T0242967

Lecture Notes in Computer Science   11029

Commenced Publication in 1973
Founding and Former Series Editors:
Gerhard Goos, Juris Hartmanis, and Jan van Leeuwen

More information about this series at http://www.springer.com/series/7409

Sven Hartmann · Hui Ma
Abdelkader Hameurlain · Günther Pernul
Roland R. Wagner (Eds.)

Database and Expert Systems Applications

29th International Conference, DEXA 2018
Regensburg, Germany, September 3–6, 2018
Proceedings, Part I

 Springer

Editors
Sven Hartmann
Clausthal University of Technology
Clausthal-Zellerfeld
Germany

Hui Ma
Victoria University of Wellington
Wellington
New Zealand

Abdelkader Hameurlain
Paul Sabatier University
Toulouse
France

Günther Pernul
University of Regensburg
Regensburg
Germany

Roland R. Wagner
Johannes Kepler University
Linz
Austria

ISSN 0302-9743   ISSN 1611-3349 (electronic)
Lecture Notes in Computer Science
ISBN 978-3-319-98808-5   ISBN 978-3-319-98809-2 (eBook)
https://doi.org/10.1007/978-3-319-98809-2

Library of Congress Control Number: 2018950662

LNCS Sublibrary: SL3 – Information Systems and Applications, incl. Internet/Web, and HCI

This Springer imprint is published by the registered company Springer Nature Switzerland AG
The registered company address is: Gewerbestrasse 11, 6330 Cham, Switzerland

Preface

This volume contains the papers presented at the 29th International Conference on Database and Expert Systems Applications (DEXA 2018), which was held in Regensburg, Germany, during September 3–6, 2018. On behalf of the Program Committee, we commend these papers to you and hope you find them useful.

Database, information, and knowledge systems have always been a core subject of computer science. The ever-increasing need to distribute, exchange, and integrate data, information, and knowledge has added further importance to this subject. Advances in the field will help facilitate new avenues of communication, to proliferate interdisciplinary discovery, and to drive innovation and commercial opportunity.

DEXA is an international conference series that showcases state-of-the-art research activities in database, information, and knowledge systems. The conference and its associated workshops provide a premier annual forum to present original research results and to examine advanced applications in the field. The goal is to bring together developers, scientists, and users to extensively discuss requirements, challenges, and solutions in database, information, and knowledge systems.

DEXA 2018 solicited original contributions dealing with any aspect of database, information, and knowledge systems. Suggested topics included, but were not limited to:

- Acquisition, Modeling, Management, and Processing of Knowledge
- Authenticity, Privacy, Security, and Trust
- Availability, Reliability, and Fault Tolerance
- Big Data Management and Analytics
- Consistency, Integrity, Quality of Data
- Constraint Modeling and Processing
- Cloud Computing and Database-as-a-Service
- Database Federation and Integration, Interoperability, Multi-Databases
- Data and Information Networks
- Data and Information Semantics
- Data Integration, Metadata Management, and Interoperability
- Data Structures and Data Management Algorithms
- Database and Information System Architecture and Performance
- Data Streams and Sensor Data
- Data Warehousing
- Decision Support Systems and Their Applications
- Dependability, Reliability, and Fault Tolerance
- Digital Libraries and Multimedia Databases
- Distributed, Parallel, P2P, Grid, and Cloud Databases
- Graph Databases
- Incomplete and Uncertain Data
- Information Retrieval

- Information and Database Systems and Their Applications
- Mobile, Pervasive, and Ubiquitous Data
- Modeling, Automation, and Optimization of Processes
- NoSQL and NewSQL Databases
- Object, Object-Relational, and Deductive Databases
- Provenance of Data and Information
- Semantic Web and Ontologies
- Social Networks, Social Web, Graph, and Personal Information Management
- Statistical and Scientific Databases
- Temporal, Spatial, and High-Dimensional Databases
- Query Processing and Transaction Management
- User Interfaces to Databases and Information Systems
- Visual Data Analytics, Data Mining, and Knowledge Discovery
- WWW and Databases, Web Services
- Workflow Management and Databases
- XML and Semi-structured Data

Following the call for papers, which yielded 160 submissions, there was a rigorous review process that saw each submission refereed by three to six international experts. The 35 submissions judged best by the Program Committee were accepted as full research papers, yielding an acceptance rate of 22%. A further 40 submissions were accepted as short research papers.

As is the tradition of DEXA, all accepted papers are published by Springer. Authors of selected papers presented at the conference were invited to submit substantially extended versions of their conference papers for publication in the Springer journal *Transactions on Large-Scale Data- and Knowledge-Centered Systems (TLDKS)*. The submitted extended versions underwent a further review process.

The success of DEXA 2018 was the result of collegial teamwork from many individuals. We wish to thank all authors who submitted papers and all conference participants for the fruitful discussions.

We are grateful to Xiaofang Zhou (The University of Queensland) for his keynote talk on "Spatial Trajectory Analytics: Past, Present, and Future" and to Tok Wang Ling (National University of Singapore) for his keynote talk on "Data Models Revisited: Improving the Quality of Database Schema Design, Integration and Keyword Search with ORA-Semantics."

This edition of DEXA also featured three international workshops covering a variety of specialized topics:

- BDMICS 2018: Third International Workshop on Big Data Management in Cloud Systems
- BIOKDD 2018: 9th International Workshop on Biological Knowledge Discovery from Data
- TIR 2018: 15th International Workshop on Technologies for Information Retrieval

We would like to thank the members of the Program Committee and the external reviewers for their timely expertise in carefully reviewing the submissions. We are grateful to our general chairs, Abdelkader Hameurlain, Günther Pernul, and

Roland R. Wagner, to our publication chair, Vladimir Marik, and to our workshop chairs, A Min Tjoa and Roland R. Wagner.

We wish to express our deep appreciation to Gabriela Wagner of the DEXA conference organization office. Without her outstanding work and excellent support, this volume would not have seen the light of day.

Finally, we like to thank Günther Pernul and his team for being our hosts during the wonderful days in Regensburg.

July 2018   Sven Hartmann
                                                                       Hui Ma

Organization

General Chairs

Abdelkader Hameurlain   IRIT, Paul Sabatier University, Toulouse, France
Günther Pernul   University of Regensburg, Germany
Roland R. Wagner   Johannes Kepler University Linz, Austria

Program Committee Chairs

Hui Ma   Victoria University of Wellington, New Zealand
Sven Hartmann   Clausthal University of Technology, Germany

Publication Chair

Vladimir Marik   Czech Technical University, Czech Republic

Program Committee

Slim Abdennadher   German University, Cairo, Egypt
Hamideh Afsarmanesh   University of Amsterdam, The Netherlands
Riccardo Albertoni   Institute of Applied Mathematics and Information Technologies - Italian National Council of Research, Italy
Idir Amine Amarouche   University Houari Boumediene, Algeria
Rachid Anane   Coventry University, UK
Annalisa Appice   Università degli Studi di Bari, Italy
Mustafa Atay   Winston-Salem State University, USA
Faten Atigui   CNAM, France
Spiridon Bakiras   Hamad bin Khalifa University, Qatar
Zhifeng Bao   National University of Singapore, Singapore
Ladjel Bellatreche   ENSMA, France
Nadia Bennani   INSA Lyon, France
Karim Benouaret   Université Claude Bernard Lyon 1, France
Benslimane Djamal   Lyon 1 University, France
Morad Benyoucef   University of Ottawa, Canada
Catherine Berrut   Grenoble University, France
Athman Bouguettaya   University of Sydney, Australia
Omar Boussaid   University of Lyon/Lyon 2, France
Stephane Bressan   National University of Singapore, Singapore
Barbara Catania   DISI, University of Genoa, Italy
Michelangelo Ceci   University of Bari, Italy
Richard Chbeir   UPPA University, France

Cindy Chen   University of Massachusetts Lowell, USA
Phoebe Chen   La Trobe University, Australia
Max Chevalier   IRIT - SIG, Université de Toulouse, France
Byron Choi   Hong Kong Baptist University, Hong Kong, SAR
                               China
Soon Ae Chun   City University of New York, USA
Deborah Dahl   Conversational Technologies, USA
Jérôme Darmont   Université de Lyon (ERIC Lyon 2), France
Roberto De Virgilio   Università Roma Tre, Italy
Vincenzo Deufemia   Università degli Studi di Salerno, Italy
Gayo Diallo   Bordeaux University, France
Juliette Dibie-Barthélemy   AgroParisTech, France
Dejing Dou   University of Oregon, USA
Cedric du Mouza   CNAM, France
Johann Eder   University of Klagenfurt, Austria
Suzanne Embury   The University of Manchester, UK
Markus Endres   University of Augsburg, Germany
Noura Faci   Lyon 1 University, France
Bettina Fazzinga   ICAR-CNR, Italy
Leonidas Fegaras   The University of Texas at Arlington, USA
Stefano Ferilli   University of Bari, Italy
Flavio Ferrarotti   Software Competence Center Hagenberg, Austria
Vladimir Fomichov   School of Business Informatics, National Research
                               University Higher School of Economics, Moscow,
                               Russian Federation
Flavius Frasincar   Erasmus University Rotterdam, The Netherlands
Bernhard Freudenthaler   Software Competence Center Hagenberg, Austria
Hiroaki Fukuda   Shibaura Institute of Technology, Japan
Steven Furnell   Plymouth University, UK
Joy Garfield   University of Worcester, UK
Claudio Gennaro   ISTI-CNR, Italy
Manolis Gergatsoulis   Ionian University, Greece
Javad Ghofrani   Leibniz Universität Hannover, Germany
Fabio Grandi   University of Bologna, Italy
Carmine Gravino   University of Salerno, Italy
Sven Groppe   Lübeck University, Germany
Jerzy Grzymala-Busse   University of Kansas, USA
Francesco Guerra   Università degli Studi di Modena e Reggio Emilia, Italy
Giovanna Guerrini   University of Genoa, Italy
Allel Hadjali   ENSMA, Poitiers, France
Abdelkader Hameurlain   Paul Sabatier University, France
Ibrahim Hamidah   Universiti Putra Malaysia, Malaysia
Takahiro Hara   Osaka University, Japan
Sven Hartmann   Clausthal University of Technology, Germany
Wynne Hsu   National University of Singapore, Singapore

Wilfred Ng	Hong Kong University of Science and Technology, Hong Kong, SAR China
Javier Nieves Acedo	IK4-Azterlan, Spain
Mourad Oussalah	University of Nantes, France
George Pallis	University of Cyprus, Cyprus
Ingrid Pappel	Tallinn University of Technology, Estonia
Marcin Paprzycki	Polish Academy of Sciences, Warsaw Management Academy, Poland
Oscar Pastor Lopez	Universitat Politecnica de Valencia, Spain
Francesco Piccialli	University of Naples Federico II, Italy
Clara Pizzuti	Institute for High Performance Computing and Networking (ICAR)-National Research Council (CNR), Italy
Pascal Poncelet	LIRMM, France
Elaheh Pourabbas	National Research Council, Italy
Claudia Raibulet	Università degli Studi di Milano-Bicocca, Italy
Praveen Rao	University of Missouri-Kansas City, USA
Rodolfo Resende	Federal University of Minas Gerais, Brazil
Claudia Roncancio	Grenoble University/LIG, France
Massimo Ruffolo	ICAR-CNR, Italy
Simonas Saltenis	Aalborg University, Denmark
N. L. Sarda	I.I.T. Bombay, India
Marinette Savonnet	University of Burgundy, France
Florence Sedes	IRIT, Paul Sabatier University, Toulouse, France
Nazha Selmaoui	University of New Caledonia, New Caledonia
Michael Sheng	Macquarie University, Australia
Patrick Siarry	Université Paris 12 (LiSSi), France
Gheorghe Cosmin Silaghi	Babes-Bolyai University of Cluj-Napoca, Romania
Hala Skaf-Molli	Nantes University, France
Bala Srinivasan	Retried, Monash University, Australia
Umberto Straccia	ISTI - CNR, Italy
Maguelonne Teisseire	Irstea - TETIS, France
Sergio Tessaris	Free University of Bozen-Bolzano, Italy
Olivier Teste	IRIT, University of Toulouse, France
Stephanie Teufel	University of Fribourg, Switzerland
Jukka Teuhola	University of Turku, Finland
Jean-Marc Thevenin	University of Toulouse 1 Capitole, France
A Min Tjoa	Vienna University of Technology, Austria
Vicenc Torra	University of Skövde, Sweden
Traian Marius Truta	Northern Kentucky University, USA
Theodoros Tzouramanis	University of the Aegean, Greece
Lucia Vaira	University of Salento, Italy
Ismini Vasileiou	University of Plymouth, UK
Krishnamurthy Vidyasankar	Memorial University of Newfoundland, Canada
Marco Vieira	University of Coimbra, Portugal
Junhu Wang	Griffith University, Brisbane, Australia

Wendy Hui Wang	Stevens Institute of Technology, USA
Piotr Wisniewski	Nicolaus Copernicus University, Poland
Ming Hour Yang	Chung Yuan Christian University, Taiwan
Yang, Xiaochun	Northeastern University, China
Yanchang Zhao	CSIRO, Australia
Qiang Zhu	The University of Michigan, USA
Marcin Zimniak	Leipzig University, Germany
Ester Zumpano	University of Calabria, Italy

Additional Reviewers

Valentyna Tsap	Tallinn University of Technology, Estonia
Liliana Ibanescu	AgroParisTech, France
Cyril Labbé	Université Grenoble-Alpes, France
Zouhaier Brahmia	University of Sfax, Tunisia
Dunren Che	Southern Illinois University, USA
Feng George Yu	Youngstown State University, USA
Gang Qian	University of Central Oklahoma, USA
Lubomir Stanchev	Cal Poly, USA
Jorge Martinez-Gil	Software Competence Center Hagenberg, Austria
Loredana Caruccio	University of Salerno, Italy
Valentina Indelli Pisano	University of Salerno, Italy
Jorge Bernardino	Polytechnic Institute of Coimbra, Portugal
Bruno Cabral	University of Coimbra, Portugal
Paulo Nunes	Polytechnic Institute of Guarda, Portugal
William Ferng	Boeing, USA
Amin Mesmoudi	LIAS/University of Poitiers, France
Sabeur Aridhi	LORIA, University of Lorraine - TELECOM Nancy, France
Julius Köpke	Alpen Adria Universität Klagenfurt, Austria
Marco Franceschetti	Alpen Adria Universität Klagenfurt, Austria
Meriem Laifa	Bordj-Bouarreridj University, Algeria
Sheik Mohammad Mostakim Fattah	University of Sydney, Australia
Mohammed Nasser Mohammed Ba-hutair	University of Sydney, Australia
Ali Hamdi Fergani Ali	University of Sydney, Australia
Masoud Salehpour	University of Sydney, Australia
Adnan Mahmood	Macquarie University, Australia
Wei Emma Zhang	Macquarie University, Australia
Zawar Hussain	Macquarie University, Australia
Hui Luo	RMIT University, Australia
Sheng Wang	RMIT University, Australia
Lucile Sautot	AgroParisTech, France
Jacques Fize	Cirad, Irstea, France

María del Carmen   Technological Institute of Aragón, Spain
  Rodríguez-Hernández
Ramón Hermoso   University of Zaragoza, Spain
Senen Gonzalez   Software Competence Center Hagenberg, Austria
Ermelinda Oro   High Performance and Computing Institute of the
                              National Research Council (ICAR-CNR), Italy
Shaoyi Yin   Paul Sabatier University, France
Jannai Tokotoko   ISEA University of New Caledonia, New Caledonia
Xiaotian Hao   Hong Kong University of Science and Technology,
                              Hong Kong, SAR China
Ji Cheng   Hong Kong University of Science & Technology,
                              Hong Kong, China
Radim Bača   Technical University of Ostrava, Czech Republic
Petr Lukáš   Technical University of Ostrava, Czech Republic
Peter Chovanec   Technical University of Ostrava, Czech Republic
Galicia Auyon Jorge   ISAE-ENSMA, Poitiers, France
  Armando
Nabila Berkani   ESI, Algiers, Algeria
Amine Roukh   Mostaganem University, Algeria
Chourouk Belheouane   USTHB, Algiers, Algeria
Angelo Impedovo   University of Bari, Italy
Emanuele Pio Barracchia   University of Bari, Italy
Arpita Chatterjee   Georgia Southern University, USA
Stephen Carden   Georgia Southern University, USA
Tharanga Wickramarachchi   U.S. Bank, USA
Divine Wanduku   Georgia Southern University, USA
Lama Saeeda   Czech Technical University in Prague, Czech Republic
Michal Med   Czech Technical University in Prague, Czech Republic
Franck Ravat   Université Toulouse 1 Capitole - IRIT, France
Julien Aligon   Université Toulouse 1 Capitole - IRIT, France
Matthew Damigos   Ionian University, Greece
Eleftherios Kalogeros   Ionian University, Greece
Srini Bhagavan   IBM, USA
Monica Senapati   University of Missouri-Kansas City, USA
Khulud Alsultan   University of Missouri-Kansas City, USA
Anas Katib   University of Missouri-Kansas City, USA
Jose Alvarez   Telecom SudParis, France
Sarah Dahab   Telecom SudParis, France
Dietrich Steinmetz   Clausthal University of Technology, Germany

Abstracts of Keynote Speakers

Data Models Revisited: Improving the Quality of Database Schema Design, Integration and Keyword Search with ORA-Semantics (Extended Abstract)

Tok Wang Ling[1], Mong Li Lee[1], Thuy Ngoc Le[2], and Zhong Zeng[3]

[1] Department of Computer Science, School of Computing, National University of Singapore
{lingtw,leeml}@comp.nus.edu.sg
[2] Google Singapore
le.thuy.ngoc@gmail.com
[3] Data Center Technology Lab, Huawei
zengzhong4@huawei.com

Introduction

Object class, relationship type, and attribute of object class and relationship type, are three basic concepts in the Entity Relationship Model. We call them ORA-semantics. In this talk, we highlight the limitations of the common database models such as the relational and XML data model. One serious common limitation of these database models is their inability to capture and explicitly represent object classes and relationship types together with their attributes in their schema languages. In fact, these data models have no concepts of object class, relationship type, and their attribute.

Without using ORA-semantics in databases, the quality of important database tasks such as relational and XML database schema design, data and schema integration, and relational and XML keyword query processing are low, and serious problems may arise. We show the reasons that lead to these problems, and demonstrate how ORA-semantics can be used to improve the result quality of these database tasks significantly.

Limitations of Relational Model

In the *relational model*, functional dependencies (FDs) and multivalued dependencies (MVDs) are integrity constraints; many of which are artificially imposed by organization or database designers. These constraints have no semantics, and cannot be automatically discovered by data mining techniques.

FDs and MVDs are used to remove redundancy and obtain normal form relations in database schema design. During normalization, we must cover the given set of FDs (i.e., the closure of the set of FDs remain unchanged), and we want to remove all MVDs. However, MVDs are relation sensitive, and it is very difficult to detect them. The existence of MVDs in a relation is because some unrelated multivalued attributes (of an object class or a relationship type) are wrongly grouped in the relation [10]. Key in relation is not the same as OID of object class. There is no concept of ORA-semantics in the relational model.

ORA Semantics in Database Schema Design

There are three common approaches for relational database schema design:

a. Decomposition. This approach is based on the Universal Relation Assumption (URA) that a database can be represented by a universal relation which contains all the attributes of the database and this relation is then decomposed into smaller relations in some good normal forms such as 3NF, BCNF, 4NF, etc. in order to remove redundant data using the given FDs and MVDs. The process is non-deterministic, and the relations obtained depend on the order of FDs and MVDs chosen for decomposition, which may not cover the given set of FDs.

b. Synthesis [1]. This approach is based on the assumption that a database can be described by a given set of attributes and a given set of functional dependencies. It also assumes URA, and a set of 3NF and BCNF relations is synthesized based on the given set of dependencies. The process is non-deterministic, and depends on the order of the redundant FDs found to generate 3NF relations. It does not consider MVDs and does not guarantee reconstructibility.

c. ER Approach. An ER diagram (ERD) is first constructed based on the database specification and requirements, and then normalized to a normal form ERD. The normal form ERD is then translated to a set of normal form relations together with a set of additional constraints that exist in the ERD but cannot be represented in the relational schema [11]. Multivalued attributes of object classes and relationships will be in separated relations. Users do not need to consider MVDs which are relation sensitive. ERD can use relaxed URA, i.e. only object identifier names must be unique, which is much more convenient than using URA.

Both the decomposition and synthesis approaches cannot handle complex relationship types such as recursive relationship type, ISA relationship, and multiple relationship types defined among 2 or more object classes. They also do not have the concept of ORA-semantics and have many problems and short comings. Other problems and issues that arise when using decomposition and synthesis methods to design a database include

(i) How to find a given set of FDs in a relational database? Can we use some data mining techniques to find FDs and MVDs in a relational database?

(ii) If a relation is not in BCNF, can we always normalize it to a set of BCNF relations?

(iii) If a relation is not in 4NF, is there a non-loss decomposition of the relation into a set of 4NF relations which cover all the given FDs?

(iv) 3NF and BCNF relations are defined on individual relations, rather than on the whole database. Hence, they cannot detect redundancy among relations of the database and may contain global redundant attributes [13].

In contrast, the ER approach captures ORA-semantics and avoids the problems of the decomposition method and synthesis method.

ORA Semantics in Data and Schema Integration

In data and schema integration, entity resolution (or object identification) is widely studied. However, this problem is still not well solved and cannot be handled fully automatically, e.g., we cannot automatically identify authors of papers completely in DBLP.

Besides entity resolution, we need to consider relationship resolution which aims to identify different relationship types between/among same object classes. We also need to differentiate between primary key vs object identifier (OID), local OID vs global OID, system generated OID vs manually designed OID, local FD vs global FD, semantic dependency vs FD/MVD constraint, structural conflicts [9], as well as schematic discrepancy [3] among schemas. All these concepts are related to ORA-semantics and they have a big impact on the quality of the integrated database and schema.

The challenge to achieve a good quality integration remains. Since the ER model can capture ORA-semantics, it is more promising to use the ER approach for data and schema integration.

ORA Semantics in Relational Keyword Search

Methods for *relational keyword search* [4, 5] can be broadly classified into two categories: data graph approach and schema graph approach. In the data graph approach, the relational database is modeled as a graph where each node represents a tuple and each edge represents a foreign key-key reference. An answer to a keyword query is typically defined as a minimal connected subgraph which contains all the keywords. This graph search is equivalent to the Steiner tree problem, which is NP-complete.

In schema graph approach, the database schema is modeled as a schema graph where each node represents a relation and each edge represents a foreign key-key constraint between two relations. Based on the schema graph, a keyword query is translated into a set of SQL statements that join the relations with tuples matching the keywords.

We identify the serious limitations of existing relational keyword search, which include incomplete answers, meaningless answers, inconsistent answers, and user difficulty in understanding the answers when they are represented as Steiner trees, etc.

In addition, the answers returned depend on the normal form of the relational database, i.e., database schema dependence. We can improve the correctness and completeness of relational keyword search by exploiting ORA-semantics because these semantics enable us to detect duplication of objects and relationships and address the above mentioned limitations [16].

We extend keyword queries by allowing keywords that match the metadata, i.e., relation name and attribute name. We also extend keyword queries with group-by and aggregate functions including sum, max, min, avg, count, etc. In order to process these extended keyword queries correctly, we use ORA-semantics to detect duplication of objects and relationships. Without using ORA-semantics, the results of aggregate functions may be computed wrongly. For more details, see [15, 17].

Limitations of XML Data Model

The *XML data model* also cannot capture ORA-Semantics [2, 12]. The constraints on the structure and content of XML can be described by DTD or XML Schema. The ID in DTD is not the same as the object identifier, ID attribute is OID of the object class, but OID of an object class may not be able to declare as ID, and a multivalued attribute of object class cannot be represented directly as an attribute in DTD/XML Schema. IDREF is not the same as foreign key to key reference in RDB. IDREF has no type.

DTD/XML Schema can only represent the hierarchical structures with simple constraints; they have no concept on ORA-semantics. The parent-child relationship in XML may not represent relationship type; relationship type (especially n-ary) is not explicitly captured in DTD/XML Schema. They cannot distinguish between attribute of object class vs attribute of relationship type.

ORA Semantics in XML Keyword Search

Existing approaches to *XML keyword search* are structure-based because they mainly rely on the exploration of the structure of XML data. These approaches can be classified as tree-based and graph-based search. Tree-based search is used when an XML document is modeled as a tree, i.e. without ID references (IDREFs), while graph-based search is used for XML documents with IDREFs.

Almost all tree-based approaches are based on some variations of LCA (Least Common Ancestor) semantics such as SLCA, MLCA, VLCA, and ELCA [14]. Given the lack of awareness of semantics in XML data, LCA-based methods do not exploit hidden ORA-semantics in data-centric XML document. This causes serious problems in processing LCA-based XML keyword queries, such as returning meaningless answers, duplicated answers, incomplete answers, missing answers, and inconsistent answers.

We can use ORA-semantics to improve the correctness and completeness of XML keyword search by detecting duplication of objects and relationships. We introduce the concepts of object tree, reversed object tree, and relative of objects to address the above mentioned problems of XML keyword search [6, 8]. We also extend XML keyword queries by considering keywords that match the metadata, i.e., tag names of XML data, and with group-by and aggregate functions [7].

Conclusion

In summary, the schemas of relational model and XML data model cannot capture the ORA-semantics which exist in the ER model. We highlight the serious problems on the quality of some database tasks due to the lack of knowledge on ORA-semantics in the relational model and XML data model. However, programmers must know the ORA-semantics of the database in order to write SQL and XQuery programs correctly. ORA-SS data model [2, 12] is designed to capture ORA-semantics in XML data.

We conclude this talk with suggestions on further research on data and schema integration, keyword query search in relational databases and XML databases such as data model independent keyword query search, and the use of ORA-semantics in NoSQL and big data applications.

References

1. Bernstein, P.A.: Synthesizing third normal form relations from functional dependencies. Trans. Database Syst. (1976)
2. Dobbie, G., Wu, X., Ling, T.W., Lee, M.L.: Ora-ss: an object-relationship-attribute model for semistructured data. Technical report, National University of Singapore (2000)
3. He, Q., Ling, T.W.: Extending and inferring functional dependencies in schema transformation. In: ACM CIKM (2004)
4. Hristidis, V., Papakonstantinou, Y.: Discover: keyword search in relational databases. In: VLDB (2002)
5. Hulgeri, A., Nakhe, C.: Keyword searching and browsing in databases using banks. In: IEEE ICDE (2002)
6. Le, T.N., Bao, Z., Ling, T.W.: Schema-independence in xml keyword search. In: Yu, E., Dobbie, G., Jarke, M., Purao, S. (eds.) ER 2014. LNCS, vol. 8824. Springer, Cham (2014)
7. Le, T.N., Bao, Z., Ling, T.W., Dobbie, G.: Group-by and aggregate functions in XML keyword search. In: DEXA (2014)
8. Le, T.N., Wu, H., Ling, T.W., Li, L., Lu, J.: From structure-based to semantics-based: towards effective XML keyword search. In: Ng, W., Storey, V.C., Trujillo J.C. (eds.) ER 2013. LNCS, vol. 8217. Springer, Heidelberg (2013)
9. Lee, M.L., Ling, T.W.: Resolving structural conicts in the integration of entity relationship schemas. In: ER (1995)
10. Ling, T.W.: An analysis of multivalued and join dependencies based on the entity-relationship approach. Data Knowl. Eng. (1985)
11. Ling, T.W.: A normal form for entity-relationship diagrams. In: ER (1985)
12. Ling, T.W., Lee, M.L., Dobbie, G.: Semistructured Database Design. Springer, New York (2005)

13. Ling, T.W., Tompa, F.W., Kameda, T.: An improved third normal form for relational databases. Trans. Database Syst. (1981)
14. Xu, Y., Papakonstantinou, Y.: Efficient keyword search for smallest LCAs in XML databases. In: ACM SIGMOD (2005)
15. Zeng, Z., Bao, Z., Le, T.N., Lee, M.L., Ling, T.W.: Expressq: identifying keyword context and search target in relational keyword queries. In: ACM CIKM (2014)
16. Zeng, Z., Bao, Z., Lee, M.L., Ling, T.W.: A semantic approach to keyword search over relational databases. In: Ng, W., Storey, V.C., Trujillo, J.C. (eds.) ER 2013. LNCS, vol. 8217. Springer, Heidelberg (2013)
17. Zeng, Z., Lee, M.L., Ling, T.W.: Answering keyword queries involving aggregates and group by on relational databases. In: EDBT (2016)

Spatial Trajectory Analytics: Past, Present and Future (Extended Abstract)

Xiaofang Zhou

School of Information Technology and Electrical Engineering,
The University of Queensland, Australia
zxf@itee.uq.edu.au

Trajectory computing involves a wide range of research topics centered around spatiotemporal data, including data management, query processing, data mining and recommendation systems, and more recently, data privacy and machine learning. It can find many applications in intelligent transport systems, location-based systems, urban planning and smart city.

Spatial trajectory computing research has attracted an extensive amount of effort from researchers in database and data mining communities. In 2011 we edited a booked to introduce the basic concepts and main research topics and progresses at that time in spatial trajectory computing [6]. This area has been developed at a very rapid and still accelerating speed, driven by the availability of massive volumes of both historical and real-time streaming trajectory data from many sources such as GPS devices, smart phones and social media applications. Major businesses also start to treat spatial trajectory data as enterprise data to support all business units that require location and movement intelligence. Trajectory data have now been embedded into traffic navigation and car sharing services, mobile apps and online social network applications, leading to more sophisticated time-dependent queries [3] and millions of concurrent queries that have not been considered in previous spatial query processing research. New computing platforms and new computational and analytics tools such as machine learning [4] have also contributed the current surge of research effort in this area. As trajectory data can reveal highly unique information about individuals [1], there are new research opportunities to address the both sides of the problem: to protect user's location and movement privacy and to link users from different trajectory datasets.

There are strong industry demands to manage and process extremely large amount of trajectory data for a diversified range of applications. Our community has developed a quite comprehensive spectrum of solutions in the past to address different aspects of trajectory analytics problems. There is an urgent need now to develop flexible and powerful trajectory data management systems with proper support from data acquisition, management to analytics. Such a system should cater for the hierarchical nature of spatial data [2] such that analytics can be applied at the right level to generate meaningful results (for example, trajectory similarity analysis can only be done using calibrated data [5]). This is the future direction of spatial trajectory computing research.

References

1. de Montjoye, Y.-A., Hidalgo, C.A., Verleysen, M., Blondel, V.D.: Unique in the crowd: the privacy bounds of human mobility. Sci. Rep. **3**, 1376 (2013). EP 03
2. Kuipers, B.: The spatial semantic hierarchy. Artif. Intell. **119**(1–2), 191–233 (2000)
3. Li, L., Hua, W., Du, X., Zhou, X.: Minimal on-road time route scheduling on time-dependent graphs. PVLDB **10**(11), 1274–1285 (2017)
4. Lv, Z., Xu, J., Zheng, K., Yin, H., Zhao, P., Zhou, X.: LC-RNN: a deep learning model for traffic speed prediction. In: IJCAI (2018)
5. Su, H., Zheng, K., Huang, J., Wang, H., Zhou, X.: Calibrating trajectory data for spatio-temporal similarity analysis. VLDB J. **24**(1), 93–116 (2015)
6. Zheng, Y., Zhou, X.: Computing with Spatial Trajectories. Springer, New York (2011)

Contents – Part I

Data Semantics

Cloud Data Processing

Time Series Data

Social Networks

Contents – Part II

Data Streams

Information Networks and Algorithms

Database System Architecture and Performance

Learning

Emerging Applications

Data Mining

Privacy

Text Processing

Big Data Analytics

Scalable Vertical Mining for Big Data Analytics of Frequent Itemsets

Carson K. Leung[1]([✉])[iD], Hao Zhang[1], Joglas Souza[1], and Wookey Lee[2]

[1] University of Manitoba, Winnipeg, MB, Canada
kleung@cs.umanitoba.ca
[2] Inha University, Incheon, South Korea

Abstract. Advances in technology and the increasing growth of popularity on Internet of Things (IoT) for many applications have produced huge volume of data at a high velocity. These valuable big data can be of a wide variety or different veracity. Embedded in these big data are useful information and valuable knowledge. This leads to data science, which aims to apply big data analytics to mine implicit, previously unknown and potentially useful information from big data. As a popular data analytic task, frequent itemset mining discovers knowledge about sets of frequently co-occurring items in the big data. Such a task has drawn attention in both academia and industry partially due to its practicality in various real-life applications. Existing mining approaches mostly use serial, distributed or parallel algorithms to mine the data horizontally (i.e., on a transaction basis). In this paper, we present an alternative big data analytic approach. Specifically, our scalable algorithm uses the MapReduce programming model that runs in a Spark environment to mine the data vertically (i.e., on an item basis). Evaluation results show the effectiveness of our algorithm in big data analytics of frequent itemsets.

Keywords: Data mining · Knowledge discovery · Frequent patterns
Vertical mining · Big data · Spark

1 Introduction

In the current era of big data, high volumes of a wide variety of valuable data of different veracity are produced at a high velocity in various modern applications. Embedded in these big data are useful information and knowledge. This calls for *data science* [6,9]—which aims to apply *data analytics* and data mining techniques for the discovery of implicit, previously unknown, and potentially useful information knowledge from big data—are in demand. From business intelligence (BI) viewpoint, the discovered knowledge usually leads to actionable decisions in business. As "a picture is worth a thousand words", visual representation of the discovered information also helps to easily interpret and comprehend the knowledge. This explains why *data and knowledge visualization*, together with *visual analytics* [14,15], are also in demand.

© Springer Nature Switzerland AG 2018
S. Hartmann et al. (Eds.): DEXA 2018, LNCS 11029, pp. 3–17, 2018.
https://doi.org/10.1007/978-3-319-98809-2_1

Characteristics of these big data can be described by 3V's, 5V's, 7V's, and even 42V's [25]. Some of the well-known V's include the following:

1. *variety*, which focuses on differences in types, contents, or formats of data (e.g., key-value pairs, graphs [2,11,12]);
2. *velocity*, which focuses on the speed at which data are collected or generated (e.g., dynamic streaming data [7]);
3. *volume*, which focuses on the quantity of data (e.g., huge volumes of data [16]);
4. *value*, which focuses on the usefulness of data (e.g., information and knowledge that can be discovered from the big data [5,13]);
5. *veracity*, which focuses on the quality of data (e.g., precise data, uncertain and imprecise data [3,24]);
6. *validity*, which focuses on interpretation of data and discovered knowledge from big data [13]; and
7. *visibility*, which focuses on visualization of data and discovered knowledge from big data [4,14].

To process these big data, *frequent itemset mining*—as an important data mining task—finds frequently co-occurring items, events, or objects (e.g., frequently purchased merchandise items in shopper market basket, frequently collocated events). Since the introduction of the frequent itemset mining problem [1], numerous frequent itemset mining algorithms [17,19] have been proposed. For instance, the *Apriori algorithm* [1] applies a generate-and-test paradigm in mining frequent itemsets in a level-wise bottom-up fashion. As it requires K database scans to discover all frequent itemsets (where K is the maximum cardinality of discovered itemsets). The *FP-growth algorithm* [10] addresses this disadvantage of the Apriori algorithm and improves efficiency by using an extended prefix-tree structure called Frequent Pattern tree (FP-tree) to capture the content of the transaction database. Unlike the Apriori algorithm, FP-growth scans the database *twice*. However, as many smaller FP-trees (e.g., for $\{a\}$-projected database, $\{a, b\}$-projected database, $\{a, b, c\}$-projected database,...) need to be built during the mining process, FP-growth requires lots of memory space. Algorithms like *TD-FP-Growth* [27] and *H-mine* [22] avoid building and keeping multiple FP-trees at the same time during the mining process. During the mining process, instead of recursively building sub-trees, TD-FP-Growth keeps updating the global FP-tree by adjusting tree pointers. Similarly, the H-mine algorithm uses a hyperlinked-array structure called H-struct to capture the content of the transaction database. Consequently, a disadvantage of both TD-FP-Growth and H-mine is that many of the pointers/hyperlinks need to be updated during the mining process. Besides these algorithms that mine frequent itemsets horizontally (i.e., using a transaction-centric approach to find what k-itemset is supported by, or contained in, a transaction), frequent itemsets can also be mined vertically (i.e., using an item-centric approach to count the number of transactions supporting or containing the itemsets). Three notable vertical frequent itemset mining algorithms are VIPER [26], Eclat [28] and dEclat [29].

To handle big data, parallel mining algorithms [18,21,23,30] have been proposed to mine frequent itemsets horizontally in parallel. For instance, a parallel Eclat algorithm called Peclat [20] was proposed in DEXA 2015, which uses the concepts of a mixed sets, for opportunistic mining of frequent itemsets. However, computation of mixed sets can be time-consuming.

This paper presents an alternative. Specifically, we present a *Scalable Vertical (SVT) algorithm* that analyzes and mines big data for frequent itemsets vertically. *Key contributions of our paper* include the design and development of the SVT algorithm. Moreover, the algorithm also reduces the communication cost and balances workload among workers when running in an Apache Spark environment.

The remainder of this paper is organized as follows. Next two sections present related work and background. Section 4 presents our frequent itemset mining algorithm called SVT. Evaluation and conclusions are given in Sects. 5 and 6, respectively.

2 Related Works

2.1 Serial Frequent Itemset Mining

Besides the well-known algorithms—such as Apriori [1], FP-growth [10] TD-FP-Growth [27] and H-mine [22]—that mine frequent itemsets horizontally (i.e., using a transaction-centric approach to find what k-itemset is supported by, or contained in, a transaction), frequent itemsets can also be mined vertically (i.e., using an item-centric approach to count the number of transactions supporting or containing the itemsets). Three notable vertical frequent itemset mining algorithms are VIPER [26], Eclat [28] and dEclat [29]. Like the Apriori algorithm, Eclat also uses a levelwise bottom-up paradigm. With Eclat, the database is treated as a collection of item lists. Each list for an item x keeps IDs of transactions (i.e., *tidset*) containing x. The length of the list for x gives the support of 1-itemset $\{x\}$. By taking the intersection of lists for two frequent itemsets α and β, the IDs of transactions containing $(\alpha \cup \beta)$ can be obtained. Again, the length of the resulting (intersected) list gives the support of the itemset $(\alpha \cup \beta)$. Eclat works well when the database is sparse. However, when the database is dense, these item lists can be long.

As an extension to Eclat, dEclat also uses a levelwise bottom-up paradigm. Unlike Eclat (which uses tidset), dEclat uses diffset which is the set difference between tidsets of two related itemsets. Specifically, the *diffset* of an itemset $X = Y \cup \{z\}$ is defined as the difference between the tidset of X and the tidset of Y. To start mining a transaction database TDB, dEclat computes the diffset of 1-itemset $\{x\}$ by taking the complement of the tidset of $\{x\}$, i.e., $diffset(\{x\}) = tidset(TDB) - tidset(\{x\}) = \{t_i | x \notin t_i \subseteq TDB\}$. For TDB containing n transactions, the support of 1-itemset $\{x\}$ is $n - |diffset(\{x\})|$. By taking the set difference between $diffset(W \cup \{z\})$ and $diffset(Y)$ where W is a $(k-1)$-prefix of a k-itemset $Y = W \cup \{y\}$, the support of k-itemset $(Y \cup \{z\})$ can be computed by subtracting the cardinality of $(Y \cup \{z\})$ from the support of Y.

dEclat works well when the database is dense. However, when the database is sparse, these diffsets can be long. Moreover, the computation of diffset may not be too intuitive.

Alternatively, VIPER represents the item lists in the form of bit vectors. Each bit in a vector for a domain item x indicates the presence (bit "1") or absence (bit "0") of transaction containing x. The number of "1" bits for x gives the support of 1-itemset $\{x\}$. By computing the dot product of vectors for two frequent itemsets α and β, the vector indicating the presence of transactions containing $(\alpha \cup \beta)$ can be obtained. Again, the number of "1" bits of this vector gives the support of the resulting itemset $(\alpha \cup \beta)$. VIPER works well when the database is dense. However, when the database is sparse, lots of space may be wasted because the vector contains lots of 0s.

2.2 Distributed and Parallel Frequent Itemset Mining

To speed up the mining process of serial algorithms, several distributed and parallel mining algorithms [21,30] have been proposed. For instance, YAFIM [23] is a parallel version of the Apriori algorithm, whereas PFP [18] is a parallel version of the FP-growth algorithm. While these parallel algorithms run faster than their serial counterparts, they also inherit disadvantages of their serial counterparts. Specifically, YAFIM requires K sets of MapReduce functions to scan the database K times for the discovery of all frequent itemsets (where K is the maximum cardinality of discovered itemsets). PFP builds many smaller FP-trees during the mining process. Hence, it requires lots of memory space. Moreover, as PFP focuses on query recommendation, it does not take into account load balancing. This problem is worsened when datasets are skewed.

In DEXA 2015, a parallel Eclat algorithm called Peclat [20] was proposed. The algorithm applies the concepts of a mixed sets for opportunistic mining of frequent itemsets. During the mining process, the mixed set of a frequent itemset X are computed based on two components—namely, (i) the tidset of X and (ii) the diffset of X.

3   Background

Over the past few years, researchers have been using the Spark framework for managing and mining big data partially because of the following advantages of using the Spark framework. First, in a Spark framework, (i) the *driver program* serves as a resource distributor and a result collector, (ii) the *cluster manager* can be considered as a built-in driver program, and (iii) *worker nodes* serve as computing units handling sub-tasks. Second, Spark uses an elastic structure— the *resilient distribute dataset (RDD)*—which can be distributed across different nodes. Third, to speed up the mining process, Spark stores intermediate results in *memory* (instead of disk as in the Hadoop framework). Fourth, Spark also extends the MapReduce framework to support more complicated computations like interactive queries and stream processing. For instance, the Spark framework provides users with the following action and transformation operators:

- map(f), which returns a new RDD formed by passing each item of the source through the function f:*Item* $\mapsto$ *Item* that maps each input item into a single output item.
- flatMap(f), which returns a new RDD formed by passing each item of the source through the function f:*Item* $\mapsto$ *SeqOfItems* that maps each input item into a sequence of 0 or more output items.
- filter(f), which returns a new RDD formed by selecting those items of the source satisfying the Boolean function f:*Item* $\mapsto$ {TRUE, FALSE}.
- collect(), which is usually used after filter(f) to return all items in the RDD as an array at the driver program.
- reduceByKey(f), which returns a dataset of key-value pairs after the values for each key are aggregated using f:$(V, V) \mapsto V$.

In addition, the "shuffle" operator redistributes the data, and the "merge" operator merges one accumulator with another same-type accumulator into one.

4   Our SVT Algorithm

Our Scalable VerTical mining algorithm SVT aims to be memory-efficient as we only needs to store either tidsets or diffsets for any itemset (cf. Peclat stores both tidset and diffset to compute mixset for each itemset). The SVT starts with tidsets then switches to diffsets depending on the data densities. Hence, our SVT algorithm can be used for datasets of different densities. Moreover, with the load balancing and communication reduction, SVT is also time-efficient.

Let us give an overview of our SVT algorithm, which consists of the following three key phases:

1. Find frequent distributed singletons by performing the following actions:
   (a) serializing the datasets and distributing the serialized sub-datasets to workers;
   (b) calculating frequencies in the driver node; and
   (c) transforming into a vertical datasets in which items are sorted in descending-frequency order.
2. Build parallel equivalence classes by performing the following actions:
   (a) computing the proper size of prefix;
   (b) mapping datasets into independent equivalence classes; and
   (c) distributing equivalence classes to workers.
3. Mine local equivalence classes in parallel by performing the following actions:
   (a) mining datasets vertically in each worker; and
   (b) collecting results from workers to the driver.

4.1   Phase 1: Finding the Global Frequent Singletons Among All Distributed Datasets

In this first key phase, data are first serialized and distributed from the driver (i.e., master) to workers. The input transaction database is partitioned into equally sub-datasets called *shards* (by applying a "flatMap" function) and distributed among the workers. The shards in workers are in the form of ⟨item ID, transaction ID⟩-pairs.

After the work is evenly distributed, each worker works simultaneously. SVT then finds all frequent singletons (i.e., 1-itemsets) by applying the "reduceByKey" and "filter" functions, which counts the number of local singletons and groups the same singletons together to find the items having frequency higher than or equal to the user-specified frequency threshold *minsup*.

Most computation is observed to happen among workers. Hence, as an enhancement to *reduce communication cost*, SVT provides users an option to request each worker to calculate and send its local ⟨item ID, support⟩-pairs to the driver.

After aggregating all the keys (i.e., item ID), the driver filters out global infrequent singletons, and keeps those that satisfy the user-specified *minsup*. It then broadcasts the resulting list of frequent 1-itemsets to each processing unit (i.e., worker) for further process.

Moreover, as the mining process uses a vertical data representation, SVT also converts datasets from the usual horizontal format into an equivalent vertical format. To accelerate the conversion process, a local hash table is generated in each partition. Each domain item x and the number of transactions containing x are both stored as a ⟨item ID, support⟩-pair in the hash table. Algorithm 1 gives a skeleton of this first key phase, and Example 1 illustrates this phase.

Algorithm 1. Key Phase 1 of SVT: Find frequent distributed singletons

---

parallelize(DB)
for transaction T_i in transactions **do**
    flatMap(T_i) $\mapsto \{item_k{:}T_i\}$
end for
for all workers **do**
    $C_1 \leftarrow$ (reduceByKey $\{item_k{:} T_i\} \mapsto \{item_k{:}sup(item_k)\}$)
end for
$L_1 \leftarrow C_1$.filter($item_k$, if $sup(item_k) \geq minsup$)
$L_1 \leftarrow L_1$.sortBy($L_1.sup$)
broadcast(L_1)

Example 1. Let us consider the transaction database TDB as shown in Table 1. Suppose there are three workers for this illustrative example. (For real-life applications, SVT uses more workers.) Our SVT algorithm first serializes the transaction database by equally dividing the database into three parts for distribution to workers:

Table 1. Transaction database TDB in a horizontal format.

T_1	$\{a, b, c, d, f, g, i, m\}$
T_2	$\{a, b, c, f, m, o\}$
T_3	$\{b, f, h, j, o\}$
T_4	$\{b, c, k, p, s\}$
T_5	$\{a, c, e, f, l, m, n, p\}$

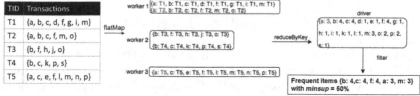

Fig. 1. In Phase 1, SVT (a) serializes the datasets and distributes them to workers, then (b) calculates frequencies in the driver node.

1. transactions T_1 and T_2 in Worker 1,
2. transactions T_3 and T_4 in Worker 2, and
3. transaction T_5 in Worker 3.

After serialization, each worker stores one part of datasets, as shown in Fig. 1. With the "flatMap" function, each worker emits a list of key-value pairs. Specifically,

- Worker 1 emits a list of key-value pairs $\{a{:}T_1,\ b{:}T_1,\ c{:}T_1,\ d{:}T_1,\ f{:}T_1,\ g{:}T_1,$ $i{:}T_1,\ m{:}T_1,\ a{:}T_2,\ b{:}T_2,\ c{:}T_2,\ f{:}T_2,\ m{:}T_2,\ o{:}T_2\}$;
- Worker 2 emits a list of key-value pairs $\{b{:}T_3,\ f{:}T_3,\ h{:}T_3,\ j{:}T_3,\ o{:}T_3,\ b{:}T_4,$ $c{:}T_4,\ k{:}T_4,\ p{:}T_4,\ s{:}T_4\}$; and
- Worker 3 emits a list of key-value pairs $\{a{:}T_5,\ c{:}T_5,\ e{:}T_5,\ f{:}T_5,\ l{:}T_5,\ m{:}T_5,$ $n{:}T_5,\ p{:}T_5\}$.

These workers send out their lists of key-value pairs to the driver node, as shown in Fig. 1. With the "reduceByKey" function, the driver node combines those values belonging to the same keys. Consequently, $\{a{:}3,\ b{:}4,\ c{:}4,\ d{:}1,\ e{:}1,\ f{:}4,\ g{:}1,$ $h{:}1,\ i{:}1,\ j{:}1,\ k{:}1,\ l{:}1,\ m{:}3,\ n{:}1,\ o{:}2,\ p{:}2,\ s{:}1\}$ is resulted.

As an enhancement, SVT provides users an option to request each worker to calculate and send its local ⟨item ID, support⟩-pairs to the driver. With this option, Worker 1 sends out a list of ⟨item ID, support⟩-pairs $\{a{:}2,\ b{:}2,\ c{:}2,\ d{:}1,$ $f{:}2,\ g{:}1,\ i{:}1,\ m{:}2,\ o{:}1\}$. Similarly, Worker 2 sends out a list $\{b{:}2,\ c{:}1,\ f{:}1,\ h{:}1,$ $j{:}1,\ k{:}1,\ o{:}1,\ p{:}1,\ s{:}1\}$, and Worker 3 sends out a list $\{a{:}1,\ c{:}1,\ e{:}1,\ f{:}1,\ l{:}1,\ m{:}1,$ $n{:}1,\ p{:}1\}$. Note that, as these lists of ⟨item ID, support⟩-pairs sent by workers to the driver are much shorter than the original lists of ⟨item ID, transaction ID⟩-pairs, communication cost is reduced. Moreover, with the "reduceByKey" function, the driver node can easily sums up those values belonging to the same

Table 2. Transformed transaction database TDB′ in a vertical format.

$tidset(\{b\})$	$\{T_1, T_2, T_3, T_4\}$
$tidset(\{c\})$	$\{T_1, T_2, T_4, T_5\}$
$tidset(\{f\})$	$\{T_1, T_2, T_3, T_5\}$
$tidset(\{a\})$	$\{T_1, T_2, T_5\}$
$tidset(\{m\})$	$\{T_1, T_2, T_5\}$

keys. Consequently, $\{a{:}3,\ b{:}4,\ c{:}4,\ d{:}1,\ e{:}1,\ f{:}4,\ g{:}1,\ h{:}1,\ i{:}1,\ j{:}1,\ k{:}1,\ l{:}1,\ m{:}3,\ n{:}1,\ o{:}2,\ p{:}2,\ s{:}1\}$ is resulted.

Afterwards, by applying the "filter" function to the list of key-value sum pairs, SVT finds that only $b{:}4$, $c{:}4$, $f{:}4$, $a{:}3$ and $m{:}3$ (in descending frequency order) are frequent when $minsup = 50\%$. This frequent-item list is then defined as a broadcasting variable, and each worker stores a copy of it. See Fig. 1.

Finally, at the end of this first key phase, the input transaction database TDB is transformed from the horizontal database to a vertical database TDB′ containing only frequent singletons and their associated transaction IDs. See Table 2.   □

4.2 Phase 2: Computing the Proper Size of Equivalence Classes that Can Fit into Workers' Memory

After finding the global frequent 1-itemsets, our SVT algorithm computes the size of equivalence classes (k-itemsets) that can fit into the memory of the working machines. A critical step in this phase is to balance the workload among workers. The size of equivalence classes may vary among different scenarios based on the density of the dataset and the capacity of the computation environment (e.g., workers' memory). Unlike existing approaches that use a fixed number for the proper size of prefix, *SVT uses a dynamic value based on the current maximum load*.

Once the proper size of prefix is determined, SVT then maps datasets into independent equivalence class. As an enhancement to *reduce communication cost*, SVT provides users an option to remap long names (or item ID) into shorter ones.

Afterwards, SVT distributes the equivalence classes to all the workers by applying the "map", "shuffle" and "merge" functions. To elaborate, each equivalence class is packed into ⟨prefix of the equivalence class *EC*, list of candidate itemsets in *EC*⟩-pairs for distribution. When distributing the equivalence class to each worker:

1. if the worker already has a list of equivalence-class itemsets, then it is necessary to merge with previous transactions for every itemsets in these two equivalence classes;
2. otherwise (i.e., when the worker does not have a list of equivalence-class itemsets), then the worker just needs to build one with current itemsets.

At the end, each partition keeps one branch of the itemsets, which have the same prefix. Algorithm 2 shows a skeleton of this second key phase, and Example 2 illustrates this phase.

Algorithm 2. Key Phase 2 of SVT: Build parallel equivalence class

---
for all workers **do**
   $C_k = \langle L_{k-1}, L_{k-1} \rangle$
end for
L_k.reduceByKey($item_k$:T_i) $\mapsto$ {$item_k$:$sup(item_k)$}
$L_k \leftarrow L_k$.filter($item_k$, if $sup(item_k) \geq minsup$)
$m_k \leftarrow$ maximum size of candidate itemsets with same prefix
if sizeof(m_k) $\leq$ memory of single worker **then**
   $L_k \rightarrow EQ_k$
else
   compute L_{k+1}
end if

Example 2. Continue with Example 1. In Phase 2(a), our SVT determines the proper size of prefix based on factors like (i) the number of workers and (ii) system load (e.g., CPU, memory). To do so, SVT generates candidate 2-itemsets from the vertical database returned by Phase 1:

– Worker 1 emits a list of ⟨2-itemset X, $tidset(X)$⟩-pairs {bc:T_1T_2, bf:T_1T_2, ba:T_1T_2, bm:T_1T_2, cf:T_1T_2, ca:T_1T_2, cm:T_1T_2, fa:T_1T_2, fm:T_1T_2, am:T_1T_2};
– Worker 2 emits {bf:T_3, bc:T_4}; and
– Worker 3 emits {cf:T_5, ca:T_5, cm:T_5, fa:T_5, fm:T_5, am:T_5}.

By applying the "reduceByKey" and "filter" functions, the driver node combines those values belonging to the same keys to generate global candidate 2-itemsets and keeps only those frequent ones. Consequently, {bc:3, bf:3, cf:3, ca:3, cm:3, fa:3, fm:3, am:3} is resulted. With this result, the best size of equivalence class for this example happens to be 2 (representing 2-itemsets). Consequently, the proper size of prefix for the equivalence classes shown in Fig. 2 is 1 (representing prefix 1-itemsets).

With the "map" function, SVT computes a list of key-value pairs by performing inner products (i.e., dot products) of the *frequent* itemsets mined from the previous levels. As the prefix is the key in the key-value pairs and value is a list of frequent candidates (with their corresponding tidsets), the results are as shown in Fig. 3:

– Worker 1 emits a list of ⟨prefix, [suffix|frequency]⟩-pairs {b:[$cf|T_1T_2$], c:[$fam|T_1T_2$], f:[$am|T_1T_2$] a:[$m|T_1T_2$]};
– Worker 2 emits {b:[$f|T_3$, $c|T_4$]}; and
– Worker 3 emits {c:[$fam|T_5$], f:[$am|T_5$], a:[$m|T_5$]}.

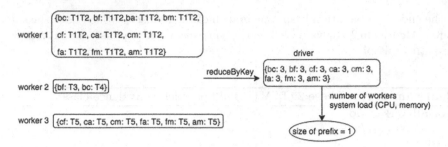

Fig. 2. In Phase 2, SVT (a) computes the proper size of prefix.

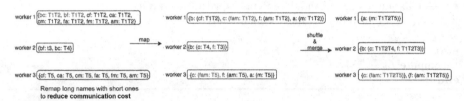

Fig. 3. In Phase 2, SVT (b) maps datasets into independent equivalence class and (c) distributes equivalence class to workers.

Afterwards, with the "shuffle" and "merge" functions, SVT distributes key-value pairs as equivalence classes to different workers. As shown in Fig. 3,

- Worker 1 gets $\{a:[m|T_1T_2T_5]\}$;
- Worker 2 gets $\{b:[c|T_1T_2T_4, f|T_1T_2T_3]\}$; and
- Worker 3 gets $\{c:[fam|T_1T_2T_5], f:[am|T_1T_2T_5]\}$.

Note that some worker (e.g., Worker 3) contains more than one equivalence class, which is computed based on the workload capacity of each worker. Moreover, the above results represent $1 + (1 + 1) + (3 + 2) = 8$ itemsets:

- $a:[m|T_1T_2T_5]$ represents itemset $\{a, m\}$, which appears in transactions T_1, T_2 and T_5;
- $b:[c|T_1T_2T_4]$ represents itemset $\{b, c\}$, which appears in transactions T_1, T_2 and T_4;
- $f:[c|T_1T_2T_3]$ represents itemset $\{f, c\}$, which appears in transactions T_1, T_2 and T_3;
- $c:[fam|T_1T_2T_5]$ represents itemsets $\{c, f\}, \{c, a\}$ and $\{c, m\}$, which appear in transactions T_1, T_2 and T_5; and
- $f:[am|T_1T_2T_5]$ represents itemsets $\{f, a\}$ and $\{f, m\}$, which appear in transactions T_1, T_2 and T_5.

This data structure is compact because the common prefix only appears once (e.g., prefix "c" only appears once for three itemsets). □

4.3 Phase 3: Distributing the Equivalence Classes to Different Workers

The final key phase of SVT is to distribute the original transaction dataset and to store the transactions as frequent equivalence classes in different units. With the "map" function, the mappers apply *hybrid vertical mining* on each partition without the need of any additional information from other workers. Unlike the traditional vertical mining algorithms like Eclat or dEclat, our SVT algorithm does not choose just a single strategy. Instead, it chooses different strategies based on the densities of datasets. Specifically, SVT first captures transaction IDs (i.e., tidsets), which consumes less time in calculating the support. SVT then computes differences among the sets of transaction IDs (i.e., diffsets). The switching from one strategy to another is based on the densities of datasets:

1. If the dataset is dense, SVT switches from using transaction IDs to using diffsets early.
2. If the dataset is sparse, SVT uses transaction IDs for longer period of mining time before it switches to diffsets.

Our analytical and empirical evaluation results suggest SVT to switch from using tidsets to using diffsets when the frequency of the subset is at least half of that of the superset. Another benefit of the switch is that, as each worker performs the vertical mining simultaneously, each worker may choose a different strategy based on the current system load. As another benefit, *SVT only needs to scan the database once* in the entire mining process.

Once vertical mining is performed by each worker, the results (i.e., frequent itemsets) are collected from these workers to the driver. Algorithm 3 shows a skeleton of this third and final key phase of SVT, and Example 3 illustrates this phase.

Algorithm 3. Key Phase 3 of SVT: Mine local equivalence class

---
$\quad$ **for all** C_i, C_j in equivalence class EQ_k **do**
$\quad\quad$ $C_{ij} = C_i \cap C_j$
$\quad\quad$ $sup(C_{ij}) = |C_{ij}|$
$\quad$ **end for**
$\quad$ **if** $sup(C_{ij}) \leq 2 \times sup(C_k)$ **then**
$\quad\quad$ vertical mining using tidsets and equivalence class
$\quad$ **else**
$\quad\quad$ vertical mining using diffsets
$\quad$ **end if**

Example 3. Let us continue with Examples 1 and 2. As the following equivalence classes $\{a:[m|T_1T_2T_5]\}$, $\{b:[c|T_1T_2T_4, f|T_1T_2T_3]\}$ and $\{c:[fam|T_1T_2T_5], f:[am|T_1T_2T_5]\}$ are distributed to Workers 1, 2 and 3 respectively, SVT then

computes the next level of frequent patterns with equivalence class transformations. The results are $\{b, c, f\}:T_1T_2T_3$, $\{c, f, a\}:T_1T_2T_5$, $\{c, f, m\}:T_1T_2T_5$,
$\{c, a, m\}:T_1T_2T_5$ and $\{f, a, m\}:T_1T_2T_5$.

When the support of $\{c, f, a\} \geq 2\times sup(\{c, f\})$, our SVT algorithm switches
from tidsets to diffsets. Specifically, SVT computes $diffset(\{c, f, a, m\})$ =
$\{c, f, m\}$ $-\{c, f, a\}$ = $\emptyset$ and thus $sup(\{c, f, a, m\})$ = 3. At this level,
$diffset(\{c, f, a, m\})$ requires less space than $tidset(\{c, f, a, m\})$. $\square$

5   Evaluation

We compared our SVT algorithm with existing algorithms like YAFIM [23],
PFP [18] and MREclat [30]. All these four algorithms were implemented and
run in a Spark environment with (a) five workers having 20 GB of memory and
an 8-core Intel Xenon CPU and (b) a driver having 8 GB of memory and a 4-
core Intel CPU. All machines are running Linux and Spark 2.0.1. We used both
synthetic datasets generated by the Synthetic Dataset Generator [8] and real-
life datasets (e.g., accidents, mushrooms, retails) from UCI ML Repository and
FIMI Repository.

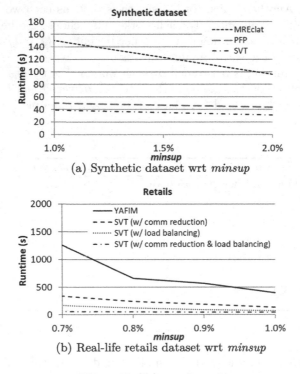

(a) Synthetic dataset wrt *minsup*

(b) Real-life retails dataset wrt *minsup*

Fig. 4. Experimental result

First, we compared the runtime of the SVT algorithm using a synthetic dataset t20i6d100k having 100,000 transactions with an average of 20 items per itemset and an average cardinality of frequent itemsets being 6 (i.e., 6-itemsets). Figure 4(a) shows that our SVT algorithms ran faster than existing algorithms like PFP and MREclat.

Similarly, we compared the runtime of the SVT algorithm using different real-life datasets from UC Irvine Machine Learning Repository. Figure 4(b) shows the results for a retail dataset with more than 1M transactions and more than 46K distinct domain items. Again, our SVT algorithms was shown to run faster than existing algorithm like YAFIM.

In addition, Fig. 4(b) also shows the benefits on load balancing and communication reduction in the vertical mining process. Specifically, communication reduction helps lower the runtime. Load balancing further reduces the runtime. SVT with both communication reduction and load balancing led to a low runtime.

Moreover, we evaluated the runtime of our SVT algorithm with increasing *minsup*. The results show that, when *minsup* increased, the runtime decreased as expected. We also evaluated the scalability of SVT with increasing number of transactions. The results show that our SVT algorithm was scalable with respect to the size of transaction databases.

6 Conclusions

In this paper, we present a scalable vertical algorithm called SVT to "vertically" mine frequent itemsets from big transaction data in a Spark environment. Our SVT algorithm is time-efficient because it (a) balances the workload by dynamically distributing work among workers based on the current system load and (b) reduces communication costs by keeping main computation among workers and only transferring results to the driver. Moreover, our SVT algorithm is also space-efficient because it (a) dynamically switches from tidset representation to diffset representation of itemset in the vertical mining process and (b) compresses data by remapping long item names or item IDs to shorter ones. Evaluation results show the scalability and effectiveness of our SVT algorithms in big data analytics of frequent itemsets—especially, vertical mining frequent itemsets from big data.

As ongoing and future work, we are exploring further enhancements in reducing the computation cost and memory consumption, as well as speeding up the mining process. Moreover, we are also conducting more exhaustive experiments on our SVT algorithm.

Acknowledgements. This project is partially supported by NSERC (Canada) and University of Manitoba.

References

1. Aggarwal, R., Srikant, R.: Fast algorithms for mining association rules. In: VLDB 1994, pp. 487–399 (1994)
2. Arora, N.R., Lee, W., Leung, C.K.-S., Kim, J., Kumar, H.: Efficient fuzzy ranking for keyword search on graphs. In: Liddle, S.W., Schewe, K.-D., Tjoa, A.M., Zhou, X. (eds.) DEXA 2012, Part I. LNCS, vol. 7446, pp. 502–510. Springer, Heidelberg (2012). https://doi.org/10.1007/978-3-642-32600-4_38
3. Braun, P., Cuzzocrea, A., Jiang, F., Leung, C.K.-S., Pazdor, A.G.M.: MapReduce-based complex big data analytics over uncertain and imprecise social networks. In: Bellatreche, L., Chakravarthy, S. (eds.) DaWaK 2017. LNCS, vol. 10440, pp. 130–145. Springer, Cham (2017). https://doi.org/10.1007/978-3-319-64283-3_10
4. Braun, P., Cuzzocrea, A., Keding, T.D., Leung, C.K., Pazdor, A.G.M., Sayson, D.: Game data mining: clustering and visualization of online game data in cyber-physical worlds. Proc. Comput. Sci. **112**, 2259–2268 (2017)
5. Brown, J.A., Cuzzocrea, A., Kresta, M., Kristjanson, K.D.L., Leung, C.K., Tebinka, T.W.: A machine learning system for supporting advanced knowledge discovery from chess game data. In: IEEE ICMLA 2017, pp. 649–654 (2017)
6. Chen, Y.C., Wang, E.T., Chen, A.L.P.: Mining user trajectories from smartphone data considering data uncertainty. In: Madria, S., Hara, T. (eds.) DaWaK 2016. LNCS, vol. 9829, pp. 51–67. Springer, Cham (2016). https://doi.org/10.1007/978-3-319-43946-4_4
7. Cuzzocrea, A., Jiang, F., Leung, C.K., Liu, D., Peddle, A., Tanbeer, S.K.: Mining popular patterns: a novel mining problem and its application to static transactional databases and dynamic data streams. In: Hameurlain, A., Küng, J., Wagner, R., Cuzzocrea, A., Dayal, U. (eds.) Transactions on Large-Scale Data- and Knowledge-Centered Systems XXI. LNCS, vol. 9260, pp. 115–139. Springer, Heidelberg (2015). https://doi.org/10.1007/978-3-662-47804-2_6
8. Fournier-Viger, P., Gomariz, A., Gueniche, T., Soltani, A., Wu, C., Tseng, V.S.: SPMF: a Java open-source pattern mining library. JMLR **15**(1), 3389–3393 (2014)
9. Gan, W., Lin, J.C.-W., Fournier-Viger, P., Chao, H.-C.: Mining recent high-utility patterns from temporal databases with time-sensitive constraint. In: Madria, S., Hara, T. (eds.) DaWaK 2016. LNCS, vol. 9829, pp. 3–18. Springer, Cham (2016). https://doi.org/10.1007/978-3-319-43946-4_1
10. Han, J., Pei, J., Yin, Y.: Mining frequent patterns without candidate generation. In: ACM SIGMOD 2000, pp. 1–12 (2000)
11. Hoi, C.S.H., Leung, C.K., Tran, K., Cuzzocrea, A., Bochicchio, M., Simonetti, M.: Supporting social information discovery from big uncertain social key-value data via graph-like metaphors. In: Xiao, J., Mao, Z.-H., Suzumura, T., Zhang, L.-J. (eds.) ICCC 2018. LNCS, vol. 10971, pp. 102–116. Springer, Cham (2018). https://doi.org/10.1007/978-3-319-94307-7_8
12. Islam, M.A., Ahmed, C.F., Leung, C.K., Hoi, C.S.H.: WFSM-MaxPWS: an efficient approach for mining weighted frequent subgraphs from edge-weighted graph databases. In: Phung, D., Tseng, V.S., Webb, G.I., Ho, B., Ganji, M., Rashidi, L. (eds.) PAKDD 2018. LNCS (LNAI), vol. 10939, pp. 664–676. Springer, Cham (2018). https://doi.org/10.1007/978-3-319-93040-4_52
13. Leung, C.K.: Big data analysis and mining. In: Encyclopedia of Information Science and Technology, 4th edn, pp. 338–348 (2018)

14. Leung, C.K.: Data and visual analytics for emerging databases. In: Lee, W., Choi, W., Jung, S., Song, M. (eds.) Proceedings of the 7th International Conference on Emerging Databases. LNEE, vol. 461, pp. 203–213. Springer, Singapore (2018). https://doi.org/10.1007/978-981-10-6520-0_21
15. Leung, C.K., Carmichael, C.L., Johnstone, P., Xing, R.R., Yuen, D.S.H.: Interactive visual analytics of big data. In: Ontologies and Big Data Considerations for Effective Intelligence, pp. 1–26 (2017)
16. Leung, C.K.-S., Jiang, F.: Big data analytics of social networks for the discovery of "following" patterns. In: Madria, S., Hara, T. (eds.) DaWaK 2015. LNCS, vol. 9263, pp. 123–135. Springer, Cham (2015). https://doi.org/10.1007/978-3-319-22729-0_10
17. Leung, C.K.-S., MacKinnon, R.K.: Balancing tree size and accuracy in fast mining of uncertain frequent patterns. In: Madria, S., Hara, T. (eds.) DaWaK 2015. LNCS, vol. 9263, pp. 57–69. Springer, Cham (2015). https://doi.org/10.1007/978-3-319-22729-0_5
18. Li, H., Wang, Y., Zhang, D., Zhang, M., Chang, E.Y.: PFP: parallel FP-growth for query recommendation. In: ACM RecSys 2008, pp. 107–114 (2008)
19. Liu, J., Li, J., Xu, S., Fung, B.C.M.: Secure outsourced frequent pattern mining by fully homomorphic encryption. In: Madria, S., Hara, T. (eds.) DaWaK 2015. LNCS, vol. 9263, pp. 70–81. Springer, Cham (2015). https://doi.org/10.1007/978-3-319-22729-0_6
20. Liu, J., Wu, Y., Zhou, Q., Fung, B.C.M., Chen, F., Yu, B.: Parallel Eclat for opportunistic mining of frequent itemsets. In: Chen, Q., Hameurlain, A., Toumani, F., Wagner, R., Decker, H. (eds.) DEXA 2015, Part I. LNCS, vol. 9261, pp. 401–415. Springer, Cham (2015). https://doi.org/10.1007/978-3-319-22849-5_27
21. Moens, S., Aksehirli, E., Goethals, B.: Frequent itemset mining for big data. In: IEEE BigData 2013, pp. 111–118 (2013)
22. Pei, J., Han, J., Lu, H., Nishio, S., Tang, S., Yang, D.: H-Mine: hyper-structure mining of frequent patterns in large databases. In: IEEE ICDM 2001, pp. 441–448 (2001)
23. Qiu, H., Gu, R., Yuan, C., Huang Y.: YAFIM: a parallel frequent itemset mining algorithm with Spark. In: IEEE IPDPS 2014 Workshops, pp. 1664–1671 (2014)
24. Rahman, M.M., Ahmed, C.F., Leung, C.K., Pazdor, A.G.M.: Frequent sequence mining with weight constraints in uncertain databases. In: ACM IMCOM 2018, Article no. 48 (2018)
25. Shafer, T.: The 42 V's of big data and data science (2017). https://www.kdnuggets.com/2017/04/42-vs-big-data-data-science.html
26. Shenoy, P., Bhalotia, J.R., Bawa, M., Shah, D.: Turbo-charging vertical mining of large databases. In: ACM SIGMOD 2000, pp. 22–33 (2000)
27. Wang, K., Tang, L., Han, J., Liu, J.: Top down FP-growth for association rule mining. In: Chen, M.-S., Yu, P.S., Liu, B. (eds.) PAKDD 2002. LNCS (LNAI), vol. 2336, pp. 334–340. Springer, Heidelberg (2002). https://doi.org/10.1007/3-540-47887-6_34
28. Zaki, M.J.: Scalable algorithms for association mining. IEEE TKDE **12**(3), 372–390 (2000)
29. Zaki, M.J., Gouda, K.: Fast vertical mining using diffsets. In: KDD 2003, pp. 326–335 (2003)
30. Zhang, Z., Ji, G., Tang, M.: MREclat: an algorithm for parallel mining frequent itemsets. In: CBD 2013, pp. 177–180 (2013)

ScaleSCAN: Scalable Density-Based Graph Clustering

Hiroaki Shiokawa[1,2(✉)], Tomokatsu Takahashi[3], and Hiroyuki Kitagawa[1,2]

[1] Center for Computational Sciences, University of Tsukuba, Tsukuba, Japan
{shiokawa,kitagawa}@cs.tsukuba.ac.jp
[2] Center for Artificial Intelligence Research,
University of Tsukuba, Tsukuba, Japan
[3] Graduate School of Systems and Information Engineering,
University of Tsukuba, Tsukuba, Japan
shihakata@kde.cs.tsukuba.ac.jp

Abstract. How can we efficiently find clusters (*a.k.a.* communities) included in a graph with millions or even billions of edges? Density-based graph clustering SCAN is one of the fundamental graph clustering algorithms that can find densely connected nodes as clusters. Although SCAN is used in many applications due to its effectiveness, it is computationally expensive to apply SCAN to large-scale graphs since SCAN needs to compute all nodes and edges. In this paper, we propose a novel density-based graph clustering algorithm named *ScaleSCAN* for tackling this problem on a multicore CPU. Towards the problem, ScaleSCAN integrates *efficient node pruning methods* and *parallel computation schemes* on the multicore CPU for avoiding the exhaustive nodes and edges computations. As a result, ScaleSCAN detects exactly same clusters as those of SCAN with much shorter computation time. Extensive experiments on both real-world and synthetic graphs demonstrate that the performance superiority of ScaleSCAN over the state-of-the-art methods.

Keywords: Graph mining · Density-based clustering
Manycore processor

1 Introduction

How can we efficiently find clusters (*a.k.a.* communities) included in a graph with millions or even billions of edges? Graph is a fundamental data structure that has helped us to understand complex systems and schema-less data in the real-world [1,7,13]. One important aspect of graphs is cluster structures where nodes in the same cluster have denser edge-connections than nodes in the different clusters. One of the most successful clustering method is density-based clustering algorithm, named *SCAN*, proposed by Xu *et al.* [20]. The main concept of SCAN is that densely connected nodes should be in the same cluster; SCAN excludes nodes with sparse connections from clusters, and SCAN classifies them

S. Hartmann et al. (Eds.): DEXA 2018, LNCS 11029, pp. 18–34, 2018.
https://doi.org/10.1007/978-3-319-98809-2_2

as either hubs or outliers. In contrast to most traditional clustering algorithms such as graph partitioning [19], spectral algorithm [14], and modularity-based method [15] that only study the problem of the cluster detection and so ignore hubs and outliers, SCAN successfully finds not only clusters but also hubs and outliers. As a result, SCAN has been used in many applications [5,12].

Although SCAN is effective in finding highly accurate results, SCAN has a serious weakness; it requires high computational costs for large-scale graphs. This is because SCAN has to find all clusters prior to identifying hubs and outliers; it finds densely connected subgraphs as clusters. It then classifies the remaining non-clustered nodes into hubs or outliers. This clustering procedure entails exhaustive density evaluations for all adjacent node pairs included in the large-scale graphs. Furthermore, in order to evaluate the density, SCAN employs a criteria, called *structural similarity*, that incurs a set intersection for each edge. Thus, SCAN requires $O(m^{1.5})$ in the worst case [3].

Existing Approaches and Challenges: To address the expensive time-complexity of SCAN, many efforts have been made for the recent few years, especially in the database and data mining communities. One of the major approaches is nodes/edge pruning: *SCAN++* [16] and *pSCAN* [3] are the most representative methods. Although these algorithms certainly succeeded in reducing the time complexity of SCAN for the real-world graphs, the computation time for large-scale graphs (*i.e.* graphs with more than 100 million edges) is still large. Thus, it is a challenging task to improving the computational efficiency for the structural graph clustering. Especially, most of existing approaches perform as a single-threaded algorithms; they do not fully exploit parallel computation architectures but this is time-consuming.

Our Approaches and Contributions: We focus on the problem of speeding up SCAN for large-scale graphs. We present a novel parallel-computing algorithm, *ScaleSCAN*, that is designed to efficiently perform on shared memory architectures with the multicore CPU. The modern multicore CPU equips a lot of physical cores on a chip, and each core highlights vector processing units (VPUs) for powerful data-parallel processing, *e.g.,* SIMD instructions. Thus, ScaleSCAN employs *thread-parallel algorithm* and *data-parallel algorithm* in order to fully exploit the performance of the multicore CPU. In addition, we also integrates existing node pruning techniques [3] and our parallel algorithm. By pruning unnecessary nodes in the parallel computation manner, we attempt to achieve further improvement of the clustering speed. As a result, ScaleSCAN has the following attractive characteristics:

1. **Efficient:** Compared with the existing approaches [3,16,18], ScaleSCAN achieves high speed clustering by using the above approaches for density computations; ScaleSCAN can avoid computing densities for the whole graph.
2. **Scalable:** ScaleSCAN shows near-linear speeding up as increasing of the number of threads. ScaleSCAN is also scalable to the dataset size.
3. **Exact:** While our approach achieves efficient and scalable clustering, it does not to sacrifice the clustering accuracy; it returns exact clusters as SCAN.

Our extensive experiments showed that ScaleSCAN runs ×500 faster than SCAN without sacrificing the clustering quality. Also, ScaleSCAN achieved from ×17.3 to ×90.2 clustering speed improvements compared with the state-of-the-art algorithms [3,18]. In specific, ScaleSCAN can compute graphs, which have more than 1.4 billion edges, within 6.4s while SCAN did not finish even after 24h. Even though SCAN is effective in enhancing application quality, it has been difficult to apply SCAN to large-scale graphs due to its performance limitations. However, by providing our scalable approach that suits the identification of clusters, hubs and outliers, ScaleSCAN will help to improve the effectiveness of a wider range of applications.

Organization: The rest of this paper is organized as follows: Sect. 2 describes a brief background of this work. Section 3 introduces our proposed approach ScaleSCAN, and we report the experimental results in Sect. 4. In Sect. 5, we briefly review the related work, and we conclude this paper in Sect. 6.

2   Preliminary

We first briefly review the baseline algorithm SCAN [20]. Then, we introduce the data-parallel computation scheme that we used in our proposal.

2.1   The Density-Based Graph Clustering: SCAN

The density-based graph clustering SCAN [20] is one of the most popular graph clustering method; it successfully detects not only clusters but also hubs and outliers unlike traditional algorithms. Given an unweighted and undirected graph $G = (V, E)$, where V is the set of nodes and E is the set of edges, SCAN detects not only the set of clusters $\mathbb{C}$ but also the set of hubs H and outliers O at the same time. We denote the number of nodes and edges in G by $n = |V|$ and $m = |E|$, respectively.

SCAN extracts clusters as the sets of nodes that have dense internal connections; it identifies the other non-clustered nodes as hubs or outliers. Thus, prior to identifying hubs and outliers, SCAN finds all clusters in a given graph G. SCAN assigns two adjacent nodes into a same cluster according to how strong the two nodes are densely connected with each other through their shared neighborhoods. Let N_u be a set of neighbors of node u, so called *structural neighborhood* defined in Definition 1, SCAN evaluates *structural similarity* between two adjacent nodes u and v defined as follows:

Definition 1 (Structural neighborhood). *The structural neighborhood of a node u, denoted by N_u, is defined as $N_u = \{v \in V | (u, v) \in E\} \cup \{u\}$.*

Definition 2 (Structural similarity). *The structural similarity $\sigma(u, v)$ between node u and v is defined as $\sigma(u, v) = |N_u \cap N_v| / \sqrt{d_u d_v}$, where $d_u = |N_u|$ and $d_v = |N_v|$.*

Algorithm 1. Baseline algorithm: SCAN(G, ϵ, μ) [20]

1: **for each** edge $(u, v) \in E$ **do**
2:     Compute $\sigma(u, v)$ by Definition 2;
3: $\mathbb{C} = \emptyset$;
4: **for each** unvisited node $u \in V$ **do**
5:     $C = \{u\}$;
6:     **for each** unvisited node $v \in C$ **do**
7:         **if** $|N_v^\epsilon| \geq \mu$ **then**
8:             $C = C \cup N_v^\epsilon$;
9:             Mark v as visited;
10:    **if** $|C| \geq 2$ **then**
11:       $\mathbb{C} = \mathbb{C} \cup C$;

We denote nodes u and v are *similar* if $\sigma(u, v) \geq \epsilon$; otherwise, the nodes are *dissimilar*.

SCAN detects a special class of node, called *core node*, that plays as the seed of a cluster, and SCAN then expands the cluster from the core node. Given a similarity threshold $\epsilon \in \mathbb{R}$ and a minimum size of a cluster $\mu \in \mathbb{N}$, core node is a node that has μ neighbors with a structural similarity that exceeds the threshold ϵ.

Definition 3 (Core node). *Given a similarity threshold $0 \leq \epsilon \leq 1$ and an integer $\mu \geq 2$, a node u is a core node iff $|N_u^\epsilon| \geq \mu$. Note that N_u^ϵ, so called ϵ-neighborhood of u, is defined as $N_u^\epsilon = \{v \in N_u | \sigma(u, v) \geq \epsilon\}$.*

When node u is a core node, SCAN assigns all nodes in N_u^ϵ to the same cluster as node u, and it expands the cluster by checking whether each node in the cluster is a core node or not.

Definition 4 (Cluster). *Let a node u be a core node that belongs to a cluster $C \in \mathbb{C}$, the cluster C is defined as $C = \{w \in N_v^\epsilon | v \in C\}$, where C is initially set to $C = \{u\}$.*

Finally, SCAN classifies non-clustered nodes (*i.e.* nodes that belong to no clusters) as hubs or outliers. If a node u is not in any clusters and its neighbors belong to two or more clusters, SCAN regards node u as a hub, and it is an outlier otherwise. Given the set of clusters, it is straightforward to obtain hubs and outliers in $O(n + m)$ time. Hereafter, we thus focus on only extracting the set of clusters in G.

Algorithm 1 overviews the pseudo code of SCAN. SCAN first evaluates structural similarities for all edges in G, and then constructs clusters by traversing all nodes. As proven in [3], Algorithm 1 is essentially based on the problem of triangle enumeration on G since each node $w \in \{N_u \cap N_v\} \backslash \{u, v\}$ forms a triangle with u and v when we compute $\sigma(u, v) = |N_u \cap N_v| / \sqrt{d_u d_v}$. This triangle enumeration basically involves $O(\alpha(G) \cdot m)$, where $\alpha(G)$ is the arboricity of G such that $\alpha(G) \leq \sqrt{m}$. Thus, the time complexity of SCAN is $O(m^{1.5})$ and is *worst-case optimal* [3].

2.2   Data-Parallel Instructions

In our proposed method, we employ the data-parallel computation schemes [17] for improving clustering speed. Thus, we briefly introduce the data-parallel instructions.

Data-parallel instructions are the fundamental instructions included in modern CPUs (*e.g.*, SSE, AVX, AVX2 in x86 architecture). By using the data-parallel instructions, we can perform the same operation on multiple data elements simultaneously. CPU usually loads only one element into for each CPU register in *non*-data-parallel computation scheme, whereas the data-parallel instructions enables to load *multiple elements* for each CPU register, and simultaneously perform an operation on the loaded elements.

The maximum number of elements that can be loaded on a register is determined by the size of the register and each element. For example, if a CPU supports 126-bit wide registers, we can load four 32-bit integers for each register. Also, CPUs with AVX2 and AVX-512 enable to perform eight and 16 integers simultaneously since the CPUs have 256-bit and 512-bit wide registers, respectively.

3   Proposed Method: ScaleSCAN

Our goal is to find exactly the same clustering results as those of SCAN from large-scale graphs within short runtimes. In this section, we present details of our proposal, ScaleSCAN. We first overview the ideas underlying ScaleSCAN and then give a full description of our proposed approaches.

3.1   Overview

The basic idea underlying ScaleSCAN is to reduce the computational cost for the structural similarity computations from algorithmic and parallel processing perspectives. Specifically, we first integrate the node pruning algorithms [3] into massively parallel computation scheme on the modern multicore CPU. We then propose the data-parallel algorithm for each structural similarity computation for further improving the clustering efficiency. By combining the node pruning and parallel computing nature, we design ScaleSCAN so as to compute only necessary pairs of nodes.

Algorithm 2 shows the pseudocode of ScaleSCAN. For efficiently detecting nodes that can be pruned, we maintain two integer values sd (*similar-degree*) [3] and ed (*effective-degree*) [3]. Formally, sd and ed are defined as follows:

Definition 5 (Similar-degree). *The similar-degree of node u, denoted $sd[u]$, is the number of neighbor nodes in N_u that have been determined to be structure-similar to node u, i.e., $\sigma(u,v) \geq \epsilon$ for $v \in N_u$.*

Definition 6 (Effective-degree). *The effective-degree of node u, denoted $ed[u]$, is d_u minus the number of neighbor nodes in $N[u]$ that have been determined to be not structure-similar to node u, i.e., $\sigma(u,v) < \epsilon$ for $v \in N_u$.*

Algorithm 2. Proposed algorithm: ScaleSCAN(G, ϵ, μ)

▷ **Step 0: Initialization**
1: **for each** node $u \in V$ **do in thread-parallel**
2:    $sd[u] \leftarrow 0$, and $ed[u] \leftarrow d_u$;

▷ **Step 1: Pre-pruning**
3: **for each** edge $(u, v) \in E$ **do in thread-parallel**
4:    Get $L[(u, v)]$ by Definition 7;
5:    **if** $L[(u, v)] \neq unknown$ **then** UpdateSdEd($L[(u, v)]$);
6: $E^{\text{unknown}} \leftarrow \{(u, v) \in E | L[(u, v)] = unknown\}$

▷ **Step 2: Core detection**
7: **for each** $(u, v) \in E^{\text{unknown}}$ **do in thread-parallel**
8:    **if** $sd[u] < \mu$ and $ed[u] \geq \mu$ **then**
9:       $L[(u, v)] \leftarrow$ PStructuralSimilarity($(u, v), \epsilon$);
10:       UpdateSdEd($L[(u, v)]$);
11: $E^{\text{core}} \leftarrow \{(u, v) \in E | sd[u] \geq \mu \text{ and } sd[v] \geq \mu\}$;

▷ **Step 3: Cluster construction**
12: **for each** $(u, v) \in E^{\text{core}}$ **do in thread-parallel**
13:    **if** find(u) $\neq$ find(v) **then**
14:       **if** $L[(u, v)] = unknown$ **then** $L[(u, v)] \leftarrow$ PStructuralSimilarity($(u, v), \epsilon$);
15:       **if** $L[(u, v)] = similar$ **then** cas_union(u, v);

16: $E^{\text{border}} \leftarrow \{(u, v) \in E \backslash E^{\text{core}} | sd[u] \geq \mu \text{ or } sd[v] \geq \mu\}$;
17: **for each** $(u, v) \in E^{\text{border}}$ **do in thread-parallel**
18:    **if** find(u) $\neq$ find(v) **then**
19:       **if** $L[(u, v)] = unknown$ **then** $L[(u, v)] \leftarrow$ PStructuralSimilarity($(u, v), \epsilon$);
20:       **if** $L[(u, v)] = similar$ **then** cas_union(u, v);

In the beginning of ScaleSCAN shown in Algorithm 2 (Lines 1–2), ScaleSCAN first initializes sd and ed for all nodes. By comparing the two values sd and ed, we determine whether a node should be prune or not in the thread-parallel manner. We describe the details of the node pruning techniques based on sd and ed in Sect. 3.3.

After the initialization, the algorithm consists of three main thread-parallel steps: (Step 1) pre-pruning, (Step 2) core detection, and (Step 3) cluster construction. In the pre-pruning, ScaleSCAN first reduces the size of given graph G in the thread-parallel manner; it prunes edges from E what are obviously either similar or dissimilar without computing the structural similarity. Then, ScaleSCAN extracts all core nodes in the core detection step that is the most time-consuming part in the density-based graph clustering. In order to reduce the computation time for the core detection, ScaleSCAN combines the nodes pruning techniques proposed by Chang *et al.* [3] and the thread-parallelization using the multicore processor. In addition, for further improving the efficiency of the core detection step, we also propose a novel structural similarity computation technique, named PStructuralSimilarity, by using the data-parallel instructions. Finally, in the cluster construction step, ScaleSCAN finds clusters based on

---

Algorithm 3. UpdateSdEd($L[(u, v)]$)

1: **if** $L[(u, v)] = similar$ **then**
2:      $sd[u] \leftarrow sd[u] + 1$ with atomic operation;
3:      $sd[v] \leftarrow sd[v] + 1$ with atomic operation;
4: **else if** $L[(u, v)] = dissimilar$ **then**
5:      $ed[u] \leftarrow ed[u] - 1$ with atomic operation;
6:      $ed[v] \leftarrow ed[v] - 1$ with atomic operation;

Definition 4 by employing union-find tree shown in Sect. 3.4. In the following sections, we describe the details of each thread-parallel step.

3.2   Pre-pruning

In this step, ScaleSCAN reduces the size of graph G by removing $(u, v) \in E$ what can be either $\sigma(u, v) \geq \epsilon$ or $\sigma(u, v) < \epsilon$ without computing the structural similarity defined in Definition 2. Specifically, let $(u, v) \in E$, we always have $\sigma(u, v) \geq \epsilon$ when $\frac{2}{\sqrt{d_u d_v}} \geq \epsilon$ since $|N_u \cap N_v| \geq 2$ from Definition 1. Meanwhile, we also have $\sigma(u, v) < \epsilon$ when $d_u < \epsilon^2 d_v$ (or $d_v < \epsilon^2 d_u$), because if $d_u < \epsilon^2 d_v$ then $\sigma(u, v) < \frac{d_u}{\sqrt{d_u d_v}} < \epsilon$. Clearly, we can check both $\frac{2}{\sqrt{d_u d_v}} \geq \epsilon$ and $d_u < \epsilon^2 d_v$ (or $d_v < \epsilon^2 d_u$) in $O(1)$. Thus, we can efficiently remove such edges from a given graph.

Based on the above discussion, we maintain *edge similarity label* $L[(u, v)]$ for each edge $(u, v) \in E$; an edge (u, v) takes one of the three edge similarity labels, *i.e.*, *similar*, *dissimilar*, and *unknown*.

Definition 7 (Edge similarity label). *Let $(u, v) \in E$, ScaleSCAN assigns the following edge similarity label $L[(u, v)]$ for (u, v):*

$$L[(u, v)] = \begin{cases} similar & (if \ \frac{2}{\sqrt{d_u d_v}} \geq \epsilon) \\ dissimilar & (if \ d_u < \epsilon^2 d_v \ or \ d_v < \epsilon^2 d_u) \\ unknown & (Otherwise) \end{cases} \tag{1}$$

If an edge (u, v) is determined to have $\sigma(u, v) \geq \epsilon$ or $\sigma(u, v) < \epsilon$, we assign $L[(u, v)]$ as *similar* or *dissimilar*, respectively; otherwise, we label the edge as *unknown*. If $L[(u, v)] = unknown$, we can not verify the edge becomes $\sigma(u, v) \geq \epsilon$ or not without computing its structural similarity. Thus, we compute the structural similarity only for $E^{\text{unknown}} = \{(u, v) \in E | L[(u, v)] = unknown\}$ in the subsequent procedure.

The pseudocode of the pre-pruning step is shown in Algorithm 2 (Lines 3–6). In this step, we assign each edge to each thread on the multicore CPU. For each edge (u, v) (Line 3), we first apply Definition 7, and obtain the edge similarity label $L[(u, v)]$ (Line 4). If $L[(u, v)] \neq unknown$, we invoke UpdateSdEd($L[(u, v)]$) (Line 5) for updating sd and ed values according to $L[(u, v)]$ (Lines 1–6 in Algorithm 3). Note that sd and ed are shared by all threads, and thus UpdateSdEd($L[(u, v)]$) has a possibility to cause write conflicts. Hence, to avoid the write conflicts, we use atomic operation (*e.g.,* omp atomic in OpenMP) for updating sd and ed values (Lines 2–3 and Lines 5–6 in Algorithm 3). After the pre-pruning procedure, we extract a set of edges $E^{unknown}$ whose edge similarity label are *unknown* (Line 6).

3.3   Core Detection

As we described in Sect. 2, core detection step is the most time-consuming part since the original algorithm SCAN needs to compute all edges in E. Thus, to speed up the core detection step, we propose a thread-parallel algorithm with the node pruning and data-parallel similarity computation method PStructural-Similarity.

(1) Thread-Parallel Node Pruning: The pseudocode of the thread-parallel node pruning is shown in Algorithm 2 (Lines 7–12). Algorithm 2 (Lines 7–12) detects all core nodes included in G by using the node pruning technique in the thread-parallel manner. As shown in (Line 7) in Algorithm 2, we first assign each edge in $E^{unknown}$ to each thread. In the threads, we compute the structural similarity only for the nodes such that (1) they have not been core or non-core, and (2) they have a possibility to be a core node. Clearly, if $sd[u] \geq \mu$ then node u satisfies the core node condition shown in Definition 3, and also if $ed[u] < \mu$ then node u never satisfies the core node condition; otherwise, we need to compute structural similarities between node u and its neighbor nodes to determine whether node u is core node or not. Hence, once we determine node u is either core or non-core, we stop to compute structural similarities between node u and its neighbor nodes (Line 6). Meanwhile, in the case of $sd[u] < \mu$ and $ed[u] \geq \mu$ (Line 6), we compute structural similarities for node u by PStructuralSimilarity (Line 7), and we finally update sd and ed by UpdateSdEd according to $L[(u,v)]$ (Line 8).

(2) Data-Parallel Similarity Computation: In the structural similarity computation, we propose a novel algorithm PStructuralSimilarity for further improving the efficiency of the core detection step. As we described in Sect. 2.2, each physical core on the modern multicore CPU equips the data-parallel instructions [17] (*e.g.,* SSE, AVX, AVX2 in x86 architecture); data-parallel instructions enable to compute multiple data elements simultaneously by using a single instruction. Our proposal, PStructuralSimilarity, reduces the computation time consumed in the structural similarity computations by using such data-parallel instructions.

Algorithm 4 shows the pseudocode of PStructuralSimilarity. For ease of explanation, we hereafter suppose that 256-bit wide registers are available, and we use 32-bit integer for representing each node in Algorithm 4. That is, we can pack eight nodes into each register. In addition, we suppose that nodes in N_u are stored in ascending order, and we denote $N_u[i]$ to specify i-th element in N_u. Given an edge (u,v) and the parameter ϵ, Algorithm 4 returns whether

$L[(u, v)] = $ *similar* or *dissimilar* based on the structural similarity $\sigma(u, v)$. In the structural similarity computations, the set intersection (*i.e.*, $|N_u \cap N_v|$) is obviously the most time-consuming part since it requires $O(\min\{d_u, d_v\})$ for obtaining $\sigma(u, v) = \frac{|N_u \cap N_v|}{\sqrt{d_u d_v}}$ while the other part (*i.e.*, $\sqrt{d_u d_v}$) can be done in $O(1)$. Hence, in PStructuralSimilarity, we employ the data-parallel instructions to improve the set intersection efficiency.

Algorithm 4 (Lines 6–11) shows our *data-parallel set intersection* algorithm that is consisted of the following three phases:

Phase 1. We load α and β nodes from N_u and N_v as blocks, respectively, and pack the blocks into the 256-bit wide registers, $\mathbf{reg}_u$ and $\mathbf{reg}_v$ (Lines 7–8). Since we need to compare all possible $\alpha \times \beta$ pairs of nodes in the data-parallel manner, we should select α and β so that $\alpha \times \beta = 8$. That is, we have only two choices: $\alpha = 8$ and $\beta = 1$, or $\alpha = 4$ and $\beta = 2$. Thus, we set $\alpha = 8$ and $\beta = 1$ if d_u and d_v are significantly different, otherwise $\alpha = 4$ and $\beta = 2$ (Lines 2–5). dp_load_permute permute nodes in the blocks in the order of permutation arrays π_α and π_β.

Example. If we have sets of loaded nodes $\{u_1, u_2, u_3, u_4\}$ and a permutation array $\pi_\alpha = [4, 3, 2, 1, 4, 3, 2, 1]$, dp_load_permute($\pi_\alpha, \{u_1, u_2, u_3, u_4\}$) loads $[u_4, u_3, u_2, u_1, u_4, u_3, u_2, u_1]$ into $\mathbf{reg}_u$. Also, dp_load_permute($\pi_\beta, \{v_1, v_2\}$) loads $[v_2, v_2, v_2, v_2, v_1, v_1, v_1, v_1]$ into $\mathbf{reg}_v$ for $\{v_1, v_2\}$ and $\pi_\beta = [2, 2, 2, 2, 1, 1, 1, 1]$.

Phase 2. We compare the $\alpha \times \beta$ pairs of nodes by dp_compare in the data-parallel manner. dp_compare compares each pair of nodes in the corresponding position of $\mathbf{reg}_u$ and $\mathbf{reg}_v$. If each pair of nodes has same node it then outputs 1, otherwise 0.

Example. Let $\mathbf{reg}_u = [u_4, u_3, u_2, u_1, u_4, u_3, u_2, u_1]$ and $\mathbf{reg}_v = [v_2, v_2, v_2, v_2, v_1, v_1, v_1, v_1]$, where $u_1 = v_1$ and $u_2 = v_2$, dp_compare outputs $[0, 0, 1, 0, 0, 0, 0, 1]$.

Phase 3. We update the blocks (Lines 10–11) and repeat these phases until we can not load any blocks from N_u or N_v (Line 6).

After the termination, we count the number of common nodes $\triangle$ by (Line 12) in Algorithm 4. Finally, we obtain $L[(u, v)]$ based on $\triangle \geq \epsilon \sqrt{d_u d_v}$ or not (Lines 13–16).

3.4   Cluster Construction

ScaleSCAN finally constructs clusters in the thread-parallel manner. For efficiently maintaining clusters, we use *union-find tree* [4], which can efficiently keep set of nodes partitioned into disjoint clusters. The union-find tree supports two fundamental operations: find(u) and union(u, v). find(u) is an operation to check which cluster does node u belong to, and union(u, v) merges two clusters, which are node u and v belong to. It is known that each operation can be done in $\Omega(A(n))$ where A is Ackermann function, thus we can check and merge clusters efficiently.

Algorithm 4. PStructuralSimilarity$((u, v), \epsilon)$

▷ **Step 0: Initialization**
1: $\triangle \leftarrow 0$, $p_u \leftarrow 0$, $p_v \leftarrow 0$, and $\mathbf{reg}_{add} \leftarrow$ dp_load($[0, 0, 0, 0, 0, 0, 0, 0]$);
2: **if** $d_u > 2d_v$ (or $d_v > 2d_u$) **then**
3:     $\alpha = 8$, $\beta = 1$, $\pi_\alpha \leftarrow [1, 2, 3, 4, 5, 6, 7, 8]$, and $\pi_\beta \leftarrow [1, 1, 1, 1, 1, 1, 1, 1]$;
4: **else**
5:     $\alpha = 4$, $\beta = 2$, $\pi_\alpha \leftarrow [4, 3, 2, 1, 4, 3, 2, 1]$, and $\pi_\beta \leftarrow [1, 1, 1, 1, 2, 2, 2, 2]$;

▷ **Step 1: Data-parallel set intersection**
6: **while** $p_u < d_u$ and $p_v < d_v$ **do**
7:     $\mathbf{reg}_u \leftarrow$ dp_load_permute($\pi_\alpha, [N_u[p_u], \cdots, N_u[p_u + \alpha - 1]]$);
8:     $\mathbf{reg}_v \leftarrow$ dp_load_permute($\pi_\beta, [N_u[p_v], \cdots, N_u[p_v + \beta - 1]]$);
9:     $\mathbf{reg}_{add} \leftarrow$ dp_add($\mathbf{reg}_{add}$, dp_compare($\mathbf{reg}_u, \mathbf{reg}_v$));
10:     **if** $N_u[p_u + \alpha - 1] \geq N_v[p_v + \beta - 1]$ **then** $p_v \leftarrow p_v + \beta$;
11:     **if** $N_u[p_u + \alpha - 1] \leq N_v[p_v + \beta - 1]$ **then** $p_u \leftarrow p_u + \alpha$;

▷ **Step 2: Edge similarity label assignment**
12: $\triangle \leftarrow \triangle +$ dp_horizontal_add($\mathbf{reg}_{add}$);
13: **if** $\triangle < \epsilon\sqrt{d_u d_v}$ **then** $\triangle \leftarrow \triangle + |\{N_u[p_u], \ldots, N_u[d_u]\} \cap \{N_v[p_v], \cdots, N_v[d_v]\}|$;
14: **if** $\triangle \geq \epsilon\sqrt{d_u d_v}$ **then** $L[(u, v)] = similar$;
15: **else** $L[(u, v)] = dissimilar$;
16: **return** $L[(u, v)]$;

Algorithm 2 (Lines 12–20) shows our parallel cluster construction. We first constructs clusters by using only core nodes (Lines 12–15), and then we attach non-core nodes to the clusters (Lines 16–20). Recall that this clustering process is done in the thread-parallel manner. For avoiding conflicts among multiple threads, we thus propose a multi-threading aware union operation, cas_union(u, v). can_union employs *compare-and-swap (CAS)* atomic operation [8] before merging two clusters.

4   Experimental Analysis

We conducted extensive experiments to evaluate the effectiveness of ScaleSCAN. We designed our experiments to demonstrate that:

- **Efficient and Scalable:** ScaleSCAN outperforms the state-of-the-art algorithms pSCAN and SCAN-XP by over one order of magnitude for all datasets. Also, SacaleSCAN is scalable to the number of threads and edges (Sect. 4.2).
- **Effectiveness:** The key techniques of ScaleSCAN, parallel node-pruning and data-parallel similarity computation, are effective for improving the clustering speed on large-scale graphs (Sect. 4.3).
- **Exactness:** Regardless of parallel nodes pruning techniques, ScaleSCAN always returns exactly same clustering results as those of SCAN (Sect. 4.4).

Table 1. Statistics of real-world datasets

Dataset name	# of nodes	# of edges	Data source
DB	317,080	1,049,866	com-DBLP [9]
LJ	4,847,571	68,993,773	soc-livejournal1 [9]
OK	3,072,441	117,185,083	com-orkut [9]
FS	65,608,366	141,874,960	com-friendster [9]
WB	118,142,155	1,019,903,190	webbase-2001 [2]
TW	41,652,230	1,468,365,182	twitter-2010 [2]

4.1 Experimental Setup

We compared ScaleSCAN with the baseline method SCAN [20], the state-of-the-art *sequential* algorithm pSCAN [3], and the state-of-the-art *thread-parallel* algorithm SCAN-XP [18]. All algorithms were implemented in g++ using -O3 option[1]. All experiments were conducted on a CentOS server with an Intel(R) Xeon(R) E5-2690 2.60 GHz GPU and 128 GB RAM. The CPU has 14 physical cores, we thus used threads for up to 14 in the experiments. Since each physical core equips 256-bit wide registers, 256-bit wide data-parallel instructions were also available. Unless otherwise stated, we used default parameters $\epsilon = 0.4$ and $\mu = 5$.

Datasets: We evaluated the algorithms on six real-world graphs, which are downloaded from the Stanford Network Analysis Platform (SNAP) [9] and the Laboratory for Web Algorithmics (LAW) [2]. Table 1 summarizes the statistics of real-world datasets. In addition to the real-world graphs, we also used synthetic graphs generated by LFR benchmark [6], which is considered as the *de facto standard* model for generating graphs. The settings will be detailed later.

4.2 Efficiency and Scalability

Efficiency: In Fig. 1, we evaluated the clustering speed on the real-world graphs through wall clock time by varying ϵ. In this evaluation we used 14 threads for the thread-parallel algorithms, *i.e.*, ScaleSCAN and SCAN-XP. Note that SCAN did not finish its clustering for WB and TW with in 24 h, so we omitted the results from Fig. 1. Overall, ScaleSCAN outperforms SCAN-XP, pSCAN, and SCAN. On average, ScaleSCAN achieves $\times 17.3$ and $\times 90.2$ faster than the state-of-the-art methods SCAN-XP and pSCAN, respectively; also, ScaleSCAN is approximately $\times 500$ faster than the baseline method SCAN. In particular, ScaleSCAN can compute TW with 1.4 billion edges within 6.4 s. Although pSCAN slightly improves its efficiency as ϵ increases, these improvements are negligible.

In Fig. 2, we also evaluated the clustering speeds on FS by varying the parameter μ. As well as Fig. 1, we used 14 threads for ScaleSCAN and SCAN-XP. We

---

[1] We opened our source codes of ScaleSCAN on our website.

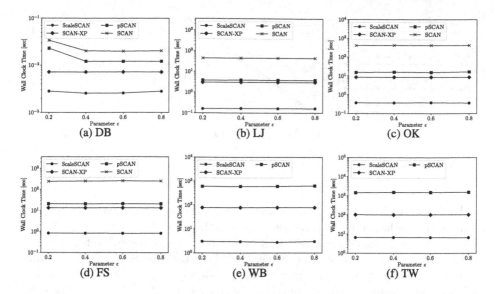

Fig. 1. Runtimes of each algorithm by varying ϵ.

omitted the results for the other datasets since they show very similar results to Fig. 2. As shown in Fig. 2, ScaleSCAN also outperforms the other algorithms that we examined even though ScaleSCAN and pSCAN slightly increase runtimes as μ increases.

Scalability: We assessed scalability tests of ScaleSCAN in Fig. 3a and b by increasing the number of threads and edges, respectively. In Fig. 3a, we used the real-world dataset TW. Meanwhile, in Fig. 3b, we generated four synthetic datasets by using LFR benchmark; we varied the number of nodes from 10^5 to 10^8 with the average degree 30. As we can see from Fig. 3, the runtimes of ScaleSCAN has near-linear in terms of the number of threads and edges. These results verify that ScaleSCAN is scalable for large-scale graphs.

4.3   Effectiveness of the Key Techniques

As mentioned in Sect. 3.3, we employed thread-parallel node pruning and data-parallel similarity computation to prune unnecessary computations. In the following experiments, we examined the effectiveness of the key techniques of ScaleSCAN.

Thread-Parallel Node Pruning. ScaleSCAN prunes nodes that have already been determined as core or non-core nodes in the thread-parallel manner. As mentioned in Sect. 3.3, ScaleSCAN specifies the nodes to be pruned by checking the two integer values sd and ed; ScaleSCAN prunes a node u from its subsequent procedure if $sd[u] \geq \mu$ or $ed[u] < \mu$ since it is determined as core or non-core, respectively.

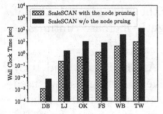
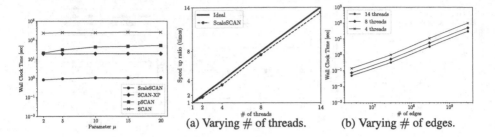

Fig. 2. Runtimes by varying μ on FS.

Fig. 3. Scalability test.

(a) Varying # of threads.

(b) Varying # of edges.

To show the effectiveness, we compared the runtimes of ScaleSCAN with and without the node-pruning techniques. We set the number of threads as 14 for each algorithm. Figure 4 shows the wall clock time of each algorithm for the real-world graphs. Figure 4 shows that ScaleSCAN is faster than ScaleSCAN without the node pruning by over one order of magnitude for all datasets. These results indicate that the node pruning significantly contributes the efficiency of ScaleS-CAN even though it requires several synchronization (*i.e.*, atomic operations) among threads for maintaining sd and ed.

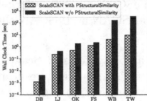
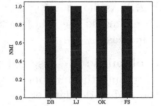

Fig. 4. Effects of the node pruning.

Fig. 5. Effects of PStruc-turalSimilarity.

Fig. 6. Evaluate exact-ness of ScaleSCAN.

Data-Parallel Similarity Computation. As shown in Algorithm 4, ScaleS-CAN computes the structural similarity by using the data-parallel algorithm PStructuralSimilarity. That is, ScaleSCAN compares two neighbor node sets N_u and N_v whether they share same nodes or not in the data-parallel manner. In order to confirm the impact of the data-parallel instructions, we evaluated the running time of a variant of ScaleSCAN that did not use PStructuralSimilarity for obtaining $\sigma(u, v)$.

Figure 5 shows the wall clock time comparisons between ScaleSCAN with and without using PStructuralSimilarity. As shown in Fig. 5, PStructuralSimilarity achieved significant improvements in several datasets, *e.g.* DB, OK, WB, and TW. On the other hand, the improvements seems to be moderated in LJ and FS. More specifically, ScaleSCAN is ×20 faster than ScaleSCAN without PStruc-turalSimilarity on average for DB, OK, WB and TW. Meanwhile, ScaleSCAN is limited to approximately ×2 improvements in LJ and FS.

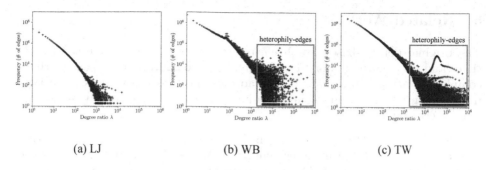

(a) LJ   (b) WB   (c) TW

Fig. 7. Distribution of degree ratio $\lambda_{(u,v)}$

For further discussing about this point, we measured the degree ratio $\lambda_{(u,v)} = \max\{\frac{d_u}{d_v}, \frac{d_v}{d_u}\}$ of each edge $(u,v) \in E$ for LJ, WB, and TW. Figure 7 shows the distributions of the degree ratio for each dataset; horizontal and vertical axis show the degree ratio $\lambda_{(u,v)}$ and the number of edges with the corresponding ratio. In Fig. 7, we can observe that WB has large number of edges with large $\lambda_{(u,v)}$ values while LJ does not have such edges. This indicates that, differ from LJ, edges in WB prefer to connect nodes with different size of degree. Here, let us say an edge with large $\lambda_{(u,v)}$ value as *heterophily-edge*, PStructuralSimilarity can perform efficiently if a graph has many heterophily-edges. This is because that, as shown in Algorithm 4 (Lines 2–3), we can load a lot of nodes from N_u (or N_v) to the 256-bit wise registers at the same time since we set $\alpha = 8$ and $beta = 1$ for the heterophily-edges. In addition, by setting such imbalanced α and $beta$, PStructuralSimilarity is expected to terminate earlier since the while loop in Algorithm 4 (Lines 6–11) stops when $p_u \geq d_u$ or $p_v \geq d_v$. As a result, PStructuralSimilarity thus performs efficiently for the heterophily-edges.

We observed that large-scale graphs tend to have a lot of heterophily-edges because their structure grows more complicated when the graphs become more larger. For example, TW shown in Fig. 7c has a peak around $\lambda_{(u,v)} = 10^5$ values (heterophily-edges), and ScaleSCAN gains large improvements on this dataset (Fig. 5). Thus, these results imply that our approach is effective for large-scale graphs.

4.4 Exactness of the Clustering Results

Finally, we experimentally confirm the exactness of clustering results produced by ScaleSCAN. In order to measure the exactness, we employed the information-theoretic metric, *NMI (normalized mutual information)* [11], that returns 1 if two clustering results are completely same, otherwise 0. In Fig. 6, we compared the clustering results produced by the original method SCAN and our proposed method ScaleSCAN. Since SCAN did not finish in WB and TW within 24 h, we omitted the results from Fig. 6. As we can see from Fig. 6, ScaleSCAN shows 1 for all conditions we examined. Thus, we experimentally confirmed that ScaleSCAN produces exactly same clustering results as those of SCAN.

5   Related Work

The original density-based graph clustering method SCAN requires $O(m^{1.5})$ times and it is known as worst-case optimal [3]. To address the expensive time-complexity, many efforts have been made for the recent few years, especially from sequential and parallel computing perspectives. Here, we briefly review the most successful algorithms.

Sequential Algorithms. One of the major approaches for improving clustering speed is the node/edge pruning techniques: SCAN++ [16] and pSCAN [3] are the representative algorithms. SCAN++ is designed to handle the property of real-world graphs; a node and its two-hop-away nodes tend to have lots of common neighbor nodes since real-world graphs have high clustering coefficients [16]. Based on this property, SCAN++ effectively reduces the number of structural similarity computations. Chang *et al.* proposed pSCAN that employs a new paradigm based on the observations in real-world graphs [3]. By following the observations, pSCAN employs several the nodes pruning techniques and their optimizations for reducing the number of structural similarity computations. To the best of our knowledge pSCAN is the state-of-the-art sequential algorithm that achieves high performance and exact clustering results at the same time. However, SCAN++ and pSCAN ignore the thread-parallel and the data-parallel computation schemes, and thus their performance improvements are still limited. Our work is different from these algorithms in that provides not only the node pruning techniques but also both thread-parallel and data-parallel algorithms. Our experimental analysis in Sect. 4 show that ScaleSCAN is approximately ×90 faster clustering than pSCAN.

Parallel Algorithms. In a recent few years, several thread-parallel algorithms have been proposed for improving the clustering speed of SCAN. To the best of our knowledge, AnySCAN [10], proposed by Son *et al.* in 2017, is the first solution that performs SCAN algorithm on the multicore CPUs. Similar to SCAN++ [16], they applied randomized algorithm in order to avoid unnecessary structural similarity computations. By performing the randomized algorithm in the thread-parallel manner, AnySCAN achieved almost similar efficiency on the multicore CPU compared with pSCAN [3]. Although AnySCAN is scalable on large-scale graphs, it basically produces approximated clustering results due to its randomized algorithm nature.

Takahashi *et al.* recently proposed SCAN-XP [18] that exploits massively parallel processing hardware for the density-based graph clustering. As far as we know, SCAN-XP is the state-of-the-art parallel algorithm that achieves the fastest clustering without sacrificing clustering quality for graphs with millions or even billions of edges. However, different from our proposed method ScaleSCAN, SCAN-XP does not have any node pruning techniques; it need to compute all nodes and edges included in a graph. As shown in Sect. 4, our ScaleSCAN is much faster than SCAN-XP; ScaleSCAN outperforms SCAN-XP by over one order of magnitude for the large datasets.

6   Conclusion

We developed a novel parallel algorithm ScaleSCAN for density-based graph clustering using the multicore CPU. We proposed thread-parallel and data-parallel approaches that combines parallel computation capabilities and efficient node pruning techniques. Our experimental evaluations showed that ScaleSCAN outperforms the state-of-the-art algorithms over one order of magnitude even though ScaleSCAN does not sacrifice its clustering qualities. The density-based graph clustering is now a fundamental graph mining tool to current and prospective applications in various disciplines. By providing our scalable algorithm, it will help to improve the effectiveness of future applications.

Acknowledgement. This work was supported by JSPS KAKENHI Early-Career Scientists Grant Number JP18K18057, JST ACT-I, and Interdisciplinary Computational Science Program in CCS, University of Tsukuba.

References

1. Arai, J., Shiokawa, H., Yamamuro, T., Onizuka, M., Iwamura, S.: Rabbit order: just-in-time parallel reordering for fast graph analysis. In: Proceedings of the 2016 IEEE International Parallel and Distributed Processing Symposium, pp. 22–31 (2016)
2. Boldi, P., Vigna, S.: The webgraph framework I: compression techniques. In: Proceedings of the 13th International Conference on World Wide Web, pp. 595–601 (2004)
3. Chang, L., Li, W., Qin, L., Zhang, W., Yang, S.: pSCAN: fast and exact structural graph clustering. IEEE Trans. Knowl. Data Eng. **29**(2), 387–401 (2017)
4. Cormen, T.H., Leiserson, C.E., Rivest, R.L., Stein, C.: Introduction to Algorithms. The MIT Press, Cambridge (2009)
5. Ding, Y., et al.: atBioNet–an integrated network analysis tool for genomics and biomarker discovery. BMC Genom. **13**(1), 1–12 (2012)
6. Fortunato, S., Lancichinetti, A.: Community detection algorithms: a comparative analysis. In: Proceedings of the 4th International ICST Conference on Performance Evaluation Methodologies and Tools, pp. 27:1–27:2 (2009)
7. Fujiwara, Y., Nakatsuji, M., Shiokawa, H., Ida, Y., Toyoda, M.: Adaptive message update for fast affinity propagation. In: Proceedings of the 21st ACM SIGKDD International Conference on Knowledge Discovery and Data Mining, pp. 309–318 (2015)
8. Herlihy, M.: Wait-free synchronization. ACM Trans. Program. Lang. Syst. **13**(1), 124–149 (1991)
9. Leskovec, J., Krevl, A.: SNAP Datasets: Stanford Large Network Dataset Collection, June 2014. http://snap.stanford.edu/data
10. Mai, S.T., Dieu, M.S., Assent, I., Jacobsen, J., Kristensen, J., Birk, M.: Scalable and interactive graph clustering algorithm on multicore CPUs. In: Proceedings of the 33rd IEEE International Conference on Data Engineering, pp. 349–360 (2017)
11. Manning, C.D., Raghavan, P., Schütze, H.: Introduction to Information Retrieval. Cambridge University Press, New York (2008)

12. Naik, A., Maeda, H., Kanojia, V., Fujita, S.: Scalable Twitter user clustering approach boosted by personalized PageRank. In: Kim, J., Shim, K., Cao, L., Lee, J.-G., Lin, X., Moon, Y.-S. (eds.) PAKDD 2017. LNCS, vol. 10234, pp. 472–485. Springer, Cham (2017). https://doi.org/10.1007/978-3-319-57454-7_37
13. Sato, T., Shiokawa, H., Yamaguchi, Y., Kitagawa, H.: FORank: fast ObjectRank for large heterogeneous graphs. In: Companion Proceedings of the the Web Conference, pp. 103–104 (2018)
14. Shi, J., Malik, J.: Normalized cuts and image segmentation. IEEE Trans. Pattern Anal. Mach. Intell. **22**(8), 888–905 (2000)
15. Shiokawa, H., Fujiwara, Y., Onizuka, M.: Fast algorithm for modularity-based graph clustering. In: Proceedings of the 27th AAAI Conference on Artificial Intelligence, pp. 1170–1176 (2013)
16. Shiokawa, H., Fujiwara, Y., Onizuka, M.: SCAN++: efficient algorithm for finding clusters, hubs and outliers on large-scale graphs. Proc. Very Large Data Bases **8**(11), 1178–1189 (2015)
17. Solihin, Y.: Fundamentals of Parallel Multicore Architecture, 1st edn. Chapman & Hall/CRC, Boca Raton (2015)
18. Takahashi, T., Shiokawa, H., Kitagawa, H.: SCAN-XP: parallel structural graph clustering algorithm on Intel Xeon Phi coprocessors. In: Proceedings of the 2nd International Workshop on Network Data Analytics, pp. 6:1–6:7 (2017)
19. Wang, L., Xiao, Y., Shao, B., Wang, H.: How to partition a billion-node graph. In: Proceedings of the IEEE 30th International Conference on Data Engineering, pp. 568–579 (2014)
20. Xu, X., Yuruk, N., Feng, Z., Schweiger, T.A.J.: SCAN: a structural clustering algorithm for networks. In: Proceedings of the 13th ACM SIGKDD International Conference on Knowledge Discovery and Data Mining, pp. 824–833 (2007)

Sequence-Based Approaches to Course Recommender Systems

Ren Wang and Osmar R. Zaïane[✉] [iD]

University of Alberta, Edmonton, Canada
{ren5,zaiane}@cs.ualberta.ca

Abstract. The scope and order of courses to take to graduate are typically defined, but liberal programs encourage flexibility and may generate many possible paths to graduation. Students and course counselors struggle with the question of choosing a suitable course at a proper time. Many researchers have focused on making course recommendations with traditional data mining techniques, yet failed to take a student's sequence of past courses into consideration. In this paper, we study sequence-based approaches for the course recommender system. First, we implement a course recommender system based on three different sequence related approaches: process mining, dependency graph and sequential pattern mining. Then, we evaluate the impact of the recommender system. The result shows that all can improve the performance of students while the approach based on dependency graph contributes most.

Keywords: Recommender systems · Dependency graph
Process mining

1 Introduction

After taking some courses, deciding which one to take next is not a trivial decision. A recommendation of learning resources relies on a recommender system (RS), a technique and software tool providing suggestions of items valuable for users [14]. The typical approaches to recommend an item are based on ranking some other items similar to another item a user or a customer has already taken, purchased, or liked. These are called Content-based recommender systems [3]. However, recommending a course simply based on similarity with previously taken courses may not be the right thing to do. In practice, in addition to course prerequisite constraints, when the curriculum is liberal, students typically chose courses where their friends are, or based on their friends suggestions (i.e. ratings). Collaborative filtering [16] is another approach for recommender systems that could be used to recommend courses. It relies on the wisdom of the crowd, -i.e. the learners that are similar to the current students in terms of courses taken or "liked". However, the exact sequence these courses are taken is not considered. The order and succession of courses is indeed relevant in choosing the next course to take. The questions students may ask include but are not restricted to: how

© Springer Nature Switzerland AG 2018
S. Hartmann et al. (Eds.): DEXA 2018, LNCS 11029, pp. 35–50, 2018.
https://doi.org/10.1007/978-3-319-98809-2_3

can I finish my study as soon as possible? Is it more advantageous to take course A before B or B before A? What is the best course for me to take this semester? Will it improve my GPA if I take this course? Answering such questions to both educators and students can greatly enhance the educational experience and process. However, very few course RS (CRS) currently take advantage of this unique sequence characteristic.

Recommender systems are widely used in commercial systems and while rarely deployed in the learning environments, their use in the e-learning context has already been advocated [9,24]. The overall goal of most RS in education is to improve students' performance. This goal can be achieved in diverse ways by recommending various learning resources [18]. A common idea is to recommend papers, books and hyperlinks [6,8,17]. Course enrollment can also be recommended [5,10]. However, most RS only apply content-based or collaborative filtering approaches, and none have considered exploiting the order of how students take courses. This missing link is what this paper tries to address.

The goal of our paper is to investigate a sequence-based CRS and show that it is possible. We study three sequence-based approaches to build this RS using process mining, dependency graphs, and sequential pattern mining.

2 CRS Based on Process Mining

2.1 Review of Process Mining

Process mining (PM) is an emerging technique that can discover the real sequence of various activities from an event log, compare different processes and ultimately find the bottlenecks of an existing process and hence improve it [20]. To be specific, PM consists of extracting knowledge from event logs recorded by an information system and discovering business process from these event logs (process discovery), comparing processes and finding discrepancies between them (Process Conformance), and providing suggestions for improvements in these processes (Process Enhancement).

Some attempts have already been made to exploit the power of PM in curriculum data. For instance, authors of one section in [15] indicate that it can be used in educational data. However, the description is too general and not enough examples are given. The authors of [19] point out the significant benefit in combining educational data with PM. The main idea is to model a curriculum as a coloured Petri net using some standard patterns. However, most of the contribution is plain theory and no real experiment is conducted. Targeted curriculum data and thereby curriculum mining is explored in [11]. Similar with the three components of PM, it clearly defines three main tasks of curriculum mining, which are curriculum model discovery, curriculum model conformance checking and curriculum model extensions. The authors explain vividly how curriculum mining can answer some of the questions that teachers and administrators may ask. However, no RS is built upon it.

2.2   Implementation of a CRS Based on Process Mining

We recommend courses to a student that successful students who have a similar course path have taken. Our course data are different from typical PM data at least in the following three aspects: First, the order of the activities is not rigidly determined. Students are quite free to take the courses they like and they do not follow a specific order. Granted that there are restrictions such as prerequisite courses or the courses we need to take in order to graduate, these dependencies are relatively rare compared with the number of courses available. Second, the dependency length is relatively short. In the course history data, we do not have a long dependency. We may have a prerequisite requirement, e.g., we must take CMPUT 174 and CMPUT 204 first in order to take CMPUT 304, but such dependency is very short. Third, the type of activities in the sequence are not singletons. Data from typical PM problems are sequence of single activities, while in our case they are a sequence of sets. Students can take several courses in the same term, which makes it more difficult to represent in a graph.

For these reasons, we do not attempt to build a dependency graph, and proceed directly to conformance checking. The intuition behind our algorithm is to recommend the path that successful students take, i.e., to recommend courses taken by the students who are both successful and similar to our students who need help. We achieve this by the steps in Algorithm 1.

Algorithm 1. Algorithm of CRS based on PM

Input :
    Logs L of finished students course history
    Student stu who needs course recommendations
Execute :
 1: Find all high GPA students from L as HS
 2: Set candidate courses $CC = \emptyset$
 3: **for all** $stuHGPA$ in HS **do**
 4:    Apply Algorithm 2 to compute the similarity sim between stu and $stuHGPA$
 5:    **if** sim is greater than a certain threshold **then**
 6:        Add courses that $stuHGPA$ take next to CC
 7:    **end if**
 8: **end for**
 9: Rank CC based on selected metrics
10: Recommend the top courses from CC to stu

In Algorithm 1 we first find the history of all past successful students. We assume success is measured based on final GPA. Other means are of course possible. From this list we only keep the successful students who are similar to the current student based on some similarity metric, and retain the courses they took as candidate courses to recommend. These are finally ranked and the top are recommended. The ranking is explained later.

The method we use to compute the similarity between two students is highlighted in Algorithm 2. It is an improved version of the casual footprint approach for conformance checking in PM. Instead of building a process model, we apply or method directly on the sequence of sets of courses to build the footprint tables. In addition, we define some new relations among activities, courses in our case, due to the special attributes of course history and the sequence of set.

- Direct succession: $x \rightarrow y$ iff x is directly followed by y
- Indirect succession: $x \rightarrow\rightarrow y$ iff x is indirectly followed by y
- Reverse direct succession: $x \leftarrow y$ iff y is directly followed by x
- Reverse indirect succession: $x \leftarrow\leftarrow y$ iff y is indirectly followed by x
- Same term: $x \parallel y$ iff x and y are in the same term
- Other: $x \# y$ for Initialization or if x and y have the same name.

With the relation terms defined, we can proceed to our improved version of the footprint algorithm which computes the similarity of two course history sequences.

Algorithm 2. Algorithm of computing the similarity of two course history sequences

---

Input :
    Course history sequence of the first student s_1
    Course history sequence of the first student s_2
Output :
  1: Truncate the longer sequence to the same length with the shorter sequence
  2: Build two blank footprint tables that map between s_1 and s_2
  3: Fill out two footprint tables based on s_1 and s_2
  4: Calculate the total elements and the number of elements that are different
  5: Compute the similarity
  6: Return the similarity of s_1 and s_2

In most cases, finished students' course histories are much longer than the current students'. To eliminate this difference we truncate the longer sequence to the same length of the shorter sequence. The next step is to build a one-to-one mapping of all courses in both sequences. Our CRS computes the above defined relations based on the two sequences and fills the relations in the footprint table separately. Lastly, our CRS calculates $differenceCount$ which is the number of elements in the footprint tables that s_1 differs from s_2, and $totalCount$ which is the total number of elements in one footprint table. $similarity$ is then:

$$similarity = 1 - \frac{differenceCount}{totalCount}. \tag{1}$$

3 CRS Based on Dependency Graph

3.1 Review of Dependency Graph (DG)

A primitive method to discover DG from event data is stated in [1]. The dependency relation is based on the intuition that for two activities A and B, if B

follows A but A does not follow B, then B is dependent on A. If they both follow each other in the data, they are independent. In fact, this simple intuitive idea lays the foundation for many process discovery algorithms in PM. These are, however, more advanced, as they use Petri nets [13] to deal with concurrency and satisfy other criteria, such as the Alpha Algorithm [21], the heuristic mining approach [23], and the fuzzy mining approach [7]. These approaches are, however, not quite suitable for our task. Our method here is based on [4]. The authors developed an approach of recommending of learning resources for users based on users' previous feedback. It learns a DG by users' ratings. Learners are required to give a rating or usefulness of the resources they used. The database evolves by filtering learning objects with low ratings as time goes by. The dependencies are discovered based on these ratings, positive or negative, using an association rule mining approach.

3.2   Implementation of a CRS Based on Dependency Graph

The method in [4] is to recommend resources to learners based on what learners have seen and rated. It creates dependencies between items i and items j only if an item j is always positively rated immediately upon appearing *after* an always positively rated i when it is *before* j, and independent or ignored otherwise. Resource j is dependent on i in the pair (i, j) based on ratings.

Admittedly, the approach is simple but has drawbacks (i.e. linear, no context used, and ignores noise), but we propose to adapt it to make it more suitable to our case of courses, and improved it as follows. We cannot ask students to rate all the courses they have taken, as these may not be very reliable for building dependencies. The indicator we built our dependencies upon is the mark obtained by students in courses. A good mark for course i before a good mark of course j often implies course i is the prerequisite or positively influencer of course j. Moreover, instead of using a universal notion of positive and negative as for the ratings, A positive mark in a course or a negative mark is defined relative to a student. A $B+$ may be a good mark in general, but for a successful student whose mark is A on average, $B+$ is not that good. Moreover, we use association rule mining parameters *support* (indicating frequency) and *confidence* (indicating how often a rule has been found to be true) to threshold pairs of courses with positive marks, and thus reduce potential noise.

Algorithm 3 outlines our approach with the above rationale. The CRS first learns dependencies from the finished students' course history. For a student who needs recommendations, the CRS checks the previous course history of this student and compares this history with the dependencies the CRS has learned. A ranking of the candidate courses constitutes the final recommendation.

Algorithm 3. Algorithm of CRS based on DG

Input :
    Logs L of finished students course history
    Student stu who needs course recommendations
Execute :
  1: Convert all marks of courses from L to positive or negative signs. The standard
     may differ based on GPA to make it relative to individual students
  2: Build the projected dataset of positive courses P_{i+} and negative courses P_{i-} with
     the highlighted modification. Remove courses in P_{i-} from P_{i+}
  3: Set candidate courses $CC = \emptyset$
  4: Add to CC courses in P_{i+} whose prerequisites are finished
  5: Rank CC based on selected metrics
  6: Recommend the top courses from CC to stu

---

4 CRS Based on Sequential Pattern Mining

4.1 Implementation of a CRS Based on Sequential Pattern Mining

Sequential pattern mining (SPM) consists of discovering frequent subsequences in a sequential database [2]. There are many algorithms for SPM but we adopt the widely used PrefixSpan [12] because of its recognized efficiency.

SPM was introduced and is typically used in the context of market basket analysis. The sequences in the database are the progression of items purchased together each time a purchaser comes back to a store, and SPM consists of predicting the next items that are likely to be purchased at the next visit. Students take few courses each term. There is no order of courses in a specific term, yet the courses of different terms do follow a chronological order. The analogy with market basket analysis is simple. A semester for a student is a store visit, and the set of courses taken during a semester are the items purchased together during one visit. Just like frequent sequence patterns of items bought by customers can be found, so can frequent sequence patterns of courses taken by students.

Our CRS Algorithm 4 based on SPM works as follows. Since we only want to find the sequential patterns of positive courses, i.e., sequences of courses taken by students with good outcome, we first filter all the course records and only keep a course record when the mark is A or $A+$. Here $A+$ and A are taken as reference examples. Note that a course deleted in one sequence of a student may be selected in another sequence for another student. For instance, a student who took CMPUT 101 and received an A then this course is kept in this student's sequence. If another student who also took CMPUT 101 but received a B this course is filtered from their sequence. After this step, the course records left in students history are all either A or $A+$. The second step in the algorithm is to treat these courses like the shopping items and process them with PrefixSpan [12] to find all the sequential patterns of courses. Among the course sequential patterns we find, some are long, while some are short. Ideally, we want to recommend courses from the most significant patterns.

Algorithm 4. Algorithm of CRS based on SPM

Input :
    Logs L of finished students course history
    Student stu who needs course recommendations

Execute :
1: Filter all the course records of L with a predefined course mark standard as FL
2: Find all the course sequential patterns SP from FL with PrefixSpan [12].
3: **for all** Sequential pattern p from SP **do**
4:     Compute the number of elements num of this sequential pattern that is also contained in stu's course history
5:     Add the next course of this p to the Hashtable HT where the key is num
6: **end for**
7: Rank courses from HT's highest key as candidate courses CC based on selected metrics
8: Recommend the top courses from CC to stu

Suppose we have a student who needs course recommendations and has already taken courses 174, 175, and 204. We have discovered a short frequent pattern $s_1 = \langle 174, 206 \rangle$ while another long frequent pattern s_2 we discovered is $\langle 174, 175, 204, 304 \rangle$. A more intuitive recommendation should be 304 because the student has already finished three courses in s_2.

Based on this intuition, the courses we recommend are the next unfinished elements from the sequential patterns that spm the longest common elements with our student's current course history. By this algorithm, the course we recommend for our example student earlier will be course 304 since the length of common elements of s_2 and this student is three, longer than one which is of s_1.

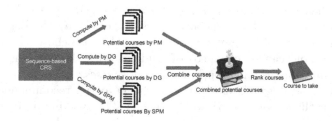

Fig. 1. The overall workflow of our CRS that combines all 3 sequence-based algorithms

In addition to the three approaches for CRS, PM-based, DG-based, and SPM-based, we combine all of our three sequence-based methods into one comprehensive one. We call it "Comprehensive" in our experiments. Since each of them produces a potential list of recommended courses, it is straight forward to combine the result of potential courses of all three methods and rank the result. The overall structure of this approach is shown in Fig. 1.

5   Ranking Results

All methods previously mentioned focus more on student's course performance, which we approximate with the GPA. Of course, other learning effectiveness measure alternatives exist. Since the quickness of a program before graduation is also of concern to many learners who would like to graduate as soon as possible, we also consider the length of sequences of courses before graduation in our recommendation. To do this, we incorporate this notion in the ranking of the candidate courses before taking the top to recommend.

The sequence of some courses and the number of courses and the compulsory courses to graduate are dictated by the school or department program. These requirements can be obtained from the school guidelines. Most of these programs, however, are liberal not enforcing most constraints and contain many electives. These optional courses can be further considered in two aspects: First, these courses may be very important that many students decide to take them even though they are not in the mandatory list. We can compute the percentage of students who take a specific course and rank courses based on this percentage from high to low. It could be a must for students who want to graduate as soon as possible if the percentage of students who take this course is above a certain threshold. The second aspect to distinguish courses that can speed up graduation is their relationship with the average duration before graduation. For one course, we can compute the average time needed to graduate by students who take this specific course. We do this for all the courses and rank them based on the average graduation time from low to high, the lower the number the faster a student graduates, i.e. the likelier it contributes to the acceleration of graduation. In short, there are three attributes we consider: First, the course is mandatory from the department's guideline; Second, is the percentage of students who take this course; Third, is the average time before graduation by students who take this course. The second category can actually be merged into the first category since they both indicate how crucial a course is, either by the department or the choice of students. We combine the courses that are chosen by more than 90% (this threshold can be changed) of students with the compulsory courses specified by educators as one group we call *key courses*.

This "agility strategy" is used to rank the potential recommended courses selected by our three sequence-based algorithms. This ranking process is always the last step of these three sequential based algorithms. To be more exact, after selecting a few courses in the potential course list by one of the three sequence-based approaches, there are three methods to rank them with this "agility" algorithm.

1. No "agility": Rank courses merely on the GPA contribution of courses.
2. Semi "agility": Always rank key courses that are in the potential course list first. The key course list and the non-key course list will be ranked based on each course's GPA contribution respectively.
3. Full "agility": Always rank key courses that are in the potential course list first. The key course list and the non-key course list will be ranked based on each course's average graduation time by students who take this course.

6 Experiments

6.1 Data Simulator

The Computing Science Department of the University of Alberta collects for each semester and for each student the courses they register in and the final mark they obtain. While there are prerequisites for courses and other strict constraints, the rules are not enforced and are thus often violated, giving a plethora of paths to graduation. This history for many years, constituting the exact needed event log, is readily available. However, such data cannot be used for research purposes or for publication even though anonymized due to lack of ethical approval. Indeed, we would need inaccessible consent from alumni learners. It is hopeless to gather the consent of all past students, and impractical to start collecting written consent from new students as it would require years to do so. We were left with alternative to simulate historic curriculum data for proof of concept and publication, and use real data for local implementation. For this paper we opted for the simulation of the event log. A simulator was developed to mimic the behaviours of undergraduate students with different characters in higher education. The simulator encompasses the dynamic course directory and the rules of enrollment, as well as student behaviour such as performance and diligence in following guideline rules. The detail of the simulator simulating arriving and graduating students one semester at a time can be found in [22].

6.2 Result Analysis

In this section we compare the performance of our CRS based on different sequence-based algorithms. We want to see which sequence-based algorithm performs better, whether the "speedup" algorithm works, and what additional insights our CRS can provide. Moreover, we add one more approach to all experiments, which is called "comprehensive" that combines all results from the three methods. If not otherwise specified, the parameters of each algorithm are the ones that performed best. The numbers presented in each table and figure for this section are the average scores of their corresponding experiment three times since the simulation is stochastic.

The first experiment is to compare the performance of different sequence-based approaches at different student stages. "Different stages" means when do students use our CRS. For example, "Year 4" means students only begin to take courses recommended by our CRS in the fourth year, while "Year 1" means students start using our CRS from the first year. Table 1 with its corresponding Fig. 2 shows the result of this experiment: 200 students' average GPAs varied by the year of starting CRS in different approaches. The blue line in the middle is our baseline 3.446 which is the average GPA if students do not take any recommendations. From Table 1 and Fig. 2 we can observe the following. Firstly, we can see a substantial effect for students who use our CRS in the first two years. This steady increase indicates students can benefit more if they start using our CRS earlier in their study. Secondly, the performance of CRS for all methods is about the same with the baseline if students only start to use our CRS in the fourth year, which means it may be too late to improve a student's GPA even with the help of a CRS. Other than Year 4, our CRS does have a positive impact. Thirdly, CRS based on DG outperforms all in nearly all scenarios while other approaches are equally matched. Note that the comprehensive approach does not outperform others. Our interpretation is that by combining the candidate courses from all three methods, it obtains too many candidates and cannot perform well if the candidates are not ranked properly. As to why CRS based on DG performs best, it may be due to the intrinsic attribute of our data simulator. The mark generation part of our simulator considers course prerequisites, which may favour the DG algorithm. Thus, other approaches may outweigh DG if we are dealing with real data.

Table 1. 200 students' average GPAs varied by the year CRS is used by different approaches

Approach	Year 4	Year 3	Year 2	Year 1
PM	3.453	3.516	3.569	3.588
DG	3.433	3.529	3.617	3.652
SPM	3.447	3.498	3.545	3.602
Comprehensive	3.441	3.512	3.564	3.593

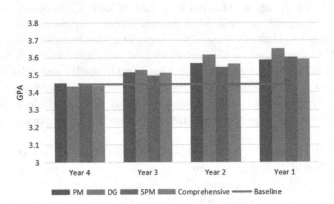

Fig. 2. 200 students' average GPAs varied by the year CRS is used by different approaches (Color figure online)

The next experiment is to check whether increasing the training data in the number of students would lead to a better performance of our CRS. Table 2 and Fig. 3 demonstrate 200 students' average GPAs varied by the number of training students of CRS in different approaches. We can see that, as the training data

Table 2. 200 students' average GPAs varied by the number of training students of CRS in different approaches

Approach	500	1000	1500
PM	3.513	3.57	3.586
DG	3.535	3.607	3.639
SPM	3.528	3.581	3.598
Comprehensive	3.522	3.582	3.597

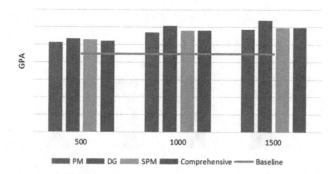

Fig. 3. 200 students' average GPAs varied by the number of training students of CRS in different approaches

size increases from 500 to 1000, the performance of our CRS improves. However, when this size further increases from 1000 to 1500, the performance of our CRS does not improve significantly. We than fixed the training data size to 1500 in all our experiments. This can be explained by the fact that the number of courses in a program is finite and small (even though dynamic) and all important dependencies are already expressed in a relatively small training dataset.

Besides improving students' performance in grades, our CRS can also speed up students' graduation process by ranking the candidate courses selected by sequence algorithms properly. Table 3 and Fig. 4 show the effect of using the full "agility" ranking setting to recommend courses based on DG to 200 students. Same as the first experiment in this section, Year X means students start to use our CRS from year X. We can see a remarkable decrease in the number of terms needed to graduate if students start using our CRS from the third year. However, after that, such change is not very notable. Since the pivotal fact to graduate fast is to take all key courses as soon as possible, our explanation is that taking key courses from the third year is timely. There is no particular need to focus on key courses in the first two years. Note that although the graduation time improvement of our CRS is only in a decimal level, it is already quite a boost considering students only need to study 12 terms in normal scenarios.

Table 3. 200 students' average graduation terms varied by the year of starting CRS based on DG with the full "agility" setting

Starting Year	Average graduation terms
Year 4	11.917
Year 3	11.615
Year 2	11.567
Year 1	11.532

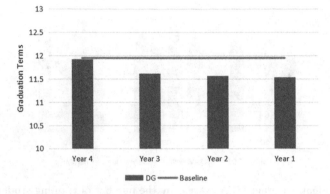

Fig. 4. 200 students' average graduation terms varied by the year of starting CRS based on DG with the full "agility" setting

Other than recommending courses, our CRS may provide some insights to educators and course counselors. We previously mentioned computing courses' GPA contribution and graduation time contribution. A course's GPA contribution is the average GPA of students who take this course, while a course's graduation time contribution is the average time before graduation of students who take this course. These indicators are used to rank the candidate courses obtained by sequence-based algorithms. Yet, these indicators themselves may have values. Table 4 demonstrates the top 5 GPA contribution courses and graduation time contribution courses. One interesting finding is course CMPUT 201. This course is not one of the preferred courses in our simulator but is a prerequisite course for many courses. A preferred course is a course that will have a very high probability to be taken in a particular term because it is the "right" course for that term. Being a prerequisite course but not a preferred course means that, CMPUT 201 has to be taken in order to perform well in other courses but many students do not take it. Thus, finding this course actually means that our CRS found an important course that is not in the curriculum but is necessary for students to succeed. Sometimes it is risky to force to do so. For example, CMPUT 275 is in the top position in the GPA contribution list, but we cannot know whether this course causes students to succeed or successful students like to take it. Nevertheless, this contribution list would still provide some insights to educators and course counselors if it is trained on real students' data and is carefully interpreted.

Table 4. The top 5 GPA contribution courses and graduation time contribution courses

Ranking	Top GPA courses	Top time courses
1	CMPUT 275	CMPUT 301
2	CMPUT 429	CMPUT 274
3	CMPUT 350	CMPUT 300
4	CMPUT 333	CMPUT 410
5	CMPUT 201	CMPUT 366

Finally, our CRS can assist educators and administrators to gain deep insights on course relations and thus improve the curriculum. Figure 5 (Left) shows the DG of courses with edge colours representing discovery sources (green = imposed and confirmed; blue = expected but not found; red = new discovered). It combines the prerequisite relations used by our simulator and the dependencies discovered by our DGA. On one hand, we can consider the prerequisite course relations used by our simulator as the "current curriculum" or behaviours we expect to see from students. On the other hand, the courses' prerequisite relations discovered by our CRS based on .the DG algorithm can be deemed as the prerequisite relations in reality or the actual behaviours by students. Many dependencies used by our simulator are found by our DG algorithm (green edges) like 204⇒304, which means that these rules are successfully carried out by students. Some dependencies used by our simulator are not found in the data (blue edges) like 175⇒229 because the students did not actually follow them, which indicates there are some discrepancies between what we expect from students and what students really do. Administrators may want to check why this happens. There are also some dependencies found by our DG algorithm but are not in the rules for our simulator (red edges), such as 304⇒366 and 272⇒415. These dependencies indicate some relations among courses unknown and unexpected to administrators but are performed by students. Educators and administrators may want to consider to add these new found prerequisites to the curriculum in the future if these are indicative of good overall performance in terms of learning objectives.

Figure 5 (Right) shows the paths of successful students (GPA above 3.8) filtered from the 1500 training students with the weight of edges representing the number of students. The thick edges mean many successful students have gone through these paths and they should be considered when trying to improve the curriculum. All in all, the benefits of these findings can be considerable when sequences of courses are taken into account.

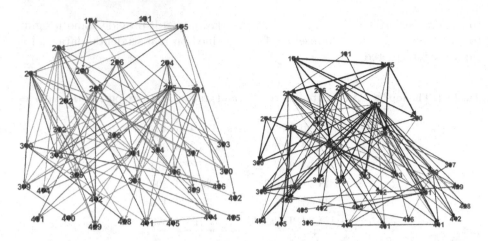

Fig. 5. Left: the DG of courses with edge colours representing discovery sources (green = imposed and confirmed; blue = expected but not found; red = new discovered). Right: the paths of successful students filtered from the 1500 training students with the weight of edges representing the number of students. (Color figure online)

7   Conclusions and Future Work

We built a course recommender system to assist students choose suitable courses in order to improve their performance. This recommender is based on three different methods yet all three are related to the sequence of taken course. We considered conformance checking of process mining as a first approach, recommending courses to a student that successful students, who have a similar a course path, have taken. We have also suggested a new approach based on dependency graphs modeling deep prerequisite relationships, by recommending courses whose prerequisites are finished. We also advocated a third method based on sequential pattern mining discovering frequent sequential course patterns of successful students. Finally, we combined all the approaches in a comprehensive method and proposed ranking methods to favour reducing the program length.

We conduct several experiments to evaluate our course recommender systems and to find the best recommendation approach. All three approaches can improve students' performance in different scales. The best recommendation method is based on the dependency graph, and the number of recommended courses accepted by students have a positive correlation with the performance. Moreover, the course recommender system we build can speed up students' graduation if set properly, and provide some useful insights for educators and course counselors.

References

1. Agrawal, R., Gunopulos, D., Leymann, F.: Mining process models from work-flow logs. In: Schek, H.-J., Alonso, G., Saltor, F., Ramos, I. (eds.) EDBT 1998. LNCS, vol. 1377, pp. 467–483. Springer, Heidelberg (1998). https://doi.org/10.1007/BFb0101003
2. Agrawal, R., Srikant, R.: Mining sequential patterns. In: Proceedings of the 11th International Conference on Data Engineering, pp. 3–14. IEEE (1995)
3. Burke, R.: Hybrid web recommender systems. In: Brusilovsky, P., Kobsa, A., Nejdl, W. (eds.) The Adaptive Web. LNCS, vol. 4321, pp. 377–408. Springer, Heidelberg (2007). https://doi.org/10.1007/978-3-540-72079-9_12
4. Cummins, D., Yacef, K., Koprinska, I.: A sequence based recommender system for learning resources. Aust. J. Intell. Inf. Process. Syst. **9**(2), 49–57 (2006)
5. García, E., Romero, C., Ventura, S., De Castro, C.: An architecture for making recommendations to courseware authors using association rule mining and collaborative filtering. User Model. User-Adap. Interact. **19**(1–2), 99–132 (2009)
6. Ghauth, K.I., Abdullah, N.A.: Learning materials recommendation using good learners' ratings and content-based filtering. Educ. Technol. Res. Dev. **58**(6), 711–727 (2010)
7. Günther, C.W., van der Aalst, W.M.P.: Fuzzy mining – adaptive process simplification based on multi-perspective metrics. In: Alonso, G., Dadam, P., Rosemann, M. (eds.) BPM 2007. LNCS, vol. 4714, pp. 328–343. Springer, Heidelberg (2007). https://doi.org/10.1007/978-3-540-75183-0_24
8. Luo, J., Dong, F., Cao, J., Song, A.: A context-aware personalized resource recommendation for pervasive learning. Cluster Comput. **13**(2), 213–239 (2010)
9. Manouselis, N., Drachsler, H., Vuorikari, R., Hummel, H., Koper, R.: Recommender systems in technology enhanced learning. In: Ricci, F., Rokach, L., Shapira, B., Kantor, P.B. (eds.) Recommender Systems Handbook, pp. 387–415. Springer, Boston, MA (2011). https://doi.org/10.1007/978-0-387-85820-3_12
10. O'Mahony, M.P., Smyth, B.: A recommender system for on-line course enrolment: an initial study. In: Proceedings of the 2007 ACM Conference on Recommender Systems, pp. 133–136. ACM (2007)
11. Pechenizkiy, M., Trcka, N., De Bra, P., Toledo, P.: CurriM: curriculum mining. In: International Conference on Educational data Mining, pp. 216–217 (2012)
12. Pei, J., et al.: PrefixSpan: mining sequential patterns efficiently by prefix-projected pattern growth. In: Proceedings of the 17th International Conference on Data Engineering. IEEE (2001)
13. Peterson, J.L.: Petri Net Theory and the Modeling of Systems, vol. 132. Prentice-Hall, Englewood Cliffs (1981)
14. Ricci, F., Rokach, L., Shapira, B.: Introduction to recommender systems handbook. In: Ricci, F., Rokach, L., Shapira, B., Kantor, P.B. (eds.) Recommender Systems Handbook, pp. 1–35. Springer, Boston, MA (2011). https://doi.org/10.1007/978-0-387-85820-3_1
15. Romero, C., Ventura, S., Pechenizkiy, M., Baker, R.S.: Handbook of Educational Data Mining. CRC Press, Boca Raton (2010)
16. Sarwar, B., Karypis, G., Konstan, J., Riedl, J.: Item-based collaborative filtering recommendation algorithms. In: Proceedings of the 10th International Conference on World Wide Web, pp. 285–295. ACM (2001)
17. Tang, T.Y., McCalla, G.: Smart recommendation for an evolving e-learning system. In: Workshop on Technologies for Electronic Documents for Supporting Learning, AIED (2003)

18. Thai-Nghe, N., Drumond, L., Krohn-Grimberghe, A., Schmidt-Thieme, L.: Recommender system for predicting student performance. Proc. Comput. Sci. 1(2), 2811–2819 (2010)
19. Trcka, N., Pechenizkiy, M.: From local patterns to global models: towards domain driven educational process mining. In: 9th International Conference on Intelligent Systems Design and Applications (ISDA), pp. 1114–1119. IEEE (2009)
20. van der Aalst, W.M.: Process Mining: Discovery, Conformance and Enhancement of Business Processes, vol. 136. Springer, Heidelberg (2011). https://doi.org/10.1007/978-3-642-19345-3
21. van der Aalst, W.M., Weijters, A., Maruster, L.: Workflow mining: discovering process models from event logs. IEEE Trans. Knowl. Data Eng. 16(9), 1128–1142 (2004)
22. Wang, R.: Sequence based approaches to course recommender systems. Master's thesis, University of Alberta, March 2017
23. Weijters, A., van der Aalst, W.M., De Medeiros, A.A.: Process mining with the heuristics miner-algorithm. Technische Universiteit Eindhoven, Technical Report WP, 166, 1–34 (2006)
24. Zaïane, O.R.: Building a recommender agent for e-learning systems. In: Proceedings International Conference on Computers in Education, pp. 55–59. IEEE (2002)

Data Integrity and Privacy

BFASTDC: A Bitwise Algorithm for Mining Denial Constraints

Eduardo H. M. Pena[1]([✉]) and Eduardo Cunha de Almeida[2]

[1] Federal University of Technology, Toledo, Paraná, Brazil
eduardopena@utfpr.edu.br
[2] Federal University of Paraná, Curitiba, Brazil
eduardo@inf.ufpr.br

Abstract. Integrity constraints (ICs) are meant for many data management tasks. However, some types of ICs can express semantic rules that others ICs cannot, or vice versa. Denial constraints (DCs) are known to be a response to this expressiveness issue because they generalize important types of ICs, such as functional dependencies (FDs), conditional FDs, and check constraints. In this regard, automatic DC discovery is essential to avoid the expensive and error-prone task of manually designing DCs. FASTDC is an algorithm that serves this purpose, but it is highly sensitive to the number of records in the dataset. This paper presents BFASTDC, a bitwise version of FASTDC that uses logical operations to form the auxiliary data structures from which DCs are mined. Our experimental study shows that BFASTDC can be more than one order of magnitude faster than FASTDC.

Keywords: Data profiling · Denial constraints · Integrity constraints

1 Introduction

Production databases often generate large and disordered datasets which become challenging to explore over time. Sometimes analysts will spend more time looking for relevant and clean data than they will do producing useful insights [1]. A research field that helps with this challenge is data profiling: the set of activities to gather statistical and structural properties, i.e, metadata, about datasets [2].

Data profiling research continually focus on developing efficient methods to discover integrity constraints (ICs) satisfied by datasets [2]. ICs validate the integrity and consistency of real-world entities that are represented in data and, although were initially devised for database schema design, are commonly used in other data management tasks, such as data integration [3] and data cleaning [4]. Well known exemplars of ICs include attribute dependencies (e.g, functional dependencies (FDs)), which express semantic relationships for data. Notice, however, that attribute dependencies may not be able to express important rules that hold in data, as shown by the examples below.

© Springer Nature Switzerland AG 2018
S. Hartmann et al. (Eds.): DEXA 2018, LNCS 11029, pp. 53–68, 2018.
https://doi.org/10.1007/978-3-319-98809-2_4

Consider an instance of relation, *employees*, as shown in Table 1. An FD could state that *(1) employees' names identify their manager.* A check constraint could state that *(2) employees' salaries must be greater than their bonus.* Denial constraints (DCs) [5,6] could state rules 1–2, and more expressive ones, for example, *(3) if two employees are managed by the same person, the one earning a higher salary has a higher bonus.* Thus, DCs are able to express many business rules, and subsume other types of ICs [6].

Table 1. An instance of the relation *employees.*

	Name	Manager	Salary	Bonus
t_0	John	Jim	$1000	$300
t_1	Brad	Frank	$1000	$400
t_2	Jim	Mark	$3000	$1100
t_3	Paul	Jim	$1200	$400

DCs define sets of predicates that databases must satisfy to prevent attributes from taking combinations of values considered semantically inconsistent. For example, the FD (1) mentioned earlier can be defined as a sequence of (in)equality predicates: if two tuples of *employees* agree on *Name* ($t_x.Name = t_y.Name$), then, they cannot disagree on *Managers* ($t_x.Manager \neq t_y.Manager$). Notice that predicates of DCs are easily expressed by SQL queries and, therefore, DCs can be readily used with commercial databases.

DCs have been adopted as the IC language in various scenarios [5,7]. Particularly, they have received considerable attention in data cleaning (violation of DCs usually indicates that data is dirty). Holoclean [7] and LLUNATIC [8] are examples of cleaning tools that use DCs. However, they assume DCs to be user-provided. Designing DCs is challenging because it requires expensive domain expertise that is not always available. Furthermore, DCs may become obsolete as business rules and data evolve. To overcome these limitations, DC-based cleaning tools (or any other DC-dependent solution) should also provide mechanisms to discover DCs holding on sample data.

Discovering DCs is nontrivial because the search space for DCs grows exponentially with the number of predicates. Predicates are defined over attributes, tuples and operators. For example, the *Salary* attribute in the relation *employees* define six predicates with the form $\{t_x.Salary\ w_o\ t_y.Salary\}$, $w_o \in \mathcal{W} : \{=, \neq, <, \leq, >, \geq\}$. Additionally, predicates can be defined over different attributes (e.g., $\{t_x.Salary\ w_o\ t_y.Bonus\}$). The predicate space $\mathbf{P}$ is the set of all predicates defined for a relation, and there are $2^{|\mathbf{P}|}$ DC candidates because a DC may be any subset of $\mathbf{P}$. Thus, checking DC candidates against every tuple combination of a relation instance becomes impractical [6].

Chu et al. [6] introduce important properties for DCs, and present a discovery algorithm called FASTDC. The algorithm uses the predicate space to compute sets of predicates that tuple pairs satisfy, namely, the evidence set. FASTDC then reduces the problem of discovering DCs to the problem of finding minimal covers for the evidence set. Unfortunately, a dominant computational cost of FASTDC is computing the evidence set. The algorithm needs to test every pair of tuples of the relation instance on every predicate in **P**; therefore, its performance is highly dependent on the number of records.

In this paper, we present a new algorithm that improves DC discovery by changing how the evidence set is built. Our algorithm, BFASTDC, is a *bitwise* version of FASTDC that exploits *bit-level* operations to avoid unnecessary tuple comparisons. BFASTDC builds associations between attribute values and lists of tuple identifiers so that different combinations of these associations indicate which tuple pairs satisfy predicates. To frame evidence sets, BFASTDC operates over auxiliary bit structures that store predicate satisfaction data. This allows our algorithm to use simple logical operations (e.g., conjunctions and disjunctions) to imply the satisfaction of remaining predicates. In addition, BFASTDC can use two modifications described in [6] to discover approximate and constant DCs. These DCs variants let the discovery process to work with data containing errors (e.g., integrated data from multiple sources). In our experiments, BFASTDC produced considerable improvements on DCs discovery performance.

Organization. Section 2 discusses the Related Work. Section 3 reviews the definition of DCs and the DC discovery problem. Section 4 describes the BFASTDC Algorithm. Section 5 presents our Experimental Study. Finally, Sect. 6 concludes this paper.

2   Related Work

Most works on IC discovery have focused on attribute dependencies. Liu et al. [9] present a comprehensive review of the topic. Papenbrock et al. [10] have looked into the experimental comparison of various FD discovery algorithms.

Dependency discovery algorithms usually employ strategies to reduce the number of candidate dependencies they must check. For example, Tane [11] is an FD discovery algorithm that uses a level-wise approach to traverse the attribute-set lattice of a relation. Supersets of attributes from level $k + 1$ of the lattice are pruned as Tane validates FDs from level k. FastFD [12] compares tuple pairs to build difference sets: the set of attributes in which two tuple differ. It uses depth-first search to find covers of difference sets and then derives valid FDs.

As data may be inconsistent, discovery algorithms need to, somehow, avoid returning unreliable ICs. Fan et al. [13] describe CTane and FastCFD to discovering conditional FDs, that is, FDs enforced by constants patterns. Conditional dependencies are particularly useful when working with integrated data because some dependencies may hold only on portions of the data [13]. Approximate discovery is another approach to avoid overfitting ICs [9,14]. For this matter, ICs are allowed to be approximately satisfied by a dataset. Liu et al. [9] also present a discussion on satisfaction metrics for approximate discovery algorithms.

As opposed to dependency discovery, for which many algorithms were devised [9,10], there are only two algorithms for discovering DCs: Hydra [15] and FASTDC [6]. Hydra can only detect exact variable DCs (DCs that are neither approximate nor contains constant predicates). The principle of the algorithm is to avoid comparing redundant tuple pairs, i.e, tuple pairs satisfying the same predicate set. It generates preliminary DCs from a sample of tuple pairs and identifies the tuple pairs violating those DCs. Hydra then derives exact DCs from the evidence set built upon the combination of the sample and tuple pairs violating the preliminary DCs. Because Hydra eliminates the need for checking every pair of tuple, it is not able to count how many times a predicate set is satisfied by a dataset. This counting feature is precisely what enables FASTDC to discover approximate DCs. The inspiration for FASTDC comes from FastFD-FastCFD, and is twofold: pairwise comparison of tuples for extracting evidence from datasets; depth-first search for finding covers for the evidence and deriving valid ICs. As described in [6], simple modifications in FASTDC enable the algorithm to also discover DCs with constant predicates. BFASTDC is designed to avoid the exhaustive tuple pairs comparison of FASTDC, but keeping the ability to discover exact, approximate and constant DCs.

3   Background

Consider a relational database schema $\mathcal{R}$ and a set of operators $\mathcal{W} : \{=, \neq, <, \leq, >, \geq\}$. A DC [5,6] has the form $\varphi : \forall t_x, t_y, ... \in r, \neg(P_1 \wedge ... \wedge P_m)$, where $t_x, t_y, ...$ are tuples of an instance of relation r of R, and $R \in \mathcal{R}$. A predicate P_i is a comparison atom with either the form $v_1 w_o v_2$ or $v_1 w_o c$: v_1, v_2 are variables $t_{id}.A_j$, $A_j \in R$, $id \in \{x, y, ...\}$, c is a constant from A_j's domain, and $w_o \in \mathcal{W}$.

Example 1. *The ICs (1), (2) and (3) from Sect. 1 can be expressed as the following DCs:* $\varphi_1 : \neg(t_x.Name = t_y.Name \wedge t_x.Manager \neq t_y.Manager), \varphi_2 : \neg(t_x.Salary < t_x.Bonus), \varphi_3 : \neg(t_x.Manager = t_y.Manager \wedge t_x.Salary > t_y.Salary \wedge t_x.Bonus < t_y.Bonus).*

An instance of relation r satisfies a DC φ if at least one predicate of φ is false, for every pair of tuples of r. In other words, the predicates of φ cannot be all true at the same time. We follow the conventions of [6] for DC discovery. We consider there is only one relation in $\mathcal{R}$, and discover DCs involving at most two tuples because they suffice to represent most rules used in practice. Allowing more tuples in a single DC would unnecessarily incur a much bigger predicate spaces for the DC discovery [6].

Table 2 shows the inverse, $\overline{w}_o$, and implication, $I(w_o)$, of the operators $w_o \in \mathcal{W}$. The inverse of a predicate $P : v_1 w_o v_2$ has the form $\overline{P} : v_1 \overline{w}_o v_2$, which is the logical complement of P. The set of predicates implied by P is $I(P) = \{P' \mid P' : v_1 w'_o v_2, \forall w'_o \in I(w_o)\}$. Every $P' \in I(P)$ is true if P is true. BFASTDC is designed to use these properties in the form of bitwise operations so that implied and inversed predicates can be transitively evaluated.

Table 2. Inverse and implied operators.

w_o	$=$	$\neq$	$<$	$\leq$	$>$	$\geq$
$\overline{w}_o$	$\neq$	$=$	$\geq$	$>$	$\leq$	$<$
$I(w_o)$	$=,\leq,\geq$	$\neq$	$<,\leq,\neq$	$\leq$	$>,\geq,\neq$	$\geq$

We follow the problem definition of [6] to discover minimal DCs. A DC φ_1 on r is minimal if there does not exist a φ_2 such that both φ_1 and φ_2 are satisfied by r, and the predicates of φ_2 are a subset of φ_1. Chu et al. [6] also describe additional properties for DCs and an inference system that helps eliminating non-minimal DCs. An in-depth discussion on the theoretical aspects of DCs and other ICs can be found in [5, 16].

3.1   DC Discovery

The first step to discover DCs is to set the predicate space $\mathbf{P}$ from which DCs are derived. Experts can define predicates for attributes based on the database structure. One could also use approaches, such as [17], for mining associations between attributes. Predicates on categorical attributes use operators $\{=,\neq\}$, and predicates on numerical attributes $\{=,\neq,<,>,\leq,\geq\}$. Figure 1 illustrates a predicate space for the relation *employees* from Sect. 1.

$P_1 : t_x.Name = t_y.Name$	$P_2 : t_x.Name \neq t_y.Name$	$P_3 : t_x.Name = t_x.Manager$
$P_4 : t_x.Name \neq t_x.Manager$	$P_5 : t_x.Manager = t_y.Manager$	$P_6 : t_x.Manager \neq t_y.Manager$
$P_7 : t_x.Salary = t_y.Salary$	$P_8 : t_x.Salary \neq t_y.Salary$	$P_9 : t_x.Salary < t_y.Salary$
$P_{10} : t_x.Salary \leq t_y.Salary$	$P_{11} : t_x.Salary > t_y.Salary$	$P_{12} : t_x.Salary \geq t_y.Salary$
$P_{13} : t_x.Bonus = t_y.Bonus$	$P_{14} : t_x.Bonus \neq t_y.Bonus$	$P_{15} : t_x.Bonus < t_y.Bonus$
$P_{16} : t_x.Bonus \leq t_y.Bonus$	$P_{17} : t_x.Bonus > t_y.Bonus$	$P_{18} : t_x.Bonus \geq t_y.Bonus$

Fig. 1. Example of predicate space for *employees*.

The satisfied predicate set $\mathbf{Q}_{t_\mu,t_\nu}$ of an arbitrary pair of tuples $(t_\mu, t_\nu) \in r$ is a subset $\mathbf{Q} \subset \mathbf{P}$ such that for every $P \in \mathbf{Q}$, $P(t_\mu, t_\nu)$ is true. The set of satisfied predicate sets of r is the evidence set $\mathbf{E}_r = \{\mathbf{Q}_{t_\mu,t_\nu} \mid \forall(t_\mu, t_\nu) \in r\}$. Different tuple pairs may return the same predicate set, hence, each $\mathbf{Q} \in \mathbf{E}_r$ is associated with an occurrence counter.

A cover for $\mathbf{E}_r$ is a set of predicates that intersects with every satisfied predicate set of $\mathbf{E}_r$, and it is minimal if none of its subsets equally intersects with $\mathbf{E}_r$. The authors of FASTDC demonstrate that minimal covers of $\mathbf{E}_r$ represent the predicates of minimal DCs [6]. Thus, the DC discovery problem becomes finding covers for evidence set $\mathbf{E}_r$.

FASTDC uses a depth-first search (DFS) strategy to find minimal covers for $\mathbf{E}_r$. Predicates of $\mathbf{P}$ are recursively arranged to form the branches of the search tree. To optimize the search, predicates that cover more elements of the evidence set are added to the path first. As minimal covers are discovered, unnecessary branches of the DFS are pruned with the inference system. Any path of the tree is a candidate cover that identifies a set of elements $\mathbf{E}_{path} \subset \mathbf{E}_r$ not yet covered. When a candidate cover includes a predicate P, elements that contain P are removed from its corresponding $\mathbf{E}_{path}$. The search stops for a branch when there are no more predicates in $\mathbf{E}_{path}$. The candidate cover is minimal if satisfies minimality property and $\mathbf{E}_{path}$ is empty.

The authors of FASTDC also present two modifications for their algorithm: A-FASTDC and C-FASTDC.

A-FASTDC is an algorithm for discovering approximate DCs, that is, DCs whose number of violations is bounded. The algorithm uses the same evidence set $\mathbf{E}_r$ as FASTDC, but modify the minimal cover search to work with approximation levels ϵ. In short, the search prioritizes predicates that appear in the most frequent predicate sets of $\mathbf{E}_r$. The search stops for branches of the search tree when their predicates cover frequent predicate sets. This means that the frequency of the predicate sets that were not used in the search are below a threshold $\epsilon |r| (|r| - 1)$. This approximate approach is only possible because the evidence set $\mathbf{E}_r$ counts the number of times a predicate set appears in the dataset.

C-FASTDC discovers DCs with constant predicates. It builds a constant predicate space from attribute domains and then follows an Apriori approach to identify τ-frequent constant predicate sets. A constant predicate set $\mathbf{C}$ is τ-frequent if $\frac{|sup(\mathbf{C}, r)|}{|r|} \geq \tau$, where $sup(\mathbf{C}, r)$ is the set of tuples of r that satisfy all predicates of $\mathbf{C}$ [6]. As τ-frequent predicate sets $\mathbf{C}$ are identified, FASTDC discovers the variable predicates holding on $sup(\mathbf{C}, r)$ and outputs DCs that are combinations of $\mathbf{C}$ and the variable predicates.

Challenge. FASTDC builds the evidence set by evaluating every predicates of the predicate space $\mathbf{P}$ on every pair of tuples of r. This computation requires $|\mathbf{P}| \times |r|^2$ predicate evaluations, of which at least half return false if we consider groups of predicates $\{P, \overline{P}, ...\}$. We next describe how BFASTDC reduces this computational cost.

4   The BFASTDC Algorithm

BFASTDC operates *at the bit level* and takes advantage of the inversion and implication properties presented in Table 2. The computational cost of our approach grows as a function of the number of predicates that evaluate to true, and is potentially smaller than FASTDC. We next describe how to set simple data structures to represent predicate satisfaction.

4.1   Data Structures

Attribute-Values Maps. Attribute values are organized as entries $\langle k, l \rangle$, where key k is an element of the set of values in attribute A_j, and l is a list of tuple

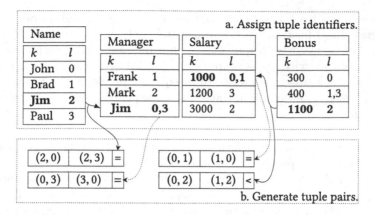

Fig. 2. Organizing attribute values: (a) assign tuple identifiers; (b) generate permutations (dashed line arrows)/Cartesian products (solid line arrows).

identifiers such that $\forall id \in l$ then $t_{id}[A_j] = k$. Procedure SEARCH(A_j, k) finds the list l for k in A_j. PREDECESSORS(A_j, k) is defined for numerical attributes. It returns the set L_2 consisting of the lists SEARCH(A_j, k_2) associated with the values k_2 smaller than k. Notice that SEARCH(A_j, k) and PREDECESSORS(A_j, k) may return $\varnothing$ if they find no tuple identifiers associated with k. Figure 2a depicts the assignment of tuples identifiers for *employees*. In the example, a key "**Jim**" from attribute *Name* is inputted to SEARCH$(Manager, \mathbf{Jim})$; and a key **1100** from attribute *Bonus* is inputted to PREDECESSORS$(Salary, \mathbf{1100})$.

Bit Vectors. A bit vector B is associated with a predicate P to represent the relationship between P and the tuple pairs that satisfy P. Notice that a relation instance of size $|r|$ generates tuple pairs: $(t_0, t_0), (t_0, t_1), ..., (t_{|r|}, t_{|r|})$. Function (1) below returns a unique identifier λ for a given pair of tuples (t_μ, t_ν) of r. Bit vector B holds 1 at position λ only if λ corresponds to a pair of tuples that satisfy P, otherwise B holds 0.

$$\lambda(t_\mu, t_\nu, r) = (|r| \mu) + \nu \qquad (1)$$

Example 2. *Consider the predicate $P_5 : t_x.Manager = t_y.Manager$, and the relation employees from Sect. 1. In the sample, Predicate P_5 is satisfied by the following tuple pairs: (t_0, t_3) and (t_3, t_0). From Function (1), considering the size of the instance $|empolyees| = 4$, with $\lambda(t_0, t_3, employees)$ and $\lambda(t_3, t_0, employees)$ we get tuple pairs identifiers $\lambda = 3$ and $\lambda = 12$. These λ are the indexes for which the bit vector B_5, holds true.*

4.2 Building Bit Vectors

Before describing the strategies to efficiently obtain indexes λ, we add some remarks regarding the possible forms of predicates.

Predicates involve one or two attributes, conventionally $\{A_1\}$ and $\{A_1, A_2\}$; and can be defined for two, (t_x, t_y), or one tuple, (t_x, t_x). We denote P_α and P_β to distinguish between two-tuple and single-tuple predicates, respectively. Let P^{w_o} be a predicate with the operator w_o, $w_o \in \mathcal{W} : \{=, \neq, <, \leq, >, \geq\}$. Hence, $P_\alpha^{w_1} : t_x.A_1 = t_y.A_1$ exemplify a two-tuple equality predicate on attribute $\{A_1\}$, $P_\beta^{w_2} : t_x.A_1 \neq t_x.A_2$ exemplify a single-tuple inequality predicate on attributes $\{A_1, A_2\}$, and so on. To ease notation for (in)equality predicates, when $o = 1$ and $o = 2$, we assume $\widehat{P}_\alpha \equiv P_\alpha^{w_1}, \widetilde{P}_\alpha \equiv P_\alpha^{w_2}$ and $\widehat{P}_\beta \equiv P_\beta^{w_1}, \widetilde{P}_\beta \equiv P_\beta^{w_2}$.

Logical operations are enough to set some of the bit vectors, but they require auxiliary bitmasks to prevent bit vectors B from holding incorrect values. Let exponentiation denote bit repetition, e.g., $10^3 = 1000$. A bitmask $mask_{st} = (z_1, ..., z_{|r|})$, where $z_n = 10^{|r|}$, helps operations on single-tuple predicates as they are not related to pair of tuples (t_μ, t_ν) if $t_\mu \neq t_\nu$. Similarly, a bitmask $mask_{tt} = (z_1, ..., z_{|r|})$, where $z_n = 01^{|r|}$, helps operations on two-tuple predicates as they are not related to pair of tuples (t_μ, t_ν) if $t_\mu = t_\nu$.

Next, we describe four strategies that arrange the set of bit vectors $\mathbf{B}$ associated with the predicate space $\mathbf{P}$. Every $B \in \mathbf{B}$ is filled with 0's at the start.

1. Predicates Involving One Categorical Attribute. Consider a predicate of the form $\widehat{P}_\alpha : \{t_x.A_1 = t_y.A_1\}$, and its associated bit vector $\widehat{B}_\alpha$. Given an entry $\langle k, l \rangle$ of A_1 where $|l| > 1$, permutations of two elements taken from l represent tuple pairs (t_μ, t_ν) that satisfy $\widehat{P}_\alpha$. From Function (1), these permutations generate tuple pair identifiers λ at which bit vector $\widehat{B}_\alpha$ is set to one, i.e, $\widehat{B}_{\alpha,\lambda} \leftarrow 1$. Figure 2b illustrates some tuple pairs arranged for *employees*. For entry $\langle \mathbf{Jim}, \{0, 3\} \rangle$ from attribute *Manager*, tuple pairs $(0, 3)$ and $(3, 0)$ do satisfy a two-tuple equality predicate involving the attribute. The above process repeats for every entry of A_1.

Consider a predicate $\widetilde{P}_\alpha : \{t_x.A \neq t_y.A\}$, and its associated bit vector $\widetilde{B}_\alpha$. Observe that $\widetilde{B}_\alpha$ is the logical complement of $\widehat{B}_\alpha$. Therefore, $\widetilde{B}_\alpha$ derives from a disjunction ($\vee$) followed by an exclusive-or operation ($\oplus$) : $\widetilde{B}_\alpha \leftarrow (\widehat{B}_\alpha \vee mask_{tt}) \oplus \widehat{B}_\alpha$.

2. Predicates Involving Two Categorical Attributes. Suppose that we want to find associations from attribute values of *Name* to attribute values of *Manager* in *employees*. Entries $\langle \mathbf{Jim}, \{2\} \rangle$ of *Name* and $\langle \mathbf{Jim}, \{0, 3\} \rangle$ of *Manager* generate an equality association, which is represented by the Cartesian product $\{(2, 0), (2, 3)\}$. Formally, consider an entry $\langle k_1, l_1 \rangle$ taken from attribute A_1 and a list of tuple identifiers l_2 such that $l_2 \leftarrow \text{SEARCH}(A_2, k_1)$. Cartesian products $l_1 \times l_2$ represent tuple pair identifiers (t_μ, t_ν) that either satisfy a predicate $\widehat{P}_\alpha : \{t_x.A_1 = t_y.A_2\}$ or $\widehat{P}_\beta : \{t_x.A_1 = t_x.A_2\}$. Given λ corresponding to $(t_\mu, t_\nu) \in l_1 \times l_2$: if $t_\mu \neq t_\nu$ then $\widehat{B}_{\alpha,\lambda} \leftarrow 1$; otherwise, $\widehat{B}_{\beta,\lambda} \leftarrow 1$. The above process runs for every entry of A_1.

Computing $\widetilde{B}_\alpha \leftarrow (\widehat{B}_\alpha \vee mask_{tt}) \oplus \widehat{B}_\alpha$ solves $\widetilde{P}_\alpha$. As for $\widetilde{P}_\beta$, it is sufficient to compute $\widetilde{B}_\beta \leftarrow (\widehat{B}_\beta \vee mask_{st}) \oplus \widehat{B}_\beta$.

3. Predicates Involving One Numerical Attribute. Numerical attributes additionally require predicates with the operators $\{<, \leq, >, \geq\}$. Given an entry $\langle k_1, l_1 \rangle$ in A_1, the set L_2 such that $L_2 \leftarrow \text{PREDECESSORS}(A_1, k_1)$ and lists of tuple identifiers $l_2 \in L_2$, the Cartesian product of every $l_1 \times l_2$ represent tuple pairs (t_μ, t_ν) that satisfy a predicate with the *less than* operator, $P_\alpha^{w_3}$. The tuple pair identifiers λ for which $B_\alpha^{w_3}$ holds one come from the products generated for every entry from A_1.

Bit vectors $\widehat{B}_\alpha$ and $\widetilde{B}_\alpha$ are set using permutations (strategy one). The predicates with the remaining operators are solved from $\widehat{B}_\alpha$ and $B_\alpha^{w_3}$. Predicate with *less than or equals* operator is given by: $B_\alpha^{w_4} \leftarrow (B_\alpha^{w_3} \wedge \widehat{B}_\alpha)$, with *greater than*: $B_\alpha^{w_5} \leftarrow \overline{B}_\alpha^{w_4}$, and *greater than or equals*: $B_\alpha^{w_6} \leftarrow (B_\alpha^{w_5} \wedge \widehat{B}_\alpha)$.

4. Predicates Involving Two Numerical Attributes. Bit vectors for single and two-tuple predicates $\{\widehat{B}_\alpha, \widetilde{B}_\alpha, \widehat{B}_\beta, \widetilde{B}_\beta\}$ are set using Cartesian products from attributes A_1 and A_2 (strategy two). In the same spirit, a slight modification on strategy three is sufficient to set order predicates involving two attributes. Cartesian products $l_1 \times l_2$ are generated such that $\langle k_1, l_1 \rangle$ is taken from A_1 and each $l_2 \in L_2$ is taken from $\text{PREDECESSORS}(A_2, k_1)$. These products generate tuple pair identifiers λ that either satisfy $B_\alpha^{w_3}$ or $B_\beta^{w_3}$. The logical operations described earlier are applied on $\{\widehat{B}_\alpha, \widetilde{B}_\alpha, B_\alpha^{w_3}, \widehat{B}_\beta, \widetilde{B}_\beta, B_\beta^{w_3}\}$ to solve the remaining predicates.

4.3 Fitting Bit Vectors into Memory

The length of bit vectors grows as a function of the relation instance size. A single bit vector would occupy 400 Mb for a relation with 20 k tuples. To avoid running out of memory and to handle large relation instances, BFASTDC splits B into smaller chunks: $B = \sum_{s \in S} b_s$. The number of chunks is given by $|S| = \lceil |r|^2 / \omega \rceil$, where ω defines a maximum chunk size. The chunk size ω is related to the amount of available memory and bounds the range that chunk b_s operates.

Let b_s be a chunk being evaluated in turn s. Assume that a list of tuple pair identifiers $\Lambda = \{\lambda_1, ..., \lambda_c, ..., \lambda_{|\Lambda|}\}$, $\lambda_c < \lambda_{c+1}$, acknowledges B_{λ_c} to be true. The only portion of B in memory is b_s, so λ_c can be used to set b_{s,λ_c} only if it is in the range covered by b_s. If not, list Λ is skipped and the last λ_c used in Λ is marked. The list Λ can be iterated from λ_{c+1} in the next time it is acquired because tuple pair identifier λ_c will never be in the range of subsequent chunks b_{s+1}. Figure 3a illustrates tuple pair identifiers on setting bit chunks. For better visualization, it considers only a subset of the predicate space $\mathbf{P}$ of Fig. 1.

4.4 Assembling the Evidence Set

Each bit vector $B \in \mathbf{B}$ represents *the set of tuple pairs that satisfy a predicate P*. Conversely, each element in the evidence set, $E \in \mathbf{E}_r$, is *the satisfied predicate set of a pair of tuples*. Our algorithm uses the same DFS strategy as FASTDC to search for minimal covers, hence, we need to transpose $\mathbf{B}$ into $\mathbf{E}_r$.

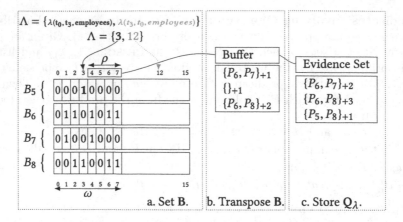

Fig. 3. Evidence set generation: (a) Fill chunks of size $\omega = 8$; (b) Transpose chunks to buffer of size $\rho = 4$; (c) Insert the buffer content into evidence set and update the predicate sets counters (denoted by the $\{\}_{+c}$ notation).

Consider $i = 0, ..., |\mathbf{P}|$, chunks of bit vectors $B_1 = \{b_{1,1}, ..., b_{1,S}\}, ..., B_{|\mathbf{P}|} = \{b_{|\mathbf{P}|,1}, ..., b_{|\mathbf{P}|,S}\}$, and $\mathbf{B} = \{B_1, ..., B_{|\mathbf{P}|}\}$. Chunks $b_{i,s}$ are transposed all at once (see Fig. 3). The evidence set is built by inserting satisfied predicate sets $\mathbf{Q}_{t_\mu, t_\nu}$ into set $\mathbf{E}_r$ (see Fig. 3c). We can assume that $\mathbf{E}_r = \{\mathbf{Q}_\lambda \mid \forall \lambda \in r\}$ because λ is a unique identifier for pair of tuples $t_\mu, t_\nu \in r$. If $b_{i,s,\lambda} = 1$, then $P_j \in \mathbf{Q}_\lambda$. Notice that BFASTDC only need to iterate over $b_{i,s}$ at indices λ that are set to true.

There are ω satisfied predicate sets $\mathbf{Q}$ to insert into $\mathbf{E}_r$ at each turn s. Given, $1 < \rho < \omega$, we have found that using a buffer holding ρ elements $\mathbf{Q}$ saves memory and decreases overall running time. If $b_{i,s,\lambda} = 1$, and λ is out of the buffer range, we skip iteration $b_{i,s}$ until the next round (similarly to chunks range scheme). At this stage, the predicate set counters of $\mathbf{E}_r$ are updated for further approximate discovery. Figure 3b illustrates a buffer operation.

4.5   Implementation Details

Hash-based dictionaries group entries of categorical attributes. Building them is linear since insertions on hash-based dictionaries are constant in time. Lookup operations are also performed in constant-time. BFASTDC uses sorted arrays to group entries of numerical attributes because they support operations $\{<, \leq, >, \geq\}$. Given a numerical entry $\langle k, l \rangle$, k and l are stored separately, into position h of two different arrays. A numerical entry is realigned by pairing both arrays with the same index h. For sorting, we have adapted the *Quicksort* algorithm to return the list of tuple identifiers for each distinct attribute value. Numerical entries are sorted according to k, which allows BFASTDC to use binary search[1]. Finally, chunks and buffers are implemented as simple bitsets.

---

[1] We have adapted binary search for procedure PREDECESSORS(A_j, k).

5   Experimental Study

In this section, we present our experimental study of BFASTDC. We compare BFASTDC with FASTDC to evaluate the scalability of our algorithm in the number of tuples and predicates. We also evaluate the performance of the algorithms on discovering approximate and constants DCs. Finally, we evaluate the effects that different sizes of chunks and buffers produce on the execution of BFASTDC.

5.1   Experimental Setup

Implementation and Hardware. We implemented FASTDC and BFASTDC using Java programming language version 1.8. The algorithms use the same implementations of predicate space building and minimal cover search. To perform the experiments, we used a machine with a 3.4 GHz Core i7, 8 MB of L3 cache, 8 GB of memory, running Linux. The algorithms run in main memory after dataset loading.

Datasets and Predicate Space. We used both synthetic and real-life datasets[2]: *Tax* and *Stock*. *Tax* is a synthetic compilation of personal information that includes fifteen attributes to represent addresses and tax-records. *Stock* gathers data from historical S&P 500 stocks in the form of a relation with seven attributes. We used *Tax* and *Stock* in our experiments because these datasets have already been used to evaluate DC discovery [6]. With regard to predicate spaces, we defined single and two-tuple predicates on: categorical attributes using operators $\{=, \neq\}$; numerical attributes using operators $\{=, \neq, <, >, \leq, \geq\}$. We defined predicates involving two different attributes provided that the values of the two attributes were in the same order of magnitude.

5.2   Results and Discussion

In the first four experiments, we fixed chunk and buffer size of BFASTDC to 4000 kb and 12 kb, respectively. These parameters are discussed in the fifth experiment. Furthermore, we report the average runtime of five runs for each experiment. We consider a running time limit of 48 h for all runs.

Exp-1: Scalability in the Number of Tuples. We varied the number of tuples from 10,000 to 1,000,000 for *Tax*, and from 10,000 to 122,000 for *Stock*. Keeping the size of the predicate spaces constant for both datasets ($|\mathbf{P}| = 50$), we measured the running time in seconds of FASTDC and BFASTDC. Figure 4 shows their scaling behavior (Y axis are in log scale). The running time of both algorithms increases in a quadratic trend as we add more tuples in their input. However, the running time for BFASTDC were at least one order of magnitude

---

[2] Available at: http://da.qcri.org/dc/.

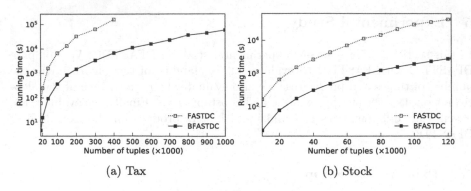

(a) Tax   (b) Stock

Fig. 4. Scalability of BFASTDC and FASTDC in the number of tuples.

smaller than the running time for FASTDC. To process 400,000 tuples of *Tax* (see Fig. 4a), FASTDC took a little more than 2656 min. In contrast, BFASTDC processed the same input in approximately 110 min; an improvement ratio of approximately 24 times. FASTDC was not able to process more than 400,000 tuples of *Tax* within the running time limit. In turn, BFASTDC processed the entire *Tax* dataset (one million tuples) in approximately 16 h. BFASTDC was also faster than FASTDC when running over *Stock* (see Fig. 4b). It processed the full dataset in approximately 47 min, while FASTDC took more than 12 h to reach completion.

Exp-2: Scalability in the Number of Predicates. Fixing the algorithms input on the first 20,000 tuples of *Tax* and *Stock*, we varied the number of predicates from 10 to 60. The attributes for which predicates were added to the predicate spaces were chosen at random. As shown in Fig. 5 (Y axis are in log scale), the running time of the algorithms increases exponentially w.r.t. the number of predicates. In addition, the BFASTDC running time improvements over FASTDC degrades when the search for minimal covers includes larger predicate spaces.

Exp-3: Approximate DC Discovery. For this experiment, we kept the number of tuples and the size of predicate space constant ($|r| = 20,000$ and $|\mathbf{P}| = 50$) for both datasets. We gradually increased the approximation levels ϵ from 10^{-6} to 2×10^{-5}. Figure 6 shows the running time for the approximate versions of BFASTDC and FASTDC (Y axis are in log scale). Despite their small improvements, the running time for both algorithms, for either *Tax* or *Stock*, remains in their original order of magnitude provided that only approximation levels differ. Indeed, varying the approximation levels did not impact on the algorithms' running time as much as varying the number of tuples or predicates did.

Exp-4: Constant DC Discovery. We used the same number of tuples and predicate space size as we did in experiment three. Then, we gradually increased

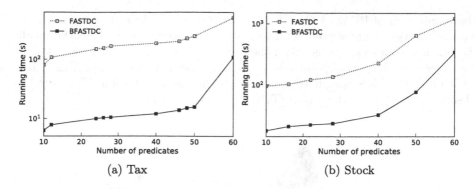

Fig. 5. Scalability BFASTDC and FASTDC in the number of predicates.

the frequency threshold τ from 0.1 to 0.5. Figure 7 shows the running time that each algorithm took to discover constant DCs (Y axis are in log scale). The algorithms are sensitive to threshold τ. For *Tax*, smaller thresholds τ resulted in longer running times. As for *Stock*, FASTDC and BFASTDC returned within virtually the same running time because there were no constant predicates to be considered by the variant portion of the algorithms.

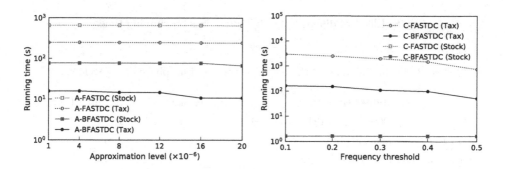

Fig. 6. Approximate DC discovery.   **Fig. 7.** Constant DC discovery.

Exp-5: BFASTDC Parameters. We report this experiment using only *Tax* dataset because the same behavior and very similar parameters were seen for *Stock*. Fixing $|\mathbf{P}| = 50$, and $|r| = 100,000$, we varied chunk size ω from 250 kb to 64,000 kb, and buffer size ρ from 5 kb to 19 kb. Figure 8 shows that the running time does not improve as we rashly increase the size of chunks or buffers. For example, configurations where $\omega < 10000$ kb and $\rho < 14$ kb produced better results if compared to configurations with higher values. The best setting was $\omega = 4000$ kb and $\rho = 12$ kb.

To better understand this result, we monitored the cache activities in the evidence set building phase of BFASTDC. Table 3 shows some ratios between the monitoring of BFASTDC in its best setting and BFASTDC running in two extreme settings. The setting with bigger ω and ρ suffers from L1 cache invalidation (i.e., chunks are bigger than the cache line leading to cache misses). But, we observe an inflection point when accessing the last level cache (LLC): bigger chunks need less concurrent access with less cache pollution. Therefore, we observe a sweet-spot where BFASTDC can be cache-efficient.

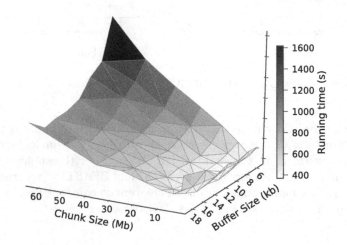

Fig. 8. Effect of different chunk/buffer sizes on running time.

Table 3. Cache behavior of the evidence set building phase of BFASTDC.

Chunk (ω) and buffer (ρ) sizes	LLC misses	L1 misses	Running time
Baseline: $\omega = 4000\,\text{kb}$, $\rho = 12\,\text{kb}$	1	1	1
Low extreme: $\omega = 250\,\text{kb}$, $\rho = 5\,\text{kb}$	2.868	0.621	1.577
High extreme: $\omega = 64000\,\text{kb}$, $\rho = 19\,\text{kb}$	1.445	2.104	2.322

Discussion. Our experiments confirm our earlier hypothesis: there is no need to check every predicate for every pair of tuples. With its attribute values organization, BFASTDC tracks bit vectors only for tuple pairs that do satisfy predicates. The bitwise representation of predicate satisfaction makes it possible to use logical operations, which are optimized in all modern CPU architectures. Such operations are cache-dependent because bit vectors are packed into processor words for processing. That is why there was an inflection point in the last experiment where the bigger the chunk and buffer sizes were, the worse the cache usage, and, therefore, the higher the running time. Experiment one demonstrates the effectiveness of BFASTDC in building the evidence set and the deep impact it had on the overall DC discovery performance. The improvements were seen in the subsequent experiments: BFASTDC was faster than FASTDC in approximate and constant DC discovery. Because of the exponential nature of the DFS used for minimal cover search, the two algorithms did not scale well with the number of predicates. Future studies could investigate not only algorithmic improvements for this phase, but how approximate discovery fits in there.

6   Conclusions

We presented BFASTDC, a bitwise, instance-driven algorithm for mining minimal DCs from relational data. BFASTDC improves the evidence set building phase of FASTDC based on two key principles: (i) it combines tuple identifiers from related values and avoids testing every pair of tuples on every predicate, and (ii) it exploits the implication relation between predicates to operate at *bit level*. BFASTDC was up to 24 times faster than FASTDC in our experimental study. In addition, BFASTDC is able to work with noisy datasets when it is modified to discover approximate and constant DCs. For those reasons, we believe BFASTDC can be a valuable part of DC-dependent tools. Future research should improve minimal covers search and evaluate the quality of the discovered DCs on real use cases.

References

1. Kandel, S., Paepcke, A., Hellerstein, J.M., Heer, J.: Enterprise data analysis and visualization: an interview study. IEEE TVCG **18**(12), 2917–2926 (2012)
2. Abedjan, Z., Golab, L., Naumann, F.: Profiling relational data: a survey. VLDB J. **24**(4), 557–581 (2015)
3. Ayat, N., Afsarmanesh, H., Akbarinia, R., Valduriez, P.: Pay-as-you-go data integration using functional dependencies. In: Quirchmayr, G., Basl, J., You, I., Xu, L., Weippl, E. (eds.) CD-ARES 2012. LNCS, vol. 7465, pp. 375–389. Springer, Heidelberg (2012). https://doi.org/10.1007/978-3-642-32498-7_28
4. Fan, W.: Data quality: from theory to practice. SIGMOD Rec. **44**(3), 7–18 (2015)
5. Bertossi, L.: Database Repairing and Consistent Query Answering. Morgan & Claypool Publishers, San Rafael (2011)
6. Chu, X., Ilyas, I.F., Papotti, P.: Discovering denial constraints. Proc. VLDB Endow. **6**(13), 1498–1509 (2013)
7. Rekatsinas, T., Chu, X., Ilyas, I.F., Ré, C.: Holoclean: holistic data repairs with probabilistic inference. PVLDB Endow. **10**(11), 1190–1201 (2017)
8. Geerts, F., Mecca, G., Papotti, P., Santoro, D.: That's all folks!: LLUNATIC goes open source. PVLDB **7**, 1565–1568 (2014)
9. Liu, J., Li, J., Liu, C., Chen, Y.: Discover dependencies from data - a review. IEEE TKDE **24**(2), 251–264 (2012)
10. Papenbrock, T., et al.: Functional dependency discovery: an experimental evaluation of seven algorithms. PVLDB **8**(10), 1082–1093 (2015)
11. Huhtala, Y., Kärkkäinen, J., Porkka, P., Toivonen, H.: TANE: an efficient algorithm for discovering functional and approximate dependencies. Comput. J. **42**(2), 100–111 (1999)

12. Wyss, C., Giannella, C., Robertson, E.: FastFDs: a heuristic-driven, depth-first algorithm for mining functional dependencies from relation instances extended abstract. In: Kambayashi, Y., Winiwarter, W., Arikawa, M. (eds.) DaWaK 2001. LNCS, vol. 2114, pp. 101–110. Springer, Heidelberg (2001). https://doi.org/10. 1007/3-540-44801-2_11
13. Fan, W., Geerts, F., Li, J., Xiong, M.: Discovering conditional functional dependencies. IEEE TKDE **23**(5), 683–698 (2011)
14. Caruccio, L., Deufemia, V., Polese, G.: Relaxed functional dependencies - a survey of approaches. IEEE TKDE **28**(1), 147–165 (2016)
15. Bleifuß, T., Kruse, S., Naumann, F.: Efficient denial constraint discovery with hydra. Proc. VLDB Endow. **11**(3), 311–323 (2017)
16. Fan, W., Geerts, F.: Foundations of Data Quality Management. Morgan & Claypool Publishers, San Rafael (2012)
17. Zhang, M., Hadjieleftheriou, M., Ooi, B.C., Procopiuc, C.M., Srivastava, D.: On multi-column foreign key discovery. PVLDB **3**(1–2), 805–814 (2010)

BOUNCER: Privacy-Aware Query Processing over Federations of RDF Datasets

Kemele M. Endris[1]([✉]), Zuhair Almhithawi[2], Ioanna Lytra[2,4], Maria-Esther Vidal[1,3], and Sören Auer[1,3]

[1] L3S Research Center, Hanover, Germany
{endris,auer}@L3S.de
[2] University of Bonn, Bonn, Germany
s6zualmh@uni-bonn.de, lytra@cs.uni-bonn.de
[3] TIB Leibniz Information Centre for Science and Technology, Hanover, Germany
Maria.Vidal@tib.eu
[4] Fraunhofer IAIS, Sankt Augustin, Germany

Abstract. Data provides the basis for emerging scientific and inter-disciplinary data-centric applications with the potential of improving the quality of life for the citizens. However, effective data-centric applications demand data management techniques able to process a large volume of data which may include sensitive data, e.g., financial transactions, medical procedures, or personal data. Managing sensitive data requires the enforcement of privacy and access control regulations, particularly, during the execution of queries against datasets that include sensitive and non-sensitive data. In this paper, we tackle the problem of enforcing privacy regulations during query processing, and propose BOUNCER, a privacy-aware query engine over federations of RDF datasets. BOUNCER allows for the description of RDF datasets in terms of RDF molecule templates, i.e., abstract descriptions of the properties of the entities in an RDF dataset and their privacy regulations. Furthermore, BOUNCER implements query decomposition and optimization techniques able to identify query plans over RDF datasets that not only contain the relevant entities to answer a query, but that are also regulated by policies that allow for accessing these relevant entities. We empirically evaluate the effectiveness of the BOUNCER privacy-aware techniques over state-of-the-art benchmarks of RDF datasets. The observed results suggest that BOUNCER can effectively enforce access control regulations at different granularity without impacting the performance of query processing.

1 Introduction

In recent years, the amount of both open data available on the Web and private data exchanged across companies and organizations, expressed as Linked Data, has been constantly increasing. To address this new challenge of effective

S. Hartmann et al. (Eds.): DEXA 2018, LNCS 11029, pp. 69–84, 2018.
https://doi.org/10.1007/978-3-319-98809-2_5

and efficient data-centric applications built on top of this data, data management techniques targeting sensitive data such as financial transactions, medical procedures, or various other personal data must consider various privacy and access control regulations and enforce privacy constraints once data is being accessed by data consumers. Existing works suggest the specification of Access Control ontologies for RDF data [5,12] and their enforcement on centralized or distributed RDF stores (e.g., [2]) or federated RDF sources (e.g., [8]). Albeit expressive, these approaches are not able to consider privacy-aware regulations during the whole pipeline of a federated query engine, i.e., during source selection, query decomposition, planning, and execution. As a consequence, efficient query plans cannot be devised in a way that privacy-aware policies are enforced.

In this paper, we introduce a privacy-aware federated query engine, called BOUNCER, which is able to enforce privacy regulations during query processing over RDF datasets. In particular, BOUNCER exploits RDF molecule templates, i.e., abstract descriptions of the properties of the entities in an RDF dataset in order to express privacy regulations as well as their automatic enforcement during query decomposition and planning. The novelty of the introduced approach is (1) the granularity of access control regulations that can be imposed; (2) the different levels at which access control statements can be enforced (at source level and at mediator level) and (3) the query plans which include physical operators that enforce the privacy and data access regulations imposed by the sources where the query is executed. The experimental evaluation of the effectiveness and efficiency of BOUNCER is conducted over the state-of-the-art benchmark BSBM for a medium size RDF dataset and 14 queries with different characteristics. The observed results suggest the effective and efficient enforcement of access control regulations during query execution, leading to minimal overhead in time incurred by the introduced access policies.

The remainder of the article is structured as follows. We motivate the privacy-aware federated query engine BOUNCER using a real case scenario from the medical domain in Sect. 2. In Sect. 4, we introduce the BOUNCER access policy model and in Sect. 5 we formally define the query decomposition and query planning techniques applied inside BOUNCER and present the architecture of our federated engine. We perform an empirical evaluation of our approach and report on the evaluation results in Sect. 6. Finally, we discuss the related work in Sect. 7 and conclude with an outlook on future work in Sect. 8.

2 Motivating Example

We motivate our work using a real-world use case from the biomedical domain where data sources from clinical records and genomics data have been integrated into an RDF graph. For instance, Fig. 1 depicts two RDF subgraphs or RDF molecules [7]. One RDF molecule represents a patient and his/her clinical information provided by source (S1), while the other RDF molecule models the results of liquid biopsy available in a research institute (S2). The privacy policy enforced at the hospital data source states that *projection (view)* of values is

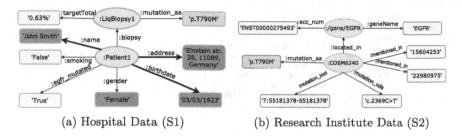

(a) Hospital Data (S1)   (b) Research Institute Data (S2)

Fig. 1. Motivating Example. Federation of RDF data sources S1 and S2. (a) An RDF molecule representing a lung cancer patient; thicker arrows correspond to controlled properties. (b) An RDF molecule representing the results of a liquid biopsy of a patient. Servers at the hospital can perform join operations.

not permitted. Properties name, date of birth, and address of a patient (thicker arrows in Fig. 1) are controlled, i.e., query operations are not permitted. Furthermore, it permits a *local join operation* (on premises of the hospital data server) of properties, such as ex:mutation_aa - peptide sequence changes that are studied for a patient, ex:targetTotal - percentage of circulating tumor DNA in the blood sample of liquid biopsy, ex:egfr_mutated - whether the patient has mutations that lead to EGFR over-expression, and ex:smoking - whether the patient is a smoker or not. Suppose a user requires to collect the Pubmed ID, mutation name, the genomic coordinates of the mutation and accession numbers of the genes associated with non-smoking lung cancer patients whose liquid biopsy has been studied for somatic mutations that involve EGFR gene amplification (over-expression). Figure 2a depicts a SPARQL query that represents this request; it is composed of 11 triple patterns. The first five triple patterns are executed against S1 while the last six triple patterns are evaluated over S2.

Existing federated query engines are able to generate query plans over these data sources. Figure 2b shows a query execution plan generated by FedX [11] federated query engine for the given query. FedX decomposes the query into two subqueries that are sent to each data source. FedX uses a nested loop join operator to join results from both sources. This operator pushes down the join operation to the data sources by binding the join variables of the right operand of the operator with values extracted from the left operand. First, triple patterns from $t1-t5$ are executed on S1, extracting values for the variables ?mutation_aa, ?lbiop, ?targetTotal, and ?patient. Then, the shared variable, ?mutation_aa, is bound and the triple patterns $t6-t11$ are executed over S2. However, executing this plan yields no answer since the privacy-policy of the hospital does not allow projection of values from the first subquery. Figure 2c shows the query execution plan generated by ANAPSID [1] federated query engine. ANAPSID creates a bushy plan where join operation is performed using GJoin operator (special type of symmetric hash join operator). This operator executes the left and right operands and makes join on the federated engine. In order to check whether the results returned from the subqueries on the left and

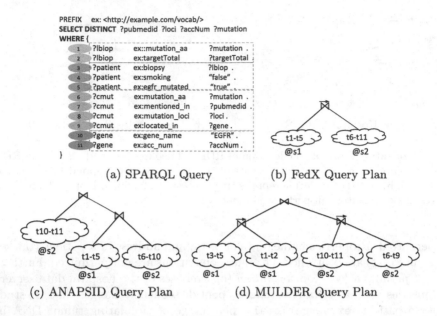

Fig. 2. Motivating Example. (a) A SPARQL query composed of four star-shaped subqueries accessing controlled and public data from S1 and S2. (b) FedX generates a plan with two subqueries. (c) ANAPSID decomposed the query into three subqueries. (d) MULDER identifies a plan with four star-shape subqueries. None of the query plan respects privacy policies of S1 and S2.

right operand can be joined, the values of shared variables from both operands have to be checked by ANAPSID, which requires extracting all values for all variables in both sources. This ignores the privacy policy enforced which yields no answer for the given query. The MULDER [7] federated query engine generates a bushy plan and decomposes the query by identifying matching RDF Molecule Templates (PRDF-MTs) as a subquery, as shown in Fig. 2d. PRDF-MT is a template that represents a set of RDF molecules that share the same RDF type (rdf:type). MULDER assigns nested hash join operator to join triple patterns $t3-t5$ associated with **Patient** PRDF-MT and triple patterns $t1-t2$ that are associated with **Liquid_Biopsy** PRDF-MT. Like in FedX, this operator extracts values for join and projection variables from the left operand, and then binds them to the same variables of the right operand. Like FedX and ANAPSID plans, the MULDER plan also ignores the privacy policy enforced at the hospital data source, which would yield an empty query answer. All of these federated engines fail to answer the query, because they ignore the privacy policy of the data sources during query decomposition as well as query execution plan generation (e.g., wrong join ordering). Also, MULDER ignores the privacy policy of the hospital during query decomposition and splits the triple patterns from this source. This leads to trying to extract results on the federation system which is not possible because of the restrictions enforced by the hospital. In addition to the join order problem, ANAPSID selects a wrong join operator which requires data from S1 to be projected for the restricted properties, i.e., $t1-t5$. In this paper, we present BOUNCER a privacy-aware federated query engine able to identify plans that respect the above-mentioned privacy and access control policies.

3 Problem Statement and Proposed Solution

In this section, we formalize the problem of privacy-aware query decomposition over a federation of RDF data sources. First we define a set of privacy-aware predicates that represent the type of operations that can be performed over an RDF dataset according to the access regulations of the federation.

Definition 1 (Privacy-Aware Operations). *Given a federated query engine* $\mathcal{M}$, *a federation* $\mathcal{F}$ *of RDF datasets* D, *and a dataset* D_i *in* D. *Let* p_{ij} *be an RDF property with domain the RDF class* C_{ij}. *The set of operations to be executed by* $\mathcal{M}$ *against* $\mathcal{F}$ *is defined as follows:*

- *join_local(D_i, p_{ij}, C_{ij}) - this predicate indicates that the join operation on property p_{ij} can be performed on the dataset D_i.*
- *join_fed(D_i, p_{ij}, C_{ij}) - this predicate indicates that the join operation on property p_{ij} can be performed by $\mathcal{M}$. The truth value of join_fed(D_i, p_{ij}, C_{ij}) implies to the truth value of join_local(D_i, p_{ij}, C_{ij}).*
- *project(D_i, p_{ij}, C_{ij}) - this predicate indicates that the values of the property p_{ij} can be projected from dataset D_i. The truth value of project(D_i, p_{ij}, C_{ij}) implies to the truth value of join_fed(D_i, p_{ij}, C_{ij}).*

Definition 2 (Access Control Theory). *Given a federated query engine* $\mathcal{M}$, *a set of RDF datasets* $D = \{D_1, \ldots, D_n\}$ *of a federation* $\mathcal{F}$. *An Access Control Theory is defined as the set of privacy-aware operations that can be performed on property* p_{ij} *of RDF class* C_{ij} *over dataset* D_i *in* D.

The access control theory for the federation described in our running example of Fig. 2a can be defined as a conjunction of the following operations:

- *join_local(s1, ex:mutation_aa, Liquid_Biopsy),*
- *join_local(s1, ex:biopsy, Patient), project(s2, ex:located_in, Mutation),*
- *join_local(s1, ex:targetTotal, Liquid_Biopsy), project(s2, ex:acc_num, Gene),*
- *join_local(s1, ex:smoking, Patient), join_local(s1, ex:egfr_mutated, Patient),*
- *project(s2, ex:mutation_aa, Mutation), project(s2, ex:gene_name, Gene),*
- *project(s2, ex:mutation_loci, Mutation), project(s2, ex:mentioned_in, Mutation).*

Note that the RDF properties :name, :gender, :address, and :birthdate of the Patient RDF class do not have operations defined in the access control theory. In our approach this fact indicates that these properties are controlled and any operation on these properties performed by the federated engine is forbidden.

Property 1. Given a property p_{ij} of an RDF class C_i from a dataset D_i in a federation $\mathcal{F}$ and an access control theory T. If there is no privacy-aware predicate in T that includes p_{ij}, then p_{ij} is a *controlled property* and no federation engine can perform operations over p_{ij} against D_i.

A basic graph pattern (BGP) in a SPARQL query is defined as a set of triple patterns $\{t_1, \ldots, t_n\}$. A BGP contains one or more triple patterns that involve a variable being projected from the original SELECT query. We call these triple patterns *projected triple patterns*, denoted as $PTP = \{t_1, \ldots, t_m\}$ such that $PTP \subseteq BGP$. A BGP includes at least one star-shaped subquery (SSQ), i.e., $BGP = \{SSQ_1, \ldots, SSQ_n\}$. A star-shaped subquery is a set of triple patterns that share the same subject variable or object [13]. Furthermore, an SSQ may contain zero or more triple patterns that involve a variable which is being projected from the original SELECT query. We call these triple patterns *projected triple patterns of an SSQ*, denoted as $PTS = \{t_1, \ldots, t_k\}$ where $PTS_i \subseteq SSQ_i$. Let PRJ be a set of triple patterns that involve a variable being projected from the original SELECT query, then projected triple patterns of a BGP, is a subset of PRJ, i.e., $PTP \subseteq PRJ$ and a projected triple pattern of SSQ_i is a subset of PTP, i.e., $PTS_i \subseteq PTP$. For example, in our running example, there is only one BGP, $BGP_1 = \{t_1, \ldots, t_{11}\}$, for which projected variables belong to triple patterns, $PRJ = \{t_6, t_7, t_8, t_{11}\}$. Projected triple patterns of BGP_1 are the same as PRJ, $PTP_{BGP_1} = \{t_6, t_7, t_8, t_{11}\}$, since there is only one BGP. Furthermore, BGP_1 can be clustered into four start-shaped subqueries, $SSQs_{BGP_1} = \{SSQ_{1=\{t_1-t_2\}}, SSQ_{2=\{t_3-t_5\}}, SSQ_{3=\{t_6-t_9\}}, SSQ_{4=\{t_{10}-t_{11}\}}\}$. Out of four $SSQs$ of BGP_1, only the last two $SSQs$ have triple patterns that are also in the projected triple patterns, i.e., $PTS_{SSQ_1} = \emptyset$, $PTS_{SSQ_2} = \emptyset$, $PTS_{SSQ_3} = \{t_6, t_7, t_8\}$, $PTS_{SSQ_4} = \{t_{11}\}$.

Property 2. Given a SPARQL query Q such that a variable $?v$ is associated with a property p of a triple pattern t in a BGP and $?v$ is projected in Q. Suppose an access control theory T regulates the access of the datasets in D of the federation $\mathcal{F}$. A federation engine $\mathcal{M}$ accepts Q iff there is a privacy-aware operation $project(D_i, p, C)$ in T for at least an RDF dataset D_i in D.

A privacy-aware query decomposition on a federation is defined. This formalization states the conditions to be met by a decomposition in order to be evaluated over a federation by enforcing their access regulations.

Definition 3 (Privacy-Aware Query Decomposition). *Let BGP be a basic graph pattern, PTP a set of projected triple patterns of a BGP, T an access control theory, and $D = \{D_1, \ldots, D_n\}$ a set of RDF datasets of a federation $\mathcal{F}$. A privacy-aware decomposition P of BGP in D, $\gamma(P|BGP, D, T, PTP)$, is a set of decomposition elements, $\Phi = \{\phi_1, \ldots, \phi_k\}$, such that ϕ_i is a four-tuple, $\phi_i = (SQ_i, SD_i, PS_i, PTS_i)$, where:*

- *SQ_i is a subset of triple patterns in BGP, i.e., $SQ_i \subseteq BGP$, and $SQ_i \neq \emptyset$, such that there is no repetition of triple patterns, i.e., If $t_a \in SQ_i$, then $!\exists t_a \in SQ_j : SQ_j \subset BGP \wedge i \neq j$,*

- SD_i is a subset of datasets in D, i.e., $SD_i \subseteq D$, and $SD_i \neq \varnothing$,
- PS_i is a set of privacy-aware operations that are permitted on triple patterns in SQ_i to be performed on datasets in SD_i and $PS_i \subseteq T$, and $PS_i \neq \varnothing$,
- PTS_i is a set of triple patterns in SQ_i that contains variables being projected from the original SELECT query, i.e., $PTS_i \subseteq SQ_i \wedge PTS_i \subseteq PTP$,
- The set composed of SQ_i in the decompositions $\phi_i \in \Phi$ corresponds to a partition of BGP and
- The selected RDF datasets are able to project out the attributes in the project clause of the query, i.e., $\forall t_a \in SQ_i : t_a \in PTP$, then $project(D_a, p_{aj}, C_{aj}) \in PS_i$ where $t_a = (s, p_{aj}, o)$, $D_a \in SD_i$, and $SQ_i \in \phi_i$.

After defining what is a decomposition of a query, we state the problem of finding a suitable decomposition for a query and a given set of data sources.

Privacy-Aware Query Decomposition Problem. Given a SPARQL query Q, RDF datasets $D = \{D_1, \ldots, D_m\}$ of a federation $\mathcal{F}$, and access control theory T. The *problem of decomposing* Q in D restricted by T is defined as follows. For all BGPs, $BGP = \{t_1, \ldots, t_n\}$ in Q, find a query decomposition $\gamma(P|BGP, D, T, PTP)$ that satisfies the following conditions:

- The evaluation of $\gamma(P|BGP, D, T, PTP)$ in D is *complete* according to the privacy-aware policies of the federation in T. Suppose D^* represents the maximal subset of D where the privacy policies of each RDF dataset $D_i \in D^*$ allow for projecting and joining the properties from D_i that appear in Q^1. Then the evaluation of BGP in D^* is equivalent to the evaluation of $\gamma(P|BGP, D, T, PTP)$ and the following expression holds:

$$[[BGP]]_{D^*} = [[\gamma(P|BGP, D, T, PTP)]]_D$$

- The cost of executing the query decomposition $\gamma(P|BGP, D, T, PTP)$ is *minimal*. Suppose the execution time of a decomposition P' of BGP in D is represented as $cost(\gamma(P'|BGP, D, T, PTP))$, then

$$\gamma(P|BGP, D, T, PTP) = \underset{\gamma(P'|BGP, D, T, PTP)}{\operatorname{argmin}} cost(\gamma(P'|BGP, D, T, PTP))$$

To solve this problem, we present BOUNCER, a federated query engine able to identify query decompositions for SPARQL queries and query plans that efficiently evaluate SPARQL queries over a federation. Two definitions are presented for a query plan over a decomposition. The next two functions are presented in order to facilitate the understanding of the definition of a query plan.

Definition 4 (The property function prop(*)). *Given a set of triple patterns, TPS, the function $prop(TPS)$ is defined as follows:*

$$prop(TPS) = \{p \mid (s, p, o) \in TPS \wedge p \text{ is constant}\}$$

---

[1] Predicates $project(Di, p_{ij}, C_{ij})$, $join\_fed(Di, p_{ij}, C_{ij})$ and $join\_local(Di, p_{ij}, C_{ij})$ are part of T for all properties in triple patterns in Q that can be answered by Di.

Definition 5 (The variable function var(*)). *Given a privacy-aware decomposition, Φ, the function $var(\Phi)$ is defined inductively as follows:*

1. **Base case:** $\Phi = \{\phi_1\}$, then $var(\Phi) = \{?x \mid (s,p,o) \in SQ_1$, where $\phi_1 = (SQ_1, SD_1, PS_1, PTS_1)$, $?x = s \wedge s$ is a variable $\vee ?x = o \wedge o$ is a variable$\}$
2. **Inductive case:** Let Φ_1 and Φ_2 be disjoint decompositions such that $\Phi = \Phi_1 \cup \Phi_2$ then, $var(\Phi) = var(\Phi_1) \cup var(\Phi_2)$.

Definition 6 (A Valid Plan over a Privacy-Aware Decomposition). *Given a privacy-aware decomposition $\gamma(P|BGP, D, T, PTP)$: $\Phi = \{\phi_1, \ldots, \phi_n\}$, a valid query plan, $\alpha(\Phi)$, is defined inductively as follows:*

1. **Base Case:** *If only one decomposition ϕ_1 belongs to Φ, i.e., $\Phi = \{\phi_1\}$, the plan unions all the service graph patterns over the selected RDF sources. Thus, $\alpha(\Phi) = UNION_{d_i \in SD_1}(SERVICE\ d_i\ SQ_1)$ is a valid plan[2,3], where:*
   - $\phi_1 = (SQ_1, SD_1, PS_1, PTS_1)$ *is a valid privacy-aware decomposition;*
   - *All the variables projected in the query have the permission to be projected, i.e., $\forall p_{i1} \in prop(PTS_1)$, $project(Di, pi1, Ci1) \in PS_1$.*
2. **Inductive Case:** *Let Φ_1 and Φ_2 be disjoint decompositions such that $\Phi = \Phi_1 \cup \Phi_2$. Then, $\alpha(\Phi) = (\alpha(\Phi_1) * \alpha(\Phi_2))$ is a valid plan, where:*
   (a) *$\alpha(\Phi_1)$ and $\alpha(\Phi_2)$ are valid plans.*
   (b) *The join variables appear jointly in the triple patterns of Φ_1 and Φ_2, i.e., $joinVars = var(\Phi_1) \cap var(\Phi_2)$.*
   (c) *$\mathcal{J}$ is a set of joint triple patterns involving join variables in BGP:*
      - $\mathcal{J} = \{t|variable(t) \subseteq joinVars, (t \in \Phi_{1(SQ)} \vee t \in \Phi_{2(SQ)})\}$
      - $\Phi_{1(SQ)} = \{SQ_i|\forall \phi_i \in \Phi_1,\ \phi_i = (SQ_i, SD_i, PS_i, PTS_i)\}$, *and*
      - $\Phi_{2(SQ)} = \{SQ_j|\forall \phi_j \in \Phi_2,\ \phi_j = (SQ_j, SD_j, PS_j, PTS_j)\}$.
   (d) *The operator $*$ is a JOIN operator, i.e., $\alpha(\Phi) = (\alpha(\Phi_1)\ JOIN\ \alpha(\Phi_2))$ is a valid plan, iff $\forall p_{ij} \in prop(\mathcal{J})$, $join\_fed(D_i, p_{ij}, C_{ij}) \in (\Phi_{1(PS)} \cap \Phi_{2(PS)})$, $\Phi_{1(PS)} = \{PS_i|\forall \phi_i \in \Phi_1,\ \phi_i = (SQ_i, SD_i, PS_i, PTS_i)\}$, and $\Phi_{2(PS)} = \{PS_j|\forall \phi_j \in \Phi_2,\ \phi_j = (SQ_j, SD_j, PS_j, PTS_j)\}$.*
   (e) *The operator $*$ is a DJOIN operator, i.e., $\alpha(\Phi) = (\alpha(\Phi_1)\ DJOIN\ \alpha(\Phi_2))$ is a valid plan iff $\forall p_{ij} \in prop(\mathcal{J})$, $join\_fed(D_i, p_{ij}, C_{ij}) \in \Phi_{1(PS)}$ and $join\_local(D_i, p_{ij}, C_{ij}) \in \Phi_{2(PS)}$[4].*

Next, we define the BOUNCER architecture and the main characteristics of the query decomposition and execution tasks implemented by BOUNCER.

4   BOUNCER: A Privacy-Aware Engine

Web interfaces provide access to RDF datasets, and can be described in terms of resources and properties in the datasets. BOUNCER employs privacy-aware RDF Molecule Templates for describing and enforcing privacy policies.

---

[2] For readability, $UNION_{d_i \in SD + i}$ represents SPARQL UNION operator.
[3] $SERVICE$ corresponds to the SPARQL SERVICE clause.
[4] DJOIN- is a dependent JOIN [14].

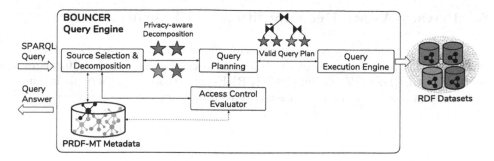

Fig. 3. BOUNCER Architecture. BOUNCER receives a SPARQL query and outputs the results of executing the SPARQL query over a federation of SPARQL endpoints. It relies on PRDF-MT descriptions and privacy-aware policies to select relevant sources, and perform query decomposition and planning. The query engine executes a valid plan against the selected sources.

Definition 7 (Privacy-Aware RDF Molecule Template(PRDF-MT)).
A privacy-aware RDF molecule template (PRDF-MT) is a 5-tuple=< WebI, C, DTP, IntraL, InterL>, where:

- *WebI – is a Web service API that provides access to an RDF dataset G via SPARQL protocol;*
- *C – is an RDF class such that the triple pattern (?s rdf:type C) is true in G;*
- *DTP – is a set of triples (p, T, op) such that p is a property with domain C and range T, the triple patterns (?s p ?o) and (?o rdf:type T) and (?s rdf:type C) are true in G, and op is an access control operator that is allowed to be performed on property p;*
- *IntraL – is a set of pairs (p, C_j) such that p is an object property with domain C and range C_j, and the triple patterns (?s p ?o) and (?o rdf:type C_j) and (?s rdf:type C) are true in G;*
- *InterL – is a set of triples (p, C_k, SW) such that p is an object property with domain C and range C_k; SW is a Web service API that provides access to an RDF dataset K, and the triple patterns (?s p ?o) and (?s rdf:type C) are true in G, and the triple pattern (?o rdf:type C_k) is true in K.*

Figure 3 depicts BOUNCER architecture. Given a SPARQL query, the source selection and query decomposition component solves the problem of identifying a privacy-aware query decomposition; they select PRDF-MTs for subqueries (SSQs) by consulting PRDF-MT metadata store and the access control evaluator component. The source selection and decomposition component is privacy-aware decomposition; it is given to the query planning component for creating a valid plan, i.e., access policies of the selected data sources should be respected. The valid plan is executed in a bushy-tree fashion by the query execution.

5   Privacy-Aware Decomposition and Execution

This section presents the privacy-aware techniques implemented by BOUNCER. They rely on the description of the RDF datasets of a federation in terms of privacy-aware RDF molecule templates (PRDF-MTs) to identify query plans that enforce data access control regulations. More importantly, these techniques are able to generate query execution plans whose operators force the execution of queries at the dataset sites in case data cannot be transferred or accessed.

5.1   Privacy-Aware Source Selection and Decomposition

The BOUNCER privacy-aware source selection and query decomposition is sketched in Algorithm 1. Given a *BGP* in a SPARQL query *Q*, BOUNCER first decomposes the query into star-shaped subqueries (SSQs), (Line 2). For instance, our running example query, in Fig. 2a, is decomposed into four SSQs, as shown in Fig. 4, i.e., SSQs around the variables ?lbiop, ?patient, ?cmut, and ?gene, respectively. The first SSQ (denoted ?lbiop-SSQ) has two triple patterns, t1–t2, the second SSQ (?patient-SSQ) is composed of three triple

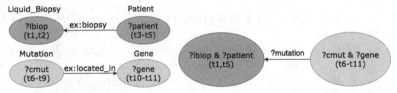

(a) Initial query decomposition   (b) Privacy-aware Query Decomposition

Fig. 4. Example of Privacy-Aware Decompositions. Decompositions for SPARQL query in the motivating example. Nodes represent SSQs and colors indicate datasets where they are executed; edges correspond to join variables. (a) Initial query decomposed into four SSQs. (b) Decomposition result where the subqueries ?lbiop-SSQ and ?patient-SSQ are composed into a single subquery to comply with the privacy policy of data source S1, while ?cmut-SSQ and ?gene-SSQ are also composed to push down the join operation to the data source S2. (Color figure online)

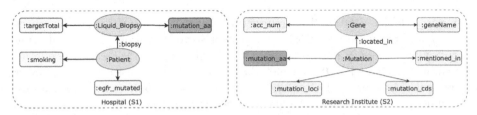

Fig. 5. Example of Privacy-aware RDF Molecule Templates (PRDF-MTs). Two PRDF-MTs for the SPARQL query in the motivating example. According to the privacy regulations the properties **:name**, **:birthdate**, and **:addresss** are controlled; they do not appear in the PRDF-MTs.

Algorithm 1. Privacy-Aware Query Decomposition: BG - Basic Graph Pattern, Q - Query, $PRMT$ - Access-aware RDF Molecule Templates

1: **procedure** DECOMPOSE(BGP, Q, $PRMT$)
2:     $SSQs \leftarrow getSSQs(BGP)$ ▷ Partition the BGP to SSQs
3:     $RES \leftarrow selectSource(PRMT, PRMT)$ ▷ RES=[(SSQ, PRMT, DataSource)]
4:     $A \leftarrow getAccessPolicies(RES)$; $\Phi \leftarrow [\ ]$; $DR \leftarrow \{\ \}$ ▷ access control statements
5:     **for** $(SSQ, RMT, p, ds, pred) \in A$ **do**
6:         **if** $p \in Query.PRJ \wedge pred\,! = project(ds, p, RMT.type)$ **then return** [ ]
7:             $DR[SSQ][PTS].append(t)\ |\ t = (s, p, o) \wedge t \in SSQ\ |\ p \in Query.PRJ$
8:             $DR[SSQ][SD].append(ds) \wedge DR[SSQ][PS].append(pred)$
9:     **end for**
10:    **for** $(SSQ_i, SD_i, PS_i, PTS_i) \in DR$ **do**
11:        $\phi_i = (SQ_i, SD_i, PS_i, PTS_i)\ |\ SQ_i \leftarrow SSQ_i$
12:        **if** $join\_local() \in PS_i$ **then**   ▷ If SSQ_i contains restricted property
13:            **for** $(SSQ_j, SD_j, PS_j, PTS_j) \in DR$ **do**
14:                **if** $SD_i \cap SD_j \neq \varnothing$ **then**
15:                    $\phi_i.extend(SSQ_j, SD_j, PS_j, PTS_j)$
16:                    $DR.remove((SSQ_j, SD_j, PS_j, PTS_j)) \wedge done \leftarrow True$
17:            **end for**
18:            **if** $NOT\ done$ **then return** [ ]
19:        **end if**
20:        $\Phi.append(\phi_i)$
21:    **end for**
22:    **return** Φ ▷ decomposed query
23: **end procedure**

patterns, t3–t5, the third SSQ (?cmut-SSQ) includes four triple patterns, and the fourth SSQ (?gene-SSQ) is composed of two triple patterns, t10–t11 (Fig. 5).

Figure 4a presents an initial decomposition with the selected PRDF-MTs for each SSQs. The subquery ?patient-SSQ is joined to the subquery ?lbiop-SSQ via ex:biopsy property. Similarly, ?cmut-SSQ is joined to ?gene-SSQ via the ex:located_in property. Given the set of properties in each SSQ and the joins between them, BOUNCER finds a matching PRDF-MT for each SSQs (Line 3), i.e., it matches the subqueries ?patient-SSQ, ?lbiop-SSQ, ?cmut-SSQ, and ?gene-SSQ to the PRDF-MTs Patient, Liquid_Biopsy, Mutation, and Gene, respectively. Once the PRDF-MTs are identified for the SSQs, BOUNCER verifies the access control policies associated with them (Line 4). A subquery SSQ associated with an PRDF-MT(s) that grants the project() permission to all of its properties is called *Independent SSQ*; otherwise, it is called *Dependent SSQ*. An SSQ in a SPARQL query Q is called *dependent* iff a property of at least one triple pattern in SSQ is associated with the privacy-aware operation join_local(). On the other hand, an SSQ is *independent* iff the privacy-aware operation project() is true for the properties of the triple patterns in SSQ.

If the value of the controlled property is in the projection list, i.e., if the property of a triple pattern in an SSQ have join_local() or join_fed() predicate, then the decomposition process exits with empty result (Line 6). Once

the SSQs are associated with PRDF-MTs, the next step is to merge the SSQs
with the same source and push down the join operation to the data source.
To comply with access control policies of a dataset, i.e., when the properties of
an SSQ have only the join_local() permission, the join operation with this
SSQ should be done at the data source. Hence, if two SSQs can be executed at
the same source, then BOUNCER decomposes them as a single subquery (SQ)
(Lines 10–21). This technique may also improve query execution time by performing join operation at the source site. Figure 4b shows a final decomposition
for our running example. ?lbiop-SSQ and ?patient-SSQ are merged because
they are dependent and the join operation can be executed at the source.

5.2 BOUNCER Privacy-Aware Query Planning Technique

Algorithm 2 sketches the BOUNCER privacy-aware query planing technique.
Given a privacy-aware decomposition Φ of a query Q, BOUNCER finds a valid
plan that respects the privacy-policy of the data sources. For each subquery in
ϕ_i a service-graph pattern is created (Lines 4 and 6) and the SPARQL UNION
operator is used whenever the subquery can be executed over more than one data
source. Then, BOUNCER selects another subquery, ϕ_j that is joinable with ϕ_i
(Line 5). If ϕ_i is composed of dependent SSQ(s) (resp., independent SSQ(s))
and ϕ_j is composed of an independent SSQ(s) (resp., dependent SSQ(s)), then a
dependent join operator (DJOIN) is selected (Lines 9–12). If both ϕ_i and ϕ_j are
merged of an independent SSQ(s), then any JOIN operator can be chosen (Lines
13–14). Finally, otherwise, an empty plan is returned indicating that there is no
valid plan for the input query (Line 16).

Algorithm 2. Query Planning over Privacy-Aware Decomposition: Φ - Privacy-
Aware query decomposition, Q - SELECT query

1: **procedure** MAKEPLAN(Φ, Q)
2:     $\alpha \leftarrow []$
3:     **for** $\phi_i \in \Phi$ **do**
4:         $\sigma_1 \leftarrow UNION_{d_i \in SD_i \wedge SD_i \in \phi_i}(SERVICE\ d_i\ SQ_i)$
5:         **for** $\phi_j \in \Phi \mid \phi_i \neq \phi_j \wedge var(SQ_i) \cap var(SQ_j) \neq \emptyset$ **do**   ▷ If joinable
6:             $\sigma_2 \leftarrow UNION_{d_j \in SD_j}(SERVICE\ d_j\ SQ_j)$
7:             $\mathcal{J} \leftarrow \{ t \mid vari(t) \subseteq [var(SQ_i) \cap var(SQ_j)] \wedge t \in [SQ_i \cup SQ_j]\}$
8:             $\rho \leftarrow prop(\mathcal{J})$ ▷ Properties of join variables
9:             **if** $\exists join\_local() \in PS_i \wedge \forall pred_{p \in \rho} \in PS_j \mid pred_{p \in \rho} \Rightarrow join\_fed()$ **then**
10:                 $\alpha.append((\sigma_2\ DJOIN\ \sigma_1))$; $joined \leftarrow True$ ▷ Dependent JOIN
11:             **if** $\exists join\_local() \in PS_j \wedge \forall pred_{p \in \rho} \in PS_i \mid pred_{p \in \rho} \Rightarrow join\_fed()$ **then**
12:                 $\alpha.append((\sigma_1\ DJOIN\ \sigma_2))$; $joined \leftarrow True$ ▷ Dependent JOIN
13:             **if** $\forall pred_{p \in \rho} \in [PS_i \cup PS_j] \mid pred_{p \in \rho} \Rightarrow join\_fed()$ **then**
14:                 $\alpha.append((\sigma_1\ JOIN\ \sigma_2))$; $joined \leftarrow True$ ▷ Independent JOIN
15:         **end for**
16:         **if** $\exists join\_local() \in PS_i \wedge NOT\ joined$ **then return** $[]$ ▷ No valid plan
17:     **end for**
18:     **return** α
19: **end procedure**

---

6 Empirical Evaluation

We study the efficiency and effectiveness of BOUNCER. First, we assess the impact of access-control policies enforcement and BOUNCER is compared to ANAPSID, FedX, and MULDER. Then, the performance of BOUNCER is evaluated. We study the following research questions: **(RQ1)** Does privacy-aware enforcement employed during source selection, query decomposition, and planning impact query execution time? **(RQ2)** Can privacy-aware policies be used to identify query plans that enhance execution time and answer completeness?

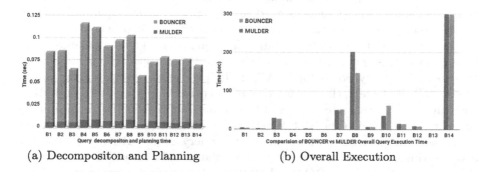

(a) Decomposition and Planning   (b) Overall Execution

Fig. 6. Decomposition and Execution Time. BOUNCER decomposition and planning are more expensive than baseline (MULDER), but BOUNCER generates more efficient plans and overall execution time is reduced.

Benchmarks: The Berlin SPARQL Benchmark (*BSBM*) generates a dataset of 200 M triples and 14 queries; answer size is limited to 10,000 per query.

Metrics: *(i)* Execution Time: Elapsed time between the submission of a query to an engine and the delivery of the answers. Timeout is set to 300 s. *(ii)* Throughput: Number of answers produced per second; this is computed as the ratio of the number of answers to execution time in seconds.

Implementation: BOUNCER privacy-aware techniques are implemented in Python 3.5 and integrated into the ANAPSID query engine. The BSBM dataset is partitioned into 8 parts (one part per RDF type) and deployed on one machine as SPARQL endpoints using Virtuoso 6.01.3127, where each dataset resides in a dedicated Virtuoso docker container. Experiments are executed on a Dell PowerEdge R805 server, AMD Opteron 2.4 GHz CPU, 64 cores, 256 GB RAM.

Experiment 1: Impact of Access Control Enforcement. The impact of privacy-aware processing techniques is studied, as well as the overhead on source selection, decomposition, and execution. In this experiment, the privacy-aware theory enables all the operations over the properties of the federation, i.e., all the operations are defined for each property and dataset. MULDER and

BOUNCER are compared; Fig. 6 reports on decomposition, planning, and execution time per query. Both engines generate the same results and BOUNCER consumes more time in query decomposition and planning. However, the overall execution time is lower in almost all queries. These results suggest that even there is an impact on query processing, BOUNCER is able to exploit privacy-aware polices, and generates query plans that speed up query execution.

Experiment 2: Impact of Privacy-Aware Query Plans. The privacy-aware query plans produced by BOUNCER are compared to the ones generated by state-of-the-art query engines. In this experiment, the privacy-aware theory enables local joins for `Person`, `Producer`, `Product`, and `ProductFeature`, and projections of the properties of `Offer`, `Review`, `ProductType`, and `Vendor`. Figure 7 reports on the throughput of each query engine. As observed, the query engines produced different query plans which allow for high performance. However, many of these plans are not valid, i.e., they do not respect the privacy-aware policies in the theory. For instance, ANAPSID produces bushy tree plans around `gjoins`; albeit efficient, these plans violate the privacy policies. FedX and MULDER are able to generate some valid plans–*by chance*– but fail in producing efficient executions. On the contrary, BOUNCER generates valid plans that in many cases increase the performance of the query engine. Results observed in two experiments suggest that efficient query plans can be identified by exploiting the privacy policies; thus, **RQ1** and **RQ2** can be positively answered.

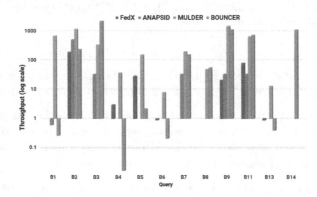

Fig. 7. Efficiency of Query Plans. Existing engines are compared based on throughput. ANAPSID plans are efficient but no valid. FedX and MULDER generate valid plans (by chance) but some are not efficient. BOUNCER generates both valid and efficient plans and overall execution time is reduced.

7   Related Work

The data privacy control problem has received extensive attention by the Database community; approaches by De Capitani et al. [6] and Bater et al. [3] are exemplars that rely on an authority network to produce valid plans. Albeit relevant, these approaches are not defined for federated systems; thus, the tasks of source selection and query decomposition are not addressed. BOUNCER also generates valid plans, but being designed for SPARQL endpoint federations, it also ensures that only relevant endpoints are selected to evaluate these valid plans. The Semantic Web community has also explored access control models for SPARQL query engines; RDF named graphs [5,8,12] and quad patterns [9] are used to enforce access control policies. Most of the work focuses on the specification of access control ontologies and enforcement on RDF data [5,12] stored in a centralized RDF store, while others explore access control specification and enforcement on distributed RDF stores [2,4] and federated query processing [8,10] scenarios. Costabello et al. [5] present SHI3LD, an access control framework for RDF stores accessed on mobile devices; it provides a pluggable filter for generic SPARQL endpoints that enforces context-aware access control at named graph level. Kirane et al. [9] propose an authorization framework that relies on stratified Datalog rules to enforce access control policies; RDF quad patterns are used to model permissions (grant or deny) on named graphs, triples, classes, and properties. Ubehauen et al. [12] propose an access control approach at the level of named graphs; it binds access control expressions to the context of RDF triples and uses a query rewriting method on an ontology for enabling the evaluation of privacy regulations in a single query. SAFE [8] is designed to query statistical RDF data cubes in distributed settings and also enables graph level access control. BOUNCER is a privacy-aware federated engine where policies are defined over RDF properties of PRDF-MTs; it also enables access control statements at source and mediator level. More important, BOUNCER generates query plans that both enforce privacy regulations and speed up execution time.

8 Conclusion and Future Work

We presented BOUNCER, a privacy-aware federated query engine for SPARQL endpoints. BOUNCER relies on privacy-aware RDF Molecule Templates (PRDF-MTs) for source description and guiding query decomposition and plan generation. Efficiency of BOUNCER was empirically evaluated, and results suggest that it is able to reduce query execution time and increase answer completeness by producing query plans that comply with the privacy policies of the data sources. In future work, we plan to integrate additional Web access interfaces, like RESTful APIs, and empower PRDF-MTs with context-aware access policies.

Acknowledgements. This work has been funded by the EU H2020 RIA under the Marie Skłodowska-Curie grant agreement No. 642795 (WDAqua) and EU H2020 Programme for the project No. 727658 (IASIS).

References

1. Acosta, M., Vidal, M.-E., Lampo, T., Castillo, J., Ruckhaus, E.: ANAPSID: an adaptive query processing engine for SPARQL endpoints. In: Aroyo, L., et al. (eds.) ISWC 2011. LNCS, vol. 7031, pp. 18–34. Springer, Heidelberg (2011). https://doi.org/10.1007/978-3-642-25073-6_2
2. Amini, M., Jalili, R.: Multi-level authorisation model and framework for distributed semantic-aware environments. IET Inf. Secur. 4(4), 301–321 (2010)
3. Bater, J., Elliott, G., Eggen, C., Goel, S., Kho, A., Rogers, J.: SMCQL: secure querying for federated databases. Proc. VLDB Endow. 10(6), 673–684 (2017)
4. Bonatti, P.A., Olmedilla, D.: Rule-based policy representation and reasoning for the semantic web. In: Antoniou, G., et al. (eds.) Reasoning Web 2007. LNCS, vol. 4636, pp. 240–268. Springer, Heidelberg (2007). https://doi.org/10.1007/978-3-540-74615-7_4
5. Costabello, L., Villata, S., Gandon, F.: Context-aware access control for RDF graph stores. In: ECAI-20th European Conference on Artificial Intelligence (2012)
6. De Capitani, S., di Vimercati, S., Foresti, S., Jajodia, S.P., Samarati, P.: Authorization enforcement in distributed query evaluation. JCS 19(4), 751–794 (2011)
7. Endris, K.M., Galkin, M., Lytra, I., Mami, M.N., Vidal, M.-E., Auer, S.: MULDER: querying the linked data web by bridging RDF molecule templates. In: Benslimane, D., Damiani, E., Grosky, W.I., Hameurlain, A., Sheth, A., Wagner, R.R. (eds.) DEXA 2017. LNCS, vol. 10438, pp. 3–18. Springer, Cham (2017). https://doi.org/10.1007/978-3-319-64468-4_1
8. Khan, Y., et al.: SAFE: SPARQL federation over RDF data cubes with access control. J. Biomed. Semant. 8(1) (2017)
9. Kirrane, S., Abdelrahman, A., Mileo, A., Decker, S.: Secure manipulation of linked data. In: Alani, H., et al. (eds.) ISWC 2013. LNCS, vol. 8218, pp. 248–263. Springer, Heidelberg (2013). https://doi.org/10.1007/978-3-642-41335-3_16
10. Kost, M., Freytag, J.-C.: SWRL-based access policies for linked data (2010)
11. Schwarte, A., Haase, P., Hose, K., Schenkel, R., Schmidt, M.: FedX: optimization techniques for federated query processing on linked data. In: Aroyo, L., et al. (eds.) ISWC 2011. LNCS, vol. 7031, pp. 601–616. Springer, Heidelberg (2011). https://doi.org/10.1007/978-3-642-25073-6_38
12. Unbehauen, J., Frommhold, M., Martin, M.: Enforcing scalable authorization on SPARQL queries. In: SEMANTiCS (Posters, Demos, SuCCESS) (2016)
13. Vidal, M.-E., Ruckhaus, E., Lampo, T., Martínez, A., Sierra, J., Polleres, A.: Efficiently joining group patterns in SPARQL queries. In: Aroyo, L., et al. (eds.) ESWC 2010. LNCS, vol. 6088, pp. 228–242. Springer, Heidelberg (2010). https://doi.org/10.1007/978-3-642-13486-9_16
14. Zadorozhny, V., Raschid, L., Vidal, M., Urhan, T., Bright, L.: Efficient evaluation of queries in a mediator for websources. In: ACM SIGMOD (2002)

Minimising Information Loss
on Anonymised High Dimensional Data
with Greedy In-Memory Processing

Nikolai J. Podlesny$^{(\boxtimes)}$, Anne V. D. M. Kayem$^{(\boxtimes)}$,
Stephan von Schorlemer$^{(\boxtimes)}$, and Matthias Uflacker$^{(\boxtimes)}$

Hasso Plattner Institute, University of Potsdam, Potsdam, Germany
Nikolai.Podlesny@student.hpi.de, Anne@mykayem.org,
{Stephan.Schorlemer,Matthias.Uflacker}@hpi.de

Abstract. Minimising information loss on anonymised high dimensional
data is important for data utility. Syntactic data anonymisation algorithms address this issue by generating datasets that are neither use-
case specific nor dependent on runtime specifications. This results in
anonymised datasets that can be re-used in different scenarios which
is performance efficient. However, syntactic data anonymisation algorithms incur high information loss on high dimensional data, making
the data unusable for analytics. In this paper, we propose an optimised
exact quasi-identifier identification scheme, based on the notion of k-
anonymity, to generate anonymised high dimensional datasets efficiently,
and with low information loss. The optimised exact quasi-identifier identification scheme works by identifying and eliminating maximal partial unique column combination (mpUCC) attributes that endanger
anonymity. By using in-memory processing to handle the attribute selection procedure, we significantly reduce the processing time required. We
evaluated the effectiveness of our proposed approach with an enriched
dataset drawn from multiple real-world data sources, and augmented
with synthetic values generated in close alignment with the real-world
data distributions. Our results indicate that in-memory processing drops
attribute selection time for the mpUCC candidates from 400s to 100s,
while significantly reducing information loss. In addition, we achieve a
time complexity speed-up of $O(3^{n/3}) \approx O(1.4422^n)$.

1   Introduction

High dimensional data holds the advantage of enabling a myriad of data analytics operations. Yet, the growth in amounts of data available has also increased
the possibilities of obtaining both direct and correlated data to describe users to
a highly fine-grained degree. Data shared with data analytics service providers
must therefore be privacy preserving to protect against de-anonymisation incidents [2,7,33,34,42], and usable to generate correct query results [1].

In contrast to their semantic counterparts, syntactic data anonymisation
algorithms such as, k-anonymity, l-diversity, and t-closeness, are better for high

© Springer Nature Switzerland AG 2018
S. Hartmann et al. (Eds.): DEXA 2018, LNCS 11029, pp. 85–100, 2018.
https://doi.org/10.1007/978-3-319-98809-2_6

dimensional data anonymisation because the anonymised datasets are not use-case specific or reliant on runtime specifications. The generated syntactic anonymised datasets can be reused for several purposes, which is performance efficient. Yet, studies of syntactic anonymisation algorithms show that the anonymisation problem is NP-hard [8,24,29], and the anonymised data is vulnerable to semantics-based attacks [23,26,38,43]. Furthermore, existing syntactic data transformation techniques like *Generalisation, Suppression*, and *Perturbation* incur high levels of information loss when applied to high dimensional datasets, which impacts negatively on query processing and on the quality of data analytics results. Semantic anonymisation algorithms, like differential privacy, alleviate information loss and de-anonymisations [4,6,9], but are designed for pre-defined use cases where knowledge of the composition of required dataset is known before runtime. Pre-processing large high dimensional datasets on a per-query basis impacts negatively on performance. Furthermore, postponing data anonymisation to runtime can enable colluding users to run multiple complimentary queries to return datasets that when combined, provide information to enable partial or even complete de-anonymisation of the original dataset [4,6,9,18]. Kifer et al. [18] address this problem with "non-interactive" differential privacy in which user queries are statistically evaluated apriori to identify and prevent collusions, but the performance issue remains.

In this paper, we propose an optimised exact quasi-identifier identification scheme, based on the notion of k-anonymity, to generate anonymised high dimensional datasets efficiently. The reason is that using a combination of quasi-identifiers and sensitive attributes protects against de-anonymisation. The optimised exact quasi-identifier identification scheme is based on optimisation techniques for the exponential and W[2]-complete search for quasi-identifiers [5], and works by prefiltering maximal partial unique column combination (mpUCC) candidates, to eliminate attributes that endanger anonymity irrespective of the use case scenario. We reduce the time complexity of the anonymisation algorithm by using in-memory processing to parallelise the attribute selection procedure. We evaluated the effectiveness of our proposed approach, based on an enriched dataset drawn from multiple real-world data sources and augmented with synthetic values generated in close alignment with the real-world data distributions. Our results indicate that for 80 columns on average, in-memory processing drops attribute selection time for the mpUCC candidates from 400s to under 100s. In addition, we achieve a theoretical speed-up of $O(3^{n/3}) \approx O(1.4422^n)$ which proves to be much faster in practice due to the prefiltering of candidates but at the same time still of exact nature.

The rest of the paper is structured as follows. We discuss general related work on data anonymisation in Sect. 2. In Sect. 3, we provide some background details on k-anonymisation focusing on how quasi-identifiers are identified, and why applying data transformation techniques such as *Generalisation*, and *Suppression* is inefficient on high dimensional data. In Sect. 4, we describe our optimised exact quasi-identifier identification scheme, and proceed in Sect. 5 to discuss results from our experiments using in-memory applications. We offer conclusions and directions for future work in Sect. 6.

2   Related Work

Syntactic data anonymisation algorithms such as k-anonymity [37], l-diversity [26], and t-closeness [23] have been studied quite extensively to prevent disclosures of sensitive personal data. In order to achieve data anonymisation syntactic data anonymisation algorithms rely on a variety of data transformation methods that include *generalisation, suppression,* and *perturbation*. On the basis of these works, one could classify methods of data transformation for anonymisation into two categories namely, *randomisation* and *generalisation*.

Randomisation algorithms alter the veracity of the data, by removing strong links between the data and an individual. This is typically achieved either by noise injections, permutations, or statistical shifting to alter the data set for anonymity [16]. For instance, in differential privacy, this is done by determining at the runtime of a query, how much noise injections to add to the resulting dataset in order to ensure the anonymity in each case [9]. Additionally, differential privacy uses the exponential mechanism to release statistical information about a dataset without revealing private details of individual data entries [27]. Furthermore, the Laplace mechanism for perturbation, supports statistical shifting in differential privacy, by employing controlled random distribution sensitive noise additions [10,20]. It is worth noting here that the discretized version [14,25] is known as matrix mechanism because both sensitive attributes and quasi-identifiers are evaluated on a per-row basis during anonymisation [22].

By contrast, *generalisation* algorithms modify dataset values according to a hierarchical model where each value progressively loses uniqueness as one moves upwards in the hierarchy. Several generalisation algorithms have been used effectively in combination with k-anonymity, l-diversity, as well as t-closeness. In k-anonymity the concept is to place each person in the data set together with at least $k - 1$ similar data records, such that there is no possibility of distinguishing between them. This is done by assimilating the $k - 1$ nearest neighbours based on their describing attributes through generalisation and suppression [37]. Generalisation is vulnerable to homogeneity and background knowledge attacks [26], which l-diversity alleviates by considering the granularity of sensitive data representations to ensure a diversity of a factor of l for each quasi-identifier within a given equivalence class (usually a size of k). Further extensions in the form of t-closeness, handle skewness and background knowledge attacks by leveraging on the relative distributions of sensitive values both in individual equivalence classes and in the entire dataset [23]. In all three anonymisation algorithms, and their extensions [3,29], generalisation and suppression are used to support data transformation [13]. *Perturbation* is conceptually similar to generalisation but instead of building groups or clusters based on attribute similarity without falsifying the data, perturbation modifies the actual attribute value to the closest similar findable value. This involves introducing an aggregated value or using a similar value in which only one value is modified instead of several to build clusters. Finding such a value is processing intensive, because all newly created values must be checked iteratively. Further work on data transformation for anonymity appears in the data mining field, with work on addressing privacy constraints in publishing anonymised datasets [12,40,41]. These methods focus on data mining tasks in specific application areas with well-defined privacy models and constraints. This is the case particularly when merging various distributed data sets to ensure privacy in each partition [45]. As mentioned before, these methods are not suited to high dimensional datasets because they operate on a per-usecase basis.

Adaptations based on a Secure Multi-party Computing (SMC) protocol have been proposed as a flexible approach on top of k-anonymity, l-diversity and t-closeness as well as heuristic optimisation to anonymise distributed and separated data silos in the medical field [19]. Furthermore, to address scalability challenges of large-scale high dimensional distributed anonymisation that emerge in the healthcare industry, Mohammed et al. [30] propose *LKC-privacy* to achieve privacy in both centralized and distributed scenarios promising scalability for anonymising large datasets. *LKC-privacy* works on the premise that acquiring background knowledge is nontrivial and therefore limits the length of quasi-identifier tuples to a predefined size. While one can argue about the practically of this approach, the main concern is the fact that *LKC-privacy* violates the basic anonymity requirements of publishing datasets in a privacy-preserving manner. Other works use a MapReduce technique based on the Hadoop distributed file system (HDFS) to boost computation capacity [46], which still does not address the issue of transforming the datasets to guarantee anonymity for high dimensional data where sensitivity is an added concern. Handling large numbers of entity describing attributes (hundreds of attributes), in a performance efficient and privacy preserving manner remains to be addressed.

3   Inefficiency of k-anonymising High Dimensional Data

In this section, we explain why standard k-anoymisation data transformation techniques like *generalisation* and *suppression* are inefficient on high dimensional data. This is to pave the way for describing our proposed approach in Sect. 4.

3.1   Notation and Definitions

Anonymity is the quality of lacking the characteristic of distinction. This is indicated through the absence of outstanding, individual, or unusual features, that separate an individual from a set of similarly characterised individuals. For example, we say that a dataset is k anonymous ($2 \leq k \leq n$, where $n \in \mathbb{Z}^+$) if and only if for all tuples in a given dataset, each the quasi-identifier of each tuple is indistinguishable from at least $k - 1$ other tuples. Expanding this definition to high dimensional data, we define the following terms.

Definition 1. *Feature*
     A feature f is a function $f : E \longrightarrow A$ mapping the set of entities $E = \{e_1, \ldots, e_m\}$ to a set A of all possible realizations of an attribute or attribute combination forming new single attributes. Additionally, $F = \{f_1, \ldots, f_n\}$ denotes a feature set.

We define *self-contained anonymity* which captures the idea of anonymity of individual records or a dataset, as follows:

Definition 2. *Self-contained Anonymity*

Let E be a set of entities. A snapshot S of E is said to be self-containing anonymous or sanitized, if no family $\mathcal{F} = \{F_1, \ldots, F_m\}$ of feature sets uniquely identifies one original entity or row.

Similar to Terrovitis [39], we do not distinguish between sensitive and non-sensitive attributes. This for two reasons, first, by observation of deanonymisation attacks (homogeneity, similarity, background knowledge, ...) we note that sensitive attributes alone are not the only basis for their success; second, defining an exhaustive set of sensitive and non-sensitive attributes is impractical for high dimensional datasets where user behaviours exhibit unique patterns that increase with the volumes of data collected on the individual.

3.2   High Dimensional Quasi-Identifier Transformation

In high dimensional datasets, *generalisation* and *suppression* are not efficient data transformation procedures for anonymisation [1]. The reason for this is that when the number of quasi-identifier attributes is very large, most of the data needs to be suppressed and generalised to achieve k-anonymity. Furthermore, methods such as k-anonymity are highly dependent on spatial locality in order to be statistically robust. This results in poor quality data for data analytics tasks. Example 1, helps to explain this point in some more depth.

Example 1. The data in Table 1a represents cases of surgery at a given hospital, with quasi-identifier "Job, Age, Sex". By *generalisation* and *suppression* Table 1a can be transformed to obtain the 2-anonymous Table 1b. If we consider that Table 1a were to be expanded at some point to include 10 new attributes in the quasi-identifier of say, "blood-type", "disease", "disease-date", "Medication", "Eye-Colour", "Blood-Pressure", "Deficiencies", "Chronic Issues", "Weight", and "Height"; one could deduce that generalising and suppressing values in such a large high dimensional dataset requires searching through all the different possible quasi-identifier combinations that can result in sensitive data exposure. In fact, as Aggrawal et al. [1] point out, preventing sensitive information exposure requires evaluating an exponential number of combinations of attribute dimensions in the quasi-identifier to prevent precise inference attacks.

We now present our time efficient approach to transforming quasi-identifier attributes to ensure adherence to k-anonymity in high dimensional datasets.

4   Optimised Exact Quasi-Identifier Selection Scheme

Our proposed optimised exact quasi-identifier selection scheme works as an in-memory application for fast quasi-identifier transformation for large high dimensional dataset anonymisation. As a first step, we identify and eliminate 1st class

Table 1. Examples given a surgery list

(a) List of Surgeries in a Hospital

Group ID	Job	Sex	Age	Surgery
A	1	Janitor	M	34	Transgender
	2	Lawyer	F	58	Plastic
	3	Mover	M	45	Urology
B	4	Lawyer	M	34	Vascular
	5	Janitor	M	25	Transgender
	6	Doctor	M	45	Plastic

(b) Generalisation and Suppression of Values to achieve 2-anonymity

Group ID	Job	Sex	Age	Surgery
A	1	Manual	M	30 - 60	Transgender
	3	Manual	M	30 - 60	Urology
B	4	Professional	M	30 - 60	Vascular
	6	Professional	M	30 - 60	Plastic
A+B	2	ANY	ANY	30 - 60	Plastic
	5	ANY	ANY	30 - 60	Transgender

identifiers, which are typically standalone attributes such as "user IDs" and "phone numbers". We then select 2nd class identifiers to ensure anonymity and minimal information loss.

4.1   Identifying 1st Class Identifiers

In selecting 1st class identifiers, we do not distinguish between sensitive and non-sensitive attributes because, classifications of sensitive and non-sensitive attributes are the primary cause of semantics-based de-anonymisations. Furthermore, growing attribute numbers in high dimensional datasets, make using sensitive attribute classifications to support anonymisation is trivial since behavior patterns are easily accessible. Instead we use 1st class identifiers to decide which attribute values to transform to reduce the number of records we eliminate from the anonymised dataset. This reduces the level of information loss and ensures anonymity. We identify 1st class identifiers on the basis of two criteria namely, attribute cardinality and classification thresholds. More formally, we define a 1st class identifier as follows:

Definition 3. *1st class identifiers*
   Let F be a set of features $F = \{f_1, \ldots, f_n\}$, where each feature is a function $f_i : E \longrightarrow A$ mapping the set of entities $E = \{e_1, \ldots, e_m\}$ to a set A of realizations of f_i. A feature f_i is called a 1st class identifier, if the function f_i is injective, i.e. for all $e_j, e_k \in E : f_i(e_j) = f_i(e_k) \implies e_j = e_k$.

To find attributes fulfilling the 1st class identifier requirement, each individual attribute has to be evaluated by counting the unique values with respect to all other entries combined with a SQL GROUP BY statement. These attributes are characterised by a high cardinality and entropy as follows:

Definition 4. *Cardinality*
   The cardinality $c \in \mathcal{Q}$ of a column or an attribute is: $c = \frac{\text{number of unique rows}}{\text{total number of rows}}$.

Definition 5. *Entropy (Kullback-Leibler Divergence)*
   Let p and q denote discrete probability distributions. The Kullback-Leibler divergence or relative entropy e of p with respect to q is: $e = \sum_i p(i) \cdot \log(\frac{p(i)}{q(i)})$.

First we compute the cardinality c and mark all columns as 1st class identifiers where the cardinality threshold is $c > 0.33$, meaning that at least every third entry is unique. This is used as a heuristic and can be configured as desired. The 1st class identifiers are suppressed from the dataset so that no direct and bijective linkages from the dataset to the original entities remain. However, one is still able to combine several attributes for re-identification. In the following section, we propose a method of identifying and removing these attribute combinations.

4.2 Identifying of 2nd Class Identifiers

We use 2nd class identifier candidates as a further evaluation step to ensure self-contained data anonymity. This is done by identifying the sets of attribute value candidates that violate the anonymity by being unique throughout the entire data set. More formally, we define 2nd class identifiers as follows:

Definition 6. *2nd class identifier*

Let $F = \{f_1, \ldots, f_n\}$ be a set of all features and $B := \mathcal{P}(F) = \{B_1, \ldots, B_k\}$ its power set, i.e. the set of all possible feature combinations.

A set of selected features $B_i \in B$, is called a 2nd class identifier, if B_i identifies at least one entity uniquely and all features $f_j \in B_i$ are not 1st class identifiers.

Assessing 2nd class identifiers is similar to finding candidates for a primary key or (maximal partial) unique column combinations (mpUCC) in the data profiling field. Unique column combinations (UCC) are tuples of columns which serve as identifier across the entire dataset, however, maximal partial UCC can be understood as identifiers for (at least) one specific row. This means one searches for the UCC for each specific row (maximal partial). We evaluate all possible combinations of columns in terms of forming the anonymised dataset, as follows: $C(n, r) = \binom{n}{r} = \frac{n!}{(r!(n-r)!)}$ where n is the population of attributes and r the subset of n. In considering 2nd class identifiers of all lengths, r must equal all potential lengths of subsets of attributes. We express this using the following equation: $C_2(n) = \sum_{r=1}^{n} \binom{n}{r} = \sum_{r=1}^{n} \frac{n!}{(r!(n-r)!)} = 2^n - 1$. For each column combination, we apply an SQL *GROUP BY* statement on the data set for the particular combination and count the number of entries for each group. If there is just one row represented for one value group, this combination may serve as mpUCC. Group statements are highly efficient in modern in-memory platforms, since through their column-wise storage and reverted indices these queries do not need to be run over the entire data set.

Even without the maximal partial criteria, and only considering unique column combinations, we note that identifying 2nd class identifiers is a NP-complete problem similar to the hidden subgroup problem (HSP) [17]. In fact, more specifically the problem is W[2]-complete which is not a fixed parameter tractable problem (FPT) [5]. This implies that there is no exact solution better than of polynomial time complexity since the number of combinations of attributes for evaluation increases exponentially [5,15,28]. As such in the next section we look at how to optimise the search strategy.

4.3   Search Optimisation

As depicted in Fig. 1 evaluating 2^n combinations of attributes is not scalable to large datasets so, instead of searching for all possible combinations with all lengths for each row (hereinafter referred to as maximal partial unique column combinations (mpUCC)), we limit the search to unique column combinations (mpmUCC) [31]. Practically, one needs to only find the minimal 2nd class identifier to prevent re-identification (see Fig. 1). We define a *Minimal 2nd Class Identifier* as follows.

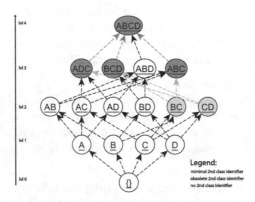

Fig. 1. Maximal partial minimal unique column combinations tree

Definition 7. *Minimal 2nd class identifier*
A 2nd class identifier $B_i \in \mathcal{P}(F)$ is called minimal, if there is no combination of features $B_j \subset B_i$ that is also a 2nd class identifier.

Example 2. Imagine a data set describing medical adherence and the drug intake behavior of patients. After potentially identifying first name, age and street name as 2nd class identifier tuple, it is clear to the reader that any additional attribute to this tuple is still a 2nd class identifier. However, a minimal 2nd class identifier contains just the minimal amount of attributes in the tuple which are needed to serve as quasi-identifier (maximal partial minimal UCC).

Therefore, the search in one branch of the search tree can be stopped as soon as a minimal 2nd class identifier is found. This is similar to Papenbrock et al.'s [31] approach to handling maximal partial UCCs. Such processing improves computation time dramatically since all super-sets can be neglected. First testing reveals that most mpmUCCs appear in the first third of the search tree but at most in the first half which still requires, due to the *symmetry of the binomial coefficient*, $\frac{2^n}{2} = 2^{n-1}$ combinations to be processed and evaluated. The symmetry and combination distribution of the binomial coefficients can be delineated by arranging the binomial coefficients to form a Pascal's triangle where each Pascal's triangle level corresponds to a n value. So, in reducing the layers and

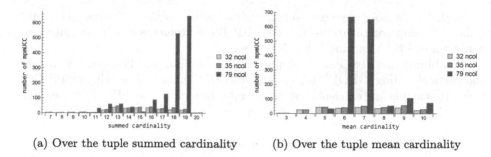

(a) Over the tuple summed cardinality   (b) Over the tuple mean cardinality

Fig. 2. Appearances of 2nd class identifier

number of combinations, we still have exponential growth. We do this by filtering the set of combinations beforehand to avoid any exponential and inefficient growth.

In the exact search for mpUCCs, the risk of compromise for each identifier type needs to be considered. As such, we prefilter column combinations by evaluating cardinality based features like the sum of their cardinality (see Fig. 2a) or its mean value (see Fig. 2b) against given thresholds. Given the observed distribution of tuple sizes regarding their elements expressed, more tuples imply more filtering at given a threshold. If no combinations are left for evaluation after filtering while the tuple length, that is up for evaluation, is incomplete with regard to the re-arranging of the binomial coefficients or while not all tree branches are covered by the already found minimal 2nd class identifiers, we decrease these thresholds successively. Having found a mpmUCC, we need to double-check its neighbors illustrated by Fig. 1. **If no sibling or parent neighbor is an (minimal) identifier**, we can stop the search for this branch.

4.4 In-Memory Applications as a Booster for 2nd Class Identifier Selections

To determine 2nd class identifiers maximal partial minimal unique column combinations (mpmUCC) are identified with the SQL GROUP BY statement. The GROUP BY is costly in traditional database systems but has the advantage of detecting mpmUCCs as well as the exact rows affected by each individual mpmUCC. This is key factor in transforming the dataset flawlessly and efficiently. Column wise databases with dictionary encoding run very efficient and fast group by statements, in comparison to traditional database benchmarks. In column-wise data storage, a GROUP BY statement does not have to read the entire dataset but rather the corresponding row saved. Additional reverse indices accelerate the access to each row further.

By handling over the task and execution of GROUP BY from the actual application to a database system, reliability and performance is gained. Vertical scaling can handle hundreds or thousands of cores in parallel without negatively impacting complexity, which is an advantage when executing several statements in parallel or in close sequence. When evaluating hundreds of thousands or millions of column combinations, the GROUP BY statements can be executed in parallel and L1–L3 caching is highly efficient.

Combining these key items, column wise, reverse indices, dictionary encoding and vertical scaling, GROUP BY statements and therefore identifying mpmUCC is highly scalable and efficient. Having a toolkit to identify mpmUCC gives us the possibility to remove all unique tuples - no matter how many attributes or which type or content they are. By removing all unique tuples, only "duplicated" ones remain which follow the original k-anonymity idea and provide sound anonymisation and therefore trustworthy data privacy. The main issue of all incidents presenting in the introduction has been, that some unique attribute combination survive the anonymisation process and may be abused for de-anonymisation. This is not possible anymore.

There are benchmarks[1] to prove that in-memory databases like HANA are up to 53% faster than the competition [11,21,32].

5   Evaluation and Results

Our experiments were conducted on a 16x Intel(R) Xeon(R) CPU E5-2697 v3 @ 2.60 GHz and 32.94 GB RAM machine, running an SAP HANA database in combination with an in-memory application based on Python[2]. The implementation platform used is the "GesundheitsCloud" application[3]. Our dataset was comprised of semi-synthetic data with 109 attributes and 1M rows that are divided into chunks of 100000 for running the benchmark multiple times with the same settings. The results are then averaged to reduce potential external noise. These real-world data include disease details and disease-disease relations, blood type distribution, drug as well as SNP and genome data and relations. The sources ranges from different data sets as part of publications [35,36,47], as well as official government websites like medicare.gov[4], US Food & Drug Administration[5], NY health data[6], Centers for Medicare & Medicaid Services[7], and many more. A list of all data sources is publicly available at github.com[8].

In processing 1st class identifiers, we need to loop over each existing attribute, group by the related column and count the rows with the same value. The sum of entities having a group count of 1 decides on its classification as 1st class identifier. Including the possibility of noise, we consider a column or attribute as a 1st class identifier, if at least 70% of its values are unique. As a consequence, attributes identified as 1st class identifiers are disregarded from further

---

[1] http://www-07.ibm.com/au/hana/pdf/S_HANA_Performance_Whitepaper.pdf.

[2] https://www.python.org/download/releases/3.0/.

[3] http://news.sap.com/germany/gesundheit-cloud/.

[4] https://www.medicare.gov/download/downloaddb.asp.

[5] https://www.fda.gov/drugs/informationondrugs/ucm142438.htm.

[6] https://health.data.ny.gov/browse?limitTo=datasets&sortBy=alpha.

[7] https://www.cms.gov/Medicare/Coding/ICD9ProviderDiagnosticCodes/codes.html.

[8] https://github.com/jaSunny/MA-enriched-Health-Data.

processing and dropped from the dataset. We then consider 2nd class identifiers. Figure 3a shows the actual number of minimal 2nd class identifiers available in the dataset, while Fig. 3b illustrates the evolution of the score for untreated data. Minor non-linear jumps can be explained by untreated 1st class identifiers that are characterised by a large number of unique values and thus large cardinality.

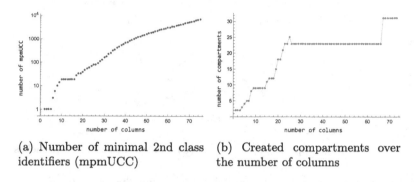

(a) Number of minimal 2nd class identifiers (mpmUCC)

(b) Created compartments over the number of columns

Fig. 3. Characteristics of the evaluation dataset

5.1   2nd Class Identifier Selection

Figure 4 shows the execution time required to identify all minimal 2nd class identifiers in comparison to the number of attributes in the quasi-identifier. The data points for each specific approach were fitted with quartic, cubic, quadratic, linear, and log curves to show the evolution of identification time over the number of present columns (attributes). Here one data point represents the time required to evaluate the entire dataset. For each x-wise step, an additional column is introduced in the dataset to visualize the time complexity. The optimising minimal 2nd class identifiers (mpmUCC) results in $O(2^{n-1})$ (see Fig. 4c). When only assessing filtered combinations, the results are illustrated and fitted in Fig. 4b. Further, Fig. 4d presents a direct comparison between all identification approaches where the effect of optimisation is clearly distinct.

5.2   Use Case Walk Through

This subsection provides an example of orchestrated transformation approaches for a predefined real-world use case provided by a large pharmaceutical company. Typical use cases involve finding drug-to-drug, gene-to-drug, drug-to-disease or disease-to-disease relationships using regression. We use the Hayden Wimmer and Loreen Powell approach [44] to investigate the effects of different transformations on such use cases. An optimal treatment composition is created by using a weighted brute force approach to transform the dataset for anonymity. In this case the time complexity is represented through an exponential interval and the decision criterion is the data score achieved for the sanitized dataset.

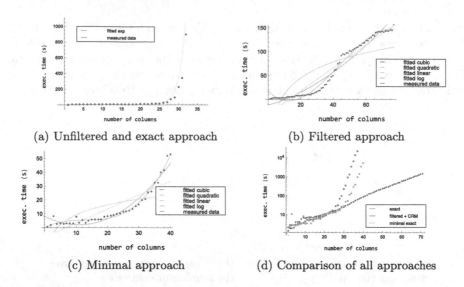

(a) Unfiltered and exact approach   (b) Filtered approach

(c) Minimal approach   (d) Comparison of all approaches

Fig. 4. Execution time for identifying all minimal 2nd class identifiers

For numerical values with a coverage of less than 50%, perturbation is used, and with more than 15% generalisation. For non-numerical values with a coverage of less than 50% suppression is used and in all other instances compartmentation as preferred treatment. For comparison, the same logistic regression function is applied to both the original, and sanitized dataset. The following case provides influencing factors for *DOID:3393*, namely "coronary artery disease" where plaque conglomerates along the inner walls of an arteries reducing the blood supply to cardiac muscles[9]. In feature selection for logistic regression, we determine height, age, blood type, weight, several single-nucleotide polymorphisms (SNPs) markers, and drug intake as interesting. Table 2 specifies the attribute coefficients as weights for influencing the probability of suffering coronary artery disease. From the original dataset, one notes that the patients age, weight and height are important factors for predicting DOID:3393. As well, blood type, drug intake, and coronary artery disease, are correlated. When perturbation or suppression are used for anonymisation, the coefficients shifts toward one feature. Compartmentation keeps most of the features, by re-weighting. The composition of weights performs the best with deviations of 10% to 20%.

This proves that information loss can be minimised without making significant compromises on privacy by combining existing (exact) anonymisation techniques. Since are no unique tuples from the original dataset, the likelihood of homogeneity and background knowledge attacks is significantly reduced.

---

[9] https://medlineplus.gov/coronaryarterydisease.html.

Table 2. Logistic regression coefficients as scaled weights for the given attributes as features

Attribute	Original coefficients	Composition coefficients	Compartmentation coefficients	Perturbation coefficients	Suppression coefficients
Age	100.00	100	32.99	0	0.05
Centimeters	49.44	37.48	100	100	6.8
drug_0	63.38	0	4.3	0	100
BloodType	33.96	8.82	45.06	0	0.05
Kilograms	50.53	62.29	24.25	0	0
snp_0	0	0	0	0	0
drug_2	0	0	6.4	0	0

6 Conclusions

Existing work has focused on optimising existing techniques based on predefined use cases through greedy or heuristic algorithms which is not adequate for high dimensional large datasets. In this paper, we have presented a hybrid approach for anonymising high dimensional datasets and presented results from experiments conducted with health data. We showed that this approach reduces the algorithmic complexity when asynchronous, use case agnostic processing is applied to the data. Additionally, we eliminate the risk of de-anonymisation by symmetric, interaction-based validations of resulting anonymous datasets because no unique attribute tuples remain. The W[2]-complete search for unique column combinations as quasi-identifiers endangering the complete anonymity of a dataset given the exponential and impractical computation efforts was studied for processing high dimensional data sets faster with cubic time complexity or exponentially at a stretching factor of 0.0889926. An optimal composition process was evaluated based on several metrics to limit increasing data quality loss (information loss) with increasing attributes in a data set. The source code, detailed implementation documentation and dataset are publicly available at github.com[10,11].

The current implementation for searching for 2nd class identifiers is based on the central processing unit (CPU), however, it would be interesting to evaluate the gains of using graphics processing units (GPU). Also, studying the effect of decoupling attributes is important for more diverse use cases besides the ones studied in this paper.

---

[10] https://github.com/jaSunny/MA-Anonymization-ETL.
[11] https://github.com/jaSunny/MA-enriched-Health-Data.

References

1. Aggarwal, C.C.: On k-anonymity and the curse of dimensionality. In: Proceedings of the 31st International Conference on Very Large Data Bases, VLDB 2005 (2005)
2. Barbaro, M., Zeller, T., Hansell, S.: A face is exposed for AOL searcher no. 4417749. New York Times **9**(2008), 8 (2006). https://www.nytimes.com/2006/08/09/technology/09aol.html
3. Bayardo, R.J., Agrawal, R.: Data privacy through optimal k-anonymization. In: Proceedings of the 21st International Conference on Data Engineering, ICDE 2005, pp. 217–228. IEEE (2005)
4. Bhaskar, R., Laxman, S., Smith, A., Thakurta, A.: Discovering frequent patterns in sensitive data. In: Proceedings of the 16th ACM SIGKDD International Conference on Knowledge Discovery and Data Mining, pp. 503–512. ACM (2010)
5. Bläsius, T., Friedrich, T., Schirneck, M.: The parameterized complexity of dependency detection in relational databases. In: LIPIcs-Leibniz International Proceedings in Informatics. Schloss Dagstuhl-Leibniz-Zentrum fuer Informatik (2017)
6. Bonomi, L., Xiong, L.: Mining frequent patterns with differential privacy. Proc. VLDB Endow. **6**(12), 1422–1427 (2013)
7. De Montjoye, Y.A., Hidalgo, C.A., Verleysen, M., Blondel, V.D.: Unique in the crowd: the privacy bounds of human mobility. Sci. Rep. **3**, 1376 (2013)
8. Dondi, R., Mauri, G., Zoppis, I.: On the complexity of the l-diversity problem. In: Murlak, F., Sankowski, P. (eds.) MFCS 2011. LNCS, vol. 6907, pp. 266–277. Springer, Heidelberg (2011). https://doi.org/10.1007/978-3-642-22993-0_26
9. Dwork, C.: Differential privacy: a survey of results. In: Agrawal, M., Du, D., Duan, Z., Li, A. (eds.) TAMC 2008. LNCS, vol. 4978, pp. 1–19. Springer, Heidelberg (2008). https://doi.org/10.1007/978-3-540-79228-4_1
10. Dwork, C., McSherry, F., Nissim, K., Smith, A.: Calibrating noise to sensitivity in private data analysis. In: Halevi, S., Rabin, T. (eds.) TCC 2006. LNCS, vol. 3876, pp. 265–284. Springer, Heidelberg (2006). https://doi.org/10.1007/11681878_14
11. Färber, F., et al.: The SAP HANA database-an architecture overview. IEEE Data Eng. Bull. **35**(1), 28–33 (2012)
12. Fienberg, S.E., Jin, J.: Privacy-preserving data sharing in high dimensional regression and classification settings. J. Priv. Confid. **4**(1), 221–243 (2012)
13. Fredj, F.B., Lammari, N., Comyn-Wattiau, I.: Abstracting anonymization techniques: a prerequisite for selecting a generalization algorithm. Procedia Comput. Sci. **60**, 206–215 (2015)
14. Ghosh, A., Roughgarden, T., Sundararajan, M.: Universally utility-maximizing privacy mechanisms. SIAM J. Comput. **41**(6), 1673–1693 (2012)
15. Ibarra, O.H.: Reversal-bounded multicounter machines and their decision problems. J. ACM (JACM) **25**(1), 116–133 (1978)
16. Islam, M.Z., Brankovic, L.: Privacy preserving data mining: a noise addition framework using a novel clustering technique. Knowl.-Based Syst. **24**(8), 1214–1223 (2011)
17. Karp, R.M.: Reducibility among combinatorial problems. In: Miller, R.E., Thatcher, J.W., Bohlinger, J.D. (eds.) Complexity of Computer Computations. IRSS, pp. 85–103. Springer, Boston (1972). https://doi.org/10.1007/978-1-4684-2001-2_9
18. Kifer, D., Machanavajjhala, A.: No free lunch in data privacy. In: Proceedings of the 2011 ACM SIGMOD, SIGMOD 2011, pp. 193–204. ACM (2011)

19. Kohlmayer, F., Prasser, F., Eckert, C., Kuhn, K.A.: A flexible approach to distributed data anonymization. J. Biomed. Inform. **50**, 62–76 (2014)
20. Koufogiannis, F., Han, S., Pappas, G.J.: Optimality of the Laplace mechanism in differential privacy (2015)
21. Lee, J., et al.: High-performance transaction processing in SAP HANA. IEEE Data Eng. Bull. **36**(2), 28–33 (2013)
22. Li, C., Miklau, G., Hay, M., McGregor, A., Rastogi, V.: The matrix mechanism: optimizing linear counting queries under differential privacy. VLDB J. **24**(6), 757–781 (2015)
23. Li, N., Li, T., Venkatasubramanian, S.: T-closeness: privacy beyond k-anonymity and l-diversity. In: 2007 IEEE 23rd ICDE, pp. 106–115, April 2007
24. Liang, H., Yuan, H.: On the complexity of t-closeness anonymization and related problems. In: Meng, W., Feng, L., Bressan, S., Winiwarter, W., Song, W. (eds.) DASFAA 2013. LNCS, vol. 7825, pp. 331–345. Springer, Heidelberg (2013). https://doi.org/10.1007/978-3-642-37487-6_26
25. Liu, F.: Generalized Gaussian mechanism for differential privacy (2016)
26. Machanavajjhala, A., Kifer, D., Gehrke, J., Venkitasubramaniam, M.: L-diversity: privacy beyond k-anonymity. ACM TKDD **1**(1), 3 (2007)
27. McSherry, F., Talwar, K.: Mechanism design via differential privacy. In: 48th IEEE Symposium Foundations of Computer Science, FOCS 2007 (2007)
28. Meyer, A.R., Stockmeyer, L.J.: The equivalence problem for regular expressions with squaring requires exponential space. In: SWAT (FOCS), pp. 125–129 (1972)
29. Meyerson, A., Williams, R.: On the complexity of optimal k-anonymity. In: Proceedings of the Twenty-Third ACM SIGMOD-SIGACT-SIGART Symposium on Principles of Database Systems, pp. 223–228. ACM (2004)
30. Mohammed, N., Fung, B., Hung, P.C., Lee, C.K.: Centralized and distributed anonymization for high-dimensional healthcare data. ACM TKDD **4**(4), 18 (2010)
31. Papenbrock, T., Naumann, F.: A hybrid approach for efficient unique column combination discovery. Proc. der Fachtagung Business, Technologie und Web (2017)
32. Plattner, H., et al.: A Course in In-Memory Data Management. Springer, Heidelberg (2013). https://doi.org/10.1007/978-3-642-55270-0
33. Polonetsky, J., Tene, O., Finch, K.: Shades of gray: seeing the full spectrum of practical data de-identification (2016)
34. Rubinstein, I., Hartzog, W.: Anonymization and risk (2015)
35. Rzhetsky, A., Wajngurt, D., Park, N., Zheng, T.: Probing genetic overlap among complex human phenotypes. Proc. Nat. Acad. Sci. **104**(28), 11694–11699 (2007)
36. Suthram, S., Dudley, J.T., Chiang, A.P., Chen, R., Hastie, T.J., Butte, A.J.: Network-based elucidation of human disease similarities reveals common functional modules enriched for pluripotent drug targets. PLoS Comput. Biol. **6**(2), 1–10 (2010)
37. Sweeney, L.: Achieving k-anonymity privacy protection using generalization and suppression. Int. J. Uncertain. Fuzziness Knowl.-Based Syst. **10**(05), 571–588 (2002)
38. Sweeney, L.: K-anonymity: a model for protecting privacy. Int. J. Uncertain. Fuzziness Knowl.-Based Syst. **10**(05), 557–570 (2002)
39. Terrovitis, M., Mamoulis, N., Kalnis, P.: Privacy-preserving anonymization of set-valued data. Proc. VLDB Endow. **1**(1), 115–125 (2008)
40. Vaidya, J., Clifton, C.: Privacy-preserving k-means clustering over vertically partitioned data. In: Proceedings of the Ninth ACM SIGKDD International Conference on Knowledge Discovery and Data Mining, pp. 206–215. ACM (2003)

41. Vaidya, J., Kantarcıoğlu, M., Clifton, C.: Privacy-preserving Naive Bayes classification. VLDB J.—Int. J. Very Large Data Bases **17**(4), 879–898 (2008)
42. Vessenes, P., Seidensticker, R.: System and method for analyzing transactions in a distributed ledger. US Patent 9,298,806, 29 March 2016
43. Wernke, M., Skvortsov, P., Dürr, F., Rothermel, K.: A classification of location privacy attacks and approaches. Pers. Ubiquit. Comput. **18**(1), 163–175 (2014)
44. Wimmer, H., Powell, L.: A comparison of the effects of k-anonymity on machine learning algorithms. In: Proceedings of the Conference for Information Systems Applied Research ISSN, vol. 2167, p. 1508 (2014)
45. Zhang, B., Dave, V., Mohammed, N., Hasan, M.A.: Feature selection for classification under anonymity constraint. arXiv preprint arXiv:1512.07158 (2015)
46. Zhang, X., Yang, L.T., Liu, C., Chen, J.: A scalable two-phase top-down specialization approach for data anonymization using mapreduce on cloud. IEEE Trans. Parallel Distrib. Syst. **25**(2), 363–373 (2014)
47. Zhou, X., Menche, J., Barabási, A.L., Sharma, A.: Human symptoms-disease network. Nat. Commun. **5**, 4212 (2014)

Decision Support Systems

A Diversification-Aware Itemset Placement Framework for Long-Term Sustainability of Retail Businesses

Parul Chaudhary[1($\boxtimes$)], Anirban Mondal[2],
and Polepalli Krishna Reddy[3]

[1] Shiv Nadar University, Greater Noida, Uttar Pradesh, India
pc230@snu.edu.in
[2] Ashoka University, Sonipat, Haryana, India
anirban.mondal@ashoka.edu.in
[3] International Institute of Information Technology,
Hyderabad, India
pkreddy@iiit.ac.in

Abstract. In addition to maximizing the revenue, retailers also aim at *diversifying* product offerings for facilitating sustainable revenue generation in the long run. Thus, it becomes a necessity for retailers to place appropriate itemsets in a limited k number of premium slots in retail stores for achieving the goals of revenue maximization and itemset diversification. In this regard, research efforts are being made to extract itemsets with high utility for maximizing the revenue, but they do not consider itemset diversification i.e., there could be duplicate (repetitive) items in the selected top-utility itemsets. Furthermore, given utility and support thresholds, the number of candidate itemsets of all sizes generated by existing utility mining approaches typically explodes. This leads to issues of memory and itemset retrieval times. In this paper, we present a framework and schemes for *efficiently* retrieving the top-utility itemsets of any given itemset size based on both revenue as well as the degree of diversification. Here, higher degree of diversification implies less duplicate items in the selected top-utility itemsets. The proposed schemes are based on *efficiently* determining and indexing the top-λ high-utility and diversified itemsets. Experiments with a real dataset show the overall effectiveness and scalability of the proposed schemes in terms of execution time, revenue and degree of diversification w.r.t. a recent existing scheme.

Keywords: Utility mining · Top-utility itemsets · Diversification
Itemset placement · Retail

1 Introduction

In retail application scenarios, the placement of items on retail store shelves considerably impacts sales revenue [1–5]. A retail store contains premium slots and non-premium slots. Premium slots are those that are easily visible as well as physically accessible to the customers e.g., slots nearer to the eye or shoulder level of the

© Springer Nature Switzerland AG 2018
S. Hartmann et al. (Eds.): DEXA 2018, LNCS 11029, pp. 103–118, 2018.
https://doi.org/10.1007/978-3-319-98809-2_7

customers; the others are non-premium slots. Furthermore, we are witnessing the trend of mega-sized retail stores, such as Walmart Supercenters, Dubai Mall and Shinsegae Centumcity Department Store (Busan, South Korea). Since these mega stores occupy more than a million square feet of retail floor space [23], they typically have **multiple blocks** of premium slots *of varying sizes* across the different aisles of the retail store.

For facilitating sustainable long-term revenue earnings, retailers not only need to *maximize the revenue*, but they also require to *diversify* their product offerings (item-sets). The issue of investigating approaches for diversifying retail businesses with the objective of long-term revenue sustainability is an active area of research. Research efforts are being made to improve diversification for real-world retail companies by collecting data about sales, customer opinions and the views of senior managers [6–8]. Hence, we can intuitively understand that diversification is critical for the long-term sustainability of businesses. As a single instance, if a retailer fails to diversify and focuses on the sales of only a few products, it may suffer huge revenue losses in case the sales of those products suddenly drop significantly. This is because consumer demand for different products is largely uncertain, volatile and unpredictable because it depends upon a wide gamut of external factors associated with the macro-environment of business. Examples of such factors include sudden economic downturn in the market, socio-cultural trends (e.g., trend towards healthier food choices), legal and regulatory changes (e.g., pulling products off retail store shelves due to public health concerns) and so on.

Regarding revenue maximization, during peak-sales periods, strategic item place-ment decisions significantly impact retail store revenue [24]. For example, the largest US retail chains witness about 30% of their annual sales during the Christmas season, and they see a good percentage of their annual sales during days such as Black Friday [24]. In such peak periods, items in the premium slots sell out quickly due to a very large number of customers. This makes it imperative for the store manager to decide quickly which high-revenue itemsets to re-stock and place in a relatively limited number of premium slots of *different sizes* across the numerous aisles of a large retail store.

Notably, diversification can cause some short-term losses in revenue for the retailer because its focus becomes spread over a larger number of products as opposed to focusing on the sales of only a few products that it specializes in selling. Thus, there is a trade-off between retail store revenue and the degree of diversification. However, as evidenced by the works in [6–8], short-term revenue losses due to diversification is generally a small price to pay for the benefits of long-term sustainable revenue earnings.

Efforts in data mining [4, 5] have focused on extracting the knowledge of frequent itemsets based on support thresholds by analyzing the customers' transactional data. Utility mining approaches [12–20] have also been proposed to identify the top-utility itemsets by incorporating the notion of item prices in addition to support. Utility mining aims at finding high-utility itemsets from transactional databases. Here, utility can be defined in terms of revenue, profits, interestingness and user convenience, depending upon the application. Utility mining approaches focus on creating repre-sentations of high-utility itemsets [13], identifying the minimal high-utility itemsets [14], proposing upper-bounds and heuristics for pruning the search space [15, 16] and using specialized data structures, such as the utility-list [17] and the UP-Tree [19], for reducing candidate itemset generation overheads. However, they do not consider itemset diversification i.e., there could be duplicate (repetitive) items in the selected top-utility itemsets. (Duplicate items occur in the selected top-utility itemsets as each itemset is preferred by different groups of customers.) Moreover, given utility and support thresholds, the number of candidate itemsets of all sizes generated by them typically explodes, thereby leading to issues of memory and itemset retrieval times.

In this paper, we investigate the placement of itemsets in the premium slots of large retail stores for achieving diversification in addition to revenue maximization. Our key contributions are a framework and schemes for *efficiently* retrieving the top-utility itemsets of any given size based on both revenue and the degree of diversification. Here, higher degree of diversification implies less duplicate items in the selected top-utility itemsets. The proposed schemes are based on *efficiently* determining and indexing the top-λ high-utility and diversified itemsets. Instead of extracting all of the itemsets of different sizes, only the top-λ high-utility itemsets corresponding to different itemset sizes are extracted. These extracted itemsets are organized in our proposed **kUI (k Utility Itemset) index** for quickly retrieving top-utility itemsets of different sizes. By setting an appropriate value of λ, we can restrict the number of candidate itemsets to be extracted, thereby avoiding candidate itemset explosion.

Overall, we propose three schemes, namely **Revenue Only (RO)**, **Diversification Only (DO)** and **Hybrid Revenue Diversification (HRD)**. The RO scheme aims at greedily maximizing the revenue of the retailer by selecting the top-λ high revenue itemsets of different retailer-specified sizes to be placed in the retail store's premium slots, but it does not consider diversification. In contrast, the DO scheme selects the top-λ itemsets for maximizing the degree of diversification, but it does not consider revenue maximization. Finally, HRD is a hybrid scheme, which selects the top-λ itemsets based on both revenue and the degree of diversification. The HRD scheme also defines the notion of a *revenue window* to limit the revenue loss due to diversification.

Our experimental results using a relatively large real dataset demonstrate that the proposed schemes could be used for efficiently determining top-utility and diversified itemsets without incurring any significant revenue losses due to diversification. The remainder of this paper is organized as follows. Section 2 reviews related works, while Sect. 3 discusses the context of the problem. Section 4 presents the proposed framework and the schemes. Section 5 reports the results of the performance evaluation. Finally, Sect. 6 concludes the paper with directions for future work.

2 Related Work

Several research efforts [9–11] have addressed the problem of association rule mining by determining frequent itemsets primarily based on support. As such, they do not incorporate any notion of utility. Furthermore, they use the downward closure property [9] i.e., the subset of a frequent itemset should also necessarily be frequent.

Given that the downward closure property is not applicable to utility mining, utility mining approaches [12–20] have been proposed for extracting high-utility patterns. The work in [12] discovers high-utility itemsets by using a two-phase algorithm, which prunes the number of candidate itemsets. Moreover, it discusses concise representations of high-utility itemsets and proposes two algorithms, namely HUG-Miner and GHUI-Miner, to mine these representations. The work in [13] proposes a representation of high-utility itemsets called MinHUIs (minimal high-utility itemsets). MinHUIs are defined as the smallest itemsets that generate a large amount of profit. The work in [15] proposes the EFIM algorithm for finding high-utility itemsets. For pruning the search space, it uses two upper-bounds called sub-tree utility and local utility. Moreover, the work in [16] discusses the EFIM-Closed algorithm for discovering closed high-utility itemsets. It uses upper-bounds for utility as well as pruning strategies.

Furthermore, the work in [17] proposes the HUI-Miner algorithm for mining high-utility itemsets. It uses a data structure, designated as the utility-list, for storing utility and other heuristic information about the itemsets, thereby enabling it to avoid expensive candidate itemset generation as well as utility computations for many candidate itemsets. The work in [18] proposed the CHUI-Miner algorithm for mining closed high-utility itemsets. In particular, the algorithm is able to compute the utility of itemsets without generating candidates. The work in [19] proposes the Utility Pattern Growth (UP-Growth) algorithm for mining high-utility itemsets. In particular, it keeps track of information concerning high-utility itemsets in a data structure called the Utility Pattern Tree (UP-Tree) and uses pruning strategies for candidate itemset generation. The work in [20] aims at finding the top-K high-utility closed patterns that are directly related to a given business goal. Its pruning strategy aims at pruning away low-utility itemsets.

Notably, none of the existing utility mining approaches [12–20] consider diversification when determining the top-utility itemsets of any given size. Hence, it is possible for the same items to repeatedly occur across the selected top-utility itemsets, thereby hindering retail business diversification and sustainable long-term revenue generation. Moreover, they are not capable of *efficiently* retrieving top-utility itemsets of varying given sizes. This is because almost all of the approaches generate a huge number of candidate high-utility itemsets of different sizes and then select the itemsets of a given size. Therefore, they suffer from efficiency and flexibility issues when trying to extract high-utility itemsets of a given size. This limits their applicability to building practically feasible applications for determining the placement of itemsets in large retail stores.

As part of our research efforts towards improving itemset placements in retail stores, our work [25] has addressed the problem of determining the top-utility itemsets when a given number of retail slots is specified as input. However, the work in [25] does not consider the important issue of diversification. Thus, the problem addressed in this paper is fundamentally different from that of the problem in [25].

A conceptual model of diversification for apparel retailers was proposed in [8]. The study in [8] also explored the nature of diversification within a successful apparel retailer in the UK and concluded that diversification benefits retailers by giving them a long-term sustainable competitive advantage over other retailers. Moreover, the study in [7] used sales data of 246 large global retail stores from different countries; its results show that retailers with a higher degree of product category diversification had better retail sales volumes. The study in [6] also reached similar conclusions regarding the benefits of diversification by exploring the retail diversification strategies of ten UK retailers through in-depth interviews with the senior management of these retailers.

3   Context of the Problem

Consider a finite set Υ of m items $\{i_1, i_2, i_3, ..., i_m\}$. We assume that each item of set Υ is physically of the same size i.e., each item consumes an equal amount of space e.g., on the shelves of the retail store. Moreover, we assume that all premium slots are of equal size and each item consumes only one slot. Each item i_j of set Υ is associated with a price ρ_j and a frequency of sales (support) σ_j. We define the net revenue NR_i of the ith item i_j as the product of its price and support i.e., $NR_i = (\rho_i * \sigma_i)$. We define an itemset of size k as a set of k distinct items $\{i_1, i_2,.., i_k\}$, where each item is an element of set Υ. We use *revenue* as an example of a utility measure. We shall use the terms revenue, net revenue and utility interchangeably. *Net revenue* of a given itemset is defined below:

Definition 1: *The net revenue of any given itemset is computed as the support of the itemset multiplied by the sum of the prices of the items in that itemset.*

Now we discuss the notion of diversification. There could be duplicate (repetitive) items in the selected top-utility itemsets as each itemset is preferred by different groups of customers. We conceptualize the degree of diversification ψ of selected top-utility itemsets as the ratio of the number of unique items across these itemsets to the total number of items in these itemsets (including duplicate items). ψ is defined as follows:

Definition 2: *Degree of diversification ψ of any given λ itemsets is the number of unique items across all of the λ itemsets divided by the total number of items in these λ itemsets.*

Given λ itemsets $\{A_1, A_2, ..., A_\lambda\}$, the value of ψ is computed as follows:

$$\psi = \frac{\left| \bigcup_{i=1}^{\lambda} A_i \right|}{\sum_{i=1}^{\lambda} |A_i|} \tag{1}$$

In Eq. 1, $0 < \psi \leq 1$. Since there is at least one unique item across all of the λ itemsets, the minimum value of ψ would always exceed 0. ψ can be at most 1 when all the items across all of the λ itemsets are unique; this is the highest possible degree of diversification. Higher values of ψ imply more diversification. As we shall see, ψ can be used as a lever to achieve diversification without incurring significant revenue loss.

Item	A	B	C	D	E	F	G	H	I
ρ	7	2	6	1	3	1	5	4	3

Itemsets: {A,D}, {A,C,G}, {A,B,C,G,H}
Unique items: {A, B, C, D, G, H} i.e., 6 items
Total number of items: 10; Ψ = 6/10= 0.6

Itemset	σ	NR
A,D	6	48
B,C,I,F	3	36
A,C,G	2	36
A,B,C,G,H	1	24
A,C,G,I	3	63

NR= σ * ρ

Fig. 1. Computation of Net Revenue (NR) and degree of diversification (Ψ)

Figure 1 shows the prices (ρ) of the items (A to I) and also depicts five itemsets with their support σ. The net revenue (NR) of the itemset {A, D} = 6 * (7 + 1) i.e., 48. Similarly, the net revenue of itemset {A, C, G, I} = 3 * (7 + 6 + 5 + 3) i.e., 63. Moreover, observe how ψ is computed for three itemsets {A, D}, {A, C, G} and {A, B, C, G, H}.

4   Proposed Framework and Schemes

In this section, we first discuss the basic idea of the proposed framework followed by three schemes for *efficiently* determining the top-utility and diversified itemsets.

4.1   Basic Idea

Transactional data of retail customers provides rich information about the purchase patterns (itemsets) of customers. Given support and utility thresholds, it is possible to extract utility patterns from a transactional database. However, as utility measures do not support downward closure property, we would need to exhaustively check all the patterns to identify the utility patterns; at low support or utility values, the number of patterns explodes. Given the limited number of premium slots, we restrict the extraction of itemsets to only a limited number λ of itemsets of each size for efficient pruning.

Regarding diversification, retailers need to expose their customers to more diversified itemsets to sustain long-term revenue earnings. As discussed earlier, diversification implies less duplicate items in the selected top-utility itemsets. A given retail store has a relatively limited number of premium slots on which the eye-balls of most customers would be likely to fall. The issue is to determine the high-utility itemsets and propose a mechanism to replace some of these high-utility itemsets with diverse itemsets without significantly degrading the utility. Such high-utility and diversified itemsets can then be placed in the premium slots. For example, a typical user buys itemsets (bundled together) such as {p_1, p_2, p_3}, {p_1, p_2}, {p_1, p_3} and {p_2, p_3}; suppose all of these are high-utility itemsets. Now if we were to place all of these high-utility itemsets in the premium slots, these itemsets would occupy 9 premium slots. Since premium slots essentially ensure good visibility to items and are limited in number, we could just place the itemset {p_1, p_2, p_3} to occupy 3 premium slots and populate the other premium slots with items (of comparable utility) albeit other than p_1, p_2 and p_3. This would avoid duplication of the items placed in the premium slots and in effect, expose customers to a more diversified set of items, while maintaining comparable utility from the perspective of the retailer. Thus, the idea allows for the efficient determination of top-utility itemsets to occupy the premium slots and enables recommendations to the retailer about the possible high-utility and diverse itemsets for placing in the premium slots.

To identify itemsets to occupy the premium slots, we propose an efficient approach to identify top-λ itemsets of different sizes and an indexing scheme, designated as the kUI index. Furthermore, we propose a diversification scheme to maximize the degree of diversification of the top-λ itemsets. Overall, we propose three schemes, namely **Revenue Only (RO)**, **Diversification Only (DO)** and **Hybrid Revenue Diversification**

(HRD). RO selects the top-λ high-revenue itemsets without considering diversification. DO maximizes the degree of diversification of the top-λ itemsets. HRD combines RO and DO to determine top-utility and diversified itemsets.

4.2   Revenue Only (RO) Scheme

RO aims to determine the top-λ high-revenue itemsets of any given size k to occupy the premium slots. Since utility measures do not follow the downward closure property, a brute-force approach would be to extract all the possible itemsets and then determine the top-λ high-revenue itemsets. However, this would be prohibitively expensive because the candidate number of itemsets would explode and also lead to memory issues.

RO extracts and maintains only the top-λ high-revenue itemsets for different itemset sizes as opposed to maintaining all the itemsets concerning different itemset sizes. We first extract the top-λ high-revenue itemsets of size 1. Based on these itemsets of size 1, we extract the top-λ high-revenue itemsets of size 2. Thus, we progressively extract the itemsets of subsequently increasing sizes. The extracted itemsets are organized in the form of the kUI (k Utility Itemset) index, where each level corresponds to itemsets of a specific size k. Given a query for determining the top-λ high-revenue itemsets of a specific size k, the kth level of the kUI index is examined for quick retrieval of itemsets.

By extracting and maintaining only the top-λ itemsets, RO restricts the number of candidate itemsets that need to be computed and subsequently maintained for building the next higher level of the index. The value of λ is specified by the retailer. If λ is set to be high, some of the top-λ itemsets would possibly have low revenue. However, if the value of λ is set too low, we may miss some itemsets with relatively high revenue. The value of λ is essentially application-dependent; we leave the determination of the optimal value of λ to future work. Now we discuss the kUI index and how to build it for use by RO.

(i) **Description of kUI Index:** kUI is a multi-level index, where each level concerns a given itemset size. At the kth level, the kUI index stores the top-η high-revenue itemsets of itemset size k. From these top-η itemsets, the top-λ itemsets will be retrieved depending upon the query, hence $\lambda < \eta$. We set the value of η based on application requirements such that queries will never request for more than the top-η itemsets. Each level corresponds to a hash bucket. For indexing itemsets of N different sizes, the index has N hash buckets i.e., one hash bucket per itemset size. Hence, a query for finding the top-λ high revenue itemsets of a given size k traverses quickly to the kth hash bucket instead of traversing through all the hash buckets corresponding to $k = \{1, 2, ..., k - 1\}$.

Now, for each level k in the kUI index, the corresponding hash bucket contains a pointer to a linked list of the top-η itemsets of size k. The entries of the linked list are of the form ($itemset$, σ, ρ, NR), where $itemset$ refers to the given itemset under consideration. Here, σ is the support of $itemset$, while ρ refers to the total price of all the items in $itemset$. NR is the product of σ and ρ, as discussed earlier in Sect. 3 (see Definition 1). Additionally, at each level of the index, the value of the degree of diversification ψ (computed based on Eq. 1 in Sect. 3) is stored for the itemsets of that level. The entries in the linked list are sorted in descending order of the value of NR to facilitate quick retrieval of the top-λ itemsets of a given size k. In case of multiple itemsets having the same value of NR, the ordering of the itemsets is performed in an arbitrary manner.

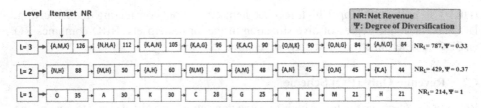

Fig. 2. Illustrative example of the kUI Index

Figure 2 depicts an illustrative example of the kUI index. Observe how the itemsets (e.g., {O}, {A}) of size 1 correspond to level 1 of the index, the itemsets of size 2 (e.g., {N, H}, {M, H}) correspond to level 2 of the index and so on. Notice how the itemsets are ordered in descending order of NR. Observe how the value of the degree of diversification ψ is maintained for the itemset size corresponding to each level of the index.

(ii) Building the kUI Index: Given the transactional database with item price values and threshold values of support, price and utility, the intuition is that items (or itemsets) with high utility (i.e., with either high support or high price) are potential candidates to be indexed under the kUI indexing scheme. First, for itemset size $k = 1$, we select only those items, whose revenue is equal to or above a given revenue threshold. The purpose of the revenue threshold is to ensure that low-revenue items (or itemsets) are efficiently pruned away from the index. Then we sort the selected items in descending order of their values of revenue and insert the top-η items into level 1 of the index. Next, we list all the combinations of the itemsets of size 2 for the items in level 1 and select only those itemsets, whose revenue is equal to or exceeds a specific revenue threshold. Among these itemsets, the top-η high-revenue itemsets are now inserted into level 2 of the index. Then, for creating itemsets of size 3, we list all the possible combinations of the items in level 1 of the kUI index and the itemsets in level 2 of the index. Among these itemsets of size 3, we select only the top-η high-revenue itemsets whose revenue exceed a given revenue threshold; then these selected itemsets are inserted into level 3.

In general, for creating level k of the index (where $k > 2$), we create itemsets of size k by combining the items from level 1 of the index and the itemsets from level $(k - 1)$ of the index. Thus, when we build the kth level of the index (where $k > 2$), only η items from level 1 and η itemsets from level $(k - 1)$ need to be examined for creating all the possible combinations of itemsets that are candidates for the kth level of the index. Notably, the value of η is only a small fraction of the total number of possible items/itemsets; this prevents the explosion in the total number of itemsets that need to be examined for building the next higher level of the index. If we were to examine *all* the possible combinations corresponding to itemsets of size 1 and itemsets of size $(k - 1)$ for building the kth level of the index, total number of combinations to be examined would explode.

Algorithm 1 depicts the creation of the kUI index. Lines 1–11 show the building of the first level of the index i.e., for itemset size of 1. In Lines 1–3, the entire set Υ of all the items is sorted in terms of support, and only those items whose support value is above mean support μ_σ are selected into set A. Here, the value of μ_σ is computed as the sum of all the support values across all the items divided by the total number of items.

Similarly, in Lines 4–6, only those items, whose price is above the mean price μ_ρ, are selected into set B. The value of μ_ρ is computed as the sum of all the price values across all the items divided by the total number of items. The rationale for selecting items with either high support or high price is to ensure that the selected items have relatively high revenue. The same items may exist in both set A and set B. Such duplicates are removed by taking the union of these two sets (see Line 7). As Lines 8–11 indicate, only the top-η items, whose net revenue either equals or exceeds the threshold revenue $TH_{NR,}$ are selected and inserted into the first level (i.e., level L_1) of the index. Here, $TH_{NR} = (\mu_{NR} + (\alpha/100) * \mu_{NR})$, where μ_{NR} is the mean revenue value across all the items in the union set i.e., it is the total revenue of all the items in the union set divided by the total number of items in that set. The parameter α is application-dependent and its value lies between 0 and 100. The purpose of the parameter α is to act as a lever to limit the number of items satisfying the revenue threshold criterion in order to effectively prune away low-revenue items from the index.

Algorithm 1: kUI index creation

Inputs: a) Set Υ of items, where σ_j, ρ_j and NR_j represent the respective support, price and net revenue of the jth item

      b) N = user-specified maximum level of the index

Output: kUI Index

Begin

/* *Building level 1 of the index* */

1. Sort the items in set Υ in descending order of σ into list Υ'
2. Compute μ_σ for the items in list Υ'/* mean support value */
3. Select items with value of σ above or equal to μ_σ into Set A
4. Sort the items in set Υ in descending order of ρ into list Υ''
5. Compute μ_ρ for the items in list Υ''/* mean price value */
6. Select items with value of ρ above or equal to μ_σ into Set B
7. Compute set X = A U B /* union of sets A and B */
8. Sort the items in set X in descending order of net revenue into list X'
9. Compute the value of TH_{NR} for the items in list X'
10. From list X', select the top-η items whose net revenue $\geq TH_{NR}$
11. Insert these top-η items into level L_1 of the index

    /* *Building the intermediate levels of the index one-by-one* */

12. for (i = 2 to N)
13.   Create combinations of itemsets of size i from level L_1 and level L_{i-1} of the index into list Y
14.   Remove duplicate itemsets from list Y
15.   Sort itemsets in list Y in descending order of net revenue
16.   Compute the value of TH_{NR} for the itemsets in list Y
17.   From list Y, select the top-η itemsets whose net revenue $\geq TH_{NR}$
18.   Insert these top-η items into level L_i of the index

End

Lines 12–18 indicate how the intermediate levels (i.e., level 2 to the maximum level N) of the kUI index are built one-by-one. In Line 13, observe how the ith level of the index is created by examining all the possible combinations of itemsets from level 1 and level (i − 1) of the index. In Line 14, all the duplicate itemsets are removed. Then in Lines 15–18, for the given level of the index, we select the top-η itemsets whose net revenue is above the value of THNR; then these top-η itemsets are inserted into that level.

4.3 Diversification Only (DO) Scheme

Although RO achieves revenue maximization, the top-utility itemsets extracted by RO can contain duplicates. Intuitively, there would be likely to be other itemsets with comparable revenue, but containing different items. By replacing some of the top-revenue itemsets extracted using RO with other itemsets, we can improve the degree of diversification in the premium slots. Thus, the idea of DO is to extract and maintain more than λ itemsets in the kUI index so that there are opportunities for replacing some of the top-λ itemsets with itemsets of comparable revenue, but containing more diversified items.

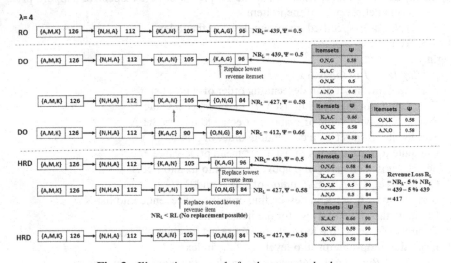

Fig. 3. Illustrative example for the proposed schemes

In the illustrative example of Fig. 3, we have selected level 3 of the example kUI index (see Fig. 2 on Page 7) to explain the notion of diversification, while determining the top-λ itemsets of size 3. For λ = 3, the itemsets selected by RO are {A, M, K}, {N, H, A}, {K, A, N} and {K, A, G}; these itemsets are sorted in descending order of revenue. Now DO will additionally consider the itemsets {O, N, G}, {K, A, C}, {O, N, K} and {A, N, O} for replacing some of the itemsets selected by RO. Here, the lowest-revenue itemset {K, A, G} is replaced by {O, N, G} to improve the degree of diversification ψ from 0.50 to 0.58. Then the next lowest-revenue itemset {K, A, N} is replaced by {K, A, C} to further improve the value of ψ from 0.58 to 0.66 and so on.

4.4   Hybrid Revenue Diversification (HRD) Scheme

RO maximizes the revenue without considering diversification, while DO maximizes the degree of diversification without taking into account the revenue. In general, there is a trade-off between the goals of revenue maximization and diversification. In other words, if we attempt to maximize the revenue, the degree of diversification will degrade and vice versa. Thus, in practice, we require a scheme, which takes into account both revenue and diversification. In particular, the scheme should be capable of improving the degree of diversification without incurring any significant revenue loss. By combining the advantages of both RO and DO, we design a hybrid scheme, designated as Hybrid Revenue Diversification (HRD) scheme. HRD uses the notion of a revenue window to limit the revenue loss due to diversification.

Now let us refer again to Fig. 3 to explain the proposed HRD scheme. Revenue (loss) window R_L is computed as, $R_L = (NR_L - a \% NR_L)$, where NR_L is the Net Revenue across the itemsets in level 3 of the index, while a is a parameter that acts as a lever to control the revenue loss due to diversification. In this example, we use $a = 5$. As in the example for DO, under HRD, the lowest-revenue itemset {K, A, G} is replaced by {O, N, G} to improve the degree of diversification ψ from 0.50 to 0.58. However, in contrast with DO, for HRD, the next lowest-utility itemset {K, A, N} cannot be replaced by {K, A, C} for further improving the degree of diversification due to the constraint of revenue loss arising from diversification being upper-limited by the revenue (loss) window.

5   Performance Evaluation

This section reports the performance evaluation. We have implemented the proposed schemes and the reference scheme [14] in Java. Our experiments use the real-world *ChainStore* dataset, which we obtained from the SPMF open-source data mining library [21]. The dataset has 46,086 items and the number of transactions in the dataset is 1,112,949. The dataset contains utility values; hence, we have used those utility values in our experiments. Table 1 summarizes the parameters of the performance study. From Table 1, observe that we set the parameter α, which controls the revenue threshold, to 30% for all our experiments. We set the total number η of top high-utility items per level of the index to 200. We set the number λ of queried top high-utility items per level of the index to 20 as the default. We also set the queried itemset size k to 4 as the default.

Table 1. Parameters of performance evaluation

Parameter	Default	Variations
Revenue threshold (α)	30%	
Total top high-utility items per level of the index (η)	200	
Queried top high-utility items per level of the index (λ)	20	40, 60, 80, 100
Queried itemset size (k)	4	2, 6, 8, 10

As reference, we adapted the recent MinFHM scheme [14]. Given a transactional database with utility information and a minimum utility threshold (*min_utility*) as input, MinFHM outputs a set of minimal high-utility itemsets having utility no less than that of *min_utility*. By scanning the database, the algorithm creates a utility-list structure for each item and then uses this structure to determine upper-bounds on the utility of extensions of each itemset. We adapted the MinFHM scheme as follows. First, we use the MinFHM scheme to generate all the itemsets across different itemset sizes (k). Second, from these generated itemsets, we extracted all the itemsets of a specific size e.g., $k = 4$. Third, from these extracted itemsets of the given size, we randomly selected any λ itemsets as the query result. We shall henceforth refer to this scheme as **MinFHM**.

Performance metrics are index build time (IBT), execution time (ET), memory consumption (MC), net revenue (NR) and the degree of diversification (ψ). IBT is the time required to build the kUI index. ET is the average execution time of a query concerning the determination of the top-λ itemsets of any given user-specified size. $ET = \frac{1}{N_c} \sum_{q=1}^{N_c} (t_f - t_o)$, where t_o is the query-issuing time, t_f is the time of the query result reaching the query-issuer, and N_C is the total number of the queries. MC is the total memory consumption of a given scheme for building its index. Given a query, the query result comprises λ itemsets. NR is the total revenue of all these λ itemsets. $NR = \sum_{j=1}^{\lambda} R_j$, where R_j is the revenue of the jth itemset. Finally, the degree of diversification ψ for the retrieved top-λ high-utility itemsets is computed as discussed in Eq. 1.

5.1 Performance of Index Creation

Figure 4 depicts the performance of index creation using the real *ChainStore* dataset. The results in Figs. 4(a) and (b) indicate that the index build time (IBT) and memory consumption (MC) increases for all the schemes with increase in the number L of the levels in the index. This occurs because building more levels of the index requires more computations as well as memory space. Our proposed schemes incur significantly lower IBT and MC than that of MinFHM because MinFHM needs to generate all of the itemsets across different itemset sizes (k). In contrast, our schemes restrict the generation of candidate itemsets by considering only the top-λ itemsets in a given index level for building the next higher levels of the index. DO incurs higher IBT and MC than RO because it needs to examine more number of itemsets for its itemset replacement strategy to improve the degree of diversification. IBT for HRD lies between that of RO and DO in terms of both IBT and MC because its notion of revenue window limits the number of itemsets to be examined for replacement as compared to that of DO.

The results in Fig. 4(c) indicate the degree of diversification provided by the different schemes at different levels of the index. Observe that the degree of diversification ψ increases for both DO and HRD essentially to their itemset replacement strategies. However, beyond a certain limit, ψ reaches a saturation point for both DO and HRD because of constraints posed by the transactional dataset. HRD provides lower values of ψ than that of DO because of the notion of the revenue loss window, which limits the degree of diversification in case of HRD. On the other hand, RO and MinFHM show considerably lower values of ψ because they do not consider diversification. In case of RO and MinFHM, the value of ψ decreases with increase in the number of levels of the index (until the saturation point of ψ is reached due to the constraints posed by the transactional data) because their focus on utility thresholds further limit the degree of diversification as the number of levels in the index (i.e., itemset sizes) increases. In other words, both RO and MinFHM only consider the high-utility itemsets as the number of levels in the index is increased, thereby increasing the possibility for items getting repeated in the selected itemsets and consequently, degrading the value of ψ.

5.2   Effect of Variations in λ

Figure 5 depicts the effect of variations in λ. The results in Fig. 5(a) indicate that as λ increases, all the schemes incur more execution time (ET) because they need to retrieve a larger number of itemsets. The proposed schemes outperform MinFHM in terms of ET due to the reasons explained for Fig. 4. DO incurs higher ET w.r.t. RO because unlike RO, it also needs to perform itemset replacements for improving the degree of diversification in the selected top-λ itemsets. HRD incurs lower ET than that of DO since it replaces a lower number of itemsets as compared to DO for diversification purposes due to its revenue loss window limit.

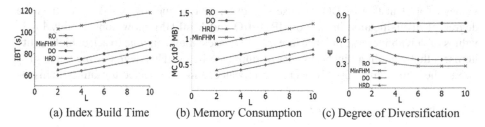

(a) Index Build Time   (b) Memory Consumption   (c) Degree of Diversification

Fig. 4.   Performance of index creation

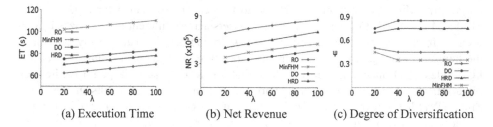

(a) Execution Time   (b) Net Revenue   (c) Degree of Diversification

Fig. 5.   Effect of variations in λ

The results in Fig. 5(b) indicate that all the schemes show higher values of net revenue (NR) with increase in λ. This occurs because as λ increases, more itemsets are retrieved as the query result for each of the schemes; an increased number of retrieved itemsets imply higher values of NR. RO shows much higher values of NR w.r.t. DO,

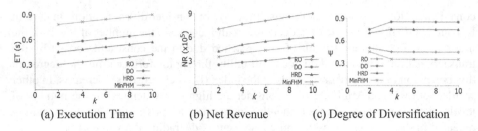

Fig. 6. Effect of variations in k

HRD and MinFHM because RO is able to directly select the top-λ high-revenue itemsets from its index. DO provides lower NR than that of RO because it trades off revenue to improve the degree of diversification. HRD provides higher NR than DO because its degree of diversification is upper-limited by the revenue loss window. MinFHM provides the lowest value of NR among all schemes because from among the itemsets (of the given size) exceeding the utility threshold, it randomly selects the λ itemsets.

The results in Fig. 5(c) indicate the degree of diversification provided by the different schemes for different values of λ. The degree of diversification ψ increases (until the saturation point is reached) for both DO and HRD essentially to their itemset replacement strategies, as explained for the results in Fig. 4(c). HRD provides lower values of ψ than that of DO because its notion of revenue loss window restricts the degree of diversification in case of HRD. RO and MinFHM show considerably lower values of ψ with increase in λ because as λ increases, they continue to select high-utility itemsets that contain a higher number of duplicate items. This degrades the value of ψ due to the same items possibly occurring repeatedly in the selected top-utility itemsets.

5.3   Effect of Variations in k

Figures 6 depict the results when we vary the queried itemset size k. The results in Fig. 6(a) indicate that as k increases, all the schemes incur more execution time (ET) because of the increased sizes of the retrieved itemsets. The proposed schemes outperform MinFHM in terms of ET due to the reasons explained for Fig. 5(a) i.e., MinFHM first needs to generate all of the itemsets across different itemset sizes before it can extract itemsets of a given queried size k. In contrast, RO can quickly determine the itemsets of any given size k by directly traversing to the corresponding level of the kUI index. DO incurs higher ET than that of RO because it performs itemset replacements for improving diversification, as explained for the results in Fig. 5(a). Since HRD has a revenue loss window limit, it performs a lower number of itemset replacements as compared to that of DO; hence, it incurs lower ET than that of DO.

The results in Fig. 6(b) indicate that all the schemes show higher values of net revenue (NR) with increase in the itemset size k because larger-sized itemsets contain more items and therefore, more revenue. RO outperforms the other schemes in terms of NR because DO and HRD lose some revenue to improve diversification, while MinFHM randomly selects from the top-utility itemsets. Furthermore, the results in Fig. 6(c) can be explained in the same manner as the results in Fig. 5(c).

6   Conclusion

Retailers typically aim not only at maximizing the revenue, but also towards diversifying their product offerings for supporting sustainable long-term revenue generation. Hence, it becomes critical for retailers to place appropriate itemsets in a limited number of premium slots in retail stores for achieving both revenue maximization as well as itemset diversification. While utility mining approaches have been proposed for extracting high-utility itemsets to support revenue maximization, they do not consider itemset diversification. Moreover, they also suffer from the drawback of candidate itemset explosion. This paper has presented a framework and schemes for *efficiently* retrieving the top-utility itemsets of any given itemset size based on both revenue and the degree of diversification. The proposed schemes efficiently determine and index the top-λ high-utility itemsets and additionally use itemset replacement strategies for improving the degree of diversification. Our experiments with a large real dataset show the overall effectiveness of the proposed schemes in terms of execution time, revenue and degree of diversification w.r.t. a recent existing scheme. In the near future, we plan to explore the relevant issues pertaining to the cost-effective integration of the proposed schemes into the existing systems of retail businesses.

References

1. Hansen, P., Heinsbroek, H.: Product selection and space allocation in supermarkets. Eur. J. Oper. Res. **3**, 474–484 (1979)
2. Yang, M.H., Chen, W.C.: A study on shelf space allocation and management. Int. J. Prod. Econ. **60–61**, 309–317 (1999)
3. Yang, M.H.: An efficient algorithm to allocate shelf space. Eur. J. Oper. Res. **131**, 107–118 (2001)
4. Chen, M.C., Lin, C.P.: A data mining approach to product assortment and shelf space allocation. Expert Syst. Appl. **32**, 976–986 (2007)
5. Chen, Y.L., Chen, J.M., Tung, C.W.: A data mining approach for retail knowledge discovery with consideration of the effect of shelf-space adjacency on sales. Decis. Support Syst. **42**, 1503–1520 (2006)
6. Hart, C.: The retail accordion and assortment strategies: an exploratory study. In: The International Review of Retail, Distribution and Consumer Research, pp. 111–126 (1999)
7. Etgar, M., Rachman-Moore, D.: Market and product diversification: the evidence from retailing. J. Mark. Channels **17**, 119–135 (2010)
8. Wigley, S.M.: A conceptual model of diversification in apparel retailing: the case of Next plc. J. Text. Inst. **102**(11), 917–934 (2011)
9. Agrawal, R., Srikant, R.: Fast algorithms for mining association rules. In: Proceedings of VLDB, pp. 487–499 (1994)
10. Han, J., Pei, J., Yin, Y.: Mining frequent patterns without candidate generation. ACM SIGMOD Rec. **29**, 1–12 (2000)
11. Pasquier, N., Bastide, Y., Taouil, R., Lakhal, L.: Discovering frequent closed itemsets for association rules. In: Proceedings of the ICDT, pp. 398–416 (1999)
12. Liu, Y., Liao, W.K., Choudhary, A.: A fast high utility itemsets mining algorithm. In: Proceedings of the International workshop on Utility-Based Data Mining, pp. 90–99 (2005)

13. Fournier-Viger, P., Wu, C.-W., Tseng, V.S.: Novel concise representations of high utility itemsets using generator patterns. In: Luo, X., Yu, J.X., Li, Z. (eds.) ADMA 2014. LNCS (LNAI), vol. 8933, pp. 30–43. Springer, Cham (2014). https://doi.org/10.1007/978-3-319-14717-8_3

14. Fournier-Viger, P., Lin, J.C.-W., Wu, C.-W., Tseng, Vincent S., Faghihi, U.: Mining minimal high-utility itemsets. In: Hartmann, S., Ma, H. (eds.) DEXA 2016. LNCS, vol. 9827, pp. 88–101. Springer, Cham (2016). https://doi.org/10.1007/978-3-319-44403-1_6

15. Zida, S., Fournier-Viger, P., Lin, J.C.-W., Wu, C.-W., Tseng, V.S.: EFIM: a highly efficient algorithm for high-utility itemset mining. In: Sidorov, G., Galicia-Haro, S.N. (eds.) MICAI 2015. LNCS (LNAI), vol. 9413, pp. 530–546. Springer, Cham (2015). https://doi.org/10.1007/978-3-319-27060-9_44

16. Fournier-Viger, P., Zida, S., Lin, J.C.-W., Wu, C.-W., Tseng, V.S.: EFIM-closed: fast and memory efficient discovery of closed high-utility itemsets. In: Perner, P. (ed.) Machine Learning and Data Mining in Pattern Recognition. LNCS (LNAI), vol. 9729, pp. 199–213. Springer, Cham (2016). https://doi.org/10.1007/978-3-319-41920-6_15

17. Liu, M., Qu, J.: Mining high utility itemsets without candidate generation. In: Proceedings of the CIKM, pp. 55–64. ACM (2012)

18. Tseng, V.S., Wu, C.W., Fournier-Viger, P., Philip, S.Y.: Efficient algorithms for mining the concise and lossless representation of high utility itemsets. IEEE TKDE 726–739 (2015)

19. Tseng, V.S., Wu, C.W., Shie, B.E., Yu, P.S.: UP-growth: an efficient algorithm for high utility itemset mining. In: Proceedings of the ACM SIGKDD, pp. 253–262. ACM (2010)

20. Chan, R., Yang, Q., Shen, Y.D.: Mining high utility itemsets. In: Proceedings of the ICDM, pp. 19–26 (2003)

21. http://www.philippe-fournier-viger.com/spmf/dataset

22. Fournier-Viger, P., Wu, C.-W., Zida, S., Tseng, V.S.: FHM: faster high-utility itemset mining using estimated utility co-occurrence pruning. In: Andreasen, T., Christiansen, H., Cubero, J.-C., Raś, Z.W. (eds.) ISMIS 2014. LNCS (LNAI), vol. 8502, pp. 83–92. Springer, Cham (2014). https://doi.org/10.1007/978-3-319-08326-1_9

23. World's Largest Retail Store. https://www.thebalance.com/largest-retail-stores-2892923

24. US Retail Industry. https://www.thebalance.com/us-retail-industry-overview-2892699

25. Chaudhary, P., Mondal, A., Reddy, P.K.: A flexible and efficient indexing scheme for placement of top-utility itemsets for different slot sizes. In: Reddy, P.K., Sureka, A., Chakravarthy, S., Bhalla, S. (eds.) BDA 2017. LNCS, vol. 10721, pp. 257–277. Springer, Cham (2017). https://doi.org/10.1007/978-3-319-72413-3_18

Global Analysis of Factors by Considering Trends to Investment Support

Makoto Kirihata[✉] and Qiang Ma

Kyoto University, Kyoto, Japan
kirihata@db.soc.i.kyoto-u.ac.jp, qiang@i.kyoto-u.ac.jp

Abstract. Understanding the factors affecting financial products is important for making investment decisions. Conventional factor analysis methods focus on revealing the impact of factors over a certain period locally, and it is not easy to predict net asset values. As a reasonable solution for the prediction of net asset values, in this paper, we propose a trend shift model for the global analysis of factors by introducing trend change points as shift interference variables into state space models. In addition, to realize the trend shift model efficiently, we propose an effective trend detection method, TP-TBSM (two-phase TBSM), by extending TBSM (trend-based segmentation method). The experimental results validate the proposed model and method.

Keywords: Factor analysis · State space model · Trend detection

1 Introduction

Recently, the Japanese government introduced the NISA (NIPPON Individual savings account) system, which encourages people to shift from savings to investments. Approximately 70% of the balance in NISA accounts is invested in investment trusts. Investment trust products are very popular and many people begin investing with investment trusts, because trust products do not require thorough knowledge of investments unlike stocks and bonds. However, there are too many similar trust products, which make determining appropriate ones for investments difficult. Revealing the factors that can be used to distinguish trust products is a considerable solution to support decisions on trust investments [3,6].

In order to support investment by considering various factors that affect the NAV (net asset value) of investment trust products, research on factor analysis has been conducted. For example, methods for quantitatively analyzing factors affecting investment trust products have been proposed. They analyze investment trust products by using text data such as monthly reports and numeric data such as NAVs of investment trusts. However, they attempt to analyze factors to explain the current situation, and they cannot be applied for predictions. In addition, some researchers report that introducing the notation of trends into a state space model is useful to improve the performance of factor analysis. However, to the best of our knowledge, there is scant work on effectively detecting

© Springer Nature Switzerland AG 2018
S. Hartmann et al. (Eds.): DEXA 2018, LNCS 11029, pp. 119–133, 2018.
https://doi.org/10.1007/978-3-319-98809-2_8

trends and analyzing factors from the global viewpoint (i.e., analyzing factors from a long-term perspective including multiple trends), which could help predict NAVs.

In this paper, we propose a trend shift model for the global analysis of factors by introducing trend change points as shift interference variables into state space models. In addition, to realize the trend shift model efficiently, we propose an effective trend detection method, TP-TBSM (two-phase TBSM), by extending TBSM (trend-based segmentation method).

The major contributions of this paper can be summarized as follows.

- We enable factor analysis across trends using a trend shift model (Sect. 3.1) and improve the accuracy of mid-term prediction (Sect. 4).
- We enable to detect flexible trends while reducing the dependence on parameters using TP-TBSM (Sect. 3.2). The experimental results demonstrate that TP-TBSM is superior to conventional methods (Sect. 4).

2   Related Work

2.1   Financial Analysis with Text Data

In order to obtain information that cannot be attained using only numerical data, many studies have analyzed text data. These studies have demonstrated outstanding results in forecasting field and market understanding [1–3].

Bollen et al. [1] proposed a method to predict the stock price by detecting the mood on Twitter. They achieved an accuracy of 86.7% in predicting the daily fluctuations in the closing values of the DJIA, and reduced the mean average percentage error more than 6%. Mahajan et al. [2] attempted to extract topics on the background of financial news using Latent Dirichlet Allocation, and discovered the topic that highly affected stock price by estimating the correlation between them. They also predicted a rise and fall in the market using extracted topics, and the average accuracy was 60%. Awano et al. [3] attempted to extract factors using the sentence structure of a monthly report on investment trust products, and developed a visualization system to support understanding of investment trust products.

These studies demonstrate that incorporating text data analysis could improve the market analysis. In this study, we use factors extracted from a monthly report of investment trust products by using the existing methods [6].

2.2   Financial Analysis with Time Series Data

Various time series analysis methods are used to study financial products and market analysis. Among them, the state space model is often used because it can flexibly build a model tailored to the purpose by incorporating various factors [4–6].

Bräuning et al. [4] used the state space model to analyze the effects of various factors on macroeconomic variables, and proposed a method to predict future values of the macroeconomic changes of the United States. Ando et al. [5] proposed a method to analyze point of sales data, which is important in marketing, using the state space model. Onishi et al. [6] quantitatively analyzed factors affecting NAV using the state space model. They extracted macro factors and micro factors from monthly reports and news, and used them in combination with numerical data such as NAV to determine the degree of influence of each factor. They concluded that considering trends could improve the accuracy.

Many other studies focused on the analysis of trends. Suzuki et al. [7] improved the accuracy of long-term prediction with non-linear prediction methods by handling trend change points. The shortcut prediction method proposed in [7] yields good results in predicting trend change points.

Chang et al. [8] proposed a method called intelligent piecewise linear representation (IPLR) for maximizing trading profit. IPLR detects a trend change point and uses it to convert time series data into a trading signal such as buying or selling. Using optimal parameters to maximize the profit learned in the neural network, it achieves better profit than rule-based transactions. Jheng-Long et al. [9] predicted buying and selling timings by using a method called TBSM together with support vector regression.

These studies show that consideration of trends and the state space model are useful for factor analysis. However, the existing trend detection methods require the specification of appropriate parameters, which is a difficult task.

3   Methodology

In this section, we first introduce a trend shift model for the global analysis of factors. Subsequently, we describe our TP-TBSM method, which detects trends automatically to realize the trend shift model efficiently.

3.1   Trend Shift Model

Generally, time series data such as stock prices are non-stationary time series whose mean and variance fluctuate with time. Therefore, it is necessary to deal with trends for analysis of such time series data. Onishi et al. [6] handled trends by delimiting data at the trend change point and constructing a state space model within it. However, as the analysis has been completed in each trend, it is not useful for future prediction. In this study, we propose a state space model incorporating the detected trend change points as slope shift interference variables. Hereafter, this model will be referred to as a trend shift model.

Assuming that the time of the i-th trend change point is τ, the slope shift interference variable can be defined as follows.

$$z_{i,t} = \begin{cases} 0 & t \leq \tau \\ t - \tau & t > \tau \end{cases} \quad (1)$$

where $z_{i,t}$ is a variable whose value increases with time changing from τ. By obtaining the regression coefficient of this variable, the slope of the trend can be estimated.

By extending the state space model proposed in [6], the trend shift model incorporating the slope shift interference variable is described as follows.

$$y_t = \mu_t + \Sigma_i \alpha_{i,t} z_{i,t} + \Sigma_k \beta_{k,t} x_{k,t} + \Sigma_m \lambda_{m,t} w_{m,t} + \epsilon_t \tag{2}$$

$$\epsilon_t \sim NID(0, \sigma_\epsilon^2) \tag{3}$$

$$\mu_{t+1} = \mu_t + \xi_t \tag{4}$$

$$\xi_t \sim NID(0, \sigma_\xi^2) \tag{5}$$

$$\alpha_{i,t+1} = \alpha_{i,t}, \ \beta_{k,t+1} = \beta_{k,t}, \ \lambda_{m,t+1} = \lambda_{m,t} \tag{6}$$

where y_t is the logarithm value of NAV at time t. μ_t represents irregular variations. $x_{k,t}$ denotes the logarithmic value of a macro variable factor k, such as the exchange rate. $w_{m,t}$ denotes a macro interference factor m, such as policy announcement; it is 0 until the event occurs, and becomes 1 after the event occurs. The parameters σ_ϵ^2, σ_ξ^2, β, λ are learned by using maximum likelihood estimations. The regression coefficients β and λ quantitatively represent the degree of influence of each factor.

3.2   TP-TBSM

We propose TP-TBSM, a method to detect trends effectively to realize the trend shift model by extending TBSM [9].

TBSM segments time series data into three kinds of trends i.e., rising, falling, and stagnating using three parameters and the point farthest from a linear function. An Example is shown in Fig. 1. In the second trend Fig. 1(a), the point where the distance from the straight line representing the trend becomes the maximum is determined. If the distance d exceeds the parameter δ_d, this point is set as a change point. If the variation is small around the change point, it is segmented into three trends (Fig. 1(b)). This judgment is made based on whether the point is included in the rectangle of X_thld and Y_thld. The second trend in (a) is segmented into three trends in (b).

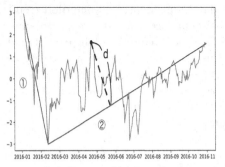
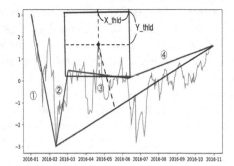

(a) Detect change points   **(b)** Detect stagnating trend

Fig. 1. TBSM (d: Distance from straight line, X_thld: Parameter of the length of trend, Y_thld: Parameter of the magnitude of variation)

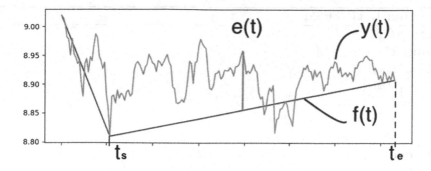

Fig. 2. Trend Error e(t) of trend (t_s, t_e)

Table 1. Symbols in TP-TBSM

Symbol	Description
$y(t)$	Time series data
(t_s, t_e)	Trend represented by a combination of points t_s and t_e
$f(t)$	Linear function representing a trend line
$e(t)$	Distance between $y(t)$ and $f(t)$
C	Set of trend change points
c_i	The i-th element of C
E	Set of trends whose trend error is large
δ_t	Parameter of the size of the minimum trend. Needs to be set
δ_d	Parameter of the magnitude of e(t). Calculated by the algorithms
δ_e	Parameters related to trend error. Calculated by the algorithms

It is difficult to determine appropriate parameters according to time series data. Therefore, we propose TP-TBSM, which relaxes the dependency on parameters. We introduce the concept of trend error, and recursively detect trends by reducing the trend error (Fig. 2).

A trend error is an average value of distance between each data point and a trend line (which can be represented by a linear function). The trend line is a straight line connecting the start and end points of the trend. The trend error is a measure showing the distance of the points from the trend line. The trend error is calculated as follows.

$$TE(y(t), t_s, t_e) = \frac{\Sigma_{t=t_s}^{t_e} e(t)}{t_e - t_s} \tag{7}$$

$$e(t) = |f(t) - y(t)| \tag{8}$$

where t_s and t_e are the start and end points of a trend respectively, y(t) is a value of time series data, and f(t) is a linear function representing a trend line.

Algorithm 1. TP-TBSM

Input: $y(t), \delta_t$
Output: C
1: $C = \{1, n\}$
2: $E = \{(1, n)\}$
3: **repeat**
4:     **if** not first iteration **then**
5:         $E = \text{Evaluation}(C, Y)$
6:     **end if**
7:     $C_{old} = C$
8:     **for** $(t_s, t_e) \in E$ **do**
9:         **if** $t_e - t_s < 2\delta_t$ **then**
10:            Go to the next trend, because the trend length is short
11:        **else**
12:            $d_{max} = \max e(t)$ in the interval $[t_s + \delta_t, t_e - \delta_t]$
13:            $\delta_d = d_{max}$
14:            C = C $\cup$ Segmentation$(y(t), \delta_t, \delta_d, t_s, t_e)$
15:        **end if**
16:    **end for**
17: **until** $C_{old} = C$
18: **return** C

In this study, a trend is considered good if $TE(y(t), t_s, t_e)$ is small. $e(t)$ is the distance between the real point $y(t)$ and the corresponding point $f(t)$ on the trend line.

As shown in Algorithm 1, the proposed method detects trends by alternately repeating two phases: evaluation and segmentation. The evaluation phase is shown in Algorithm 2, and the segmentation phase is shown in Algorithm 3. After describing these two phases, the algorithm of TP-TBSM will be explained. The symbols commonly used in the algorithms are listed in Table 1.

In the evaluation phase, we determine trends, which should be further segmented by considering their trend errors.

Step 1: Calculate the trend error for each trend and set the parameter δ_e as their average value (Line: 2–5).

Step 2: A trend whose trend error is larger than δ_e is subject to segmentation (Line: 6–10).

In the segmentation phase, we segment trends, as follows.

Step 1: Determine the point whose distance to the trend line is the maximum. Such a point is a candidate for a trend change point. We are considering the interval $[start + \delta_t, end - \delta_t]$ to ensure that the length of the trend is greater than or equal to the parameter δ_t to avoid segments that are too short (Line 1).

Algorithm 2. Evaluation

Input: C

Output: E

1: $E = \emptyset$

2: **for** $i = 1 : p$ **do**   // p: Number of trends

3:     $e\_list[i] = TE(y(t), c_i, c_{i+1})$ // e_list: List of length p

4: **end for**

5: $\delta_e = Average(e\_list)$

6: **for** $i = 1 : p$ **do**

7:     **if** $e\_list[i] > \delta_e$ **then**

8:         $E = E \cup (c_i, c_{i+1})$

9:     **end if**

10: **end for**

11: **return** E

Step 2: Determine whether to segment by using the parameter δ_d (Line 2).

Step 3: Check whether there is a stagnating trend around the trend change point. A stagnating trend indicates that the value variation in the trend is small.

  (1) As preparation for the checking, we construct a list H consisting of points whose values are close to that of the candidate trend change point (Line 3–8).

  (2) If H is sufficiently long, and more than half of the points in H have a value close to that of the candidate trend change point, we conclude that a stagnating trend exists, and thereafter divide the current trend into three sub-trends including a stagnating trend (Line 9–13).

  (3) If no stagnating trend exists, we simply segment the current trend into two sub-trends using the (candidate) trend change point (Line 15–17).

The TP-TBSM algorithm is shown in Algorithm 1.

Step 1: The start and end points of the time series data are considered as the initial trend change points, and the trend line connecting these points is considered as the initial trend (Line 1–2).

Step 2: An evaluation phase is performed. A trend with large trend error is selected and placed in the set E (Line 4–6).

Step 3: The length of the trends in E is examined. If the trend length is shorter than $2\delta_t$, we do not perform further segmentation for this trend to avoid trends shorter than δ_t (Line 8–10).

Step 4: If segmentation is possible, δ_d for segmentation is determined, and the segmentation phase is performed. The parameter δ_d is set to the maximum distance to the trend line (Line 11–16).

Step 5: Steps 2–4 are repeated until the result does not change (Line 17).

Algorithm 3. Segmentation

Input: $y(t), \delta_t, \delta_d, (t_s, t_e)$

Output: C

1: $d_{max} = \max e(t)$ in the interval $[t_s + \delta_t, t_e - \delta_d]$. Let t_d be that time
2: **if** $(d_{max} \geq \delta_d)$ **then**
3:     $p = 0$ $//\ p$:Number of points included in H
4:     **for** $t_i = (t_d - \delta_t) : (t_d + \delta_t)$ **do**
5:         **if** $|y(t_i) - y(t_d)| < \frac{\delta_d}{2}$ **then**
6:             $H[p] = i, p = p+1$ $//\ H$:Point list for a stagnating trend
7:         **end if**
8:     **end for**
9:     **if** $(H[p] - H[1] > \delta_t)$ and $(p > \frac{H[p]-H[1]}{2})$ **then**
10:         $c_a = \text{Segmentation}(y(t), \delta_t, \delta_d, t_s, H[1])$
11:         $c_b = \{H[1], H[k]\}$
12:         $c_c = \text{Segmentation}(y(t), \delta_t, \delta_d, H[k], t_e)$
13:         **return** $\{c_a, c_b, c_c\}$
14:     **else**
15:         $c_a = \text{Segmentation}(y(t), \delta_t, \delta_d, t_s, t_d)$
16:         $c_c = \text{Segmentation}(y(t), \delta_t, \delta_d, t_d, t_e)$
17:         **return** $\{c_a, c_c\}$
18:     **end if**
19: **end if**
20: **return** $\{t_s, t_e\}$

Figure 3 shows an example of detecting trends by using TP-TBSM. In Fig. 3(a), each trend is evaluated using trend error. The trend error of the second trend is large. In Fig. 3(b), the point where e(t) becomes maximum is detected as the trend change point. In Fig. 3(c), it is verified whether there is a stagnation trend. There is no stagnation trend in this instance. In Fig. 3(d), segmentation is performed. This process is repeated to detect trends.

4   Experiments

First, we evaluate the usefulness of the trend shift model by comparing the trend shift with the basic state space models. Second, we construct trend shift models with different trend detection methods to evaluate our TP-TBSM method.

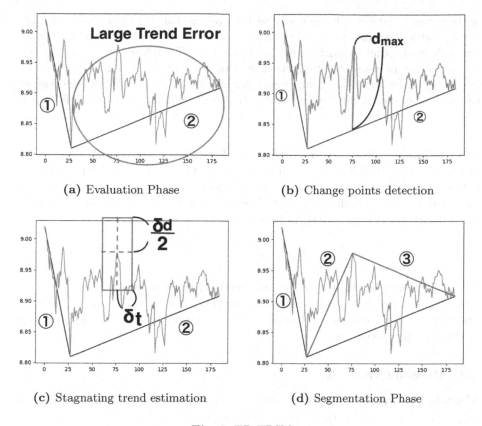

(a) Evaluation Phase

(b) Change points detection

(c) Stagnating trend estimation

(d) Segmentation Phase

Fig. 3. TP-TBSM

4.1 Outline of the Experiment

We used the data set collected by Onishi et al. [6] consisting of 13 trust products from January 4, 2016 to October 31, 2016. The data for the last 20 days are used for testing mid-term predictions, and the other data are used for learning. The 20 days will be about a month's worth of data excluding days with no NAV data such as Saturdays and Sundays. The parameter δ_t used to detect trends using TP-TBSM was also set as 20 days. We used the macro and micro factors extracted using the existing method [6].

As the state space model assumes that the standardized prediction error is independent and normal, we analyzed 13 trust products with each model and used only 11 products for further analysis. These 11 products satisfied the Ljung–Box test and the Shapiro–Wilk test with the significance level 5%.

4.2 Evaluation Measures

Average Error of Mid-term Prediction. State space models are rarely used for prediction and are often used for factor analysis. Therefore, the focus is often on how much data can be reproduced for evaluations. However, in investment trust products, accuracy of prediction is also important, and we propose a global analysis model that could be used for prediction. Therefore, in this study, the average error of mid-term prediction is used for the evaluation of the model. However, as the regression components are included in the model, it is necessary to use the observed data with respect to them, and hence, this prediction is closer to completion than pure prediction.

AIC (Akaike Information Criterion). In addition to the mid-term prediction error, the Akaike information criterion (AIC) is used for the model evaluation. Let L be the maximized log-likelihood, r be the number of unknown parameters, q be the number of initial points in a diffuse initial state, and n be the number of points; the AIC in time series is expressed as follows.

$$AIC = \frac{-2L + 2(q + r)}{n} \tag{9}$$

AIC is penalized by the number of parameters that must be estimated for maximum log likelihood. As the likelihood of the time series is based on the one-step prediction error, the model with small AIC is a simple one with the high accuracy of the one-step prediction.

4.3 Baseline Methods

Models Used for Comparison with the Trend Shift Model. We compare our trend shift model with the following existing models.

- **Local model** proposed in [6]. It is a model with $\Sigma_i \alpha_{i,t} z_{i,t}$ removed from equation (2).
- **Linear model** is a variation of the local linear trend model [10], which extends the local model by introducing a slope term. In short, the linear model modifies Eq. (3) of the trend shift model as follows.

$$\mu_{t+1} = \mu_t + \nu_t + \xi_t, \ \xi_t \sim NID(0, \sigma_\xi^2) \tag{10}$$

$$\nu_{t+1} = \nu_t \tag{11}$$

- **Trend model** is also a variation of the local linear trend model [10]. In the trend model, Eq. (3) is modified as follows.

$$\mu_{t+1} = \mu_t + \nu_t \tag{12}$$

$$\nu_{t+1} = \nu_t + \xi_t, \ \xi_t \sim NID(0, \sigma_\xi^2) \tag{13}$$

Comparative Method for TP-TBSM. To evaluate TP-TBSM, we construct trend shift models with different trend detection methods: our TP-TBSM and the dynamic programming (DP) method [6]. The method of detecting trends using DP was used by Onishi [6]. For each trend, the DP method prepares a straight line connecting the boundary points of the trend, and calculates the root mean square error by comparing with the NAV. The DP method dynamically changes the trend points to minimize the error. It is necessary to determine the number of trends.

4.4   Results and Discussion

Trend Shift Model. The local model, linear model, trend model, and trend shift model (TP-TBSM) are compared. As presented in Table 2, the average error of the mid-term prediction of the trend shift model is the smallest for eight out of 11 products. This indicates that the trend shift model could accurately estimate the influence coefficient of the factors.

In addition, the prediction errors of the local and linear models are larger for most products. These models do not fully consider the influence of trends. The error variation of the trend model is large. This is because the value of the slope term expressing the trend is largely influenced by the immediately preceding value in the trend model.

As presented in Table 3, the local model exhibits the lowest AIC value for all the products and the linear model exhibits the second lowest value. It is thought that AIC has become smaller because simple random walk is used for these two. Overfittings are caused by random walks. Further details are provided in the case study.

Upon comparing the trend model with the trend shift model, it can be observed that the trend shift model shows a smaller AIC value for eight out of 11 products, and it can be concluded that the trend shift model is a better model than the trend model.

Table 2. Average error of mid-term prediction

Product	Local	Linear	Trend	Trend shift
1	0.0121257	0.0123519	0.0360227	**0.00760425**
2	0.01692213	0.01192982	0.02064795	**0.00807945**
3	0.0263603	0.0187761	0.01338155	**0.0096469**
4	**0.0091394**	0.01260265	0.0132246	0.0095815
5	0.02357305	0.0224242	**0.01857475**	0.01863415
6	0.01534265	0.0147249	**0.0112846**	0.0217712
7	0.019291	0.0176646	0.0425186	**0.01261265**
8	0.01504885	0.02040415	0.027809	**0.00924125**
9	0.0211324	0.01933145	0.03037835	**0.01348215**
10	0.01992795	0.01959765	0.0345798	**0.012273**
11	0.01532515	0.01736685	0.0213364	**0.00893415**

Table 3. AIC

Product	Local	Linear	Trend	Trend Shift
1	**−4.052401**	−3.969067	−3.764244	−3.814294
2	**−4.212589**	−4.127697	−3.918999	−3.968967
3	**−3.723839**	−3.644144	−3.416982	−3.509882
4	**−3.730298**	−3.64922	−3.416481	−3.584658
5	**−3.365685**	−3.284338	−3.0366	−3.217069
6	**−4.281146**	−4.194311	−4.109473	−3.82778
7	**−4.076699**	−3.991273	−3.76433	−3.830447
8	**−3.960386**	−3.876909	−3.623052	−3.839674
9	**−4.133614**	−4.047846	−3.821848	−3.887454
10	**−4.133536**	−4.047639	−3.820015	−3.744925
11	**−4.193313**	−4.1072	−3.888612	−3.868926

TP-TBSM. The results (average error of mid-term predication and AIC) of the trend shift models constructed based on DP and TP-TBSM are compared. The parameter δ_t of TP-TBSM was set to 5, 0, 15, and 20.

As presented in Table 5, the model based on TP-TBSM achieved better results in terms of AIC than the model based on DP. The number of trends in DP is fixed at 9, whereas TP-TBSM detects different numbers of trends.

As presented in Table 4, the smaller the parameter δ_t, the better the result of the mid-term prediction. In addition, the prediction error of TP-TBSM is smaller than that of DP for almost all the products. In short, the TP-TBSM method could flexibly determine the number of trends and achieve better results in terms of AIC and prediction error.

Case Study. We discuss the effect of the trend shift model on the product 11.

Table 4. Average error of mid-term prediction. "error" denotes the failed prediction.

Product	DP	TP-TBSM(5)	TP-TBSM(10)	TP-TBSM(15)	TP-TBSM(20)
1	0.015405445	0.00886747	**0.006761714**	0.008439111	0.00760425
2	0.018084755	0.009292621	**0.007683202**	0.009119887	0.00807945
3	0.013712115	**0.07039487**	0.01172114	0.009646895	0.009646895
4	0.01047508	error	**0.009581494**	**0.009581494**	**0.009581494**
5	0.017375915	**0.01237681**	0.01863413	0.01863413	0.01863413
6	error	error	0.02571607	0.02795278	**0.0217712**
7	**0.007685725**	0.01445365	0.01199136	0.01452575	0.01261265
8	0.0199715	**0.008097168**	0.0091934	0.01072543	0.00924125
9	error	error	0.01348301	**0.007741808**	0.01348215
10	0.012518285	**0.006099533**	0.006322822	0.0131806	0.012273
11	0.009192925	**0.006014863**	0.00893464	0.01202038	0.00893415

Table 5. AIC. "error" denotes the failed prediction.

Product	DP	TP-TBSM(5)	TP-TBSM(10)	TP-TBSM(15)	TP-TBSM(20)
1	−3.580444	−3.716145	−3.817783	**−3.818378**	−3.814294
2	−3.686648	−3.966148	−3.970557	**−3.971028**	−3.968967
3	−3.26868	−2.658826	−3.451055	**−3.509882**	**−3.509882**
4	−3.245881	error	**−3.584658**	**−3.584658**	**−3.584658**
5	−2.816559	−2.669863	**−3.217069**	**−3.217069**	**−3.217069**
6	error	error	−3.593841	−3.787913	**−3.82778**
7	−3.495332	−3.817393	−3.831919	**−3.832069**	−3.830447
8	−3.387825	−3.485216	−3.775669	−2.899403	**−3.839674**
9	error	error	−3.674589	−3.805849	**−3.887454**
10	−3.586449	−3.27506	−3.484745	**−3.861185**	−3.744925
11	−3.681659	−3.26022	−3.871297	−3.712011	**−3.868926**

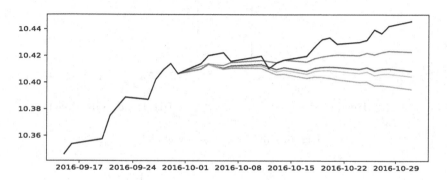

Fig. 4. Mid-term prediction for product 11; local model: blue, linear model: yellow, trend model: green, trend shift model (TP-TBSM): red (Color figure online)

The prediction of the middle term is shown in Fig. 4. The average error of the trend shift model using TP-TBSM is the smallest one among all the models. From this figure, it can be observed that the trend shift model can successfully estimate the trend. Local models and linear models do not change much since the start of prediction.

Discuss overfittings caused by random walks. μ_t of each model is shown in Fig. 5. As μ_t varies owing to random walk, larger variation of μ_t indicates that the change of NAV is random and we could not estimate the influence degrees of factors.

In the local model and linear model, μ_t significantly varies every day. In the trend model, this level term fluctuates smoothly, and hence, it is different from the change of NAV of local and linear models. Therefore, the influence of μ_t becomes small, and the variation by chance decreases. In the trend shift model,

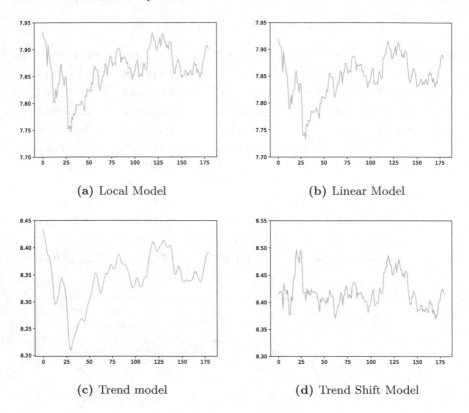

(a) Local Model

(b) Linear Model

(c) Trend model

(d) Trend Shift Model

Fig. 5. Difference in μ_t by model

the variation of μ_t is suppressed, and we may conclude that the trend shift model could reduce the effects of chance to yield better results of factor analysis.

5   Conclusion and Future Work

In this paper, we proposed a trend shift model by incorporating the trend change points into a state space model in order to quantitatively analyze factors affecting the NAV and predict future NAVs. To realize the trend shift model, we also proposed a trend detection model, i.e., TP-TBSM. In the TP-TBSM, by repeating the evaluation and segmentation phases, it is possible to reduce the dependence on the parameter, as compared with the conventional method, and to detect the trend more flexibly. The trend shift model enables global analysis across trends. From the experimental results, we observed that the trend shift model incorporating the change point detected using TP-TBSM has higher prediction accuracy than the baseline. We will carry out further extensive experiments to validate and improve our model. We also plan to extend the TP-TBSM method to multiple time series data. Another future work is to compare multiple products to support investment.

Acknowledgments. This work was partly supported by JSPS KAKENHI (16K12532).

References

1. Bollen, J., Mao, H., Zeng, X.: Twitter mood predicts the stock market. J. Comput. Sci. **2**(1), 1–8 (2011)
2. Mahajan, A., Dey, L., Haque, S.M.: Mining financial news for major events and their impacts on the market. In: Proceedings of the 2008 IEEE/WIC/ACM International Conference on Web Intelligence and Intelligent Agent Technology-Volume 01, pp. 423–426. IEEE (2008)
3. Awano, Y., Ma, Q., Yoshikawa, M.: Causal analysis for supporting users' understanding of investment trusts. In: Proceedings of the 16th International Conference on Information Integration and Web-based Applications and Services, pp. 524–528. ACM (2014)
4. Bäruning, F., Koopman, S.J.: Forecasting macroeconomic variables using collapsed dynamic factor analysis. Int. J. Forecast. **30**(3), 572–584 (2014)
5. Ando, T.: Bayesian state space modeling approach for measuring the effectiveness of marketing activities and baseline sales from POS data. In: Sixth International Conference on Data Mining (ICDM 2006), pp. 21–32. IEEE (2006)
6. Onishi, N., Ma, Q.: Factor analysis of investment trust products by using monthly reports and news articles. In: 2017 Twelfth International Conference on Digital Information Management (ICDIM), pp. 32–37. IEEE (2017)
7. Suzuki, T., Ota, M., et al.: Nonlinear prediction for top and bottom values of time series. Trans. Math. Model. Appl. **2**(1), 123–132 (2009). In Japanese
8. Chang, P.C., Fan, C.Y., Liu, C.H.: Integrating a piecewise linear representation method and a neural network model for stock trading points prediction. IEEE Trans. Syst. Man Cybern. Part C (Appl. Rev.) **39**(1), 80–92 (2009)
9. Wu, J.L., Chang, P.C.: A trend-based segmentation method and the support vector regression for financial time series forecasting. Math. Probl. Eng. **2012**, 20 p. (2012)
10. Durbin, J., Koopman, S.J.: Time series analysis by state space methods. Oxford University Press, Oxford (2012). ISBN 9780199641178

Efficient Aggregation Query Processing for Large-Scale Multidimensional Data by Combining RDB and KVS

Yuya Watari[1], Atsushi Keyaki[1], Jun Miyazaki[1($\boxtimes$)], and Masahide Nakamura[2]

[1] Department of Computer Science, School of Computing,
Tokyo Institute of Technology, Tokyo, Japan
{watari,keyaki}@lsc.cs.titech.ac.jp, miyazaki@cs.titech.ac.jp
[2] Kobe University, Kobe, Japan
masa-n@cs.kobe-u.ac.jp

Abstract. This paper presents a highly efficient aggregation query processing method for large-scale multidimensional data. Recent developments in network technologies have led to the generation of a large amount of multidimensional data, such as sensor data. Aggregation queries play an important role in analyzing such data. Although relational databases (RDBs) support efficient aggregation queries with indexes that enable faster query processing, increasing data size may lead to bottlenecks. On the other hand, the use of a distributed key-value store (D-KVS) is key to obtaining scale-out performance for data insertion throughput. However, querying multidimensional data sometimes requires a full data scan owing to its insufficient support for indexes. The proposed method combines an RDB and D-KVS to use their advantages complementarily. In addition, a novel technique is presented wherein data are divided into several subsets called grids, and the aggregated values for each grid are precomputed. This technique improves query processing performance by reducing the amount of scanned data. We evaluated the efficiency of the proposed method by comparing its performance with current state-of-the-art methods and showed that the proposed method performs better than the current ones in terms of query and insertion.

Keywords: Multidimensional data · Aggregation query
RDB · Distributed KVS

1 Introduction

In scenes including business activities, various types of data, such as product purchase data or sensor data, are generated. Accumulating and analyzing such data leads to obtaining new findings, and online analytical processing (OLAP) [1] is a type of such analysis. In OLAP, data are treated as multidimensional. Such data can be organized on a hypercube or a data cube. An analysis process is converted to an operation on the data cube, which is key to efficiently handle multidimensional data in OLAP.

© Springer Nature Switzerland AG 2018
S. Hartmann et al. (Eds.): DEXA 2018, LNCS 11029, pp. 134–149, 2018.
https://doi.org/10.1007/978-3-319-98809-2_9

In addition to this background, the rapid developments in network technology have led to an increase in the number of devices that are connected to the Internet and generation of multidimensional data. A large amount of data is generated from the backbone of what has been called the Internet of Things (IoT). Hence, analyzing the sensor data generated by IoT devices has gained prominence. One of the most useful operations that enable the analysis is an aggregation query. There are various challenges to compute such aggregation queries. Since sensor data are generated continuously and frequently, the data store must offer high insertion throughput and compute the aggregation queries by efficiently managing multidimensional data.

Several studies have focused on these challenges [2–5]. Nishimura et al. [6] proposed MD-HBase, which handles multidimensional data efficiently in a key-value store only with a one-dimensional index. The key idea behind MD-HBase is to transform multidimensional data into one-dimensional data by using a space-filing curve, which is embedded into the key-value store.

In this paper, we consider the combined advantages of current data stores, i.e., relational databases (RDBs) and distributed key-value stores (D-KVSs).

- RDBs [7] are widely used as reliable data stores in many applications. They are equipped with state-of-the-art features, such as indexes to manage complex data efficiently, transactions to protect data, and SQL to search data with complex query conditions. Multidimensional exact match queries and range queries can be processed efficiently using the indexes. However, despite the number of studies on distributed and parallel databases [8,9], RDBs do not provide good scale-out performance owing to their complex query processing capabilities such as strict transaction, indexes, and SQL.
- A key-value store (KVS) [10–13] is a simplified table-type database in which a tuple, called "row", consists of two attributes: key and value. Compared with an RDB, the data structure of a KVS is relatively simple. Thus, it is easy to decentralize data over several servers by horizontal partitioning, which is also called distributed KVS (D-KVS). In addition, most D-KVSs do not support transactions, rich query languages, and complex indexes, which adds to the bottleneck in database systems. These restrictions enable a D-KVS to provide good scale-out performance. In contrast to this advantage, most D-KVSs support an index only on a key. Therefore, it is difficult to execute flexible and complex queries because of the costs incurred in carrying out a full data scan over a large amount of data.

We also consider a precomputation technique, such as a materialized view, to reduce the computation cost required to process a multidimensional query. Using this technique, some aggregation queries can efficiently be evaluated with partial precomputed aggregation results. For example, consider a data set D that is divided into three blocks B_1, B_2, and B_3. The sum of D, sum(D), can be obtained by adding partial summation values such as sum(D) = sum(B_1) + sum(B_2) + sum(B_3). We only have to add the three partial sum values of these blocks by calculating and storing them in advance. Therefore, we can significantly reduce the cost of scanning data D.

Based on the above discussion, we propose an efficient multidimensional data store for a large amount of data by middleware that combines an RDB and

D-KVS. The proposed data store also enables the precomputation of partial aggregation results for efficient processing and optimizing multidimensional queries.

The proposed data store has two key properties. First, the raw data are stored in a D-KVS and their corresponding multidimensional indexes are stored in an RDB. The D-KVS offers high insertion throughput and the RDB provides efficient management of complex data by indexes. This approach provides better maintainability of the software of the data store because of the middleware that controls them only with their APIs. Second, the multidimensional space is divided into subspaces, which are called grids. For each grid, partial aggregation values, such as *sum*, *max*, *min*, and *number of data*, are precomputed for efficient aggregation query processing.

The remainder of the paper is organized as follows: Sect. 2 describes related work. In Sect. 3, the problem of executing aggregate operations for multidimensional data is formulated. Next, in Sect. 4, the proposed method for improving aggregation query processing performance is described. In Sect. 5, we discuss our evaluation experiments and results. Finally, we conclude the findings of this work in Sect. 6.

2   Related Work

There are many indexes for handling multidimensional data. Z-order curve [14] and Hilbert curve [15] are space-filling curves that convert multidimensional data into one-dimensional data. These curves can be used as multidimensional indexes by giving the converted value to a one-dimensional index. Tree structures, such as R-tree [16], quadtree [17] and k-d tree [18], are also commonly used for multidimensional indexes.

A k-d tree [18] is a binary search tree constructed by dividing a multidimensional space in a top-down manner. This division is conducted by a hyperplane that is perpendicular to an axis; the axis is chosen cyclically. There are several approaches to choose division points: using the median or mean value of data and center value of the hyperrectangle. Using the median value enables the k-d tree to become well balanced. The problem is that the computation cost of obtaining the median value is relatively high; however, the mean value can be calculated easily. Thus, the mean value is often used instead. We call this division *mean-value-division*. On the other hand, if the center value of the hyperrectangle is chosen, the shape of each node of the k-d tree can be kept uniform. We call this division *center-division*.

Multidimensional indexes including a k-d tree have been used in RDBs, but recent studies involved applying them to D-KVSs, such as MD-HBase [6]. MD-HBase is an improved version of HBase, which can conduct multidimensional range queries efficiently. MD-HBase transforms the multidimensional data into one-dimensional data by the Z-order curve [14], which is a space-filling curve. The transformation can be attained by assigning numbers in the order through which the curve passes. The numbers obtained are used as keys in an HBase table. In addition, MD-HBase splits the multidimensional space into several regions by a k-d tree and holds the minimum and maximum key values of each region as an index. When executing multidimensional range queries, MD-HBase finds the minimum and maximum values of the key range for a given query then conducts a range scan on HBase. At this instance, MD-HBase skips scanning some regions that do not intersect with the range of the query. This optimization skips unnecessary data scans of such regions.

MD-HBase requires the modification of the complex code in HBase to construct an index embedded in HBase. Applying the same approach to other D-KVSs is cumbersome, and its implementation and maintenance costs are quite high. In contrast, the proposed method does not require the building of a new index layer in the D-KVS. It uses only the APIs provided by an RDB and D-KVS, in which the indexes are automatically and consistently maintained by the RDB. In other words, the implementation and maintenance costs of the proposed method can be suppressed; thus, it achieves high sustainability. In a previous study [19], MD-HBase was extended to optimize the data scan. However, the query pattern must be known in advance.

Instead of MD-HBase, it is possible to use MapReduce [20] as a framework for managing large-scale data. In MapReduce, we have only to define *map* and *reduce* steps. Combining them makes it possible to easily implement highly parallelized processing. In addition to text processing, MapReduce can also be applied to aggregate operations on sensor data. SpatialHadoop [21] extends Hadoop for spatial data. It constructs multidimensional indexes such as grid, R-tree, and R+-tree. The index constructions are executed with MapReduce. Hence, it handles with static data or a snapshot of data, while the proposed method can handle dynamically and continuously generated data. MapReduce and Spatial-Hadoop are based on batch process, which leads to longer response time. In contrast, the proposed method achieves efficient aggregation query processing in both response time and throughput.

3   Problem Formulation

In our study, we assume that data are a set of points in multidimensional space. The domain of the data is called *a data space* $D \, (\in \mathbb{R}^n)$, where n is the dimensionality of the data and D is a hyperrectangle, i.e., D is expressed by a Cartesian product as follows: $D = [s_1, e_1] \times [s_2, e_2] \times \cdots \times [s_n, e_n]$, where s_i and e_i denote the start and end points in the i-th dimension of the hyperrectangle, respectively.

We consider *a partially computable aggregation operation* for multidimensional data. This operation can be defined as follows:

Definition 1 (Partially computable aggregation operation). Given a query range $Q \, (Q \subseteq D)$ and an aggregation operation $f(Q)$, which calculates an aggregation value for data within Q, f is partially computable if and only if there exists a function c that satisfies $f(Q) = c(f(G_1), f(G_2), \ldots, f(G_m))$. Here, Q is divided into hyperrectangles $G_1, G_2, \ldots$, and G_m; in other words, the following formulae hold:

$$\forall i \neq j \, (G_i \cap G_j) = \emptyset, \quad \text{and} \quad \bigcup_{i=1,\ldots,m} G_i = Q.$$

Examples of partially computable aggregation operations are *sum, count, average, minimum,* and *maximum*; *cardinality* is not a partially computable aggregation operation.

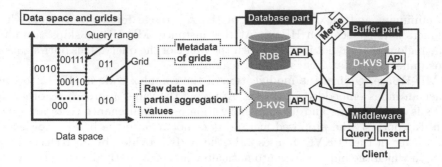

Fig. 1. Architecture of proposed data store

Our goal is to efficiently execute a partially computable aggregation operation for the data contained in a region $Q\,(Q \subseteq D)$, where Q is a hyperrectangle.

4   Proposed Method

In this section, we present an outline of our approaches for the proposed data store, which is illustrated in Fig. 1. The presented approaches reduce the amount of data to be scanned, as follows:

1. The data space is split into several hyperrectangles, which are called *grids* (the left side in Fig. 1). This split follows the algorithm of the k-d tree.
2. A partial aggregation value for each grid is precomputed.
3. Given a query (shown as a dashed line in Fig. 1), scans of the data in grids that are entirely contained in the query, are omitted because the aggregation values of such grids have already been computed. This optimization reduces the amount of data to be scanned.

For example, when we calculate the sum over the query range shown as a dashed line in Fig. 1, we first get the partial aggregation values of grids 00110 and 00111, assuming that they are 12 and 15. These values can be obtained quickly because they have already been precomputed. Then, the data contained in grid 000 are scanned and summed up, say, it is 5. Finally, the result is found by adding these three values, i.e., $12 + 15 + 5 = 32$.

As described in Sect. 1, the key feature of our method is using the advantages of both an RDB and D-KVS. There are three types of data required for our method:

– metadata of grids including their locations, sizes, and IDs;
– raw data; and
– partial aggregation values.

The size of metadata is not significantly large unless the grid size is extremely small. However, to answer a query, the number of grids that intersect with the query range must be enumerated, which is a challenging problem. To address this problem, we store the metadata in an RDB with indexes. Compared to the frequency of data insertion, grid split occurrences are relatively low. Therefore, it is reasonable to adopt the replication for the indexes, since the metadata are not frequently updated.

The size of raw data could be significantly large. When handling with sensor data, raw data and partial aggregation values must be updated frequently because such data are continuously generated. Therefore, these data should be stored in a scalable D-KVS, which can execute high insertion throughput.

By using the advantages of an RDB and D-KVS complementarily, we can address the challenges to handle a large amount of multidimensional data. In this study, we adopted PostgreSQL [22] as an RDB and HBase [23] as a D-KVS. Note that the proposed method can be implemented using any RDB and D-KVS.

4.1   Grid Splitting

As shown in Fig. 1, grid splitting follows the algorithm of the k-d tree. When the number of data entries in a grid exceeds a certain threshold, the grid is divided based on a cyclically selected axis. Let this threshold be $N_{\text{threshold}}$. The division is executed recursively until the number of data entries in the grid is less than *the grid size* (N_{size}). Note that $N_{\text{size}} \leq N_{\text{threshold}}$ always holds, which means that the number of data entries in the grid is allowed to exceed N_{size}. As a result, the frequency of grid splitting can be suppressed. We use mean-value-division and center-division as a division strategy for the k-d tree.

4.2   System Architecture

With our method, the data are stored in both an RDB and D-KVS. The architecture of our data store consists of three parts: *database, buffer*, and *middleware*, which are illustrated in Fig. 1. The database part stores three types of data—metadata of grids, raw data, and partial aggregation values. The buffer part temporarily keeps the data to be stored in the database, so that insertion throughput can be improved. The middleware accepts queries and controls the database and the buffer through their APIs for query processing.

When inserting new data, some partial aggregation values must be updated in the grids associated with them. Moreover, a grid must be split if the number of data entries in a grid becomes larger than $N_{\text{threshold}}$. Grid splitting is executed with mutual exclusion because all data must be consistent even when multiple clients simultaneously insert data into the same grid. If clients directly insert data into the database part, this costly mutual exclusion results in the degradation of data insertion throughput. To avoid this problem, clients insert data into the buffer part temporarily. Since clients do not update the database part, no mutual exclusion is needed. Moreover, the buffer is organized with the D-KVS to provide scalable insertion throughput.

Aggregation queries related to data in the buffer do not return accurate values because partial aggregation values are not precomputed. Therefore, such data must quickly be moved into the database part; this operation is referred to as a *merge* operation. The merge operation is controlled by the middleware and executed on the D-KVS servers in parallel. If the merge operation is faster than the case in which clients directly insert data into the database part, aggregation queries can return accurate results more quickly. The details of the merge operation are described in the next section.

4.3   Insert and Merge Operations

The algorithm of insertion is very simple. As described in Sect. 4.2, when a client inserts data, the data are simply inserted into an HBase table, which works as the buffer.

The merge operation is executed on multiple servers in parallel. It can cause grid splitting and updating of partial aggregation values. Figure 2 shows the flow of the merge operation, where three servers, A, B, and C, are under the merge process. Each server is responsible for merging the data based on the assigned key prefix, which uniquely maps the server to the process.

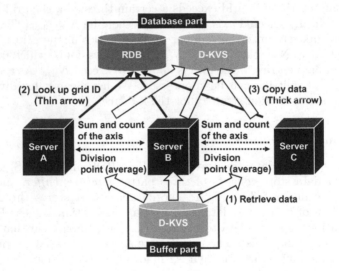

Fig. 2. Merge operation: numbers in figure correspond to those in Algorithm 1

The algorithm for the merge operation is as follows.

Algorithm 1. Merging data
1. Retrieve the data associated with a server from the buffer.
2. On PostgreSQL, search the grid ID to which the data obtained in step 1 belong.
3. Copy the data obtained in step 1 into the HBase table while adding the grid ID to its key prefix. Execute grid splitting if necessary.
4. Delete the data obtained in step 1 from the buffer.

In step 3, if the total number of data entries in the database part and buffer exceeds $N_{threshold}$, grid splitting is initiated. This split process is operated by several servers in parallel as follows. First, the master role is assigned to an arbitrary server (in Fig. 2, B is the master). The master receives the information used to determine the division point from other servers. After the division point is calculated by the master, it notifies others of the division point. Finally, the master updates partial aggregation values in HBase and the metadata in PostgreSQL while maintaining consistency by a transaction in an RDB.

Note that the master can cause a bottleneck when a large number of grids have to be split. However, this master role for each grid can be migrated to a different server to avoid a bottleneck because split processes for different grids can work independently.

4.4 Query

Given a query range $Q\,(\subseteq D)$, the aggregation query of the data within Q is processed by the middleware as follows[1].

Algorithm 2. Querying Q

---

1. Find all grids that intersect with Q by using the grid information table in PostgreSQL. Let G be a set of the obtained grids. Check if each grid range is completely included in Q.
2. Combine the partial aggregation results of the grids in G that are completely included in the query (grids 00110 and 00111 in Figure 1).
   These partial aggregation values can be obtained quickly because they are stored in HBase.
3. Scan all data in the grids in G that are partially included in the query range and aggregate the values within Q (grid 000 in Figure 1).
   We conduct a prefix scan with row keys.
4. Combine the results obtained in steps 2 and 3.

---

5 Experimental Evaluations

We conducted experiments to evaluate the proposed method.

In some experiments, we compared the proposed method to an open source implementation of MD-HBase[2]. We improved its original implementation for support of higher dimensionality and better insertion and query performance.

The experiments we conducted are as follows. We compared the insertion throughput among the proposed and current methods (Sect. 5.2). We then evaluated query performance (Sects. 5.3 and 5.4). Finally, we measured throughput with mixed read/write workloads (Sect. 5.5). In some experiments, we compared

---

[1] Our implementation uses a custom filter in HBase for a prefix scan in Step 3 of Algorithm 2, which efficiently extracts the data contained within the given query range.

[2] https://github.com/shojinishimura/Tiny-MD-HBase.

the proposed method to PostgreSQL-only and HBase-only schemes to clarify the effectiveness of combining them in the proposed method.

All experiments were conducted on a cluster with 16 PCs, each of which was equipped with an Intel Core i7-3770 CPU (3.4 GHz), 32 GB of memory, and a 2-TB HDD, running HBase 1.2.0 under CentOS 6.7. 13 PCs out of 16 operated as region servers. HBase stored data over the region servers. In addition, PostgreSQL 9.6.1 was installed on the 13 PCs, which were configured as a multi-standby replication setup.

5.1   Dataset

We used the following two datasets in our experiments.

SFB Data (Moving Objects in San Francisco Bay Area Data, 22 Million). We generated 22,352,824 points of moving objects in the San Francisco Bay Area using a network-based generator [24]. Each data entry has two attributes – latitude and longitude. We call such data *SFB data.*

Indoor Sensor Data (100 Million). We collected 2,032,918 data entries from indoor environmental sensors between January 14, 2010 and April 11, 2014. Each entry consists of 16 attributes. We extracted the entries from original data for 3 years from 2011 to 2013. Given the insufficient size of the data, we generated pseudo data by replicating the existing data by a factor of 70, giving rise to 100 million data entries from 2011 to 2031. We call the pseudo data *indoor sensor data.*

5.2   Evaluation of Insertion Throughput

To compare the insertion performance of the proposed method relative to those of MD-HBase, PostgreSQL, and HBase, we inserted *SFB data* into these systems and measured their throughputs. We used the data because they were close to large and frequently generated data with sensor devices such as automobiles. Note that the insertion throughput with the proposed method was calculated based on the elapsed time from when the client started inserting until the merge process finished.

We configured one PC in the cluster as a client for inserting data. During insertion, we varied the grid size N_{size} with the proposed method and MD-HBase as follows: $N_{size} = 50, 125, 250, 500, 1000, 2000, 4000, 8000, 16000, 32000$. The grid size in MD-HBase represents the number of data entries in a bucket used for determining the threshold for splitting. We set $N_{threshold} = N_{size} \times 10$. In addition, we used mean-value-division as a division strategy of the k-d tree.

Results. Figure 3 shows the results of insertion throughputs. Due to space limitations, we plotted some of the results. The numbers for the proposed method and MD-HBase represent the grid size N_{size}. The results indicate that the proposed method achieved higher throughput than MD-HBase and PostgreSQL for any grid size. It improved by 16.4x–39.8x and 4.0x–12.4x compared to MD-HBase

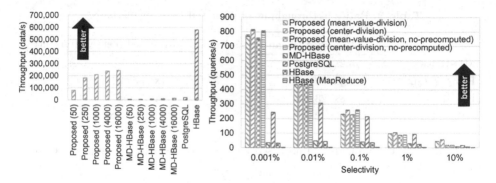

Fig. 3. Insertion throughput   **Fig. 4.** Query throughputs while varying selectivity

Table 1. Average time lags in merge process

N_{size}	50	125	250	500	1000	2000	4000	8000	16000	32000
Time lag (s)	86.8	39.7	32.9	36.2	34.1	32.0	22.4	24.0	22.0	23.7

and PostgreSQL, respectively. Note that this comparison might be overstated because the MD-HBase we used was not sufficiently optimized in terms of insertion. In contrast, the throughput of the proposed method was lower than that of HBase, which was up to around 0.4x. The merge process caused this lower insertion throughput.

We now examine this effect in more detail. There is a time lag from when data are inserted into the buffer until they are merged in the database part. Table 1 lists the average time lags. The time lag reached 22–87 s. In the merge process, an additional data access occurred since data are read from the buffer and written back to the database. This access caused a drop in insertion throughput. Improving the merge process to reduce time lag is a future task. We discuss the effect of the time lag on query processing in Sect. 5.5.

5.3 Evaluation of Query Throughput

We evaluated the query performances for the proposed method and other methods (MD-HBase, PostgreSQL, HBase, and MapReduce). We inserted *indoor sensor data* into these systems and conducted the four-dimensional range queries to measure the throughput. These data are suitable for evaluating query processing performance in high dimensional data since they have many attributes. The queries were randomly generated so that their selectivity would become 0.001, 0.01, 0.1, 1, and 10%. They were issued from 120 clients simultaneously while varying selectivity.

With the proposed method, we used both mean-value-division and center-division as the division strategies for the k-d tree and set the grid size N_{size} to the following values: N_{size} = 50, 125, 250, 500, 1000, 2000, 4000, 8000, 16000, 32000, and 64000. Also, the grid sizes in MD-HBase were N_{size} = 8000, 16000, 32000,

Table 2. Ratios in throughput of proposed to other methods

	Proposed (mean-value-division)	Proposed (center-division)
MD-HBase	3.2x–21.0x	3.5x–23.2x
PostgreSQL	1.0x–3.0x	1.1x–3.5x
HBase	3.8x–23.2x	4.1x–25.6x
HBase (MapReduce)	38.9x–241.3x	42.2x–266.3x

Table 3. Ratios of throughput of proposed method w/ precomputing to proposed one w/o precomputing

Selectivity	0.001%	0.01%	0.1%	1%	10%
Mean-value-division	1.0	1.0	1.0	1.1	2.4
Center-division	1.0	1.1	1.1	1.3	3.4

$64000, 128000, 256000, 512000$, and 1024000. These values were selected as those that demonstrate the highest query processing performance of each method based on preliminary experiments.

Results. Figure 4 depicts the query performance results. The note "no-precomputing" indicates that the precomputation of aggregation values was not available. In other words, this evaluation was for testing for simple range queries. We plotted only the best cases while changing grid sizes.

Table 2 describes the improvement rate of the throughputs. The proposed method exhibited significantly higher throughput than MD-HBase, HBase, and MapReduce. Even for PostgreSQL, the proposed method in center-division exhibited higher performance at any selectivity. Furthermore, Fig. 4 illustrates that simple range query performance of the proposed method is superior to or the same as the other methods.

Now we discuss the effects of reusing precomputed aggregation values. Table 3 shows the improvement in throughputs by reusing them. The throughput of center-division at 10% of selectivity increased 3.4x by using the precomputed values, while there was no increase at low selectivity. With the proposed method, the number of grids completely included in a query range must be large to execute queries efficiently. Such a number is proportional to the volume of the query range, which is a^n when we consider a range query as an n-dimensional hypercube whose side length is a. On the other hand, the amount of data to be scanned, which is related to execution time, depends on the number of grids partially included in the query range. This is proportional to the surface area of the query range, which is $2na^{n-1}$. Hence, it is possible to reduce the data to be scanned for a large query range. Therefore, the proposed method could obtain high throughput under 10% of selectivity.

This claim is also supported in Table 4, which shows various statistics for the proposed method. Skipped data indicates the data that are selected by the query but do not need to be scanned, i.e., they exist in a grid completely included by a

Table 4. Statistics for various selectivity ratios

Selectivity		0.001%	0.01%	0.1%	1%	10%
(a)		1,042	10,449	104,490	1,044,327	10,445,637
(b)	Mean-value-division	135,024	298,488	872,584	1,704,079	3,705,552
	Center-division	93,874	224,273	646,822	1,583,853	2,444,884
(c)	Mean-value-division	0	0	0	294,998	8,169,864
	Center-division	6	40	1,912	368,445	9,200,398
(b)/(a)	Mean-value-division	129.63	28.57	8.35	1.63	0.35
	Center-division	90.12	21.46	6.19	1.52	0.23
(c)/(a)	Mean-value-division	0.00	0.00	0.00	0.28	0.78
	Center-division	0.01	0.00	0.02	0.35	0.88

(a) # of selected data entries, (b) # of scanned data entries, (c) # of skipped data entries.

Table 5. Grid sizes that demonstrate highest throughput with proposed method

Selectivity	0.001%	0.01%	0.1%	1%	10%
Mean-value-division	4000	4000	8000	2000	2000
Center-division	4000	4000	8000	4000	2000

given query range. The "(b)/(a)" in Table 4 represents the ratio of the number of data entries in the grids which are partially included in a given query range to that of selected data entries. Similarly, the "(c)/(a)" indicates the ratio in the completely included case. Although 88% entries of the selected data did not require scanning when the selectivity was 10% in the center-division, we could not reduce the amount of data to be scanned at 0.001% of selectivity. In addition, the ratio "(b)/(a)" was much larger than 1. This means that the proposed method scanned a considerable amount of data which were not related to the query result.

In summary, increasing query range, the precomputing technique in the proposed method works more effectively and improves query processing performance.

Finally, we evaluated the effect of grid size and grid division strategy on query processing performance. Table 5 lists grid sizes that demonstrate the highest throughput. These sizes are in the range from 2000 to 8000. The best grid size for *indoor sensor data* is considered to be about 4000, although the best one cannot be obtained in advance.

In this experiment, we used mean-value-division and center-division as division strategies. From the above results, center-division yielded better performance. From "(c)/(a)" in Table 4, center-division can avoid scan more efficiently than mean-value-division. This caused the difference in throughput. Center-division keeps the shape of grids uniform compared with mean-value-division.

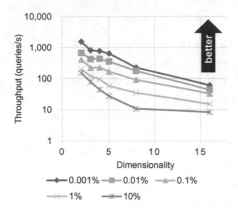
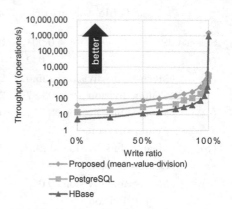

Fig. 5. Query throughput with varying dimensionality

Fig. 6. Mixed read/write workload throughput (selectivity is 10%)

5.4 Evaluation of Insertion Throughput with Varying Dimensionality

We examined how much the query processing performance of the proposed method is affected by dimensionality.

In this experiment, we used *indoor sensor data*, and inserted them into the proposed system by varying the dimensionality from $n = 2$ to 16. We executed several queries on the data while varying the selectivity, i.e., 0.001, 0.01, 0.1, 1, and 10%. In the experiment, we created an index on the first k attributes in *indoor sensor data* when the dimensionality was $n = k$.

Results. Figure 5 illustrates the results of this experiment. The vertical axis of the figure is log scale. An increase in the dimensionality had a negative impact on query throughput. As discussed in Sect. 5.3, the amount of scanned data is considered to be proportional to the surface area of the query range, which is $2na^{n-1}$ under n-dimensional space when we assume the query as a hypercube. Hence, the query performance is adversely affected by an increase in dimensionality. This theoretical analysis matches the results in Fig. 5.

The reuse of precomputed aggregation values is effective only when dimensionally is low or selectivity is high. In addition to the inefficiency in the low-selectivity case discussed in Sect. 5.3, we analyzed the reasons the throughputs decrease in higher dimensional cases. The amount of data to be scanned is proportional to $2kna^{n-1}$ when we assume that the query is a hypercube. This value is obtained by multiplying the surface area of the query by the side length of a grid k. The ratio of this value to the query volume is $2kna^{n-1}/a^n = 2kn/a$, which becomes larger as n increases. Thus, it becomes difficult to reuse the precomputed aggregation values in high dimensional data space.

5.5   Evaluation with Mixed Read/Write Workload

This section evaluates the throughput of read/write mixed workloads. We compared the throughputs of the proposed method, PostgreSQL, and HBase by changing the *write ratio*, which indicates the ratio of write operations to the entire operations. In this experiment, one operation denoted either one aggregation query (read) or insertion of one record (write). Therefore, the data size handled by a read operation is much larger than that by a write one. These operations were issued from multiple clients simultaneously.

In the experiment, we first inserted *indoor sensor data*. After that, 120 clients simultaneously issued operations at a specified write ratio, and its throughput was measured. With the proposed method, we set the grid size N_{size} to 4000, where the highest performance was expected according to Table 5.

Results. Figure 6 shows the results of this experiment. The proposed method exhibited higher throughput than PostgreSQL and HBase at most selectivity ranges and write ratios. In particular, the throughput was superior to that of PostgreSQL in all cases and significantly higher than that of HBase, except for when the write ratio was extremely high.

These results proved that the objective of this research, i.e., using an RDB and D-KVS complementarily, was sufficiently achieved. Focusing only on the results of PostgreSQL and HBase, the RDB (PostgreSQL) had higher throughput at a lower write ratio. It can handle complicated data efficiently by an index. In contrast, the D-KVS (HBase) exhibited superior performance at a higher write ratio because it can efficiently handle data insertion. The proposed method took advantage of both, which led to higher throughput.

We should note that there was a time lag between insertion and merge process. However, the adverse effects on query processing due to the time lag were sufficiently suppressed since the results indicated that the proposed method exhibited higher performance than the current methods even when the write ratio was low. Some applications require aggregation queries even to the recently inserted data. Such data might temporarily be stored in the buffer part and can properly be aggregated by our method. However, such aggregation processing to the buffer can cause slower response time than that only to the database.

6   Conclusion

We proposed a novel method for efficient aggregation query processing for large-scale multidimensional data. The proposed method combines an RDB and D-KVS with middleware, so that the advantages of both data stores can be used complimentarily. This method can also reduce the amount of data to be scanned on query processing by using the precomputed aggregation values.

We implemented our method using PostgreSQL and HBase, and evaluated the insertion and query performances by comparing it to PostgreSQL, HBase, and MD-HBase which is an existing multidimensional data store. The experimental results indicated that the proposed method exhibited the highest query throughput. The insertion throughput was also much higher than PostgreSQL and MD-HBase. In addition, the evaluation with the mixed read/write workloads showed that the proposed method was superior to PostgreSQL and HBase at any write ratio. These results obviously proved that the proposed method could utilize both an RDB and D-KVS sufficiently. We also investigated the behavior of the proposed method with various dimensional data. An increase in dimensionality resulted in a decrease in query throughput. The decrease was more prominent for queries with higher selectivity.

For future work, we will attempt to improve query performance for higher dimensional data owing to the challenges faced in using precomputed aggregation values. Besides, the estimation of the best parameters, such as grid sizes, for a given dataset is one of the most important challenges for the future.

Acknowledgements. This work was partly supported by JSPS KAKENHI Grant Numbers 15H02701, 16H02908, 17K12684, 18H03242, 18H03342, and ACT-I, JST.

References

1. Codd, E., Codd, S., Salley, C.: Providing OLAP (On-line Analytical Processing) to User-Analysts: An IT Mandate. Codd & Associates (1993)
2. Wang, J., Wu, S., Gao, H., Li, J., Ooi, B.C.: Indexing multi-dimensional data in a cloud system. In: Proceedings of the 2010 ACM SIGMOD International Conference on Management of Data, pp. 591–602. ACM (2010)
3. Zhang, X., Ai, J., Wang, Z., Lu, J., Meng, X.: An efficient multi-dimensional index for cloud data management. In: Proceedings of the First International Workshop on Cloud Data Management, pp. 17–24. ACM (2009)
4. Li, X., Kim, Y.J., Govindan, R., Hong, W.: Multi-dimensional range queries in sensor networks. In: Proceedings of the 1st International Conference on Embedded Networked Sensor Systems, pp. 63–75. ACM (2003)
5. Escriva, R., Wong, B., Sirer, E.G.: Hyperdex: a distributed, searchable key-value store. ACM SIGCOMM Comput. Commun. Rev. **42**(4), 25–36 (2012)
6. Nishimura, S., Das, S., Agrawal, D., El Abbadi, A.: $\mathcal{MD}$-hbase: design and implementation of an elastic data infrastructure for cloud-scale location services. Distrib. Parallel Databases **31**(2), 289–319 (2013)
7. Codd, E.F.: A relational model of data for large shared data banks. Commun. ACM **13**(6), 377–387 (1970)
8. Lu, H., Tan, K.L., Ooi, B.-C.: Query Processing in Parallel Relational Database Systems. IEEE Computer Society Press, Los Alamitos (1994)
9. Özsu, M.T., Valduriez, P.: Principles of Distributed Database Systems. Springer, Heidelberg (2011). https://doi.org/10.1007/978-1-4419-8834-8
10. Lakshman, A., Malik, P.: Cassandra: a decentralized structured storage system. ACM SIGOPS Oper. Syst. Rev. **44**(2), 35–40 (2010)
11. Cooper, B.F., et al.: PNUTS: Yahoo!'s hosted data serving platform. Proc. VLDB Endow. **1**(2), 1277–1288 (2008)
12. Redis: Redis. https://redis.io/
13. DeCandia, G., Hastorun, D., Jampani, M., Kakulapati, G., Lakshman, A., Pilchin, A., Sivasubramanian, S., Vosshall, P., Vogels, W.: Dynamo: Amazon's highly available key-value store. ACM SIGOPS Oper. Syst. Rev. **41**(6), 205–220 (2007)
14. Morton, G.M.: A computer oriented geodetic data base and a new technique in file sequencing. In: International Business Machines Company New York (1966)

15. Hilbert, D.: Ueber die stetige abbildung einer line auf ein flächenstück. Math. Ann. **38**(3), 459–460 (1891)
16. Guttman, A.: R-trees: a dynamic index structure for spatial searching. In: Proceedings of the 1984 ACM SIGMOD International Conference on Management of Data, SIGMOD 1984, pp. 47–57. ACM, New York (1984)
17. Finkel, R.A., Bentley, J.L.: Quad trees a data structure for retrieval on composite keys. Acta Inf. **4**(1), 1–9 (1974)
18. Bentley, J.L.: Multidimensional binary search trees used for associative searching. Commun. ACM **18**(9), 509–517 (1975)
19. Nishimura, S., Yokota, H.: Quilts: multidimensional data partitioning framework based on query-aware and skew-tolerant space-filling curves. In: Proceedings of the 2017 ACM International Conference on Management of Data, pp. 1525–1537. ACM (2017)
20. Dean, J., Ghemawat, S.: MapReduce: simplified data processing on large clusters. Commun. ACM **51**(1), 107–113 (2008)
21. Eldawy, A., Mokbel, M.F.: SpatialHadoop: a MapReduce framework for spatial data. In: 2015 IEEE 31st International Conference on Data Engineering, pp. 1352–1363, April 2015
22. Korry Douglas, S.D.: PostgreSQL: A Comprehensive Guide to Building, Programming, and Administering PostgresSQL Databases. Sams Publishing, Indianapolis (2003)
23. The Apache Software Foundation: Apache HBase. https://hbase.apache.org/
24. Brinkhoff, T.: A framework for generating network-based moving objects. GeoInformatica **6**(2), 153–180 (2002)

Data Semantics

Learning Interpretable Entity Representation in Linked Data

Takahiro Komamizu(✉)

Nagoya University, Nagoya, Japan
taka-coma@acm.org

Abstract. Linked Data has become a valuable source of factual records. However, because of its simple representations of records (i.e., a set of triples), learning representations of entities is required for various applications such as information retrieval and data mining. Entity representations can be roughly classified into two categories; (1) interpretable representations, and (2) latent representations. Interpretability of learned representations is important for understanding relationship between two entities, like why they are similar. Therefore, this paper focuses on the former category. Existing methods are based on heuristics which determine relevant *fields* (i.e., predicates and related entities) to constitute entity representations. Since the heuristics require laboursome human decisions, this paper aims at removing the labours by applying a graph proximity measurement. To this end, this paper proposes RWRDoc, an RWR (random walk with restart)-based representation learning method which learns representations of entities by weighted combinations of minimal representations of whole reachable entities w.r.t. RWR. Comprehensive experiments on diverse applications (such as ad-hoc entity search, recommender system using Linked Data, and entity summarization) indicate that RWRDoc learns proper interpretable entity representations.

Keywords: Entity representation learning
Random walk with restart · Linked data · Entity search
Entity summarization

1 Introduction

As Linked Data [3] consists of factual records about entities in RDF (Resource Description Framework) [1] where each record is called triple, $\langle subject, predicate, object \rangle$, which expresses relationship between two entities or property of an entity, entity representation is crucial for various applications on Linked Data. Examples of the applications include ad-hoc entity search [18] and entity summarization [4,7,23], which directly utilize entity representations. Recommender systems with knowledge graph [2,13,16] and information retrieval with entities [19,22] are examples of other applications which indirectly utilize entity representations. Entity representations of existing methods can be roughly classified

© Springer Nature Switzerland AG 2018
S. Hartmann et al. (Eds.): DEXA 2018, LNCS 11029, pp. 153–168, 2018.
https://doi.org/10.1007/978-3-319-98809-2_10

into two categories; (1) interpretable representations, and (2) latent representations. Interpretability of learned representations is important for understanding relationship between two entities, like why they are similar. Therefore, this paper focuses on the former category of entity representations.

Basic idea of existing interpretable entity representations is that an entity is described by closely related texts and entities. One of the simplest entity representation is to include directly connected texts in Linked Data, e.g., literals of `rdfs:label` and `rdfs:comment`. Fielded documentation technique [14] is an extended idea of the simplest method, which heuristically selects informative predicates and consider texts at their objects are more important to be included into the representations. Moreover, the fielded documentation approaches can be extended from single predicates (e.g., `rdfs:label`) to a sequence of multiple predicates (e.g., (`dbo:birthPlace`, `rdfs:label`)).

Although existing interpretable entity representation learning methods are considerably reasonable approaches, there are two major concerns: (1) Determining appropriate sequences of predicates (or fields) is cumbersome. (2) There is no evidential proximity for reasonable lengths of predicate sequences. Large varieties of vocabularies make the determination harder. Therefore, to include descriptive texts in the "neighbouring" entities is an extended idea of the first. However, defining neighbouring entities is not straightforward. Shorter hops could be reasonable choices, but there is no evidence for the number of hops (or proximity).

This paper tackles with the aforementioned concerns by exploiting random walk with restart (RWR) [24, 26] as a proximity measurement between entities. Taking random walk into account is an idea to introduce random sampling of surrounding entities with respect to reachability. Simple random walk takes all reachable entities into account by random jump, however, closer entities should be more relevant. Therefore, "with restart" characteristics (which occasionally stops random walk and restart from source vertices) is adequate to realize this.

Based on the idea above, this paper proposes an RWR-based entity representation learning, **RWRDoc** for entities on Linked Data (introduced in Sect. 2). RWRDoc is a three-step method: (1) minimal entity representation for obtaining self-descriptive contents of entities, (2) RWR to measure proximities between entities, and (3) learning representations of entities as weighted combination of minimal representations of all entities with respect to the proximities.

RWRDoc is a beneficial approach comparing with the existing work in terms of generality, effectiveness, and interpretability. RWRDoc is not dependent on any heuristics of fields, therefore, it is a general approach which is applicable for any dataset of Linked Data. Experimental evaluations indicate the applicability of RWRDoc for various applications of entity representations including ad-hoc entity search, entity summarization, and recommender systems (Sect. 3).

Contributions

– This paper proposes **RWRDoc**, a random walk with restart based interpretable entity representation learning which takes minimal representations of all reachable entities into account according with RWR-based proximities.

- RWRDoc is **non-heuristic** approach unlike existing works, that is, RWRDoc does not require human assistances such as pre-defined sequences of predicates with importance metrics and proximity constraints.
- This paper demonstrates the **effectiveness** and **interpretability** of RWR-Doc by testing on various applications in the experiments.

2   RWRDoc: RWR-Based Documentation

RWRDoc is a random walk with restart (RWR)-based entity representation learning method. Basic idea of RWRDoc is, for an entity, entities with high proximity to the entity are highly relevant and descriptive to the entity. For example, Toyotomi Hideyoshi[1] who is a Japanese general in the Sengoku period who is known as a general who launches the invasions of the Joseon dynasty[2]. However, description of him represented by dbo:abstract does not include the historical fact, furthermore, other texts reachable within one predicate do not contain it as well. The fact is reachable from his entry through dbo:subject and contents in dbo:Japanese_invasions_of_Korea_(1592-98), and the fact is not reachable from most of other entities. It is not reasonable to say dbo:subject predicate is always important since it includes broader kinds of facts. This suggests reachability-based proximity is appropriate.

RWRDoc regards Linked Data dataset as a *data graph* G defined as follows:

Definition 1 (Data Graph). *Given Linked Data dataset, data graph G is a graph $G = (V, E)$, where set $V = R \cup L \cup B$ of vertices are union of set R of entities, set L of literals, and set B of blank nodes, and set $E \subseteq V \times P \times V$ of labeled edges between vertices with predicates in P as labels.* $\square$

This paper regards all resources represented by URIs (Uniform Resource Identifier) in Linked Data dataset as entities, thus they are included in R.

RWR [24] is a random walk-based reachability calculation method. RWR assigns reachability values from starting vertex to each vertex. Therefore, RWR vector $\mathbf{z}_u$ of entity u (which is a vector of length $|R|$) is calculated as follows:

$$\mathbf{z}_u = d \cdot \mathbf{z}_u \cdot A + (1 - d) \cdot \mathbf{s}$$

where A is a $|R| \times |R|$ adjacency matrix which represents network composed on entities R, $\mathbf{s}$ is a vector with length $|R|$ for restart that only item corresponding with u is 1, 0 otherwise, and d is a dumping factor (d is experimentally set to 0.4). A is derived from an induced subgraph G' of the data graph G. $G' = (R, E')$ is consists of set $R \subseteq V$ of entities as vertices and set $E' \subseteq R \times R$ of edges which are links between entities in R regardless of predicates.

In this paper, representation $\mathbf{x}_u$ of entity u (which is $|W|$-length vector, where W is a vocabulary set) is defined as a linear combination of *minimal* representations (each of them is represented by $\mathbf{m}_v$ where $v \in R$ which is also $|W|$-length

---

[1] http://dbpedia.org/resource/Toyotomi_Hideyoshi.

[2] http://dbpedia.org/resource/Japanese_invasions_of_Korea_(1592-98).

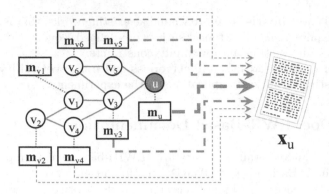

Fig. 1. RWRDoc overview: RWR-based representation generation of entity u. To make representation $\mathbf{x_u}$ of u, minimal representations ($\mathbf{m_{v_1}} \ldots \mathbf{m_{v_6}}$ and $\mathbf{m_u}$) of reachable vertices ($v_1 \ldots v_6$) are combined with respect to RWR scores (drawn by thickness of dashed arrows).

vector) of entities (including u) with respect to proximity scores from u. Figure 1 depicts the idea, that entities are represented as vertices u and $v_1, v_2, \ldots, v_6$, and corresponding minimal representations are associated with vertices (dotted lines). For entity u in the figure, representation $\mathbf{x_u}$ of u is the weighted summation of the minimal representations of entities where each weight is expressed by thickness of dashed arrows. The following provide formal definitions of minimal entity representation (Definition 2) and entity representation (Definition 3).

Definition 2 (Minimal Entity Representation). *Minimal representation* $\mathbf{m_v}$ *of entity* $v \in R$ *is a* $|W|$*-length vector of terms on literals within one hop.* $\square$

In this paper, the minimal entity representation of an entity is a TFIDF vector based on texts within one predicate away. Note that RWRDoc does not necessarily require TFIDF vectors, any vector representation is acceptable if their dimensions are shared among entities. Firstly, the following SPARQL query is executed to obtain texts of entities.

```
SELECT ?entity ?vals
WHERE {?entity ?p ?vals .
        FILTER isLiteral(?vals).}
```

Listing 1. SPARQL query for getting texts for each entity.

Secondly, the texts for entities compose bags of words, and TFIDF vectors for entities are calculated using them as follows:

$$\mathbf{m}_v = \Big(tf(t, v) \cdot idf(t, R) \Big)_{t \in W}$$

Algorithm 1. RWRDoc

Input: $G = (V, E)$: LD dataset
Output: $\mathbf{X}$: Learned Representation Matrix
1: Minimal Representation Matrix $\mathbf{M}$, RWR Matrix $\mathbf{Z}$
2: $G' \leftarrow DataGraph(G)$ ▷ Prepare data graph G' for RWR computation.
3: **for** $v \in R$ **do**
4:     $\mathbf{M}[v] \leftarrow TFIDF(v, G)$ ▷ Calculate TFIDF vector for entity v.
5:     $\mathbf{Z}[v] \leftarrow RWR(v, G')$ ▷ Calculate RWR for source entity v.
6: **end for**
7: $\mathbf{X} = \mathbf{Z} \cdot \mathbf{M}$

---

where R is a set of entities and W is a vocabulary set. $tf(t, v)$ is a term frequency of term t in the bag of words of v and $idf(t, R)$ is an inverse document frequency of t over all bags of words of entities R.

Entity representation $\mathbf{x}_u$ of entity u is represented as linear combination of representations of entities. $\mathbf{x}_u = \sum_{v \in R} z_{u,v} \cdot \mathbf{m}_v$ where $z_{u,v} \in \mathbf{z}_u$ is a proximity value from u to v. To simplify the computation, let $\mathbf{M}$ be a minimal representation matrix, which is a $|R| \times |W|$ matrix and each row corresponds with the minimal representation $\mathbf{m}_v$ of entity v. Therefore, the linear combination above can be rewritten as $\mathbf{x}_u = \mathbf{z}_u \cdot \mathbf{M}$. Consequently, entity representation $\mathbf{x}_u$ of entity u is defined as follows:

Definition 3 (Entity Representation). *Entity representation $\mathbf{x}_u$ of entity u is represented as linear combination of representations of entities as follows:*

$$\mathbf{x}_u = \mathbf{z}_u \cdot \mathbf{M}$$

where $\mathbf{z}_u$ is an RWR vector of u and $\mathbf{M}$ is a minimal representation matrix. □

Let $\mathbf{Z}$ be an RWR matrix, which is a $|R| \times |R|$ matrix where each row corresponds with RWR vector $\mathbf{z}_v$ from entity v. Then, entity representation learning process can be represented as matrix multiplication of $\mathbf{Z}$ and $\mathbf{M}$. Let $\mathbf{X}$ be an entity representation matrix, which is the result of the multiplication, that is, $\mathbf{X} = \mathbf{Z} \cdot \mathbf{W}$. Consequently, $\mathbf{X}$ is a $|R| \times |W|$ matrix where each row corresponds with entity representation $\mathbf{x}_u$ of entity u as calculated in Definition 3.

Algorithm 1 summarizes the procedure of RWRDoc for a given LOD dataset G. The first step of the algorithm (line 2) prepares the data graph G' from G. Then, the next step computes a minimal representation $\mathbf{m}_v$ and an RWR vector $\mathbf{z}_v$ for each entity v, and they are stored into corresponding matrices (i.e., $\mathbf{M}$ for minimal representations and $\mathbf{Z}$ for RWR vectors). Finally, representation matrix $\mathbf{X}$ is computed from $\mathbf{Z}$ and $\mathbf{M}$. RWRDoc Implementation in this paper employs a TFIDF vectorizer in scikit-learn[3] and, for calculating RWR, TPA algorithm [26] which is a quick calculation of approximate RWR values.

---

[3] http://scikit-learn.org/stable/modules/generated/sklearn.feature_extraction.text.TfidfVectorizer.html.

3   Experimental Evaluation

Experimentation of this paper attempts to investigate *generality, effectiveness* and *interpretability* of RWRDoc. Generality stands for its applicability to various applications related with entity documentation including entity documents themselves and document-based entity similarity. Effectiveness stands for qualities on the applications comparing with baseline approaches and the state-of-the-art. Interpretability stands for user-understandability of the learned representations comparing with a naïve baseline.

The application scenarios in this experiment are as follows: ad-hoc entity search (Sect. 3.1), recommender system with entities (Sect. 3.2), and entity summarization (Sect. 3.3). Ad-hoc entity search tests the expressive power of RWR-Doc for keyword search. Recommender system with entities checks capability of RWRDoc for entity similarity. Entity summarization observes interpretability of representations from RWRDoc. Each applications uses DBpedia 2015_10 dataset[4] as Linked Data dataset. Testing datasets and competitors are explained in the individual sections.

3.1   Ranking Quality on Ad-hoc Entity Search

Ad-hoc entity search [18] is a task for finding entities in Linked Data for given keyword queries. Basic strategy is to design vector representations of entities and queries, then find similar entities in terms of the representations with queries. To measure the similarities as discussed in information retrieval communities, various approaches have been applied to the ad-hoc entity search task, for example, BM25, language modeling, and fielded extensions of them.

RWRDoc is a representation learning method of entities and it is expected to have widely expressive information from reachable entities, therefore, more accurate search results are expected. To examine this expectation, this experiment compares RWRDoc-based ad-hoc entity search with the state-of-the-art presented in a representative benchmark, DBpedia-Entity v2 [8][5].

This paper follows the evaluation methodology in the benchmark, each ad-hoc entity search method is evaluated by their ranking quality. For given queries, each method returns ranked lists of entities, and with the gold standard in the benchmark, the lists are evaluated by NDCG (normalized discounted cumulative gain) [9] for top-10 and top-100 results. NDCG measures how the given ranking is close to ideal ranking, formal definition of NDCG is as follows:

$$DCG_k = \sum_{i=1}^{k} \frac{2^{rel_i} - 1}{\log_2(i + 1)} \tag{1}$$

$$NDCG_k = \frac{DCG_k}{IDCG} \tag{2}$$

---

[4] http://downloads.dbpedia.org/2015-10/.
[5] https://github.com/iai-group/DBpedia-Entity.

NDCG is based on DCG calculated as Eq. 1 where k is a rank position and rel_i is a true relevance score of i-th entity in the ranking (i.e., 1 for relevant and 0 for non-relevant in this experiment). Then, NDCG for p is calculated as Eq. 2 where $IDCG$ is calculated as the ideal ranking, that is, all relevant entities are on the top of the ranking. To rank entities with RWRDoc, similarities between entities and queries are calculated by standard cosine similarity.

Table 1 displays the results of ad-hoc entity search task. Note that results for the state-of-the-arts are quoted from the benchmark paper [8], since experimental settings are identical to this paper. The results are divided into five sections which indicate results for four different types of queries (i.e., 'SemSearch ES' for named entity queries, 'INEX-LD' for keyword queries, 'ListSearch' for queries seeking a list of entities, and 'QALD-2' for natural language questions) and an overall result ('Total'). Besides, for each type of queries, there are two subsections @10 and @100, respectively. In the table, the best scores for each column are highlighted as bold and underlined. Additionally, RWRDoc, has a *Residual* row which represents the residual from the second best if RWRDoc is the best or the best if RWRDoc is not.

Table 1. Ad-hoc entity search results. Model indicates task types of queries, and top-k indicates the selected k values (10 or 100). Each cell contains an NDCG value for corresponding condition. For each column, the best score is boldface and underlined, and the proposed method has residual from the best if it is not the best or the second best if it is.

Model	SemSearch ES		INEX-LD		ListSearch		QALD-2		Total	
top-k	@10	@100	@10	@100	@10	@100	@10	@100	@10	@100
BM25	0.2497	0.4110	0.1828	0.3612	0.0627	0.3302	0.2751	0.3366	0.2558	0.3582
PRMS	0.5340	0.6108	0.3590	0.4295	0.3684	0.4436	0.3151	0.4026	0.3905	0.4688
MLM-all	0.5528	0.6247	0.3752	0.4493	0.3712	0.4577	0.3249	0.4208	0.4021	0.4852
LM	0.5555	0.6475	0.3999	0.4745	0.3925	0.4723	0.3412	0.4338	0.4182	0.5036
SDM	0.5535	0.6672	0.4030	0.4911	0.3961	0.4900	0.3390	0.4274	0.4185	0.5143
LM + ELR	0.5554	0.6469	0.4040	0.4816	0.3992	0.4845	0.3491	0.4383	0.4230	0.5093
SDM + ELR	0.5548	0.6680	0.4104	0.4988	0.4123	0.4992	0.3446	0.4363	0.4261	0.5211
MLM-CA	0.6247	0.6854	0.4029	0.4796	0.4021	0.4786	0.3365	0.4301	0.4365	0.5143
BM25-CA	0.5858	0.6883	0.4120	0.5050	0.4220	0.5142	0.3566	0.4426	0.4399	0.5329
FSDM	0.6521	0.7220	0.4214	0.5043	0.4196	0.4952	0.3401	0.4358	0.4524	0.5342
BM25F-CA	0.6281	0.7200	**0.4394**	**0.5296**	**0.4252**	0.5106	**0.3689**	0.4614	**0.4605**	0.5505
FSDM+ELR	**0.6563**	**0.7257**	0.4354	0.5134	0.4220	0.4985	0.3468	0.4456	0.4590	0.5408
RWRDoc	0.5877	0.7215	0.4189	**0.5296**	0.4119	**0.5845**	0.3346	**0.5163**	0.4348	**0.5643**
Residual	-6.86%	-0.42%	-2.05%	0%	-1.33%	$+7.03\%$	-3.43%	$+5.49\%$	-2.57%	$+1.38\%$

The table indicates that RWRDoc performs the best in the total performance for top-100 ranking, however, earlier rankings (i.e., top-10) are 2.57% worse on average than the second best. This indicates that RWRDoc brings up relevant entities from out of top-100 to top-100, therefore, top-100 ranking results by RWRDoc have more relevant entities than others. Consequently, RWRDoc increase recall but lack of ranking capability.

Finding 1. RWR-based entity representation learning is effective to collect relevant terms for each entity from surrounding entities. However, in order to obtain higher ranking quality, similarity computations and ranking functions should take more sophisticated approaches.

3.2   Accuracy on Recommender Systems

Linked Data is expected to be auxiliary information to improve recommender systems [2,13]. Linked Data provides semantic relationships between entities such as music artists in a similar genre. Semantic relationships can be a help to estimate users' preferences which do not appear on rating information.

Basic idea of existing works [2,13] is that users prefer entity e_1 if they like another entity e_2 which is semantically similar to e_1. For this experiment, one baseline (TFIDF) and two representative methods (PPR [13] and PLDSD [2]) are selected as competitors. *TFIDF* models each entity as a minimal representation (Definition 2) and calculates semantic similarities between entities by cosine similarity between representations. *PPR* measures semantic similarities between entities by personalized PageRank. In particular, PPR first calculates personalized PageRank vector for each entity, then calculates cosine similarity between vectors of entities as semantic similarity. Note that dumping factor of PPR is set to the same value as RWRDoc for fair comparison. *PLDSD* measures semantic similarities by heuristic measurements based on commonalities of neighbours. PLDSD is an extension from LDSD [16] which measures semantic similarities by commonalities of neighbours, PLDSD extends LDSD by propagating scores in neighbouring entities.

In order to incorporate RWRDoc into recommender systems, learned representations are used for measuring semantic similarities between entities. Specifically, for each pair of entities, semantic similarity of them is calculated by cosine similarity of their representations.

This experiment examines whether entity representations by RWRDoc can measure semantic similarities of entities by applying to a recommendation task. This paper utilizes the HetRec 2011 dataset[6] which includes users' listening list of artists on Last.FM. In order to incorporate Linked Data, this experiment uses a mapping[7] [15] of artists to DBpedia entities. Since recommender system is typically modeled as ranking problem, this experiment evaluates RWRDoc and the baseline methods by ranking measurement NDCG (Eq. 2).

Figure 2 displays the evaluation result of recommender systems. The figure represents NDCG for top-k recommended artists by the comparing methods. Lines are corresponding with average NDCG scores of the methods. Dotted line indicates PPR, dashed line indicates PLDSD, dash-dot line indicates TFIDF,

---

[6] https://grouplens.org/datasets/hetrec-2011/.
[7] http://sisinflab.poliba.it/semanticweb/lod/recsys/datasets/.

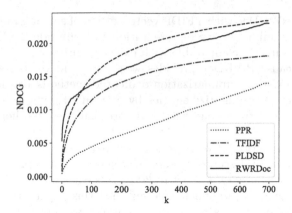

Fig. 2. Recommendation result. Lines represent average NDCG at k: dotted line indicates personalized PageRank (PPR), dashed line indicates PLDSD, dash-dot line indicates TFIDF, and solid line indicates the proposed method (RWRDoc). RWRDoc is superior to PPR and TFIDF and comparable with PLDSD. In the earlier items in the list, RWRDoc have higher quality but, in the later items, PLDSD have higher quality.

and solid line indicates the proposed method (RWRDoc). RWRDoc is, on average, superior to PPR and TFIDF and comparable with PLDSD.

The figure indicates that RWRDoc is superior to TFIDF and PPR and comparable with PLDSD. This results mean that RWRDoc provides richer semantic representations of entities than TFIDF and PPR, and the representations contribute to increase recommendation quality. While, RWRDoc is comparable with PLDSD, for the earlier recommend items, RWRDoc have more relevant items than PLDSD but for the later items, PLDSD have more relevant items. This indicates that semantic similarities based on RWRDoc entity representation is not always better than PLDSD which calculates semantic similarities by fully utilizing semantic information on Linked Data such as labels of predicates. Therefore, RWRDoc still leaves space to improving representation or similarity computation method for incorporating semantic information into account.

Finding 2. RWR-based representation learning is better performing than both of text-only representation (i.e., TFIDF) and topology-only representation (i.e., PPR). This ensures that RWR-based representation learning provides richer entity representations. On the other hand, in terms of similarity and ranking capability, RWR-based representation leaves space to improve.

3.3 Qualitative Evaluation on Entity Summarization

Entity summarization [4,7,23] is a task to describe entities in a human-readable format. Successful summary of an entity is that human judges can determine what the entity is from the summary.

This experiment attempts to show interpretability of representations which are expected to have richer vocabularies than naïve method. To show this, this paper compares RWRDoc with TFIDF vectorization of surrounding texts (which is identical with minimal entity representation in Definition 2). Unfortunately, RWRDoc is not directly comparable with existing entity summarization methods [4,7,23], because RWRDoc provides weighted term vectors as representations while the existing summarization-dedicated methods provide richer formats. These methods summarize entities by attributed texts which are derived from predicates and surrounding texts, and note that these methods have higher expressiveness than RWRDoc (to deal with such summarization of RWRDoc is a promising future direction). Consequently, this paper showcases, for each entity, a top-k list of terms in descending order of weights in the representation of the entity as its entity summary. k is set to 30 in this experiment.

To measure the goodness of entity summaries, this paper asks human judges whether terms in summaries are relevant enough to determine what are the entities. In this experiments, five voluntary human judges who are four males and one female, are in 22 to 25 y.o., and are majoring computer sciences in master courses. Every summary is checked by three judges and terms which are judges as relevant by two or more judges are regarded as relevant to the entity. Based on the judgements, RWRDoc-based summary and a baseline are evaluated in terms of precision@k (Eq. 3) which evaluates how many relevant terms are in a top-k list.

$$Precision@k = \frac{|\{relevant\ items\ in\ k\}|}{k} \tag{3}$$

Figure 3(a) showcases evaluation result of entity summarization. Lines indicates average precision@k for the comparing methods (solid line represents RWRDoc and dashed line represents TFIDF) and error bars indicate standard deviations. The figure indicates that RWRDoc achieves significantly better accuracy than TFIDF, especially in terms with high scores.

The reason why RWRDoc is superior to TFIDF is that relevant terms but not included in the minimal representations are at the top of the summaries by RWRDoc. This means that minimal representations of closer entities include descriptive facts related to the entity. Therefore, the number of relevant terms in each entity summary by RWRDoc is larger than that by TFIDF. To ensure this, Fig. 3(b) displays the average number of relevant terms in summaries with error bars for standard deviations. As expected, the number of relevant terms in summaries is larger for RWRDoc. Therefore, RWRDoc summaries entities with larger vocabularies.

To show differences of summaries by RWRDoc with those by TFIDF, Table 2 shows two examples of top-10 terms in RWRDoc documentations and TFIDF representations. Here, two examples are selected: one is *Hideyoshi Toyotomi* (see footnote 1) and the other is *Nagoya city, Japan*[8]. Table 2(a) is the top-10 term list of the former and Table 2(b) is that of the latter. The tables include relevance judgements beside the terms in *Rel.* columns, and shaded terms are only appearing either top-30 term lists of RWRDoc or TFIDF. Since RWRDoc

---

[8] http://dbpedia.org/resource/Nagoya.

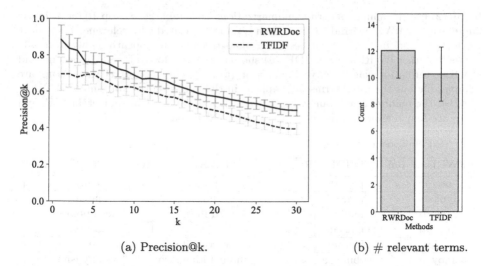

(a) Precision@k.

(b) # relevant terms.

Fig. 3. Entity summarization results, comparison between the proposed method (RWRDoc) and the baseline method (TFIDF). (a) average (lines) and standard deviations (error bars) of scores of top-k terms in summaries. (b) average (bars) and standard deviations (error bars) of the numbers of relevant terms. RWRDoc performs better than TFIDF and provide more relevant terms than TFIDF.

incorporates not only representations of surrounding entities but also those of further entities, entity representations by RWRDoc hold terms not in term lists in TFIDF. For Table 2(a), the numbers of relevant terms are comparable but the top-2 terms only appear in the entity representation of RWRDoc. For Table 2(b), the number of relevant terms of RWRDoc is larger than that of TFIDF, and there are four relevant terms only appearing in RWRDoc.

RWRDoc entity representations in Table 2 include relevant facts which are not described in the 1-hop neighouring texts. For the first example, Hideyoshi Toyotomi was a samurai in the Sengoku period in Japan and he stayed at the Momoyama castle. Table 2(a) indicates that both RWRDoc and TFIDF include the fact which is explained in his description of DBpedia. RWRDoc representation includes another fact which is not included in the TFIDF representation, that is, he launches the invasions of the Joseon dynasty. This is not directly written in his description of DBpedia but written in the relevant DBpedia entity (see footnote 2). The latter example, Nagoya city, is a city located in Aichi prefecture in Chubu region in Japan. In addition to the fact, RWRDoc documentation in Table 2(b) includes terms related to *Chunichi Doragons* which is a Japanese professional baseball team based in Nagoya, which mascot character is called *Doala*.

The results of this experiment indicate that RWRDoc successfully incorporates representations of reachable entities not only surrounding entities. The number of relevant vocabularies increases two or more within 30-term summaries

Table 2. Result samples of entity summarization. Each table shows top-10 terms in the summaries by RWRDoc and TFIDF. Each term is associated with relevance judgement (✓ for relevant) in *Rel.* column beside it. Shaded terms are appearing only in top-30 terms by either RWRDoc or TFIDF. (a) showcases terms for Hideyoshi Toyotomi and (b) lists terms for Nagoya city, Japan. For (a), the numbers of relevant terms are comparable but the top-2 terms only appear in the entity representation of RWRDoc. For (b), the number of relevant terms of RWRDoc is larger than that of TFIDF, and there are four relevant terms only appearing in RWRDoc.

(a) Hideyoshi Toyotomi

RWRDoc	Rel.	TFIDF	Rel.
joseon	✓	period	
dynasty	✓	samurai	✓
period		unifier	✓
samurai	✓	momoyama	✓
unifier	✓	ieyasu	✓
momoyama	✓	nobunaga	✓
ieyasu	✓	daimyo	✓
nobunaga	✓	liege	✓
daimyo	✓	sengoku	✓
liege	✓	legacies	

(b) Nagoya

RWRDoc	Rel.	TFIDF	Rel.
japan	✓	chky	
chky		japan	✓
chunichi	✓	metropolitan	✓
wii		largest	
metropolitan	✓	area	
chunichidragonzu	✓	kitakyushu	
doala	✓	chubu	✓
chunichi	✓	city	✓
region		honshu	✓
city	✓	aichi	✓

than TFIDF. As the number of relevant terms increases, RWRDoc achieves more appropriate summaries than TFIDF.

Finding 3. Incorporating reachable minimal representations of reachable entities increases the chance to include relevant facts into the representaitons of entities. RWR helps to give terms in relevant facts higher weights.

3.4   Remarks: Pros and Cons

Pros: RWRDoc successfully incorporates related facts for entities into entity representations by integrating minimal entity representations in terms of a graph proximity measurement, RWR. Entity representations by RWRDoc are richer representations, therefore, recall of ad-hoc entity search, accuracy of recommendation task, and quality of entity summarization are (not always significant but) better than baselines.

Cons: RWRDoc fails to incorporate relationship information between entities, since RWRDoc does not take predicate labels into account for representation learning. This is the main reason that RWRDoc cannot clearly outperform PLDSD in recommendation tasks. These experimental facts indicate that RWRDoc should take semantic relationships between entities into consideration. For similarity computations and ranking capabilities, RWRDoc seems to be not sufficient as shown in ad-hoc entity search task.

4   Related Work

Entity documentation in this paper is equivalent to representation learning of entities on Linked Data. Representation learning is a large research area ranging from vector space modeling, to deep learning based representation learning (a.k.a. graph and word embedding). Vector space modeling [14,21] is a major representation learning in ad-hoc entity search. For more complicated tasks such as question answering, more modern approach [5] employs deep learning technique to learn representations of entities.

4.1   Vector Space Model-Based Approaches

Vector space model-based representation learning is inspired from information retrieval techniques. TFIDF vectorization in Sect. 2 is one of vector space modeling. In attributed documents domain, fielded extension is an effective method, which can differentiate importances of attributes (for example, in Web page vectorization, words in *title* are more important than those in *body*). Fielded extension of entity representation is also studied [14]. Kotov [11] has provided a good overview of existing entity representations and entity retrieval models.

Existing vector space model-based approaches are reasonable, but they suffer from determination of importances of attributes (i.e., predicates in Linked Data). Fielded extension is known to outperform basic vector space modeling, but in order to apply fielded extension version of vector space modeling, the importances of predicates must be determined in advance. However, in Linked Data, determining importances of predicates is troublesome, because there are large number of predicates in Linked Data [10].

4.2   Deep Learning-Based Approaches

As deep learning techniques become popular, they are applied for various applications, in particular to Linked Data, network embedding [6,17] is an application of deep learning techniques. Network embedding is to vectorize vertices in a network based on topology of the network. Network embedding is a powerful technique that it achieves higher performance in various applications such as link prediction and vertex classification. Afterward, extending researches [12,25] have been including textual attributive information of vertices into network embedding. This extension enriches network embeddings more semantically meaningful.

Although deep learning-based techniques are powerful, there are two major drawbacks; one is human-understandability of learnt representations and computational costs. The embedded space is a latent space, therefore, dimensions of the space are not human understandable. Thus, learnt representations of entities are indeed not human understandable. Deep learning-based approach for RDF [20] is not exceptional to this, that is, it lacks the understandability of learned entities.

4.3   Advantages of RWRDoc

One of the most important feature of RWRDoc is **parameter-free** learning algorithm. It incorporates all reachable entities with respect to RWR scores, therefore, it does not suffer from the problem of setting different importances on predicates. Experimental evaluations in Sect. 3 show that RWRDoc is superior or comparable with fully-tuned heuristic vector space modeling approaches.

RWRDoc does not suffer from drawbacks on Sect. 4.2. Documentation of RWRDoc is human understandable because features are terms occurring in any description of entities. Furthermore, weights for terms in documentations properly indicate the relevancy of the terms to the entities, therefore, as shown in Sect. 3.3, the documentations can still work as summaries of entities. Moreover, the documentation algorithm of RWRDoc include RWR computation and TFIDF computation. The larger the number of vertices on Linked Data, the larger computation cost is required for RWRDoc, however, the cost is still not as large as that of deep learning algorithms.

5   Conclusion and Future Direction

This paper proposes RWRDoc, a simple and parameter-free entity documentation method. It combines representations of reachable entities in a linear combination manner. It employs random walk with restart (RWR) as a weighting method, because RWR frees parameter settings for weighting schemes. Since RWRDoc is a general purpose entity documentation method, experimental evaluation showcases its generality as well as pros and cons. Due to its rich representation of RWRDoc, it can perform well on various tasks comparing with the reasonable baselines. However, RWRDoc is still not significantly superior to the state-of-the-art on several tasks, since the state-of-the-art incorporate richer contents (e.g., predicate types) into account. This indicates that taking full advantage of Linked Data is the future direction of RWRDoc. A possible direction is that RWR can be performed on an ObjectRank manner [10] which differentiates transitivity probabilities on predicates for random walk.

Acknowledgments. This work was partly supported by JSPS KAKENHI Grant Number JP18K18056.

References

1. Resource Description Framework (RDF): Concepts and Abstract Syntax. https://www.w3.org/TR/rdf11-concepts/
2. Alfarhood, S., Labille, K., Gauch, S.: PLDSD: propagated linked data semantic distance. In: WETICE 2017, pp. 278–283 (2017)
3. Bizer, C., Heath, T., Berners-Lee, T.: Linked data - the story so far. Int. J. Semant. Web Inf. Syst. **5**(3), 1–22 (2009)

4. Cheng, G., Tran, T., Qu, Y.: RELIN: relatedness and informativeness-based centrality for entity summarization. In: Aroyo, L., et al. (eds.) ISWC 2011. LNCS, vol. 7031, pp. 114–129. Springer, Heidelberg (2011). https://doi.org/10.1007/978-3-642-25073-6_8

5. Shijia, E., Xiang, Y.: Entity search based on the representation learning model with different embedding strategies. IEEE Access **5**, 15174–15183 (2017)

6. Grover, A., Leskovec, J.: node2vec: scalable feature learning for networks. In: SIGKDD 2016, pp. 855–864 (2016)

7. Gunaratna, K., Thirunarayan, K., Sheth, A.P.: FACES: diversity-aware entity summarization using incremental hierarchical conceptual clustering. In: AAAI 2015, pp. 116–122 (2015)

8. Hasibi, F., et al.: DBpedia-entity v2: a test collection for entity search. In: SIGIR 2017, pp. 1265–1268 (2017)

9. Järvelin, K., Kekäläinen, J.: Cumulated gain-based evaluation of IR techniques. ACM Trans. Inf. Syst. **20**(4), 422–446 (2002)

10. Komamizu, T., Okumura, S., Amagasa, T., Kitagawa, H.: FORK: feedback-aware ObjectRank-based keyword search over linked data. In: Sung, W.K., et al. (eds.) AIRS 2017. LNCS, vol. 10648, pp. 58–70. Springer, Cham (2017). https://doi.org/10.1007/978-3-319-70145-5_5

11. Kotov, A.: Knowledge graph entity representation and retrieval. In: Tutorial Chapter, RuSSIR 2016 (2016)

12. Li, J., Dani, H., Hu, X., Tang, J., Chang, Y., Liu, H.: Attributed network embedding for learning in a dynamic environment. In: CIKM 2017, pp. 387–396 (2017)

13. Nguyen, P., Tomeo, P., Noia, T.D., Sciascio, E.D.: An evaluation of SimRank and personalized PageRank to build a recommender system for the web of Data. In: WWW 2015, pp. 1477–1482 (2015)

14. Nikolaev, F., Kotov, A., Zhiltsov, N.: Parameterized fielded term dependence models for ad-hoc entity retrieval from knowledge graph. In: SIGIR 2016, pp. 435–444 (2016)

15. Noia, T.D., Ostuni, V.C., Tomeo, P., Sciascio, E.D.: SPrank: semantic path-based ranking for top-N recommendations using linked open data. ACM TIST **8**(1), 9:1–9:34 (2016)

16. Passant, A.: Measuring semantic distance on linking data and using it for resources recommendations. In: AAAI Spring Symposium 2010 (2010)

17. Perozzi, B., Al-Rfou, R., Skiena, S.: DeepWalk: online learning of social representations. In: SIGKDD 2014, pp. 701–710 (2014)

18. Pound, J., Mika, P., Zaragoza, H.: Ad-hoc object retrieval in the web of data. In: WWW 2010, pp. 771–780 (2010)

19. Raviv, H., Kurland, O., Carmel, D.: Document retrieval using entity-based language models. In: SIGIR 2016, pp. 65–74 (2016)

20. Ristoski, P., Paulheim, H.: RDF2Vec: RDF graph embeddings for data mining. In: Groth, P., et al. (eds.) ISWC 2016. LNCS, vol. 9981, pp. 498–514. Springer, Cham (2016). https://doi.org/10.1007/978-3-319-46523-4_30

21. Robertson, S.E., Zaragoza, H.: The probabilistic relevance framework: BM25 and beyond. Found. Trends Inf. Retrieval **3**(4), 333–389 (2009)

22. Sartori, E., Velegrakis, Y., Guerra, F.: Entity-based keyword search in web documents. Trans. Comput. Collect. Intell. **21**, 21–49 (2016)

23. Thalhammer, A., Lasierra, N., Rettinger, A.: LinkSUM: using link analysis to summarize entity data. In: Bozzon, A., Cudre-Maroux, P., Pautasso, C. (eds.) ICWE 2016. LNCS, vol. 9671, pp. 244–261. Springer, Cham (2016). https://doi.org/10.1007/978-3-319-38791-8_14

24. Tong, H., Faloutsos, C., Pan, J.: Random walk with restart: fast solutions and applications. Knowl. Inf. Syst. **14**(3), 327–346 (2008)
25. Yang, C., Liu, Z., Zhao, D., Sun, M., Chang, E.Y.: Network representation learning with rich text information. In: IJCAI 2015, pp. 2111–2117 (2015)
26. Yoon, M., Jung, J., Kang, U.: TPA: two phase approximation for random walk with restart. CoRR abs/1708.02574 (2017). http://arxiv.org/abs/1708.02574

GARUM: A Semantic Similarity Measure Based on Machine Learning and Entity Characteristics

Ignacio Traverso-Ribón[1(✉)] and Maria-Esther Vidal[2,3]

[1] University of Cadiz, Cádiz, Spain
ignacio.traverso@uca.es
[2] L3S Research Center, Hanover, Germany
[3] TIB Leibniz Information Center for Science and Technology,
Hanover, Germany
maria.vidal@tib.eu

Abstract. Knowledge graphs encode semantics that describes entities in terms of several *characteristics*, e.g., attributes, neighbors, class hierarchies, or association degrees. Several *data-driven* tasks, e.g., ranking, clustering, or link discovery, require for determining the relatedness between knowledge graph entities. However, state-of-the-art similarity measures may not consider all the characteristics of an entity to determine entity relatedness. We address the problem of similarity assessment between knowledge graph entities and devise GARUM, a semantic similarity measure for knowledge graphs. GARUM relies on similarities of entity characteristics and computes similarity values considering simultaneously several entity characteristics. This combination can be manually or automatically defined with the help of a machine learning approach. We empirically evaluate the accuracy of GARUM on knowledge graphs from different domains, e.g., networks of proteins and media news. In the experimental study, GARUM exhibits higher correlation with gold standards than studied existing approaches. Thus, these results suggest that similarity measures should not consider *entity characteristics* in isolation; contrary, combinations of these characteristics are required to precisely determine relatedness among entities in a knowledge graph. Further, the combination functions found by a machine learning approach outperform the results obtained by the manually defined aggregation functions.

1 Introduction

Semantic Web and Linked Data communities foster the publication of large volumes of data in the form of semantically annotated knowledge graphs. For example, knowledge graphs like DBpedia[1], Wikidata or Yago[2], represent general domain concepts such as musicians, actors, or sports, using RDF vocabularies.

---

[1] http://dbpedia.org.
[2] http://yago-knowledge.org.

© Springer Nature Switzerland AG 2018
S. Hartmann et al. (Eds.): DEXA 2018, LNCS 11029, pp. 169–183, 2018.
https://doi.org/10.1007/978-3-319-98809-2_11

Additionally, domain specific communities like Life Sciences and the financial domain, have also enthusiastically supported the collaborative development of diverse ontologies and semantic vocabularies to enhance the description of knowledge graph entities and reduce the ambiguity in such descriptions, e.g., the Gene Ontology (GO) [2], the Human Phenotype Ontology (HPO) [10], or the Financial Industry Business Ontology (FIBO)[3]. Knowledge graphs encode semantics that describe entities in terms of several *entity characteristics*, e.g., class hierarchies, neighbors, attributes, and association degrees. During the last years, several semantic similarity measures for knowledge graph entities have been proposed, e.g., GBSS [15], HeteSim [22], and PathSim [24]. However, these measures do not consider all the *entity characteristics* represented in a knowledge graph at the same time in a aggregated fashion. The importance of precisely determining relatedness in data-driven tasks, e.g., knowledge discovery, and the increasing size of existing knowledge graphs, introduce the challenge of defining semantic similarity measures able to exploit all the information described in knowledge graphs, i.e., all the *characteristics* of the represented entities.

We present GARUM, a G̲rA̲ph entity R̲egression s̲U̲pported similarity M̲easure. GARUM exploits knowledge encoded in *characteristics* of an entity, i.e., ancestors or *hierarchies*, neighborhoods, associations, or *shared information*, and literals or *attributes*. GARUM receives a knowledge graph and two entities to be compared. As a result, GARUM returns a similarity value that aggregates similarity values computed based on the different *entity characteristics*; a domain-dependent aggregation function α combines similarity values specific for each *entity characteristic*. The function α can be either manually defined or predicted by a regression machine learning approach. The intuition is that knowledge represented in *entity characteristics*, precisely describes entities and allows for determining more accurate similarity values.

We conduct an empirical study with the aim of analyzing the impact of considering *entity characteristics* in the accuracy of a similarity measure over a knowledge graph. GARUM is evaluated over entities of three different knowledge graphs: The first knowledge graph describes news articles annotated with DBpedia entities; and the other two graphs describe proteins annotated with the Gene Ontology. GARUM is compared with state-of-the-art similarity measures with the goal of determining if GARUM similarity values are more correlated to the gold standards. Our experimental results suggest that: (*i*) Considering all *entity characteristics* allow for computing more accurate similarity values; (*ii*) GARUM is able to outperform state-of-art approaches obtaining higher values of correlation; and (*iii*) Machine learning approaches are able to predict aggregation functions that outperform the manually functions defined by humans.

The remainder of this article is structured as follows: Sect. 2 motivates our approach using a subgraph from DBpedia. Section 3 describes GARUM and Sect. 4 summarizes experimental results. Related work is presented in Sect. 5, and finally, Sect. 6 concludes and give insights for future work.

---

[3] https://www.w3.org/community/fibo/.

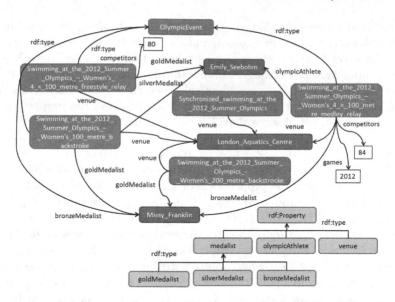

Fig. 1. Motivating Example. Two subgraphs from DBpedia. The above graph describes swimming events and entities related to these events, while the other graph represents a hierarchy of the properties in DBpedia.

2 Motivating Example

We motivate our work with a real-world knowledge graph extracted from DBpedia (Fig. 1); it describes swimming events in olympic games. Each event is related to other entities, e.g., athletes, locations, or years, using different relations or RDF *properties*, e.g., *goldMedalist* or *venue*. These RDF properties are also described in terms of the RDF property rdf:type as depicted in Fig. 1. Relatedness between entities is determined based on different *entity characteristics*, i.e., class hierarchy, neighbors, shared associations, and properties.

Consider entities *Swimming at the 2012 Summer Olympics - Women's 100 m backstroke, Swimming at the 2012 Summer Olympics - Women's 4x100 m freestyle relay*, and *Swimming at the 2012 Summer Olympics - Women's 4x100 m medley relay*. For the sake of clarity we rename them as *Women's 100 m backstroke, Women's 4x100 m freestyle*, and *Women's 4x100 m medley relay*, respectively. The entity hierarchy is induced by the *rdf:type* property, which describes an entity as instance of an RDF class. Particularly, these swimming events are described as instances of the *OlympicEvent* class, which is at the fifth level of depth in the DBpedia ontology hierarchy. Thus, based on the knowledge encoded in this hierarchy, these entities are highly similar. Additionally, these entities share exactly the same set of neighbors that is formed by the entities *Emily Seebohm, Missy Franklin*, and *London Aquatic Centre*. However, the relations with *Emily Seebohm* and *Missy Franklin* are different. *Women's 4x100 m freestyle* and *Women's 100 m backstroke* are related with *Emily Seebohm* through properties *goldMedalist* and *silverMedalist*, respectively, and with *Missy Franklin* through properties *bronzeMedalist* and *goldMedalist*. Nevertheless, *Women's 4x100 m medley relay* is related with *Missy Franklin* through the property *bronzeMedalist*, and with *Emily Seebohm* through *olympicAthlete*. Considering only the entities in these neighborhoods, they are identical since they share exactly the same set of neighbors. However, whenever properties labels and the property hierarchy are considered, we observe that *Women's 4x100 m freestyle* and *Women's 100 m backstroke* are more similar since in both events *Missy Franklin* and *Emily Seebohm* are *medalists*, while in *Women's 4x100 m medley relay* only *Missy Franklin* is *medalist*. Furthermore, swimming events are also related with attributes through datatype properties. For the sake of clarity, we only include a portion of these attributes in Fig. 1. Considering these attributes, 84 athletes participated in *Women's 4x100 m medley relay*, while only 80 participated in *Women's 4x100 m freestyle*. Finally, the node degree or *shared information* is different for each entity in the graph. Entities with a high node degree are considered abstract entities, while others with low node degree are considered specific. For instance, in Fig. 1, the entity *London Aquatic Centre* has five incident edges, while *Emily Seebohm* has four edges and *Missy Franklin* has only three incident edges. Thus, the entity *London Aquatic Centre* is less specific than *Emily Seebohm*, which is also less specific than *Missy Franklin*.

According to these observations, the similarity between two knowledge graph entities cannot be estimated only considering one *entity characteristic*. Hence, combinations of them may have to be taken into account to precisely determine relatedness between entities in a knowledge graph.

3   Our Approach: GARUM

We propose GARUM, a semantic similarity measure for determining relatedness between entities represented in knowledge graphs. GARUM considers the knowledge encoded in *entity characteristics*, e.g., hierarchies, neighborhoods, shared information, and attributes to accurately compute similarity values between entities in a knowledge graph. GARUM calculates values of similarity for each *entity characteristic* independently and combines these values to produce an aggregated similarity value between the compared entities. Figure 2 depicts the GARUM architecture. GARUM receives as input a knowledge graph G and two entities e_1, e_2 to be compared. *Entity characteristics* of the compared entities are extracted from the knowledge graph and compared as isolated elements.

Definition 1. *Knowledge graph. Given a set of entities V, a set of edges E, and a set of property labels L, a knowledge graph G is defined as $G = (V, E, L)$. An edge corresponds to a triple (v_1, r, v_2), where $v_1, v_2 \in V$ are entities in the graph, and $r \in L$ is a property label.*

Definition 2. *Individual similarity measure. Given a knowledge graph $G = (V, E, L)$, two entities e_1 and e_2 in V, and an entity characteristic $\mathcal{EC}$ of e_1 and e_2 in G, an individual similarity measure $Sim_{\mathcal{EC}}(e_1, e_2)$ corresponds to a similarity function defined in terms of $\mathcal{EC}$ for e_1 and e_2.*

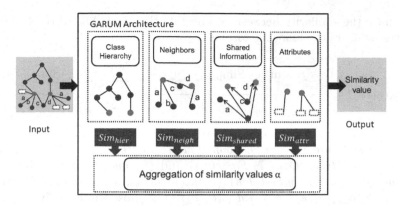

Fig. 2. The GARUM Architecture. GARUM receives a knowledge graph G and two entities to be compared (red nodes). Based on semantics encoded in the knowledge graph (blue nodes), GARUM computes similarity values in terms of class hierarchies, neighborhoods, shared information and the attributes of the input entities. Generated similarity values, $Sim_{hier}, Sim_{neigh}, Sim_{shared}, Sim_{attr}$, are combined using a function α. The aggregated value is returned as output. (Color figure online)

The hierarchical similarity $Sim_{hier}(e_1, e_2)$ or the neighborhood similarity $Sim_{neigh}(e_1, e_2)$ are examples of individual similarity measures. These individual similarity measures are combined using an aggregation function α. Next, we describe the four considered individual similarity measures.

Hierarchical Similarity: Given a knowledge graph G, a hierarchy is induced by a set of hierarchical edges $HE = \{(v_i, r, v_j) | (v_i, r, v_j) \in E \wedge \text{Hierarchical}(r)\}$. HE is a subset of edges in the knowledge graph whose property labels refer to a hierarchical relation, e.g., *rdf:type*, *rdfs:subClassOf*, or *skos:broader*. Generally, every relation that presents an entity as a generalization (ancestor) or an specification (successor) of another entity is a hierarchical relation. GARUM relies on existing hierarchical distance measures, e.g., d_{tax} [1] and d_{ps} [16] to determine the hierarchical similarity between entities; it is defined as follows:

$$Sim_{hier}(e_1, e_2) = \begin{cases} 1 - d_{tax}(e_1, e_2) \\ 1 - d_{ps}(e_1, e_2) \end{cases} \quad (1)$$

Neighborhood Similarity: The neighborhood of an entity $e \in V$ is defined as the set of relation-entity pairs $N(e)$ whose entities are at one-hop distance of e, i.e., $N(e) = \{(r, e_i) | (e, r, e_i) \in E\}$. With this definition of neighborhood, we can consider the neighbor entity and the relation type of the edge at the same time. GARUM uses the knowledge encoded in the relation and class hierarchies of the knowledge graph to compare two pairs $p_1 = (r_1, e_1)$ and $p_2 = (r_2, e_2)$. The similarity between two pairs p_1 and p_2 is computed as $Sim_{pair}(p_1, p_2) = Sim_{hier}(e_1, e_2) \cdot Sim_{hier}(r_1, r_2)$. Note that Sim_{hier} can be used with any entity of the knowledge graph, regardless of it is an instance, a class or a relation. In order to maximize the similarity between two neighborhoods, GARUM combines pair comparisons using the following formula:

$$\text{Sim}_{\text{neigh}}(e_1, e_2) = \frac{\displaystyle\sum_{i=0}^{|N(e_1)|} \max_{p_x \in N(e_2)} \text{Sim}_{\text{pair}}(p_i, p_x) + \sum_{j=0}^{|N(e_2)|} \max_{p_y \in N(e_1)} \text{Sim}_{\text{pair}}(p_j, p_y)}{|N(e_1)| + |N(e_2)|}$$

(2)

In Fig. 1, the neighborhoods of .*Women's 100 m backstroke* and *Women's 4x100 m freestyle* are {(*venue, London Aquatic Centre*), (*silverMedalist, Emily Seebohm*), (*goldMedalist, Missy Franklin*)} and {(*venue, London Aquatic Centre*), (*goldMedalist, Emily Seebohm*), (*bronzeMedalist, Missy Franklin*)}, respectively. Let $\text{Sim}_{\text{hier}}(e_1, e_2) = 1 - d_{\text{tax}}(e_1, e_2)$. The most similar pair to (*venue, London Aquatic Centre*) is itself and with similarity value of 1.0. The most similar pair to (*silverMedalist, Emily Seebohm*) is (*goldMedalist, Emily Seebohm*) with a similarity value of 0.5. This similarity value is result of the product between Sim_{hier}(*Emily Seebohm, Emily Seebohm*), whose result is 1.0, and Sim_{hier}(*goldMedalist, silverMedalist*), whose result is 0.5. Similarly, the most similar pair to (*goldMedalist, Missy Franklin*) is (*bronzeMedalist, Missy Franklin*) with a similarity value of 0.5. Thus, the similarity between neighborhoods of *Women's 100 m backstroke* and *Women's 4x100 m freestyle* is computed as $\text{Sim}_{\text{neigh}} = \frac{(1+0.5+0.5)+(1+0.5+0.5)}{3+3} = \frac{4}{6} = 0.667$.

Shared Information: Beyond the hierarchical similarity, the amount of information shared by two entities in a knowledge graph can be measured examining the human use of such entities. Two entities are considered to share information whenever they are used in a corpus similarly. Considering the knowledge graph as a corpus, the information shared by two entities x and y is directly proportional to the amount of entities that have x and y together in their neighborhood, i.e., the co-occurrences of x and y in the neighborhoods of the entities in the knowledge graph. Let $G = (V, E, L)$ be a knowledge graph and $e \in V$ an entity in the knowledge graph. The set of entities that have e in their neighborhood is defined as $Incident(e) = \{e_i | (e_i, r, e) \in E\}$. Then, GARUM computes the information shared by two entities using the following formula:

$$\textbf{Sim}_{\textbf{shared}}(e_1, e_2) = \frac{|\text{Incident}(e_1) \cap \text{Incident}(e_2)|}{|\text{Incident}(e_1) \cup \text{Incident}(e_2)|},$$

(3)

The values depends on how much informative or specific are the compared entities. For example, an entity representing *London Aquatic Centre* is included in several neighborhoods in a knowledge graph like DBpedia. This means that *London Aquatic Centre* is not a specific entity. This is reflected in the denominator of $\text{Sim}_{\text{shared}}$. Thus, abstract or non-specific entities require a greater amount of co-occurrences in order to obtain a high value of similarity. In Fig. 1, entities *Emily Seebohm, Missy Franklin,* and *London Aquatic Centre* have incident edges. *London Aquatic Centre* have five incident edges, while *Emily Seebohm* and *Missy Franklin* have four and three, respectively. *Emily Seebohm* and *Missy Franklin* co-occurs in three neighborhoods. Thus, $\text{Sim}_{\text{shared}}$ returns a value of $\frac{3}{4} = 0.75$.

London Aquatic Centre is included in five neighborhoods in sub-graph showed in Fig. 1. However, it is included in the neighborhood of each sport event located in this venue in the full graph of DBpedia.

Attributes: Entities in knowledge graphs are related with other entities and with attributes through datatype properties, e.g., temperature or protein sequence. GARUM considers only shared attributes, i.e., attributes connected to entities through the same datatype property. Given that attributes can be compared with domain similarity measures, e.g., SeqSim [23] for genes or Jaro-Winkler for strings, GARUM does not rely on a specific measure to compare attributes. Depending on the domain, users should choose a similarity measure for each type of attribute. Figure 1 depicts the entity representing *Women's 4x100 m medley relay*; it has attributes *competitors* and *games*, while *Women's 4x100 m freestyle* has only the attribute *competitors*. Thus, Sim_{attr} between these entities only considers the attribute *competitors*.

Aggregation Functions: GARUM combines four individual similarity measures and returns a similarity value that aggregates the relatedness among two compared entities. The aggregation function can be manually defined or computed by a supervised machine learning algorithm like a regression algorithm. A regression algorithm receives a set of input variables or predictors and an output or dependent variable. In the case of GARUM, the predictors are the individual similarity measures, i.e., Sim_{hier}, $\text{Sim}_{\text{neigh}}$, $\text{Sim}_{\text{shared}}$ and Sim_{attr}. The dependent variable is defined by a gold standard similarity measure, e.g., a crowd-funded similarity value. Thus, a regression algorithm produces as output a function $\alpha : X^n \rightarrow Y$, where X^n represents the predictors and Y corresponds to the dependent variable. Hence, GARUM is defined in terms of a function α:

$$\text{GARUM}(e_1, e_2) = \alpha(\text{Sim}_{\text{hier}}, \text{Sim}_{\text{neigh}}, \text{Sim}_{\text{shared}}, \text{Sim}_{\text{attr}}) \qquad (4)$$

Depending on the regression type, α can be a linear or a non-linear combination of the predictors. In both cases and regardless the used regression algorithm, α is computed by minimizing a loss function. In the case of GARUM, the loss function is the mean squared error (MSE) defined as follows:

$$\text{MSE} = \frac{1}{n} \sum_{i=1}^{n} (\hat{Y}_i - Y_i)^2, \qquad (5)$$

Y is a vector of n observed values, i.e., gold standard values, and $\hat{Y}$ is a vector of n predictions, i.e., $\hat{Y}$ corresponds to results of the computed function α. Hence, the regression algorithm implemented in GARUM learns from a training dataset how to combine the individual similarity measures by means of a function α, such that the MSE among the results produced by α and the corresponding gold standard (e.g., SeqSim, ECC) is minimized. However, gold standards are usually defined for annotation sets, i.e., sets of knowledge graph entities, instead of for pairs of knowledge graph entities. CESSM [18], and Lee50 [13] datasets are good examples of this phenomenon, where real world entities (proteins or texts) are

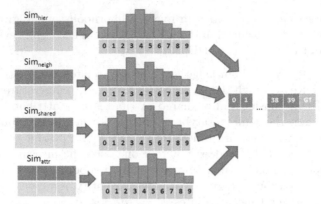

(a) Combination function for input matrices. For each matrix a 10-positions vector with the corresponding density value is generated. GT represents the ground truth.

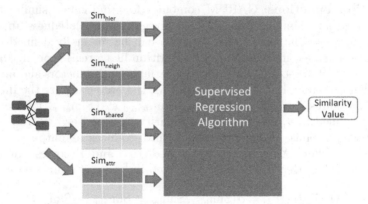

(b) Workflow of the supervised regression algorithm

Fig. 3. Training Phase of the GARUM Similarity Measure. (a) Training workflow using a regression algorithm; (b) Transformation of the input matrices into an aggregated value representing the combination of similarity measures

annotated with terms from ontologies, e.g., the Gene Ontology or the DBpedia ontology. Thus, the regression approach receives as input two sets of knowledge graph entities as showed in Fig. 3(b). Based on these sets, a similarity matrix for each individual similarity measure is computed. The output represents the aggregated similarity value computed by the estimated regression function α. Classical machine learning algorithms have a fix number of input features. However, the dimensions of the matrices depend on the cardinality of the compared sets. Hence, the matrices cannot be directly used, but a transformation to a fixed structure is required. Figure 3(a) introduces the matrix transformation. For each matrix, a density histogram with 10 bins is created. Thus, the input dimensions are fixed to $10 \times |$Individual similarity measures$|$. In Fig. 3(b), the input consists of an array with 40 features. Finally, the transformed data is used to train the regression algorithm. This algorithm learns, based on the input, how to combine the value of the histograms to minimize the MSE with respect to the ground truth (i.e., GT in Fig. 3(a)).

4 Experimental Results

We empirically evaluate the accuracy of GARUM in three different knowledge graphs. We compare GARUM with state-of-the-art approaches and measure the effectiveness comparing our results with available gold standards. For each knowledge graph, we provide a manually defined aggregation function α, as well as the results obtained using Support Vector Machines as supervised machine learning approach to compute the aggregation function automatically.

Research Questions: We aim at answering the following research questions: **(RQ1)** Does semantics encoded in *entity characteristics* improve the accuracy of similarity values between entities in a knowledge graph? **(RQ2)** Is GARUM able to outperform state-of-the-art similarity measures comparing knowledge graph entities from different domains?

Datasets. GARUM is evaluated on three knowledge graphs: Lee50[4], CESSM-2008[5], and CESSM-2014[6]. Lee50 is a knowledge graph defined by Paul et al. [15] that describes 50 news articles 8 (collected by Lee et al. [13]) with DBpedia entities. Each article has a length among 51 and 126 words, and is described on average with 10 DBpedia entities. The similarity value of each pair of news articles has been rated multiple times by humans. For each pair, we consider the average of human rates as gold standard. CESSM-2008 [18] (see footnote 5) and CESSM-2014 (see footnote 6) consist of proteins described in a knowledge graph with Gene Ontology (GO) entities. CESSM-2008 contains 13,430 pairs of proteins from UniProt with 1,039 distinct proteins, while the CESSM 2014 collection comprises 22,302 pairs with 1,559 distinct proteins. The knowledge graph of CESSM-2008 contains 1,908 distinct GO entities and the graph of 2014 includes 3,909 GO entities. The quality of the similarity measures is estimated by means the Pearson's coefficient with respect to three gold standards: SeqSim [23], Pfam [18], and ECC [5] (Table 1).

Implementation. GARUM is implemented in Java 1.8 and Python 2.7; as machine learning approaches, we used the support vector regression (SVR) implemented in the scikit-learn library[7] and a neural network of three layers implemented with the Keras[8] library, both in Python. The experimental study

---

[4] https://github.com/chrispau1/SemRelDocSearch/blob/master/data/Pincombe_ann otated_xLisa.json.
[5] http://xldb.di.fc.ul.pt/tools/cessm/index.php.
[6] http://xldb.fc.ul.pt/biotools/cessm2014/index.html.
[7] http://scikit-learn.org/stable/index.html.
[8] https://keras.io/.

Table 1. Properties of the knowledge graphs used during the evaluation.

Datasets	Comparisons	Ontology
CESSM 2008	13,430	Gene Ontology
CESSM 2014	22,302	Gene Ontology
Lee50	1,225	DBpedia

was executed on an Ubuntu 14.04 64 bits machine with CPU: Intel(R) Core(TM) i5-4300U 1.9 GHz (4 physical cores) and 8 GB RAM. To ensure the quality and correctness of the evaluation, both datasets are split following a 10-cross fold validation strategy. Apart from the machine learning based strategy, since entities (proteins and documents) are described with ontology terms from the Gene ontology or the DBpedia ontology, we manually define two aggregation strategies. Let $A \subseteq V$ and $B \subseteq V$ be set of knowledge graph entities. In the first aggregation strategy, we maximize the similarity value of sim(A, B) using the following formula:

$$sim(A, B) = \frac{\sum_{i=0}^{|A|} \max_{e_x \in B} \text{GARUM}(e_i, e_x) + \sum_{j=0}^{|B|} \max_{e_x \in A} \text{GARUM}(e_j, e_x)}{|A| + |B|}$$

In the second aggregation strategy, we perform a 1-1 maximum matching implemented with the Hungarian algorithm [11], such that each knowledge graph entity e_i in A is matched with one and only one knowledge graph entity e_j in B; the following formula of sim(A, B) is maximized:

$$sim(A, B) = \frac{2 \cdot \sum_{(e_i, e_j) \in \text{1-1 Matching}} \text{GARUM}(e_i, e_j)}{|A| + |B|}$$

The first aggregation strategy is used in knowledge graphs Lee50, while the 1-1 matching strategy is used in CESSM-2008 and CESSM-2014.

4.1   Lee50: News Articles Comparison

We compare pairwise the 50 news articles included in Lee50, and consider the knowledge encoded in the hierarchy, the neighbors, and the shared information. Knowledge encoded in attributes is not taken into account. Particularly, we define the aggregation function $\alpha(e_1, e_2)$ as follows:

$$\alpha(e_1, e_2) = \frac{\text{Sim}_{\text{hier}}(e_1, e_2) \cdot \text{Sim}_{\text{shared}}(e_1, e_2) + \text{Sim}_{\text{neigh}}(e_1, e_2)}{2} \tag{6}$$

where $\text{Sim}_{\text{hier}} = 1 - d_{\text{tax}}$.

Results in Table 2 suggest that GARUM outperforms the evaluated similarity measures in terms of correlation. Though d_{ps} obtains alone better results than

d_{tax}, its combination with the other two individual similarity measures delivers worse results. Further, we observe that the aggregation function obtained by the SVR and NN approaches outperforms the manually defined aggregation function.

Table 2. Comparison of Similarity Measures. Pearson's coefficient of similarity measures on the Lee et al. knowledge graph [13]; highest values in **bold**

Similarity measure	Pearson's coefficient
LSA [12]	0.696
SSA [7]	0.684
GED [20]	0.63
ESA [6]	0.656
d_{ps} [16]	0.692
d_{tax} [1]	0.652
GBSS$_{r=1}$ [15]	0.7
GBSS$_{r=2}$ [15]	0.714
GBSS$_{r=3}$ [15]	0.704
GARUM	**0.727**
GARUM SVR	**0.73**
GARUM NN	**0.74**

4.2 CESSM: Protein Comparison

CESSM knowledge graphs are used to compare proteins based on their associated GO annotations. GARUM considers the hierarchy, the neighborhoods, and the shared information as *entity characteristics*. In this knowledge graph, the different characteristics are combined automatically by SVR and with the following manually defined function:

$$\alpha(e_1, e_2) = \text{Sim}_{\text{hier}}(e_1, e_2) \cdot \text{Sim}_{\text{neigh}}(e_1, e_2) \cdot \text{Sim}_{\text{shared}}(e_1, e_2),$$

where $\text{Sim}_{\text{hier}} = 1 - d_{\text{tax}}$.

Table 3 reports on the correlation between state-of-the-art similarity measures and GARUM with the gold standards ECC, Pfam, and SeqSim on CESSM 2008 and 2014. The correlation is measured with the Pearson's coefficient. The top-5 values are highlighted in gray, and the highest correlation with respect to each gold standard is highlighted in bold. We observe that GARUM SVR and GARUM are the most correlated measures with respect to the three gold standard measures in both versions of the knowledge graph, 2008 and 2014. However, GARUM SVR obtains the highest correlation coefficient in CESSM 2008, while GARUM NN has the highest correlation coefficient for SeqSim in 2014[9].

---

[9] Due to the lack of training data GARUM could not be evaluated in CESSM 2014 with ECC and Pfam.

Table 3. Comparison of Similarity Measures. Pearson's correlation coefficient between three gold standards and eleven similarity measures of CESSM. The Top-5 correlations are highlighted in gray, and the highest correlation with respect to each gold standard is highlighted in *bold*. The similarity measures are: simUI (UI), simGIC (GI), Resnik's Average (RA), Resnik's Maximum (RM), Resnik's Best-Match Average (RB/RG), Lin's Average (LA), Lin's Maximum (LM), Lin's Best-Match Average (LB), Jiang & Conrath's Average (JA), Jiang & Conrath's Maximum (JM), Jiang & Conrath's Best-Match Average (JB). GARUM SVR and NN could not be executed for ECC and Pfam in CESSM 2014 due to lack of training data.

| Similarity | 2008 | | | 2014 | | |
measure	*SeqSim*	*ECC*	*Pfam*	*SeqSim*	*ECC*	*Pfam*
GI [17]	0.773	0.398	0.454	0.799	0.458	0.421
UI [17]	0.730	0.402	0.450	0.776	0.470	0.436
RA [19]	0.406	0.302	0.323	0.411	0.308	0.264
RM [21]	0.302	0.307	0.262	0.448	0.436	0.297
RB [3]	0.739	0.444	0.458	0.794	0.513	0.424
LA [14]	0.340	0.304	0.286	0.446	0.325	0.263
LM [21]	0.254	0.313	0.206	0.350	0.460	0.252
LB [3]	0.636	0.435	0.372	0.715	0.511	0.364
JA [8]	0.216	0.193	0.173	0.517	0.268	0.261
JM [21]	0.234	0.251	0.164	0.342	0.390	0.214
JB [3]	0.586	0.370	0.331	0.715	0.451	0.355
d_{tax} [1]	0.650	0.388	0.459	0.682	0.434	0.407
d_{ps} [16]	0.714	0.424	0.502	0.75	0.48	0.45
OnSim [26]	0.733	0.378	0.514	0.774	0.455	0.457
IC-OnSim [25]	0.779	0.443	0.539	0.81	0.513	0.489
GARUM	0.78	0.446	0.539	0.812	**0.515**	**0.49**
GARUM SVR	**0.86**	**0.7**	**0.7**	0.864	-	-
GARUM NN	0.85	0.6	0.696	**0.878**	-	-

5 Related Work

Several similarity measures have been proposed in the literature to determine the relatedness between knowledge graph entities; they exploit knowledge encoded in different *entity characteristics* in the knowledge graph including: hierarchies, length and amount of the paths among entities, or information content.

The measures d_{tax} [1] and d_{ps} [16] only consider hierarchies of a knowledge graph during the comparison of knowledge graph entities. These measures compute similarity values based on the relative distance of entities to their lowest common ancestor. Depending on the knowledge graph, different relation types may represent hierarchical relations. In OWL ontologies *owl:subClassOf* and *rdf:type* are considered the main hierarchical relations. However, in some knowledge graphs such as DBpedia [4], other relations like *dct:subject*, can be also regarded as hierarchical relations. PathSim [24] and HeteSim [22] among others consider only the neighbors during the computation of the similarity between two entities in a knowledge graph. They compute the similarity between two entities based on the number of existing paths between them. The similarity value is proportional to the number of paths between the compared entities. Unlike GARUM, PathSim and HeteSim do not distinguish between relation types and consider all relation types in the same manner, i.e., knowledge graphs are regarded as pairs $G = (V, E)$, where edges are not labeled. GBSS [15] considers two of the identified entity characteristics: the hierarchy and the neighbors. Unlike PathSim and HeteSim, GBSS distinguishes between hierarchical and *transversal* relations[10]; they also consider the length of the paths during the computation of the similarity. The similarity between two entities is directly proportional to the number of paths between these entities. Shorter paths have higher weight during the computation of the similarity. Unlike GARUM, GBSS does not take into account the property types that relate entities with their neighbors.

Information Content based similarity measures rely on specificity and hierarchical information [8,14,19]. These measures determine relatedness between two entities based on the Information Content of their lowest common ancestor. The Information Content is a measure to represent the generality or specificity of a certain entity in a dataset. The greater the usage frequency, the more general is the entity and lower is the respective Information Content value. Contrary to GARUM, these measures do not consider knowledge encoded in other *entity characteristics* like neighborhood. OnSim and IC-OnSim [25,26] compare ontology-based annotated entities. Though both measures rely on neighborhoods of entities and relation types, they require the execution of an OWL reasoner to obtain inferred axioms and their justifications. These justifications are taken into account for determining relatedness of two annotated entities. Thus, OnSim and IC-OnSim can be costly in terms of computational complexity. The worst case for the classification task with an OWL2 reasoner is 2NEXP-Time [9]. GARUM does not make use of justifications, which reduces significantly the execution time and allows for its use in non-OWL graphs.

6   Conclusions and Future Work

We define GARUM a new semantic similarity measure for entities in knowledge graphs. GARUM relies on knowledge encoded in *entity characteristics* to compute similarity values between entities and is able to determine automatically aggregation functions based on individual similarity measures and a supervised machine learning algorithm. Experimental results suggest that GARUM is able to outperform state-of-the-art similarity measures obtaining more accurate similarity values. Further, observed results show that the machine learning approach is able to find better combination functions than the manually defined functions.

In the future, we will evaluate the impact of GARUM in data-driven tasks like clustering or search and in to enhance knowledge graph quality, e.g., link discovery, knowledge graph integration, and association discovery.

---

[10] Transversal relations correspond to object properties in the knowledge graph.

Acknowledgements. This work has been partially funded by the EU H2020 Programme for the Project No. 727658 (IASIS).

References

1. Benik, J., Chang, C., Raschid, L., Vidal, M.-E., Palma, G., Thor, A.: Finding cross genome patterns in annotation graphs. In: Bodenreider, O., Rance, B. (eds.) DILS 2012. LNCS, vol. 7348, pp. 21–36. Springer, Heidelberg (2012). https://doi.org/10.1007/978-3-642-31040-9_3
2. Gene Ontology Consortium, et al.: Gene ontology consortium: going forward. Nucleic Acids Res. **43**(D1), D1049–D1056 (2015)
3. Couto, F.M., Silva, M.J., Coutinho, P.M.: Measuring semantic similarity between Gene Ontology terms. Data Knowl. Eng. **61**(1), 137–152 (2007)
4. Damljanovic, D., Stankovic, M., Laublet, P.: Linked data-based concept recommendation: comparison of different methods in open innovation scenario. In: Simperl, E., Cimiano, P., Polleres, A., Corcho, O., Presutti, V. (eds.) ESWC 2012. LNCS, vol. 7295, pp. 24–38. Springer, Heidelberg (2012). https://doi.org/10.1007/978-3-642-30284-8_9
5. Devos, D., Valencia, A.: Practical limits of function prediction. Prot.: Struct. Funct. Bioinform. **41**(1), 98–107 (2000)
6. Gabrilovich, E., Markovitch, S.: Computing semantic relatedness using Wikipedia-based explicit semantic analysis. In: IJCAI, vol. 7, pp. 1606–1611 (2007)
7. Hassan, S., Mihalcea, R.: Semantic relatedness using salient semantic analysis. In: AAAI (2011)
8. Jiang, J.J., Conrath, D.W.: Semantic similarity based on corpus statistics and lexical taxonomy. arXiv preprint arXiv:cmp-lg/9709008 (1997)
9. Kazakov, Y.: SRIQ and SROIQ are harder than SHOIQ. In: Description Logics. CEUR Workshop Proceedings, vol. 353. CEUR-WS.org (2008)
10. Köhler, S., et al.: The Human Phenotype Ontology project: linking molecular biology and disease through phenotype data. Nucleic Acids Res. **42**(D1), D966–D974 (2014)
11. Kuhn, H.W.: The Hungarian method for the assignment problem. Naval Res. Log. Q. **2**(1–2), 83–97 (1955)
12. Landauer, T.K., Laham, D., Rehder, B., Schreiner, M.E.: How well can passage meaning be derived without using word order? A comparison of Latent Semantic Analysis and humans. In: Proceedings of the 19th annual meeting of the Cognitive Science Society, pp. 412–417 (1997)
13. Lee, M., Pincombe, B., Welsh, M.: An empirical evaluation of models of text document similarity. In: Cognitive Science (2005)
14. Lin, D.: An information-theoretic definition of similarity. In: ICML, vol. 98, pp. 296–304 (1998)
15. Paul, C., Rettinger, A., Mogadala, A., Knoblock, C.A., Szekely, P.: Efficient graph-based document similarity. In: Sack, H., Blomqvist, E., d'Aquin, M., Ghidini, C., Ponzetto, S.P., Lange, C. (eds.) ESWC 2016. LNCS, vol. 9678, pp. 334–349. Springer, Cham (2016). https://doi.org/10.1007/978-3-319-34129-3_21
16. Pekar, V., Staab, S.: Taxonomy learning: factoring the structure of a taxonomy into a semantic classification decision. In: Proceedings of the 19th International Conference on Computational Linguistics, vol. 1, pp. 1–7. Association for Computational Linguistics (2002)

17. Pesquita, C., Faria, D., Bastos, H., Falcão, A., Couto, F.: Evaluating go-based semantic similarity measures. In: Proceedings of 10th Annual Bio-Ontologies Meeting, vol. 37, p. 38 (2007)
18. Pesquita, C., Pessoa, D., Faria, D., Couto, F.: CESSM: collaborative evaluation of semantic similarity measures. JB2009: Chall. Bioinform. **157**, 190 (2009)
19. Resnik, P., et al.: Semantic similarity in a taxonomy: an information-based measure and its application to problems of ambiguity in natural language. J. Artif. Intell. Res. (JAIR) **11**, 95–130 (1999)
20. Schuhmacher, M., Ponzetto, S.P.: Knowledge-based graph document modeling. In: Proceedings of the 7th ACM International Conference on Web Search and Data Mining, pp. 543–552. ACM (2014)
21. Sevilla, J.L., et al.: Correlation between gene expression and GO semantic similarity. IEEE/ACM Trans. Comput. Biol. Bioinform. **2**(4), 330–338 (2005)
22. Shi, C., Kong, X., Huang, Y., Yu, P.S., Wu, B.: HeteSim: a general framework for relevance measure in heterogeneous networks. IEEE Trans. Knowl. Data Eng. **26**(10), 2479–2492 (2014)
23. Smith, T.F., Waterman, M.S.: Identification of common molecular subsequences. J. Mol. Biol. **147**(1), 195–197 (1981)
24. Sun, Y., Han, J., Yan, X., Yu, P.S., Wu, T.: PathSim: meta path-based top-k similarity search in heterogeneous information networks. In: VLDB 2011 (2011)
25. Traverso-Ribón, I., Vidal, M.: Exploiting information content and semantics to accurately compute similarity of GO-based annotated entities. In: IEEE Conference on Computational Intelligence in Bioinformatics and Computational Biology, CIBCB, pp. 1–8 (2015)
26. Traverso-Ribón, I., Vidal, M.-E., Palma, G.: OnSim: a similarity measure for determining relatedness between ontology terms. In: Ashish, N., Ambite, J.-L. (eds.) DILS 2015. LNCS, vol. 9162, pp. 70–86. Springer, Cham (2015). https://doi.org/10.1007/978-3-319-21843-4_6

Knowledge Graphs for Semantically Integrating Cyber-Physical Systems

Irlán Grangel-González[1,2(✉)], Lavdim Halilaj[1,2], Maria-Esther Vidal[3,4],
Omar Rana[1], Steffen Lohmann[2], Sören Auer[3,4], and Andreas W. Müller[5]

[1] Enterprise Information Systems (EIS), University of Bonn, Bonn, Germany
{grangel,halilaj,s6omrana}@cs.uni-bonn.de
[2] Fraunhofer Institute for Intelligent Analysis and Information Systems (IAIS),
Sankt Augustin, Germany
steffen.lohmann@iais.fraunhofer.de
[3] L3S Research Center, Hanover, Germany
[4] TIB Leibniz Information Center for Science and Technology, Hanover, Germany
{maria.vidal,soeren.auer}@tib.eu
[5] Schaeffler Technologies, Herzogenaurach, Germany
andreas_w.mueller@schaeffler.com

Abstract. Cyber-Physical Systems (CPSs) are engineered systems that result from the integration of both physical and computational components designed from different engineering perspectives (e.g., mechanical, electrical, and software). Standards related to Smart Manufacturing (e.g., AutomationML) are used to describe CPS components, as well as to facilitate their integration. Albeit expressive, smart manufacturing standards allow for the representation of the same features in various ways, thus hampering a fully integrated description of a CPS component. We tackle this integration problem of CPS components and propose an approach that captures the knowledge encoded in smart manufacturing standards to effectively describe CPSs. We devise SEMCPS, a framework able to combine Probabilistic Soft Logic and Knowledge Graphs to semantically describe both a CPS and its components. We have empirically evaluated SEMCPS on a benchmark of AutomationML documents describing CPS components from various perspectives. Results suggest that SEMCPS enables not only the semantic integration of the descriptions of CPS components, but also allows for preserving the individual characterization of these components.

1 Introduction

The *Smart Manufacturing* vision aims at creating smart factories on top of the Internet of Things, Internet of Services, and Cyber-Physical Systems (CPSs). This vision is currently supported by various initiatives worldwide, including the "Industrie 4.0" activities in Germany [2], the "Factory of the Future" initiative in France and UK [27], the "Industrial Internet Consortium" in the USA as well as the "Smart Manufacturing" effort in China [19].

© Springer Nature Switzerland AG 2018
S. Hartmann et al. (Eds.): DEXA 2018, LNCS 11029, pp. 184–199, 2018.
https://doi.org/10.1007/978-3-319-98809-2_12

CPSs are complex mechatronic systems, e.g., robotic systems or smart grids [28], and are designed according to various engineering perspectives, e.g., specifications of a conveyor system usually comprise mechanical, electrical, and software viewpoints. The final design of a CPS includes the characteristics of the CPS specified in each perspective.

However, perspectives are defined independently and conflicting specifications of the same characteristics may exist [15], e.g., a software perspective may specify safety functions of a conveyor system than are not considered in the electrical viewpoint. These particularities in a perspective may generate *semantic heterogeneity*. Consequently, one of the biggest challenges for the realization of a CPS is the integration of these perspectives based on the *knowledge* encoded in each of them [3,20,21], i.e., the semantic integration of these perspectives.

Perspectives enclose core characteristics of the CPS that need to be represented in the integrated design, e.g., descriptions of a robot system's inputs and outputs and its main functionality; these characteristics correspond to *hard knowledge facts*. In addition, properties individually modeled in each perspective, as well as the resolution of the corresponding heterogeneity issues that may be caused, should be part of the final design according to how consistent they are with respect to the rest of the perspectives. These features are *uncertain* in the integrated CPS, e.g., safety issues expressed in the electrical perspective may also be included in the software perspective and vice versa. Such properties that are *totally* or *partially* covered by other perspectives can be modeled as *soft knowledge facts* in the integrated design.

Semantic heterogeneity issues that may occur in an integrated CPS have been characterized before [4,17]. Further, a number of approaches have been defined for solving such integration problems [11,21,28]. Although existing approaches support the integration of CPS perspectives based on the resolution of *semantic heterogeneity* issues, none of them is able to distinguish hard and soft knowledge facts during integration.

We devise SEMCPS, a rule-based framework that relies on Probabilistic Soft Logic (PSL) for capturing the knowledge encoded in different CPS perspectives and for exploiting this knowledge to enable a semantic integration of CPS perspectives. SEMCPS includes weighted rules representing the conditions to be met by hard and soft knowledge facts. It relies on uncertain knowledge graphs [6,13] where edges are annotated with weights to represent the knowledge of different views and to integrate this knowledge into a final design.

We evaluated the effectiveness of SEMCPS in a benchmark of real-world based CPS perspectives described using documents of the AutomationML standard. Experimental results suggest that SEMCPS accurately identifies integrated characteristics of CPSs while preserving the main individual characterization and description of the components.

The contributions of this paper are in particular:

– Formal definitions of CPS uncertain knowledge graphs and the problem of integrating CPS perspectives into a CPS uncertain knowledge graph;

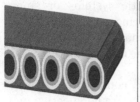 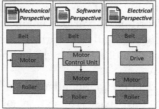 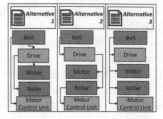

(a) Conveyor belt   (b) CPS design perspectives (c) Alternatives of a CPS design

Fig. 1. Motivating Example. Description of a conveyor belt. (a) A simple Cyber-Physical System (CPS) resulting from a multi-disciplinary engineering design. (b) The representation of the CPS according to its mechanical, electrical, and software perspectives; the CPS is defined in terms of various components and attributes in each perspective. (c) Alternatives integrate perspectives and describe final CPS designs. Each perspective solves the data integration problem differently. (Color figure online)

- SEMCPS, a PSL-based framework to capture knowledge encoded in CPS perspectives and solve *semantic heterogeneity* among CPS perspectives; and
- An empirical evaluation of the effectiveness of SEMCPS on a testbed of various perspectives describing CPSs.

The rest of the paper is structured as follows: Sect. 2 motivates the problem of integrating CPS perspectives. Section 3 provides background information and introduces the terminology relevant to our approach. Section 4 defines CPS uncertain knowledge graphs and details the integration problem tackled in this paper. Section 5 presents the SEMCPS framework, followed by its empirical evaluation presented in Sect. 6. Section 7 summarizes related work, before Sect. 8 concludes the paper and gives an outlook to future work.

2   Motivating Example

The engineering process in smart manufacturing environments combines various expertise for designing and developing a CPS, in particular skills in mechanical, electrical, and software engineering. As a result, diverse perspectives are generated for the same CPS; they may suffer of semantic heterogeneity issues caused by overlapped or inconsistent designs [22]. The goal of this collaborative design process is to produce a final design where overlapping and inconsistencies are minimized and semantic heterogeneity issues are solved [23–25]. The final design has to respect the original intent of the different perspectives; it also has to ensure that all knowledge encoded in each perspective is captured during the integration process.

Figure 1a illustrates a CPS described from different perspectives. Each perspective is defined according to an expert understanding of the domain; different elements, e.g., components, attributes, and relations may be used to describe the same CPS in each perspective. Figure 1b presents three perspectives of the CPS shown in Fig. 1a; they share some elements, e.g., `Belt`, `Motor`, and `Roller`. On the other hand, `Drive` and `Motor Control Unit` are only included in the *software* and the *electrical* perspectives, respectively. Elements that appear in all the perspectives should be included in the final integrated design of a CPS; they correspond to *hard knowledge facts*. Moreover, some elements are not part of all the perspectives, e.g., the aforementioned `Drive` and `Motor Control Unit`, causing that the granularity of the description of elements like `Belt` varies in these designs. These elements are uncertain in the final design and can be considered as *soft knowledge facts*.

Figure 1c outlines alternative integrated CPS designs. In *Alternative 1*, all the elements from three given perspectives are included: `Motor` and `Roller` are *related* to `Drive`, while `Motor Control Unit` is only *related* to `Belt`. Furthermore, because `Drive` is related to `Belt`, `Motor`, and `Roller` are also related to `Belt`. The granularity description of `Belt` is *compatible* with the software and electrical perspectives, while the properties present in all the perspectives are preserved. In contrast, neither *Alternative 2* nor *Alternative 3* describe elements at the same level of granularity. Therefore, *Alternative 1* seems to be most *complete* according to the specifications of this CPS design; however, uncertainty about the membership of elements like `Drive` and `Motor Control Unit` should be modeled. The approach we present in this work relies on knowledge graphs and allows for the representation and integration of these three alternative designs, as well as for the selection of *Alternative 1* as the final integrated design.

3 Background

A huge variety of standards, covering different aspects of smart manufacturing, are utilized to describe CPSs. For example, *OPC UA* [10] is used to describe the communication of CPSs, while *PLCOpen* [9] and *AutomationML* (AML) [8] are used for CPS programming and design, respectively. Despite the heterogeneous landscape of standards in the context of smart manufacturing, they share the commonality of containing information models to represent knowledge about the CPS and its lifecycle, from its creation until the end of its productive life. These models capture knowledge about main properties of a CPS from a particular perspective; it is represented in documents according to the specifications of the standards, e.g., using XML-based languages that includes terms representing main concepts of smart manufacturing standards, such as CPS attributes, components, relations, and datatypes. Semantic heterogeneity is caused by different viewpoints involved in CPS design, i.e., how equivalent and different concepts for the same CPS are expressed [15].

Several authors [4,17,30] have characterized forms of semantic heterogeneity that may occur in a CPS design: **(M1) Value processing:** Attributes and relations are modeled differently, e.g., using different datatypes. **(M2) Granularity:** Components modeled at various levels of detail. **(M3) Schematic differences:** Components and attributes are differently related. **(M4) Conditional mappings:** Relations between components and attributes exist only if

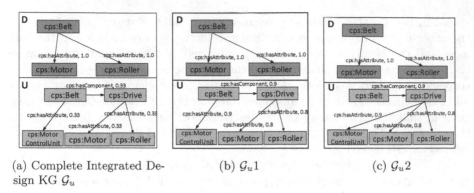

(a) Complete Integrated Design KG $\mathcal{G}_u$ (b) $\mathcal{G}_u 1$ (c) $\mathcal{G}_u 2$

Fig. 2. Uncertain KGs for CPS final design. Uncertain KGs are built based on the alternatives of the motivating example. They combine hard (D) and soft (U) knowledge facts; (a), (b) and (c) represent alternative integrated designs. (Color figure online)

certain conditions are met. **(M5) Bidirectional mappings:** Relations between components and attributes may be bidirectional. **(M6) Grouping and aggregation:** Using different relations, components and attributes can be grouped and aggregated in various ways. **(M7) Restrictions on values:** Different restrictions on the possible values of the attributes of a component are implemented.

4   Problem Statement and Solution

In this section, CPS uncertain knowledge graphs are defined. Then, the problem of integrating CPS perspectives is presented as an inference problem on uncertain knowledge graphs. PSL framework provides a practical solution to this problem.

4.1   CPS Knowledge Graphs

A knowledge graph is defined as a labeled directed graph encoded using the RDF data model [12]. Given sets I and V that correspond to URIs identifying elements in a CPS document and terms from a CPS standard vocabulary, respectively; furthermore, let L be a set of literals. A CPS Knowledge Graph $\mathcal{G}$ is a 4-tuple $\langle I, V, L, G \rangle$, where G is a set of triples of the form $(s, p, o) \in I \times V \times (I \cup L)$. Given two CPS knowledge graphs $\mathcal{G}_1 = \langle I, V, L, G_1 \rangle$, $\mathcal{G}_2 = \langle I, V, L, G_2 \rangle$ the entailment for $\mathcal{G}_1 \models \mathcal{G}_2$ is defined as the standard RDF entailment G_1 and G_2 [12], i.e., $G_1 \models G_2$. Chekol et al. [6] have shown that knowledge graphs can be extended with uncertainty; the maximum a-posteriori inference process from Markov Logic Networks (MLNs) is used to compute the interpretation of the triples in an uncertain KG that minimizes the overall uncertainty. Similarly, we define a *CPS Uncertain Knowledge Graph* as a knowledge graph where each fact is annotated with a weight in the range $[0, 1]$; weights represent uncertainty about the membership of the corresponding facts to the knowledge graph, i.e., soft knowledge facts. Moreover, we devise an entailment relation between two CPS uncertain knowledge graphs; this relation allows for deciding when a CPS uncertain knowledge graph covers the hard and soft knowledge facts of the other knowledge graph. Formally, given L, I, and V, three sets of literals, URIs identifying elements in a CPS document, and terms in a CPS standard vocabulary, respectively. A *CPS Uncertain Knowledge Graph* $\mathcal{G}_u$ is a 5-tuple $\langle I, V, L, D, U \rangle$:

- D is an RDF graph of the form $(s, p, o) \in I \times V \times (I \cup L)$. D represents a set of *hard knowledge facts*.
- U is an RDF graph where triples are annotated with weights. U is a set of *soft knowledge facts*, defined as follows:

$$U = \{(t, w) \mid t \in I \times V \times (I \cup L) \text{ and } w \in [0, 1]\}$$

- $\tau(U)$ is the set of triples in U, with $\tau(U) \cap D = \emptyset$, i.e.,

$$\tau(U) = \{t \mid (t, w) \in U\}.$$

Example 1. Figure 2b shows an Uncertain Knowledge Graph $\mathcal{G}_u1$ for Alternative 1 in Fig. 1c. Edges between blue nodes represent hard knowledge facts in D, while soft knowledge facts are modeled as edges between green nodes in U. Elements in the perspectives in Fig. 1b correspond to hard knowledge facts, e.g., elements stating that `Motor` and `Roller` are related to `Belt`. Also, the relation between `Motor Control Unit` and `Belt` is only included in one perspective; the corresponding element corresponds to a soft knowledge fact in U.

The semantics of a CPS uncertain KG $\mathcal{G}_u$ is defined in terms of the probability distribution of the values of weights of the triples in $\mathcal{G}_u$. As defined by Chekol et al. [6], the weights of the triples in $\mathcal{G}_u$ are characterized by a log-linear probability distribution. For any CPS Uncertain Knowledge Graph $\mathcal{G}_u^*$ over the same sets I, V, and L, i.e., $\mathcal{G}_u^* = \langle I, V, L, D^*, U^* \rangle$ the probability of $\mathcal{G}_u^*$ is as follows:

$$P(\mathcal{G}_u^*) = \begin{cases} \frac{1}{Z} exp \left(\displaystyle\sum_{\{(t_i, w_i) \in U : D^* \cup \tau(U^*) \models t_i\}} w_i \right) & \text{if } D^* \cup \tau(U^*) \models D \\ 0 & \text{otherwise} \end{cases} \tag{1}$$

Z is the normalization constant of the log-linear probability distribution P.

Example 2. Consider the CPS uncertain KGs depicted in Fig. 2; they represent alternate integrated designs in Fig. 1c. In Fig. 2a, we present a CPS uncertain KG $\mathcal{G}_u$ where all the elements present in the three perspectives are included in the knowledge graph D, i.e., they correspond to hard knowledge facts; additionally, the knowledge graph U includes uncertain triples representing soft knowledge facts; weights denote how many times a fact is represented in the three perspectives. For example, the relation between `Drive` and `Belt` is only included in one out of three perspectives, so, the weight is 0.3. This KG can be seen as a complete integrated design of the CPS. Furthermore, uncertain KGs in Figs. 2b and c represent alternate integrated designs; the probability of these KGs with respect to the one in Fig. 2a is computed following Eq. 1. Figure 2b presents a KG with the highest probability; it corresponds to *Alternative 1* in the motivating example where the majority of the facts in the KG are also in KG in Fig. 2a.

Definition 1. *Let $\mathcal{G}_u = \langle I, V, L, D, U \rangle$ be a CPS uncertain knowledge graph. The entailment for any $\mathcal{G}_u^* = \langle I, V, L, D^*, U^* \rangle$ $\mathcal{G}_u^* \models_u \mathcal{G}_u$ holds if $P(\mathcal{G}_u^*) > 0$.*

Example 3. Consider again the CPS uncertain KGs presented in Fig. 2, because the probability of the uncertain KGs in Figs. 2b and c with respect to the KG in Fig. 2a is greater than 0.0, we can say that the entailment relation is met, i.e., $\mathcal{G}_u^1 \models_u \mathcal{G}_u$, $\mathcal{G}_u^2 \models_u \mathcal{G}_u$, and $\mathcal{G}_u^3 \models_u \mathcal{G}_u$.

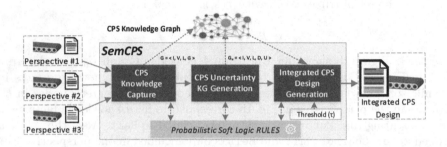

Fig. 3. The SEMCPS Architecture. SEMCPS receives documents describing a Cyber-Physical System (CPS) from various perspectives; they are represented in standards like AML. SEMCPS outputs a final design document describing the integration of the perspectives, a Knowledge Graph (KG). (1) Input documents are represented as a KG in RDF. (2) A rule-based system is used to identify heterogeneity among the perspectives represented in KG. (3) A rule-based system is utilized to solve heterogeneity and produced the final integrated CPS design.

4.2   Problem Statement

Integrating CPS perspectives corresponds to the problem of identifying a CPS Uncertain KG $\mathcal{G}_u^*$ where the probability distribution with respect to the complete integrated design $\mathcal{G}_u$ is maximized. This problem optimization is follows:

$$\underset{\mathcal{G}_u^* \models_u \mathcal{G}_u}{\operatorname{argmax}}(P(\mathcal{G}_u^*))$$

Example 4. Consider the CPS uncertain KGs shown in Fig. 2a. An optimal solution of integrating CPS perspectives is the CPS uncertain KG in Fig. 2b; this KG represents *Alternative 1* which according to Prinz [24], is the most complete representation of the CPS perspectives described in Fig. 1b.

4.3 Proposed Solution

As shown by Chekol et al. [6], solving the maximum a-posteriori inference process required to compute the probability of an uncertain KG is NP-hard in general. In order to provide a practical solution to this problem, we propose a rule-based system that relies on PSL to generate uncertain KGs that correspond to approximate solutions to the problem of integrating CPS perspectives. PSL [1,16] has been utilized as the probabilistic inference engine in several integration problems, e.g., knowledge graphs [26] and ontology alignment [5]. PSL allows for the definition of rules with an associated non-negative weight that captures the importance of a rule in a given probabilistic model. A PSL model is defined using a set of weighted rules in first-order logic, as follows:

$$SemSimComp(B, A) \land Rel(B, Y) \Rightarrow Rel(A, Y) \mid 0.9 \qquad (2)$$

SemCPS includes a set of PSL rules capturing the conditions to be met by a CPS *Uncertain KG* that solves the integration of CPS perspectives. For example, Rule 2 generates new elements in an integrated design assuming that semantically similar components are related to same attributes. Further, Rule 3 determines semantic similarity of components.

$$Component(A) \land Component(B) \land hasRefSem(A, Z) \land$$
$$hasRefSem(B, Z) \Rightarrow SemSimComp(A, B) \mid 0.8 \qquad (3)$$

The PSL program receives as input facts representing all the elements in the perspectives to be integrated, as well as their semantic references. Then, Rules 2 and 3 determine that `Drive` is a sub component of `Belt`, and that `Belt` is related to the same elements that `Drive`, i.e., `Motor` and `Roller` are related to `Belt`. Based on the weights of these rules, these facts have a high degree of membership to the integrated design. Similarly, rules are utilized for determining that `Motor Control Unit` is related to `Belt` in the integrated design. The PSL program builds the uncertain KG in Fig. 2b maximizing the probability distribution with respect to the complete integrated design in Fig. 2a.

5 The SemCPS Framework

We present SEMCPS, a framework to integrate different perspectives of a CPS. Figure 3 depicts the architectural components of SEMCPS. SEMCPS receives as input a set of documents describing a CPS in a given smart manufacturing standard and a *membership degree threshold*; the output is a final integrated design of the CPS. SEMCPS builds a CPS knowledge graph $\mathcal{G} = \langle I, V, L, G \rangle$ to capture the knowledge encoded in the CPS documents. Then, the PSL program is used to solve the heterogeneity issues existing among the elements in the different CPS perspectives; a CPS uncertain knowledge graph $\mathcal{G}_u^* = \langle I, V, L, D^*, U^* \rangle$ represents an integrated design of the CPS. Finally, the membership degree threshold is used to select the soft knowledge facts from $\mathcal{G}_u^*$ that in conjunction with the hard knowledge facts in D^* are part of the final integrated design.

Capturing Knowledge Encoded in CPS Documents. The CPS Knowledge Capture component receives as inputs documents in a given standard containing the description of the perspectives of a CPS design (cf. Sect. 2). Next, these documents are automatically transformed into RDF, by following the semantics encoded in the corresponding standard vocabulary. To this end, a set of XLST-based mapping rules are executed in the *Krextor* [18] framework to create an RDF KG using a CPS vocabulary. Consequently, the output of this component is $\mathcal{G}$, a KG comprising the input data in RDF.

Generating a CPS Uncertain Knowledge Graph. The CPS Uncertain KG Generation component creates, based on the input KG, the hard and soft knowledge facts, i.e., the uncertain KG. To achieve this goal, SEMCPS relies on the PSL rules described in Fig. 3. Next, all facts with degree of membership equal to 1.0 correspond to hard knowledge facts. The rest generated during the evaluation of the rules correspond to soft knowledge facts.

Generating a Final Integrated CPS Design. The Final Integrated CPS Design Generation component utilizes a membership degree threshold to select the facts in the CPS uncertain KG. Facts with scores below the value of the threshold are removed while the rest will be part of the final integrated design.

6   Empirical Evaluation

We empirically study the effectiveness of SEMCPS in the solution of the problem of *integrating CPS perspectives*. The goal of the experiment is to analyze the impact of: (1) the number of heterogeneity on the effectiveness of SEMCPS; and (2) the size of CPS perspectives on the efficiency of SEMCPS. Particularly, we assess the following research questions:

(RQ1) Does the type of heterogeneity among the perspectives of a CPS impact on the effectiveness of SEMCPS?

(RQ2) Does the size of the perspectives of a CPS affect the effectiveness of SEMCPS?

(RQ3) Does the degree of membership threshold impact on the effectiveness of SEMCPS?

We compare SEMCPS with the Expressive and Declarative Ontology Alignment Language (EDOAL) [29] and the Linked Data Integration Framework (SILK) [32]. Both frameworks allow for representing correspondences between the entities of different ontologies and instance data by means of rules. With the goal to compare both approaches, we created rules in EDOAL and SILK to solve heterogeneity issues between CPS perspectives[1]. For both frameworks,

---

[1] https://github.com/i40-Tools/Related-Integration-Tools.

SPARQL queries are generated based on their rules. These queries are then executed on top of the CPS perspectives after their conversion to RDF. To the best of our knowledge, real-world publicly benchmarks in the industry domain are not available. Moreover, many of the smart manufacturing standards are not even publicly accessible. This complicates the access to a full benchmark of real-world CPS documents. To address this issue, we define a generator of CPS perspectives. The generator creates CPS perspectives representing real-world scenarios and allow for the empirical evaluation of SEMCPS.

6.1 CPS Document Generator

The CPS Document generator[2] produces different perspectives of a seed real-world CPS[3]; generated perspectives include combinations of seven semantic heterogeneity described in [17]. Based on a *Poisson* distribution, a value between one and seven is selected; it simulates the number of heterogeneity that exist in each perspective. The parameter λ of the *Poisson* distribution indicates the average number of heterogeneity among perspectives; λ is set to two and simulates an average of 16 heterogeneity pair-wise perspectives. Thus, generated perspectives include components, attributes, and relations which are commonly included in real-world AutomationML documents[4].

Table 1. Testbed Description. Minimal and maximal configurations (Config.) in terms of number of elements, relations, heterogeneity, and document size

Config.	# Elements	# Relations	# M1–M7	Size (KB)
Minimal	20	8	1	5.7
Maximal	600	350	7	116.2

6.2 Experiment Configuration

Testbeds. We considered a testbed with 70 seed CPS, and two perspectives per CPS. Each perspective has in average 200 elements related using 100 relations; furthermore, in average three heterogeneity occur between the two perspectives of a CPS. Table 1 summarizes the features of the evaluated CPS perspectives. As Table 1 shows, the testbed comprises a variety of elements, relations, and heterogeneity with the aim of simulating real-world CPS designs. **Gold Standard.** The Gold Standard includes uncertain knowledge graphs–$\mathcal{G}_u$–corresponding to complete integrated designs of CPS perspectives in the testbed.

---

[2] https://github.com/i40-Tools/CPSDocumentGenerator.
[3] Source: Drath, GMA 6.16.
[4] https://raw.githubusercontent.com/i40-Tools/iafCaseStudy/master/IAF_AMLModel_journal.aml.

Metrics. A final integrated design denoted by $\mathcal{G}_u^*$, describes the output of SEM-CPS (cf. Fig. 3), i.e., facts annotated with uncertainty values lower than the degree of membership threshold are removed. The complete integrated design denoted by $\mathcal{G}_u$, corresponds to a CPS uncertain KG in the Gold Standard. We evaluate SEMCPS in terms of the following metrics: **Precision** is the fraction of the cardinality of the final integrated design produced by SEMCPS (denoted by $\mathcal{G}_u^*$) and the cardinality of the complete integrated design (denoted by $\mathcal{G}_u$). **Recall** is the fraction of the cardinality of the complete integrated design (denoted by $\mathcal{G}_u$) and the cardinality of the final integrated design (denoted by $\mathcal{G}_u^*$). **F-Measure (F1)** is the harmonic mean of Precision and Recall (Table 2).

Table 2. Metrics of precision and recall.

$$\text{Precision} = \frac{|\mathcal{G}_u^*| \cap |\mathcal{G}_u|}{|\mathcal{G}_u^*|} \quad \textbf{Recall} = \frac{|\mathcal{G}_u^*| \cap |\mathcal{G}_u|}{|\mathcal{G}_u|}$$

Implementation. The generator and SEMCPS are implemented in Java 1.8. SEMCPS also uses PSL 1.2.1. Experiments were run on a Windows 8 machine with an Intel I7-4710HQ 2.5 GHz CPU and 16 GB 1333 MHz DDR3 RAM. Results can be reproduced by using the generator along with data for the experiments[5]; SEMCPS is publicly available[6].

Table 3. Experiment 1: SemCPS Effectiveness on different types of heterogeneity. SemCPS exhibits the best performance for the increasing number of heterogeneity, i.e., from M1 to M7, e.g., EDOAL and SILK

H	SemCPS			EDOAL			SILK		
	Precision	Recall	F1	Precision	Recall	F1	Precision	Recall	F1
M1	**0.93**	**0.93**	**0.93**	0.8	0.28	0.42	0.85	0.28	0.42
M1–M2	**0.88**	**0.86**	**0.87**	0.8	0.4	0.45	0.82	0.31	0.45
M1–M3	**0.93**	**0.95**	**0.94**	0.81	0.46	0.59	0.76	0.46	0.57
M1–M4	**1.0**	**0.61**	**0.76**	0.8	0.59	0.68	0.67	0.54	0.63
M1–M5	**0.96**	**0.94**	**0.95**	0.88	0.57	0.69	0.85	0.57	0.68
M1–M6	**0.93**	**0.93**	**0.93**	0.82	0.65	0.73	0.72	0.64	0.68
M1–M7	**0.92**	**0.96**	**0.94**	0.79	0.62	0.69	0.79	0.62	0.69

Impact of the Type of Heterogeneity. To answer **RQ1**, the perspectives of 70 CPSs are considered; the membership degree threshold is set to 0.5. SEMCPS

---

[5] https://github.com/i40-Tools/HeterogeneityExampleData/tree/master/AutomationML.

[6] https://github.com/i40-Tools/SemCPS.

is executed in seven iterations. During an iteration i where $1 < i < 7$, the two perspectives of each of the 70 CPS have only heterogeneity of type $\mathbf{M}_i$; Table 3 reports on the effectiveness of *SemCPS* for each iteration in terms of the average of Precision, Recall and F1. These observed results (cf. Table 3) suggest that the behavior of SEMCPS is slightly affected by heterogeneity types. Overall, SEMCPS exhibits highest values of precision and F1 than EDOAL and SILK. Therefore, these results suggest that SEMCPS is able to identify alternatives that represent and integrate the majority of the facts in the evaluated perspectives.

Table 4. Experiment 2: SemCPS Effectiveness on based on the size of CPS perspectives. SemCPS exhibits the best performance for the increasing number of elements in perspectives, e.g., EDOAL and SILK

Elements	SemCPS			EDOAL			SILK		
	Precision	Recall	F1	Precision	Recall	F1	Precision	Recall	F1
30	**0.96**	**1.0**	**0.98**	0.97	0.70	0.81	0.97	0.70	0.81
60	**0.97**	**1.0**	**0.98**	0.97	0.48	0.64	0.97	0.48	0.64
180	**1.0**	**0.9**	**0.95**	0.88	0.91	0.88	0.86	0.91	0.89
210	**1.0**	**0.78**	**0.87**	0.87	0.77	0.81	0.87	0.77	0.81
600	**1.0**	0.95	**0.98**	0.95	**0.96**	0.96	0.95	**0.96**	0.96

Impact of the Size of CPS Perspectives. To assess **RQ2**, sizes of two perspectives of a seed CPS are changed; the experiment is run in five iterations. In iteration one, 30 elements are included in each perspective; then 60, 180, 210, and 600 elements are considered in the next iterations. For this experiment, the membership degree threshold is set to 0.5. EDOAL and SILK are executed on top of the same CPS perspectives documents. Table 4 reports on the effectiveness of SEMCPS, EDOAL, and SILK in terms of the average of precision, recall, and F1; results suggest that SEMCPS outperforms the compared frameworks. With an exception, i.e., when 210 elements are considered, SEMCPS seems not to be impacted by the number of elements.

Impact of the Degree of the Membership Threshold. To evaluate **RQ3**, SEMCPS is executed five times with a variation in the membership degree threshold from 0.5 up to 0.9. For each perspective, 210 elements are considered. As shown in Fig. 4, precision is not affected whereas recall decreases up to approximately 0.75 in the last threshold, i.e., 0.9. The membership degree threshold has lowered the performance since in every execution where the threshold is incremented, more *soft knowledge facts* are excluded from the final integrated design $\mathcal{G}_u^*$; thus lowering recall. These results suggest that the membership degree threshold impacts on the performance of SEMCPS. Note that more *soft knowledge facts* are removed from $\mathcal{G}_u^*$ whenever the values of the membership degree threshold increase. Consequently, the completeness of the results is negatively impacted and lower values of recall are observed.

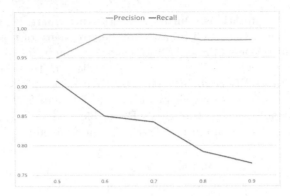

Fig. 4. Membership degree of threshold.

Overall, the observed behavior of the evaluated frameworks provides evidence about the quality of the data integration solution implemented by SEMCPS.

7   Related Work

Recently, there has been a large amount of research investigating the integration of CPS [31], as well as the use of semantic technologies for the resolution of semantic heterogeneity in related scenarios [14]. BI et al. [3] present *MSCIM*, a Mechatronics System Common Information Model to support the multi-disciplinary design in the CPS scenario. MSCIM relies on XML and XML Web services technologies to leverage the integration. Further, MSCIM utilizes the wrapper integration approach in a very generic level. Lüder et al. [20] describe a manual approach for the CPS information integration by means of AutomationML. Chen et al. [7] develop a framework for the integration of the design of CPS; requirements for each one of the disciplines involved are characterized, as well as the representation of constraints among disciplines. In [25], a method for integration of mechatronic objects design is proposed; the method combines advantages of bottom and top down approaches into a hybrid approach. Kovalenko and Euzenat [17] investigate ontology matching techniques to execute identification and integration of data in this context. A survey of existing languages for realizing this task is presented; furthermore, EDOAL is proposed for tackling the problem of semantic matching of the semantic heterogeneity for engineering documents. Sabou et al. [21] describe a semantic web-based method for data integration in multi-disciplinary engineering. The work is based on the design of a Hydro Power Plant, which is considered as a CPS. The semantics of each local data source is presented using a hybrid ontology model. Moreover, a generic ontology and three local ontologies representing data from three engineering perspectives are defined. The above mentioned approaches have potential to solve specific integration problems for CPS documents. However, isolated problems are tackled, and a general method capable of producing a final CPS integrate design considering the uncertain nature of CPS design have not been developed. On the contrary, SEMCPS combines PSL and Semantic Web technologies to effectively integrate CPS documents.

8   Conclusions and Future Work

This paper presents SEMCPS, a framework for enabling the seamless integration of descriptions of cyber-physical systems in knowledge graphs. SEMCPS combines Probabilistic Soft Logic and semantic technologies to *accurately* capture the knowledge that characterizes different types of semantic heterogeneity in CPS documents. Results of the empirical evaluation suggest that SEMCPS is able to effectively solve the problem of integrating CPS perspectives by using *Uncertain Knowledge Graphs* of smart manufacturing related standards such as AutomationML. In general, SEMCPS exhibits better performance than EDOAL and SILK when it is executed with an increasing number of semantic heterogeneity types and when an increasing number of elements are added. Finally, the effectiveness of SEMCPS seems to be impacted for higher values of the membership degree threshold. In the future, we envision to improve SEMCPS by including the semantics of OWL in combination with PSL rules. Further, we plan to extend SEMCPS to integrate documents combining information models of more than one smart manufacturing standard like OPC UA and AutomationML.

Acknowledgements. This work has partly been supported by the German Federal Ministry of Education, Research (BMBF) in the context of the project *Industrial Data Space Plus* (grant no. 01IS17031), and EU H2020 Programme for the project *BOOST 4.0* (grant no. 780732).

References

1. Bach, S.H., Broecheler, M., Huang, B., Getoor, L.: Hinge-loss markov random fields and probabilistic soft logic. J. Mach. Learn. Res. (JMLR) **18**, 1–67 (2017)
2. Bauernhansl, T., ten Hompel, M., Vogel-Heuser, B.: Industrie 4.0 in Produktion, Automatisierung und Logistik: Anwendung, Technologien, Migration. Springer, Wiesbaden (2014). https://doi.org/10.1007/978-3-658-04682-8
3. Bi, L., Jiao, Z.: An information integration framework based on XML to support mechatronics multi-disciplinary design. In: IEEE Conference on Robotics, Automation and Mechatronics, RAM, China, pp. 175–179 (2008)
4. Biffl, S., Kovalenko, O., Lüder, A., Schmidt, N., Rosendahl, R.: Semantic mapping support in AutomationML. In: ETFA, pp. 1–4. IEEE (2014)
5. Bröcheler, M., Mihalkova, L., Getoor, L.: Probabilistic similarity logic. In: Proceedings of the Twenty-Sixth Conference on Uncertainty in Artificial Intelligence, UAI 2010, Catalina Island, CA, USA, pp. 73–82 (2010)
6. Chekol, M.W., Pirrò, G., Schoenfisch, J., Stuckenschmidt, H.: Marrying uncertainty and time in knowledge graphs. In: Proceedings of the Thirty-First AAAI Conference on Artificial Intelligence, California, USA, pp. 88–94 (2017)

7. Chen, K., Bankston, J., Panchal, J.H., Schaefer, D.: A framework for integrated design of mechatronic systems. In: Wang, L., Nee, A. (eds.) Collaborative Design and Planning for Digital Manufacturing, pp. 37–70. Springer, London (2009). https://doi.org/10.1007/978-1-84882-287-0_2

8. Drath, R.: Datenaustausch in der Anlagenplanung mit AutomationML: Integration von CAEX, PLCopen XML und COLLADA. Springer, Heidelberg (2009)

9. Estévez-Estévez, E., Marcos, M., Lüder, A., Hundt, L.: PLCopen for achieving interoperability between development phases. In: Proceedings of 15th IEEE International Conference on Emerging Technologies and Factory Automation, ETFA, Spain, pp. 1–8 (2010)

10. OPC Foundation. OPC Unified Architecture Specification. Part 1: Overview and Concepts (2015)

11. Grangel-González, I., et al.: Alligator: a deductive approach for the integration of industry 4.0 standards. In: Blomqvist, E., Ciancarini, P., Poggi, F., Vitali, F. (eds.) EKAW 2016. LNCS (LNAI), vol. 10024, pp. 272–287. Springer, Cham (2016). https://doi.org/10.1007/978-3-319-49004-5_18

12. Gutierrez, C., Hurtado, C.A., Mendelzon, A.O., Pérez, J.: Foundations of semantic web databases. J. Comput. Syst. Sci. **77**(3), 520–541 (2011)

13. Huber, J., Niepert, M., Noessner, J., Schoenfisch, J., Meilicke, C., Stuckenschmidt, H.: An infrastructure for probabilistic reasoning with web ontologies. Semant. Web **8**(2), 255–269 (2017)

14. Jacoby, M., Antonić, A., Kreiner, K., Łapacz, R., Pielorz, J.: Semantic interoperability as key to IoT platform federation. In: Podnar Žarko, I., Broering, A., Soursos, S., Serrano, M. (eds.) InterOSS-IoT 2016. LNCS, vol. 10218, pp. 3–19. Springer, Cham (2017). https://doi.org/10.1007/978-3-319-56877-5_1

15. Jirkovský, V., Obitko, M., Mařík, V.: Understanding data heterogeneity in the context of cyber-physical systems integration. IEEE Trans. Ind. Inform. **13**(2), 660–667 (2017)

16. Kimmig, A., Bach, S., Broecheler, M., Huang, B., Getoor, L.: A short introduction to Probabilistic Soft Logic. In: Proceedings of the NIPS Workshop on Probabilistic Programming: Foundations and Applications, pp. 1–4 (2012)

17. Kovalenko, O., Euzenat, J.: Semantic matching of engineering data structures. In: Biffl, S., Sabou, M. (eds.) Semantic Web Technologies for Intelligent Engineering Applications, pp. 137–157. Springer, Cham (2016). https://doi.org/10.1007/978-3-319-41490-4_6

18. Lange, C.: Krextor - an extensible XML→RDF extraction framework. In: Scripting and Development for the Semantic Web (SFSW). CEUR Workshop Proceedings, vol. 449, Aachen, May 2009

19. Li, Q., Jiang, H., Tang, Q., Chen, Y., Li, J., Zhou, J.: Smart manufacturing standardization: reference model and standards framework. In: Ciuciu, I., et al. (eds.) OTM 2016. LNCS, vol. 10034, pp. 16–25. Springer, Cham (2017). https://doi.org/10.1007/978-3-319-55961-2_2

20. Lüder, A., Schmidt, N., Rosendahl, R., John, M.: Integrating different information types within AutomationML. In: Proceedings of the IEEE Emerging Technology and Factory Automation, ETFA, Spain, pp. 1–5 (2014)

21. Sabou, M., Ekaputra, F.J., Biffl, S.: Semantic web technologies for data integration in multi-disciplinary engineering. In: Biffl, S., Lüder, A., Gerhard, D. (eds.) Multi-Disciplinary Engineering for Cyber-Physical Production Systems, pp. 301–329. Springer, Cham (2017). https://doi.org/10.1007/978-3-319-56345-9_12

22. Mordinyi, R., Winkler, D., Ekaputra, F.J., Wimmer, M., Biffl, S.: Investigating model slicing capabilities on integrated plant models with AutomationML. In: Proceedings of 21st IEEE International Conference on Emerging Technologies and Factory Automation, ETFA, Germany, pp. 1–8 (2016)
23. Moser, T., Mordinyi, R., Winkler, D.: Extending mechatronic objects for automation systems engineering in heterogeneous engineering environments. In: Proceedings of IEEE 17th International Conference on Emerging Technologies & Factory Automation, ETFA, Poland, pp. 1–8 (2012)
24. Prinz, J.: Consistent merging of AutomationML documents in multiple sources scenarios. In: 4th AutomationML User Conference, Germany (2016)
25. Prösser, M., Moore, P.R., Chen, X., Wong, C., Schmidt, U.: A new approach towards systems integration within the mechatronic engineering design process of manufacturing systems. Int. J. Comput. Integr. Manuf. **26**(8), 806–815 (2013)
26. Pujara, J., Getoor, L.: Generic statistical relational entity resolution in knowledge graphs. CoRR, abs/1607.00992 (2016)
27. Ridgway, K., Clegg, C., Williams, D.: The Factory of the Future, Future Manufacturing Project: Evidence Paper 29. Foresight, Government Office for Science, London (2013)
28. Sabou, M., Ekaputra, F., Kovalenko, O., Biffl, S.: Supporting the engineering of cyber-physical production systems with the AutomationML analyzer. In: 1st International Workshop on Cyber-Physical Production Systems (CPPS), pp. 1–8. IEEE (2016)
29. Scharffe, F., Zimmermann, A.: D2. 2.10: Expressive alignment language and implementation. Deliverable D2, 2 (2007)
30. Schleipen, M., Gutting, D., Sauerwein, F.: Domain dependant matching of MES knowledge and domain independent mapping of AutomationML models. In: Proceedings of IEEE 17th International Conference on Emerging Technologies & Factory Automation, ETFA, Poland, pp. 1–7 (2012)
31. Schmidt, N., Lüder, A., Rosendahl, R., Ryashentseva, D., Foehr, M., Vollmar, J.: Surveying integration approaches for relevance in cyber physical production systems. In: 20th IEEE Conference on Emerging Technologies & Factory Automation, ETFA, Luxembourg, pp. 1–8 (2015)
32. Volz, J., Bizer, C., Gaedke, M., Kobilarov, G.: Silk - a link discovery framework for the web of data. In: Proceedings of the WWW 2009 Workshop on Linked Data on the Web, LDOW, Madrid, Spain, 20 April 2009

Cloud Data Processing

Efficient Top-k Cloud Services Query Processing Using Trust and QoS

Karim Benouaret[1]([⊠]), Idir Benouaret[2], Mahmoud Barhamgi[1],
and Djamal Benslimane[1]

[1] Univ Lyon, Université Claude Bernard Lyon 1, CNRS, LIRIS,
69622 Villeurbanne, France
{karim.benouaret,mahmoud.barhamgi,djamal.benslimane}@liris.cnrs.fr
[2] Univ Lyon, Université Jean-Monnet-Saint-Étienne, CNRS,
Laboratoire Hubert Curien, 42000 Saint-Étienne, France
idir.benouaret@univ-st-etienne.fr

Abstract. The growing number of cloud service providers has led to an exploding number of functionally similar cloud services, with a wide range choices of non-functional properties (NFPs). Thus, selecting services based on NFPs becomes of significant importance. Current approaches assume that a cloud service provider offers a single service per functionality, therefore, they do not perform well in the real-life setting where each cloud service provider offers different service plans. In contrast, in this paper, we propose an approach to select top-k cloud services that is built taking into account the real-life setting. Our approach combines the trust, determined by the reputation of the provider, and the QoS. We present different algorithms for processing such selection queries and evaluate them through a set of experiments.

Keywords: Cloud service selection · QoS · Trust · Reputation

1 Introduction

Cloud services are designed to provide easy and scalable access to applications, resources and services, and are fully managed by cloud service providers. Examples of cloud services include cloud storage, cloud backup, cloud hosting and cloud accounting services.

As cloud services are so widely used among individuals and businesses, cloud providers are competing with each other to offer similar services at different prices and performance levels, i.e., non-functional properties (NFPs). Hence, selecting services based on NFPs becomes of significant importance. NFPs of a cloud service mainly contain QoS (e.g., price and storage size) and users' feedbacks, determining the reputation of its provider, for trust concerns; i.e., the reputation of a provider represents a general opinion about how good its provided services have been rated. In other words, reputation is considered as a

© Springer Nature Switzerland AG 2018
S. Hartmann et al. (Eds.): DEXA 2018, LNCS 11029, pp. 203–217, 2018.
https://doi.org/10.1007/978-3-319-98809-2_13

collective measure of trustworthiness as illustrated by the following statement: "I trust you because of your good reputation" [7].

To illustrate this point, assume the existence of a cloud service search engine that is connected to a large number of cloud service providers. In a typical scenario, a user provides the type of the requested cloud service, subsequently, the search engine issues a query to the different providers in order to get the possible service plans. Then, the user selects the desired service. After use, he/she provides a rating, which is used to determine the reputation of the provider. However, as cloud service providers proliferate, the selection task would be very painful for the user. Therefore, a sophisticated cloud service search engine needs to identify the best candidates.

Table 1. A sample provided cloud storage services

Provider	Reputation	Service	Price ($)	Storage size (GB)
p_1	0.9	s_{11}	20	2000
		s_{12}	35	4000
p_2	0.8	s_{21}	20	2500
		s_{22}	35	4000
p_3	0.7	s_{31}	30	1000
		s_{32}	35	2000
		s_{33}	40	3000
p_4	0.6	s_{41}	40	2000
		s_{42}	45	3000
		s_{43}	50	4500
p_5	0.4	s_{51}	20	3000
		s_{52}	30	4000
		s_{53}	35	5000

Obviously, the perfect cloud service, i.e., service that dominates the other services on all NFPs, is seldom found because of the trade-offs offered by the providers. Roughly speaking, the more you pay, the more you get. Table 1 illustrates this with an example of a set of cloud service providers offering various cloud storage services. In this example, each provider has its own reputation, and provides different service plans. For instance, provider p_1 offers services s_{11} and s_{12}. The services are specified with their cost and storage size. Observe that regarding the trust, the services advertised by provider p_1 are the most favorable, while under QoS, provider p_5 offers the most interesting service plans. However, more often, users put almost the same emphasis on trust and QoS [16]. While some approaches combining the trust and QoS for cloud service selection exist; e.g., [9,16], they are not designed to the real-life setting for two main reasons. First, they assume that a provider can offer only a single service per functionality. Second, they suppose that the advertised QoS does not change over time.

As a consequence, the reputation is naturally assigned to the service, and not to the provider. However, in real-life setting, each cloud service provider offers different service plans to reach a maximum number of customers. In addition, the advertised QoS may vary over time for different reasons (e.g., for economic purposes). Thus, it makes better sense to assign the reputation to the provider, and not to its services. As a result, these approaches do not perform well in the real-life setting.

For such a critical issue, industry cloud service search engines, such as *Cloudwards*[1], follow another direction to help potential users find their desired cloud services, usually by identifying the top-k providers based on their reputation (e.g., average rating). Users are then required to probe the service plans advertised by the providers to find their desired services. However, this task remains painful for users.

To effectively deal with this problem, it is imperative to provide the results in a useful form. In this paper, we propose a method for retrieving the top-k cloud service plans, instead of the top-k cloud service providers.

Processing such a top-k query over the real-life setting is challenging, since the cloud service search engine has to interact with the different cloud service providers at query time as the advertised QoS may vary over time. Various algorithms are proposed to process top-k queries; see [6] for a survey. The most popular is Threshold Algorithm (TA) proposed by Fagin et al. [2]. Unfortunately, these algorithms are not designed for cloud service search engines. In fact, as we will see, an adaptation of TA do not perform well. In this paper, we propose a novel strategy to efficiently retrieve the top-k cloud services. We also present a naive algorithm and an adaptation of TA, and compare these techniques through a set of experiments.

To sum up, the main contributions of our work are the following:

- We introduce and formally define the problem of top-k cloud services in the real-life setting;
- We develop an adaptation of TA for computing the top-k cloud services and propose a more efficient algorithm;
- We perform an experimental study to evaluate the efficiency of these algorithms.

The remainder of this paper is organized as follows. In Sect. 2, we formally define the problem of top-k cloud services. Then, in Sect. 3, we describe the top-k cloud services query processing algorithms, and evaluate them in Sect. 4. In Sect. 5, we present related work. Finally, Sect. 6 concludes the paper.

2   Problem Definition

In this section, we present our terminology, and define the problem of top-k cloud services. For reference, Table 2 gives the used notation.

---

[1] https://www.cloudwards.net.

Table 2. Notation

Symbol	Description
$\mathcal{P}$, p_i	Set of providers, provider
$\mathcal{S}_i$, s_{ij}	Set of services provided by p_i, service provided by p_i
$\mathcal{Q}$, q_k	Set of QoS parameters, QoS parameter
$dom(q_k)$	Domain of q_k
q_k^-, q_k^+	Lower bound of $dom(q_k)$, upper bound of $dom(q_k)$
$s_{ij}.q_k$	Value of s_{ij} on q_k
$s_{ij}.q_k\prime$	Normalized value of $s_{ij}.q_k$
$s_r(p_i)$	Reputation score of p_i
$s_q(s_{ij})$	QoS score of s_{ij}
$s_g(s_{ij})$	Global score of s_{ij}
$p_i.q_k^*$	The best value of q_k
$p_i.q_k^*\prime$	Normalized value of $p_i.q_k^*$
$s_q^{max}(p_i)$	Maximal attainable QoS score of services offered by p_i
$s_g^{max}(p_i)$	Maximal attainable global score of services offered by p_i

Given a set of cloud service providers $\mathcal{P} = \{p_1, p_2, \ldots, p_n\}$, where each provider p_i offers a set of cloud services $\mathcal{S}_i = \{s_{i1}, s_{i2}, \ldots, s_{im_i}\}$ defined on a set of QoS parameters $\mathcal{Q} = \{q_1, q_2, \ldots, q_d\}$. The domain of each QoS parameter q_k is $dom(q_k)$. We use q_k^- and q_k^+ to denote respectively the lower bound and the upper bound of $dom(q_k)$ and $s_{ij}.q_k$ to denote the value of service s_{ij} on QoS parameter q_k.

Further, assume that each provider p_i has a reputation score according to the historical invocations of its provided services. The reputation score of a provider p_i can be calculated as:

$$s_r(p_i) = 1/u \sum_{\ell=1}^{u} r_{i\ell} \qquad (1)$$

Where u is the total number of evaluator users for provider p_i, and $r_{i\ell}$ represents the rating of user ℓ on provider p_i.

Now, given a cloud service s_{ij}, its QoS score is defined as follows:

$$s_q(s_{ij}) = \sum_{k=1}^{d} w_k \times s_{ij}.q_k\prime \qquad (2)$$

Where w_k is the weight associated to QoS parameter q_k, such as $w_k \in [0, 1]$ and $\sum_{k=1}^{d} w_k = 1$ and $s_{ij}.q_k\prime$ is the normalized value of $s_{ij}.q_k$. The normalization is done as follows. For negative QoS parameters, i.e., the higher the value, the lower the quality (e.g., price), the values are normalized according to Eq. 3. For positive QoS parameters, i.e., the higher the value, the higher the quality (e.g., storage size), the values are normalized according to Eq. 4.

$$s_{ij}.q_k\prime = \begin{cases} \frac{q_k^+ - s_{ij}.q_k}{q_k^+ - q_k^-} & \text{if } q_k^+ - q_k^- \neq 0 \\ 1 & \text{if } q_k^+ - q_k^- = 0 \end{cases} \tag{3}$$

$$s_{ij}.q_k\prime = \begin{cases} \frac{s_{ij}.q_k - q_k^-}{q_k^+ - q_k^-} & \text{if } q_k^+ - q_k^- \neq 0 \\ 1 & \text{if } q_k^+ - q_k^- = 0 \end{cases} \tag{4}$$

Consequently, the global score of a cloud service s_{ij} aggregating these scores is defined as follows[2]:

$$s_g(s_{ij}) = \lambda \times s_r(p_i) + (1 - \lambda) \times s_q(s_{ij}) \tag{5}$$

Where the parameter λ determines the trade-off between the two factors trust and QoS.

The result of a top-k cloud services query is the ranked list of the k cloud services with the highest global scores; we break ties arbitrarily for simplicity and ease of presentation.

We are now ready to state the problem of top-k cloud services.

Problem Statement. Given a set of cloud service providers $\mathcal{P} = \{p_1, p_2, \ldots, p_n\}$, where each provider p_i offers a set of cloud services $\mathcal{S}_i = \{s_{i1}, s_{i2}, \ldots, s_{im_i}\}$ defined on a set of QoS parameters $\mathcal{Q} = \{q_1, q_2, \ldots, q_d\}$, and a requested cloud service specified by a set of weights $\mathcal{W} = \{w_1, w_2, \ldots, w_d\}$ on $\mathcal{Q}$ and an emphasis factor λ. Return the top-k cloud services according to their global scores.

Table 3. Services' scores

s_{ij}	$s_r(p_i)$	$s_q(s_{ij})$	$s_g(s_{ij})$
s_{11}	0.9	0.7	0.80
s_{12}		0.6	0.75
s_{21}	0.8	0.75	0.775
s_{22}		0.6	0.70
s_{31}	0.7	0.4	0.55
s_{32}		0.4	0.55
s_{33}		0.4	0.55
s_{41}	0.6	0.3	0.45
s_{42}		0.3	0.45
s_{43}		0.35	0.475
s_{51}	0.4	0.8	0.6
s_{52}		0.7	0.55
s_{53}		0.7	0.55

---

[2] Our model and associated algorithms can handle other scoring functions of $s_r(p_i)$, $s_q(s_{ij})$ and $s_g(s_{ij})$.

3 Computing the Top-k Cloud Services

In this section, we present algorithms for computing the top-k cloud services. Specifically, in Sect. 3.1, we present a naive but expensive approach. Then, in Sect. 3.2, we adapt TA to our context. Finally, in Sect. 3.3, we introduce our novel algorithm for computing efficiently the top-k cloud services.

To show the flow of these algorithms, consider in the following a top-3 query, with emphasis factor $\lambda = 0.5$ (i.e., trust and QoS have the same emphasis), and assume that the weights of price and storage size parameters are 0.6 and 0.4, respectively. Table 3 shows the services of our example and their corresponding scores. Observe that the top-3 services are s_{11}, s_{21} and s_{12} (in the right order).

3.1 The Naive Algorithm

A naive strategy to compute the top-k cloud services is to interact with each provider p_i in $\mathcal{P}$ to get its different service plans $\mathcal{S}_i$, calculating their corresponding scores, and finally returning the k services with the highest scores. The pseudocode of this algorithm is depicted in Algorithm 1. Notice, however, that this algorithm is time consuming since it needs to interact with all providers and compute the scores of all services.

Algorithm 1. NA

    Input   : set of providers $\mathcal{P}$; set of weights $\mathcal{W}$; emphasis factor λ;
    Output: top-k cloud services $\mathcal{R}$;

1 **foreach** $p_i \in \mathcal{P}$ **do**
2     $\mathcal{S}_i \leftarrow$ get service plans from p_i;
3   **foreach** $s_{ij} \in \mathcal{S}_i$ **do**
4   compute $s_g(s_{ij})$;

5 $\mathcal{R} \leftarrow$ compute top-k services;
6 **return** $\mathcal{R}$;

Applying NA on our example, the scores of all services will be computed and services s_{11}, s_{21} and s_{12} will be returned.

3.2 The Threshold Algorithm Adaptation

Our first approach to compute the top-k cloud services is an adaptation of TA [2]. The algorithm is based on the following property.

Lemma 1. *Consider a top-k cloud services query and assume that at some point in time a set of k services are retrieved. Suppose that a provider p_i has a reputation score $s_r(p_i)$ such as $\lambda \cdot s_r(p_i) + 1 - \lambda$ is lower or equal than the global score of every retrieved service. Then, the services provided by p_i are not part of the top-k services.*

Proof. Let $s_{ij} \in \mathcal{S}_i$. As $s_q(s_{ij}) \leq 1$, it holds that $s_g(s_{ij}) \leq \lambda \cdot s_r(p_i) + 1 - \lambda$. Then, since $\lambda \cdot s_r(p_i) + 1 - \lambda$ is lower or equal than the global score of every retrieved service, $s_g(s_{ij})$ is lower or equal than the global score of every retrieved service. Hence, s_{ij} is not part of the top-k services – recall that we break ties arbitrarily.

Lemma 1 provides a termination condition. In fact, to exploit this property, we maintain the list of the providers sorted in non-ascending order of their reputation scores. Then, if the property holds for a given provider p_i, it also holds for the providers located after p_i, i.e., services offered by p_i and those offered by the providers that are located after p_i are not part of the top-k services.

Algorithm 2. TAA

    Input : set of providers $\mathcal{P}$ sorted in non-ascending order of reputation score; set
            of weights $\mathcal{W}$; emphasis factor λ;
    Output: top-k cloud services $\mathcal{R}$;

1 $s_g^{min} \leftarrow 0$;
2 **foreach** $p_i \in \mathcal{P}$ **do**
3     $t \leftarrow \lambda \cdot s_r(p_i) + 1 - \lambda$;
4   **if** $|\mathcal{R}| = k \wedge t \leq s_g^{min}$ **then**
5 | *break*;
6   **else**
7         $\mathcal{S}_i \leftarrow$ get service plans from p_i;
8   **foreach** $s_{ij} \in \mathcal{S}_i$ **do**
9   compute $s_g(s_{ij})$;
10   **if** $s_g(s_{ij}) > s_g^{min}$ **then**
11   **if** $|\mathcal{R}| = k$ **then**
12   remove the worst service from $\mathcal{R}$;
13   insert s_{ij} into $\mathcal{R}$;
14   update s_g^{min};
15   **else**
16   insert s_{ij} into $\mathcal{R}$;

17 **return** $\mathcal{R}$;

Algorithm 2 presents the pseudocode of the Threshold Algorithm Adaptation (TAA). The algorithm maintains the list of the providers sorted in non-ascending order of the reputation scores, and uses two variables: s_g^{min} which stores the minimal global score of the top-k services discovered so far, and a threshold t which determines the termination condition. Initially, s_g^{min} is set to 0 (line 1); t does not need to be initialized. Then, the algorithm iterates over the providers (loop in line 2). At each step, the threshold t is updated according to the current provider p_i as avowed in Lemma 1 (line 3). If the termination condition is reached (line 4) then the algorithm breaks out of for-loop (line 5) and returns the result set $\mathcal{R}$ (line 17); otherwise, TAA interacts with the provider p_i to get its different service plans $\mathcal{S}_i$ (line 7), and iterates over $\mathcal{S}_i$ (loop in line 8) for computing the scores of each service $s_{ij} \in \mathcal{S}_i$ (line 9) and updates (or fills) the current top-k services set $\mathcal{R}$ (lines 10–16). If all providers are examined (i.e., the termination condition is not reached) the result set $\mathcal{R}$ is returned (line 17).

Applying TAA on our example, the scores of the services provided by p_1, p_2, p_3 and p_4 will be computed and services s_{11}, s_{21} and s_{12} will be returned. The scores of the services provided by p_5 will not be computed as $0.4 \cdot 0.5 + 0.5 = 0.7$ is lower than $s_g(s12) = 0.75$, i.e., the termination condition will be reached.

3.3 The Double Threshold Algorithm

Hereafter, we present a novel algorithm called Double Threshold Algorithm (DTA) for computing the top-k cloud services. This algorithm leads to efficient executions by minimizing the number of computed scores. In fact, the key ideas of DTA are: (1) the use of the termination condition, previously described, and (2) the definition of upper bounds for the global scores of the services of each provider, so as the number of computed scores will be minimized.

Given a provider $p_i \in \mathcal{P}$ and a QoS parameter $q_k \in \mathcal{Q}$. Let $p_i.q_k^\star$ be the best value of q_k proposed by p_i, i.e., $p_i.q_k^\star = \min_{s_{ij} \in \mathcal{S}_i} s_{ij}.q_k$ for negative QoS parameters and $p_i.q_k^\star = \max_{s_{ij} \in \mathcal{S}_i} s_{ij}.q_k$ for positive QoS parameters. For instance, the best values of the price and the storage size regarding provider p_3 are 30 and 3000 respectively.

Then, we define the maximal attainable QoS score of any service provided by p_i as:

$$s_q^{max}(p_i) = \sum_{k=1}^{d} w_k \times p_i.q_k^{\star\prime} \tag{6}$$

Where w_k is the weight associated to QoS parameter q_k and $p_i.q_k^{\star\prime}$ is the normalized value of $p_i.q_k^\star$. The normalization is done as follows. For negative QoS parameters, the values are normalized according to Eq. 7. For positive QoS parameters, the values are normalized according to Eq. 8.

$$p_i.q_k^{\star\prime} = \begin{cases} \frac{q_k^+ - p_i.q_k^\star}{q_k^+ - q_k^-} & \text{if } q_k^+ - q_k^- \neq 0 \\ 1 & \text{if } q_k^+ - q_k^- = 0 \end{cases} \tag{7}$$

$$p_i.q_k^{\star\prime} = \begin{cases} \frac{p_i.q_k^\star - q_k^-}{q_k^+ - q_k^-} & \text{if } q_k^+ - q_k^- \neq 0 \\ 1 & \text{if } q_k^+ - q_k^- = 0 \end{cases} \tag{8}$$

Consequently, the maximal attainable global score of any service provided by p_i is defined as follows:

$$s_g(p_i)^{max} = \lambda \times s_r(p_i) + (1 - \lambda) \times s_q^{max}(p_i) \tag{9}$$

Table 4 shows the maximal attainable QoS scores and the maximal attainable global scores of any service provided by each provider of our example.

DTA is based on Lemma 1 and the following key property.

Table 4. Maximal attainable scores

p_i	$s_r(p_i)$	$s_q^{max}(p_i)$	$s_g^{max}(p_i)$
p_1	0.9	0.90	0.90
p_2	0.8	0.90	0.85
p_3	0.7	0.60	0.65
p_4	0.6	0.55	0.575
p_5	0.4	1.00	0.70

Lemma 2. *Consider a top-k cloud services query and suppose that at some point in time a set of k services are retrieved. Consider a provider p_i such as the maximal attainable global score of any of its services $s_g(p_i)^{max}$ is lower or equal than the global score of every retrieved service. Then, the services provided by p_i are not part of the top-k services.*

Proof. It is apparent since k services with higher global scores are retrieved so far – recall that we break ties arbitrarily.

Lemma 2 helps minimize the number of computed scores. In fact, if this property holds for a given provider p_i. It is unnecessary to compute the scores of the services provided by p_i.

DTA is presented in Algorithm 3. As TAA, DTA maintains the list of the providers sorted in non-ascending order of the reputation scores. DTA, uses three variables: s_g^{min} which stores the minimal global score of the top-k services discovered so far, a threshold t_p (for providers) which determines the termination condition, and a threshold t_s (for services) to exploit Lemma 2. Initially, s_g^{min} is set to 0 (line 1); t_p and t_s do not need to be initialized. Then, the algorithm iterates over the providers (loop in line 2). At each step, the threshold t_p is updated according to the current provider p_i as avowed in Lemma 1 (line 3). If the termination condition is reached (line 4) then the algorithm breaks out of for-loop (line 5) and returns the result set $\mathcal{R}$ (line 21); otherwise, DTA interacts with the provider p_i to get its different service plans $\mathcal{S}_i$ (line 7) and t_s is set to the maximal attainable global score of any service provided by p_i (line 8) in order to exploit Lemma 2. In fact, if the condition in line 9 is satisfied then $\mathcal{S}_i$ is discarded (line 10) since every service that belongs to $\mathcal{S}_i$ is not part of the top-k services according to Lemma 2; otherwise, DTA iterates over $\mathcal{S}_i$ (loop in line 12) for computing the scores of each service $s_{ij} \in \mathcal{S}_i$ (line 13) and updates (or fills) the current top-k services set $\mathcal{R}$ (lines 14–20). If all providers are examined (i.e., the termination condition is not reached) the result set $\mathcal{R}$ is returned (line 21).

Applying DTA on our example, the scores of the services provided by p_1 and p_2 will be computed. The scores of the services provided by p_3 and p_4 will not be computed since $s_g^{max}(p_3) = 0.65$ and $s_g^{max}(p_4) = 0.575$ are lower than $s_g(s12) = 0.75$. Then, services s_{11}, s_{21} and s_{12} will be returned. The scores of

---

Algorithm 3. DTA

Input : set of providers $\mathcal{P}$ sorted in non-ascending order of reputation score; set
          of weights $\mathcal{W}$; emphasis factor λ;
Output: top-k cloud services $\mathcal{R}$;

1   $s_g^{min} \leftarrow 0$;
2   **foreach** $p_i \in \mathcal{P}$ **do**
3      $t_p \leftarrow \lambda \cdot s_r(p_i) + 1 - \lambda$;
4   **if** $|\mathcal{R}| = k \wedge t_p \leq s_g^{min}$ **then**
5   *break*;
6   **else**
7         $\mathcal{S}_i \leftarrow$ get service plans from p_i;
8         $t_s \leftarrow$ compute $s_g(p_i)^{max}$;
9   **if** $|\mathcal{R}| = k \wedge t_s \leq s_g^{min}$ **then**
10   discard $\mathcal{S}_i$;
11   **else**
12   **foreach** $s_{ij} \in \mathcal{S}_i$ **do**
13   compute $s_g(s_{ij})$;
14   **if** $s_g(s_{ij}) > s_g^{min}$ **then**
15   **if** $|\mathcal{R}| = k$ **then**
16   remove the worst service from $\mathcal{R}$;
17   insert s_{ij} into $\mathcal{R}$;
18   update s_g^{min};
19   **else**
20   insert s_{ij} into $\mathcal{R}$;

21 **return** $\mathcal{R}$;

---

the services provided by p_5 will not be computed as $0.4 \cdot 0.5 + 0.5 = 0.7$ is lower than $s_g(s12) = 0.75$, i.e., the termination condition will be reached.

4   Experimental Evaluation

In this section, we evaluate the performance of the algorithms presented in Sect. 3.

Because real datasets are limited for evaluating extensive settings, we implemented a dataset generator. The providers and their offered services are generated following three distributions: (1) *correlated*, where the reputation of the providers and the QoS parameters of their offered services are positively correlated, i.e., a good reputation of a given provider increases the possibility of good QoS values of its offered services; (2) *independent*, where the reputation of the providers and the QoS values of their offered services are assigned independently; and (3) *anti-corretaled*, where the reputation of the providers and the QoS parameters of their offered services are negatively correlated, i.e., a good

Table 5. Parameters and examined values

Parameter	Values
Number of providers (n)	10K, 50K, **100K**, 500K, 1M
Number of services per provider (m)	30, 40, **50**, 60, 70
Number of QoS parameters (d)	5, 6, **7**, 8, 9
Number of requested services (k)	10, 20, **30**, 40, 50
Emphasis factor (λ)	0.1, 0.3, **0.5**, 0.7, 0.9

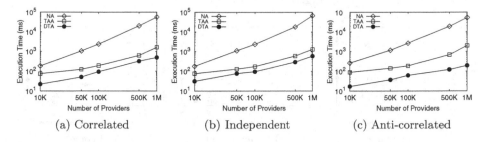

(a) Correlated (b) Independent (c) Anti-correlated

Fig. 1. Execution time vs n

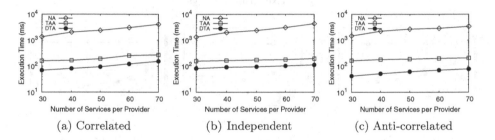

(a) Correlated (b) Independent (c) Anti-correlated

Fig. 2. Execution time vs m

reputation of a given provider increases the possibility of bad QoS values of its offered services.

The involved parameters and their examined values are summarized in Table 5. In all experimental setups, we investigate the effect of one parameter, while setting the remaining ones to their default values, shown bold in Table 5.

The algorithms were implemented in Java, and all experiments were conducted on a 3.0 GHz Intel Core i7 processor with 8 GB RAM, running Windows.

Varying n: In the first experiment, we study the impact of n. The results are shown in Fig. 1. As expected, when the n increases, the performance of all algorithms deteriorates since more providers and services have to be evaluated.

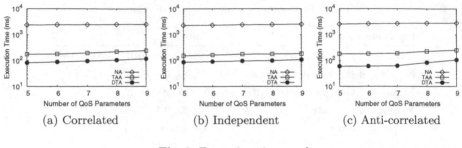

(a) Correlated   (b) Independent   (c) Anti-correlated

Fig. 3. Execution time vs d

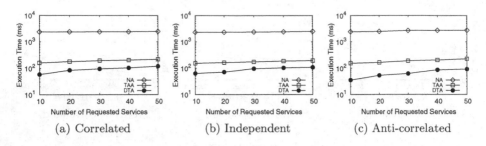

(a) Correlated   (b) Independent   (c) Anti-correlated

Fig. 4. Execution time vs k

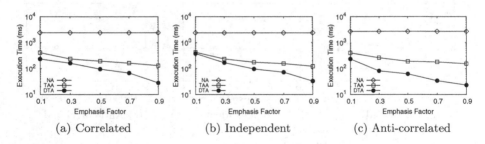

(a) Correlated   (b) Independent   (c) Anti-correlated

Fig. 5. Execution time vs λ

Varying m: In second experiment, we investigate the effect of m. Figure 2 shows the results of this experiment. The execution time of the three algorithms increases with the increase of m as more services have to be evaluated.

Varying d: In the next experiment, we consider the impact of d. The results are depicted in Fig. 3. The execution time of all algorithms increases as d increases since more time is required to computed the QoS scores of the services.

Varying k: In this experiment, we investigate the effect of k. Figure 4 shows the results of this experiment. As k increases, the execution time of the three algorithms increases, since all algorithms need to retrieve more services.

Varying λ: In the last experiment, we study the effect of λ. Figure 5 depicts the results of this experiment. Contrary to the other parameters, the performance of

NA remains stable, while TAA and DTA run faster with higher λ, since NA need to compute all scores, which is not affected by λ, while the termination condition used by both TAA and DTA is sensible to λ. Indeed, when λ increases, the global scores of services are dominated by the reputation of their providers. Thus, the termination condition is reached earlier.

Overall, the results indicate that DTA consistently outperforms both NA and TAA. In other words, the results clearly demonstrate that the optimization techniques employed by DTA significantly save the cost of computing. In addition, observe that in contrast to NA and TAA, DTA runs faster on anti-correlated datasets. This is because, in anti-correlated datasets providers with good reputation are more likely to offer services with bad QoS values. Therefore, the maximal attainable global scores of their provided services will be bad. Hence, more providers will be discarded.

5   Related Work

With the proliferation of cloud service providers and cloud services over the web, the problem of cloud service selection has received much attention in recent years.

Optimization-based approaches are proposed. In [1], the authors develop a dynamic programming algorithm for selecting cloud storage service providers that maximize the amont of surviving data, subject to a fixed budget. In [15], the authors develop a greedy algorithm for cloud service selection. The algorithm is based on a B+-Tree, which indexes cloud service provider and encodes services and user requirements. Zheng *et al.* propose in [18] a personalized QoS ranking prediction framework for cloud services based on collaborative filtering. By taking advantage of the past usage experiences of other users, their approach identifies and aggregates the preferences between pairs of services to produce a ranking of services. He *et al.* propose in [5] the use of integer programming, skyline and greedy techniques to help SaaS developers determine the optimal services. In [10], the authors propose a decision model for discrete dynamic optimization problems in cloud service selection to help organization identify appropriate cloud services by minimizing costs and risks.

Some approaches are based on simple aggregating functions. Zeng *et al.* propose in [17] algorithms for cloud service selection. The algorithms are based on a utility function, which determines the trade-off between the minimized cost and the maximized gain. In [9], the authors present a reputation-based framework for SaaS service rating and selection. The proposed service rating allows feedbacks from users. A reputation derivation model is also proposed to aggregate feedbacks into a reputation value. A selection algorithm based on a ranking function that aggregates the quality, cost, and reputation parameters is designed to assist customers in selecting the most appropriate service. In [14], the authors propose an effective service selection middleware for cloud environment. The service selection is based on ELECTRE; many parameters such as, service cost, trust, scalability, etc. are considered. Martens *et al.* propose in [11] a community platform, which assists companies and users to select appropriate cloud services.

Users have the option of evaluating individual services. The authors introduce a model for the quality assessment of cloud services. The model measures the distance between the cloud service and the user requirements in order to indicate the degree of compliance with the user requirements. The degree of compliance is computed as a weighted average function.

Other approaches use Analytic Hierarchy Process (AHP) and Analytic Network Process (ANP) techniques. Godse and Mulik propose in [4] an approach for ranking SaaS services based on AHP. The relative importance of service parameters is weighted by aggregating user preferences and domain experts' opinions. Garg *et al.* propose in [3] an AHP-based framework for ranking cloud service according to a number of performance parameters defined by the Cloud Services Measurement Initiative Consortium (CSMIC) [13]. In [8], the authors propose an AHP-based ranking method for IaaS and SaaS services. The QoS parameters are layered and categorized based on their influential relations. Mapping rules are defined in order to get the best service combination of IaaS and SaaS. In [12], the authors propose an ANP-based framework for IaaS service selection. The framework is based on a comprehensive parameters catalogue, which differentiates cloud infrastructures in a variety of dimensions: cost, benefits, opportunities and risks.

However, as mentioned in Sect. 1, these approaches are not designed to the real-life settings contrary to our work.

6   Conclusion

In this paper, we addressed the issue of finding top-k cloud services in the real-life setting. We formally defined the problem and studied its characteristics. We then presented a naive algorithm and showed how to adapt TA so that it can handle the problem of top-k cloud services in the real-life setting, and also proposed a novel algorithm. Our experimental evaluation demonstrated that our algorithm produces the best execution time for various parameter and a variety of dataset distributions.

As a future work, we intend to consider the case where the query involves multiple users, e.g., the department heads of a university that would like to obtain a software license of a cloud-based data analytics service.

References

1. Chang, C., Liu, P., Wu, J.: Probability-based cloud storage providers selection algorithms with maximum availability. In: Proceedings of the International Conference on Parallel Processing, ICPP 2012, Pittsburgh, PA, USA, 10–13 September 2012, pp. 199–208 (2012)
2. Fagin, R., Lotem, A., Naor, M.: Optimal aggregation algorithms for middleware. J. Comput. Syst. Sci. **66**(4), 614–656 (2003)
3. Garg, S.K., Versteeg, S., Buyya, R.: SMICloud: a framework for comparing and ranking cloud services. In: Proceedings of the IEEE/ACM International Conference on Utility and Cloud Computing, UCC 2011, pp. 210–218 (2011)

4. Godse, M., Mulik, S.: An approach for selecting software-as-a-service (SaaS) product. In: Proceedings of the IEEE International Conference on Cloud Computing, IEEE CLOUD, pp. 155–158 (2009)
5. He, Q., Han, J., Yang, Y., Grundy, J., Jin, H.: QoS-driven service selection for multi-tenant SaaS. In: Proceedings of the IEEE International Conference on Cloud Computing, IEEE CLOUD, pp. 566–573 (2012)
6. Ilyas, I.F., Beskales, G., Soliman, M.A.: A survey of top-k query processing techniques in relational database systems. ACM Comput. Surv. (CSUR) 40(4), 11:1–11:58 (2008)
7. Jøsang, A., Ismail, R., Boyd, C.: A survey of trust and reputation systems for online service provision. Decis. Support Syst. 43(2), 618–644 (2007)
8. Karim, R., Ding, C.C., Miri, A.: An end-to-end QoS mapping approach for cloud service selection. In: Proceedings of the IEEE World Congress on Services, IEEE SERVICES, pp. 341–348 (2013)
9. Limam, N., Boutaba, R.: Assessing software service quality and trustworthiness at selection time. IEEE Trans. Softw. Eng. (TSE) 36(4), 559–574 (2010)
10. Martens, B., Teuteberg, F.: Decision-making in cloud computing environments: a cost and risk based approach. Inform. Syst. Front. 14(4), 871–893 (2012)
11. Martens, B., Teuteberg, F., Gräuler, M.: Design and implementation of a community platform for the evaluation and selection of cloud computing services: a market analysis. In: Proceedings of the European Conference on Information Systems, ECIS 2011, p. 215 (2011)
12. Menzel, M., Schönherr, M., Tai, S.: $(MC^2)^2$: criteria, requirements and a software prototype for cloud infrastructure decisions. Softw. Pract. Exp. 43(11), 1283–1297 (2013)
13. Siegel, J., Perdue, J.: Cloud services measures for global use: the service measurement index (SMI). In: Proceedings of the Annual SRII Global Conference, SRII, pp. 411–415 (2012)
14. Silas, S., Rajsingh, E.B., Ezra, K.: Efficient service selection middleware using electre methodology for cloud environments. Inf. Technol. J. 11(7), 868 (2012)
15. Sundareswaran, S., Squicciarini, A.C., Lin, D.: A brokerage-based approach for cloud service selection. In: Proceedings of the IEEE International Conference on Cloud Computing, IEEE CLOUD, pp. 558–565 (2012)
16. Wang, H., Yu, C., Wang, L., Yu, Q.: Effective bigdata-space service selection over trust and heterogeneous QoS preferences. IEEE Trans. Serv. Comput. (TSC) (Forthcoming)
17. Zeng, W., Zhao, Y., Zeng, J.: Cloud service and service selection algorithm research. In: Proceedings of the World Summit on Genetic and Evolutionary Computation, GEC Summit, pp. 1045–1048 (2009)
18. Zheng, Z., Wu, X., Zhang, Y., Lyu, M.R., Wang, J.: QoS ranking prediction for cloud services. IEEE Trans. Parallel Distrib. Syst. TPDS 24(6), 1213–1222 (2013)

Answering Top-k Queries over Outsourced Sensitive Data in the Cloud

Sakina Mahboubi$^{(\boxtimes)}$, Reza Akbarinia, and Patrick Valduriez

INRIA and LIRMM, University of Montpellier, Montpellier, France
{Sakina.Mahboubi,Reza.Akbarinia,Patrick.Valduriez}@inria.fr

Abstract. The cloud provides users and companies with powerful capabilities to store and process their data in third-party data centers. However, the privacy of the outsourced data is not guaranteed by the cloud providers. One solution for protecting the user data is to encrypt it before sending to the cloud. Then, the main problem is to evaluate user queries over the encrypted data.

In this paper, we consider the problem of answering top-k queries over encrypted data. We propose a novel system, called BuckTop, designed to encrypt and outsource the user sensitive data to the cloud. BuckTop comes with a top-k query processing algorithm that is able to process efficiently top-k queries over the encrypted data, without decrypting the data in the cloud data centers.

We implemented BuckTop and compared its performance for processing top-k queries over encrypted data with that of the popular threshold algorithm (TA) over original (plaintext) data. The results show the effectiveness of BuckTop for outsourcing sensitive data in the cloud and answering top-k queries.

Keywords: Cloud · Sensitive data · Top-k query

1 Introduction

The cloud allows users and companies to efficiently store and process their data in third-party data centers. However, users typically loose physical access control to their data. Thus, potentially sensitive data gets at risk of security attacks, *e.g.*, from employees of the cloud provider. According to a recent report published by the Cloud Security Alliance [4], security attacks are one of the main concerns for cloud users.

One solution for protecting user sensitive data is to encrypt it before sending to the cloud. Then, the challenge is to answer user queries over encrypted data. A naive solution for answering queries is to retrieve the encrypted database from the cloud to the client, decrypt it, and then evaluate the queries over *plaintext (non encrypted)* data. This solution is inefficient, because it does not take advantage of the cloud computing power for evaluating queries.

In this paper, we are interested in processing top-k queries over encrypted data in the cloud. A top-k query allows the user to specify a number k, and the

S. Hartmann et al. (Eds.): DEXA 2018, LNCS 11029, pp. 218–231, 2018.
https://doi.org/10.1007/978-3-319-98809-2_14

system returns the k tuples which are most relevant to the query. The relevance degree of tuples to the query is determined by a *scoring function*.

Top-k query processing over encrypted data is critical for many applications that outsource sensitive data. For example, consider a university that outsources the students database in a public cloud, with non-trusted nodes. The database is encrypted for privacy reasons. Then, an interesting top-k query over the outsourced encrypted data is the following: return the k students that have the worst averages in some given courses.

There are many different approaches for processing top-k queries over plaintext data. One of the best known approaches is TA (threshold algorithm) [8] that works on sorted lists of attribute values. TA can find efficiently the top-k results because of a smart strategy for deciding when to stop reading the database. However, TA and its extensions assume that the attribute values are available as plaintext, and not encrypted.

In this paper, we address the problem of privacy preserving top-k query processing in clouds. We first propose a basic approach, called OPE-based, that uses a combination of the order preserving encryption (OPE) and the FA algorithm for privacy preserving top-k query processing.

Then, we propose a complete system, called *BuckTop*, that is able to efficiently evaluate top-k queries over encrypted data, without decrypting them in the cloud. BuckTop includes a top-k query processing algorithm that works on the encrypted data, and returns a set that is proved to contain the encrypted data corresponding to the top-k results. It also comes with an efficient filtering algorithm that is executed in the cloud and removes most of the false positives included in the set returned by the top-k query processing algorithm. This filtering is done without needing to decrypt the data in the cloud.

We implemented BuckTop, and compared its response time over encrypted data with a baseline algorithm and with TA over original (plaintext) data. The experimental results show excellent performance gains for BuckTop. For example, the results show that the response time of BuckTop over encrypted data is close to TA over plaintext data. The results also illustrate that more than 99.9% of the false positives can be eliminated in the cloud by BuckTop's filtering algorithm.

The rest of this paper is organized as follows. Section 2 gives the problem definition. Section 3 presents our basic approach for privacy preserving top-k query processing. Section 4 describes our BuckTop system and its algorithms. Section 5 reports performance evaluation results. Section 6 discusses related work, and Sect. 7 concludes.

2   Problem Definition

In this paper, we address the problem of processing top-k queries over encrypted data in the cloud.

By a top-k query, the user specifies a number k, and the system should return the k most relevant answers. The relevance degree of the answers to the query is determined by a *scoring function*. A common method for efficient top-k query processing is to run the algorithms over *sorted lists* (also called *inverted lists*) [8]. Let us define them formally.

Let D be a set of n data items, then the sorted lists are m lists $L_1, L_2, \ldots, L_m$, such that each list L_i contains every data item $d \in D$ in the form of a pair $(id(d), s_i(d))$ where $id(d)$ is the identification of d and $s_i(d)$ is a value that denotes the *local score* (attribute value) of d in L_i. The data items in each list L_i are sorted in descending order of their local scores. For example, in a relational table, each sorted list represents a sorted column of the table where the local score of a data item is its attribute value in that column.

Let f be a scoring function given by the user in the top-k query. For each data item $d \in D$ an *overall score*, denoted by $ov(d)$, is calculated by applying the function f on the local scores of d. Formally, we have $ov(d) = f(s_1(d), s_2(d), \ldots, s_m(d))$. The result of a top-k query is the set of k elements that have the highest overall scores among all elements of the database. Like many previous works on top-k query processing (*e.g.*, [8]), we assume that the scoring function is monotonic.

The sorted lists model for top-k query processing is simple and general. For example, suppose we want to find the top-k tuples in a relational table according to some scoring function over its attributes. To answer such query, it is sufficient to have a sorted (indexed) list of the values of each attribute involved in the scoring function, and return the k tuples whose overall scores in the lists are the highest.

For processing top-k queries over sorted lists, two modes of access are usually used [8]. The first is *sorted (sequential) access* that allows us to sequentially access the next data item in the sorted list. This access begins with the first item in the list. The second is *random access* by which we look up a given data item in the list.

In this paper, we consider the honest-but-curious adversary model for the cloud. In this model, the adversary is inquisitive to learn the sensitive data without introducing any modification in the data or protocols. This model is widely used in many solutions proposed for secure processing of the different queries [13].

Let us now formally state the problem which we address. Let D be a database, and $E(D)$ be its encrypted version such that each data $c \in E(D)$ is the ciphertext of a data $d \in D$, *i.e.*, $c = Enc(d)$ where $Enc()$ is an encryption function. *We assume that the database $E(D)$ is stored in one node of the cloud.*

Given a number k and a scoring function f, our goal is to develop an algorithm A, such that when A is executed over the database $E(D)$, its output contains the ciphertexts of the top-k results.

3   OPE-Based Top-k Query Processing Approach

In this section, we propose an approach, called OPE-based, that uses a combination of the order preserving encryption (OPE) [1] and the FA algorithm

[7] for privacy preserving top-k query processing. Our main contribution, called BuckTop, is presented in the next section.

Let us first explain how the local scores are encrypted. With the OPE-based approach, the local scores (attribute values) in the sorted lists are encrypted using the order preserving encryption technique. We also use a *deterministic* encryption method for encrypting the ID of data items. The *deterministic* encryption generates the same ciphertexts for two equal inputs. This allows us to do random access to the encrypted sorted lists by using the ID of data items.

After encrypting the data IDs and local scores in each sorted list, the lists are sent to the cloud.

Let us now describe how top-k queries can be answered in the cloud over the encrypted data. Given a top-k query Q with a scoring function f, the query is sent to the cloud. Then, the cloud uses the FA algorithm for processing Q as follows. It continuously performs sorted access in parallel to each sorted list, and maintains the encrypted data IDs and their encrypted local scores in a set Y. When there are at least k encrypted data IDs in Y such that each of them has been seen in each of the lists, then the cloud stops doing sorted access to the lists. Then, for each data item d involved in Y, and each list L_i, the cloud performs random access to L_i to find the encrypted local scores of d in L_i (if it has not been seen yet). The cloud sends Y to the user machine which decrypts the local scores of each item $d \in Y$, computes their overall scores, and find the final k items with the highest overall scores.

Theorem 1. *Given a top-k query with a monotonic scoring function, the OPE-based approach returns a set that includes the encrypted top-k elements.*

Proof. Let Y be the set of data items, which have been seen by top-k query processing algorithm in some lists before it stops. Let $Y' \subseteq Y$ be set of data items that have been seen in all lists. Let $d' \in Y'$ be the data item whose overall score among the data items in Y' is the minimum. In each list L_i, let s'_i be the real (plaintext) local score of d' in L_i.

We show that any data item d, which has not been seen by the algorithm under sorted access, has an overall score that is less than or equal to that of d'. In each list L_i, let s_i be the plaintext local score of d in L_i. Since d has not been seen by the top-k query processing algorithm, and the encrypted data items in the lists are sorted according to their initial order, we have $s_i \leq s'_i$, for $1 \leq i \leq m$. Since, the scoring function f is monotonic, then we have $f(s_1, \ldots, s_m) \leq f(s'_1, \ldots, s'_m)$. Thus, the overall score of d is less than or equal to that of d'. Therefore, the set Y contains at least k data items whose overall scores are greater than or equal to that of the unseen data d. □

4 BuckTop System

In this section, we present our BuckTop system. We first describe the architecture of BuckTop, and introduce our method for encrypting the data items and storing them in the cloud. Afterwards, we propose an algorithm for processing top-k queries over encrypted data, and an algorithm for filtering the false positives in the cloud.

4.1 System Architecture and Data Encryption

The architecture of BuckTop system has two main components:

- **Trusted client.** It is responsible for encrypting the user data, decrypting the results and controlling the user accesses. The security keys used for data encryption/decryption are managed by this part of the system. When a query is issued by a user, the trusted client checks the access rights of the user. If the user does not have the required rights to see the query results, then her demand is rejected. Otherwise, the query is transformed to a query that can be executed over the encrypted data.
  For example, suppose we have a relation R with attributes $att_1, att_2, \ldots, att_m$, and the user issues the following query:
  SELECT * FROM R ORDERED BY $f(att_1, \ldots, att_m)$ LIMIT k;
  This query is transformed to:
  SELECT * FROM E(R) ORDERED BY $F(E(att_1), \ldots, E(att_m))$ LIMIT k;
  where $E(R)$ and $E(att_i)$ are the encrypted name of the relation R and the attribute att_i respectively.
  Note that the trusted client component should be installed in a trusted location, $e.g.$, the machine(s) of the person/organization that outsources the data.
- **Service provider.** It is installed in the cloud, and is responsible for storing the encrypted data, executing the queries provided by the trusted client, and returning the results. This component does not keep any security key, thus cannot decrypt the encrypted data in the cloud.

Let us now present our approach for encrypting and outsourcing the data to the cloud. As mentioned before, the trusted client component of BuckTop is responsible for encrypting the user databases. Before encrypting a database, the trusted client creates sorted lists for all important attributes, $i.e.$, those that may be used in the top-k queries. Then, each sorted list is partitioned into *buckets*. There are several methods for partitioning a sorted list, for example dividing the attribute domain of the list to almost equal intervals or creating buckets with equal sizes [9]. In the current implementation of our system, we use the latter method, $i.e.$, we create buckets with almost the same size where the bucket size is configurable by the system administrator.

Let $b_1, b_2, \ldots, b_t$ be the created buckets for a sorted list L_j. Each bucket b_i has a lower bound, denoted by $min(b_i)$, and an upper bound, denoted by $max(b_i)$. A data item d is in the bucket b_i, if and only if its local score (attribute value) in the list L_j is between the lower and upper bounds of the bucket, $i.e.$, $min(b_i) \leq s_j(d) < max(b_i)$.

We use two types of encryption schemes (methods) for encrypting the data itme ids and the local scores of the sorted lists: *deterministic* and *probabilistic*. With the *deterministic* scheme, for two equal inputs, the same ciphertexts (encrypted values) are generated. We use this scheme to encrypt the ID of the data items. This allows us to have the same encrypted ID for each data item in all sorted lists.

The *probabilistic* scheme is used to encrypt the local scores (attribute values) of data items. With the *probabilistic* encryption, for the same plaintexts different ciphertexts are generated, but the decryption function returns the same plaintext for them. Thus, for example if two data items have the same local scores in a sorted list, their encrypted scores may be different. The probabilistic encryption is the strongest type of encryption.

After encrypting the data IDs and local scores of each list L_i, the trusted client puts them in their bucket (chosen based on the local score). Then, the trusted client sends the buckets of each sorted list to the cloud. The buckets are stored in the cloud according to their lower bound order. However, there is no order for the data items inside each bucket, *i.e.*, *the place of the data items inside each bucket is chosen randomly*. This prevents the cloud to know the order of data items inside the buckets.

4.2 Top-k Query Processing Algorithm of BuckTop

The main idea behind top-k query processing in BuckTop system is to use the bucket boundaries to decide when to stop reading the encrypted data from the lists.

Given a top-k query Q including a number k and a scoring function f. To answer Q, the following top-k processing algorithm is executed by the service provider component of BuckTop:

1. Let Y be an empty set;
2. Perform sorted access to the lists:
   2.1. Read the next bucket, say b_i, from each list L_i (starting from the head of the list);
   2.2. For each encrypted data d contained in the bucket b_i:
      2.2.1. Perform random access in parallel to the other lists to find the encrypted score and the bucket of d in all lists;
      2.2.2. Compute a minimum overall score for d, denoted by $min\_ovl(d)$, by applying the scoring function on the lower bound of the buckets that contain d in different lists. Formally, $min\_ovl(d) = f(min(b_1), min(b_2), \ldots, min(b_m))$, where b_i is the bucket involving d in the list L_i.
      2.2.3. Store the encrypted ID of d, its encrypted local scores, and its min_ovl score in the set Y.
   2.3. Compute a threshold TH as follows: $TH = f(min(b'_1), min(b'_2), \ldots, min(b'_m))$, where b'_i is the last bucket seen under sorted access in the L_i, for $1 < i < m$. In other words, TH is computed by applying the scoring function on the lower bounds of the last seen buckets in the lists.
   2.4. If the set Y contains at least k encrypted data items having minimum overall scores higher than TH, then stop. Otherwise, go to Step 2.1.

When the top-k query processing algorithm stops, the set Y includes the encrypted top-k data items (see the proof below). This set is sent to the trusted client that decrypts its contained data items, computes the overall scores of the items, removes the false positives (*i.e.*, the items that are in Y but not among the top-k results), and returns the top-k items to the user.

The following theorem shows that the output of BuckTop top-k query processing algorithm contains the encrypted top-k data items.

Theorem 2. *Given a top-k query with a monotonic scoring function f, the output of BuckTop top-k query processing algorithm contains the encrypted top-k results.*

Proof. Let Y be the output of the BuckTop top-k query processing algorithm, *i.e.*, the set that contains all the encrypted data items seen under sorted access when the algorithm ends. We show that each data item d that is not in Y $(d \notin Y)$, has an overall score that is less than or equal to the overall score of at least k data items in Y. Let s_i be the local score of d in the list L_i. Let b'_i be the last bucket seen under sorted access in the list L_i, *i.e.*, when the algorithm ends. Since d is not in Y, it has not been seen under sorted access in the lists. Thus, its involving buckets are after the buckets seen under sorted access by the algorithm. Therefore, we have $s_i < min(b'_i)$ for $1 \leq i \leq m$, *i.e.*, the local score of d in each list L_i is less than the lower bound of the last bucket read under sorted access in L_i. Since the scoring function is monotonic, we have $f(s_1, \ldots, s_m) < f(min(b'_1), min(b'_2), \ldots, min(b'_m)) = TH$. Thus, the overall score of d is less than TH. When the algorithm stops, there are at least k data items in Y whose minimum overall scores are greater than or equal to TH. Thus, their overall scores are at least TH. Therefore, their overall scores are greater than or equal to that of the data item d.

In the set Y returned by the top-k query processing algorithm of BuckTop, in addition to the top-k results there may be false positives. Below, we propose a filtering algorithm to eliminate most of them in the cloud, without decrypting the data items. As shown by our experimental results, our filtering algorithm eliminates most of the false positives (more than 99% in the different tested datasets). This improves significantly the response time of top-k queries, because the eliminated false positives do not need to be communicated to the trusted client and should not be decrypted by it.

In the filtering algorithm, we use the *maximum overall score*, denoted by *max_ovl* of each data item. This score is computed by applying the scoring function on the upper bound of the buckets involving the data item in the lists. The algorithm proceeds as follows:

1. Let $Y' \subseteq Y$ be the k data items in Y that have the highest minimum overall scores (*min_ovl*) among the items contained in Y.
2. Let d_{min} be the data item that has the lowest *min_ovl* score in Y'.
3. For each item $d \in Y$

3.1. Compute the maximum overall score of d, i.e., $max\_ovl(d)$, by applying the scoring function on the upper bound of the buckets involving d in the lists. Formally, let $max(b_i)$ be the upper bound of the bucket involving d in the list L_i. Then, $max\_ovl(d) = f(max(b_1), max(b_2), \ldots, max(b_m))$.

3.2. If the maximum overall score of d is less than or equal to the minimum overall score of d_{min}, then remove d from Y. In other words, if $max\_ovl(d) \leq min\_ovl(d_{min}) \Rightarrow Y = Y - \{d\}$.

Let us prove that the filtering algorithm works correctly. We first show that the minimum overall score of any data item d, i.e. $min\_ovl(d)$, which is computed based on the lower bound of its buckets, is less than or equal to its overall score. We also show that the maximum overall score of d, i.e. $max\_ovl(d)$, is higher than or equal to its overall score.

Lemma 1. *Given a monotonic scoring function f, the minimum overall score of any data item d is less than or equal to its overall score.*

Proof. The minimum overall score of a data item d is calculated by applying the scoring function on the lower bound of the buckets in which d is involved. Let b_i be the bucket that contains d in the list L_i. Let s_i be the local score of d in L_i. Since $d \in b_i$, its local score is higher than or equal to the lower bound of b_i, i.e. $min(b_i) \leq s_i$. Since f is monotonic, we have $f(min(b_1), \ldots, min(b_m)) \leq f(s_1, \ldots, s_m)$. Therefore, the minimum overall score of d is less than or equal to its overall score. $\square$

Lemma 2. *Given a monotonic scoring function f, the maximum overall score of any data item d is greater than or equal to its overall score.*

Proof. The proof can be done in a similar way as Lemma 1. $\square$

The following theorem shows that the filtering algorithm works correctly, i.e., the removed data are only false positives.

Theorem 3. *Any data item removed by the filtering algorithm cannot belong to the top-k results.*

Proof. The proof can be done by considering the fact that any removed data item d has a maximum overall score that is lower than the minimum overall score of at least k data items. Thus, by using Lemmas 1 and 2, the overall score of d is less than or equal to that of at least k data items. Therefore, we can eliminate d. $\square$

A security analysis of the BuckTop system is provided in [15].

5 Performance Evaluation

In this section, we evaluate the performance of BuckTop using synthetic and real datasets. We first describe the experimental setup, and then report the results of our experiments.

5.1   Experimental Setup

We implemented our top-k query processing system and performed our tests on real and synthetic datasets. As in some previous work on encrypted data (*e.g.*, [13]), we use the Gowalla database, which is a location-based social networking dataset collected from users locations. The database contains 6 million tuples where each tuple represents user number, time, user geographic position, etc. In our experiments, we are interested in the attribute time, which is the second value in each tuple. As in [13], we decompose this attribute into 6 attributes (year, month, day, hour, minute, second), and then create a database with the following schema R(ID, year, month, date, hour, minute, second), where ID is the tuple identifier. In addition to the real dataset, we have also generated random datasets using uniform and Gaussian distributions.

We compare our solution with the two following approaches:

- *OPE*: this is the OPE-based solution (presented in Sect. 3) that uses the order preserving encryption for encrypting the data scores.
- *TA over plaintext data*: the objective is to show the overhead of top-k query processing by BuckTop over encrypted data compared to an efficient top-k algorithm over plaintext data.

In our experiments, we have two versions of each database: (1) the plaintext database used for running TA; (2) the encrypted database used for running BuckTop and OPE.

In our performance evaluation, we study the effect of several parameters: (1) n: the number of data items in the database; (2) m: the number of lists; (3) k: the number of required top items; (4) $bsize$: the number of data items in the buckets of BuckTop. The default value for n is 2M items. Unless otherwise specified, m is 5, k is 50, and $bsize$ is 20. In our tests, the default database is the synthetic uniform database.

In the experiments, we measure the following metrics:

- **Cloud top-k time:** the time required by the service provider of BuckTop in the cloud to find the set that includes the top-k results, *i.e.*, the set Y.
- **Response time:** the total time elapsed between the time when the query is sent to the cloud and the time when the k decrypted results are returned to the user. This time includes the cloud top-k time, the filtering, and the result post-processing in the client (*e.g.*, decryption).
- **Filtering rate:** the number of false positives eliminated by the filtering algorithm in the cloud.

We performed our experiments using a node with 16 GB of main memory and Intel Core i7-5500 @ 2.40 Ghz as processor.

5.2   Effect of the Number of Data Items

In this section, we compare the performance of TA over plaintext data with BuckTop and OPE over encrypted data, while varying the number of data items, *i.e.*, n.

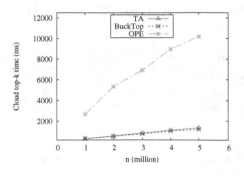

Fig. 1. Cloud top-k time vs. number of database tuples

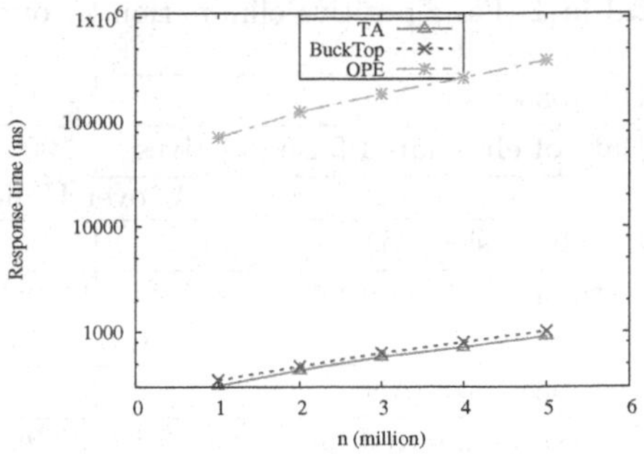

Fig. 2. Response time vs. number of database tuples

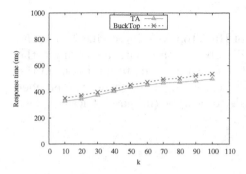

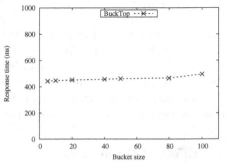

Fig. 3. Response time vs. k

Fig. 4. Response time vs. bucket size

Figure 1 shows how cloud top-k time evolves, with increasing n, and the other parameters set as default values described in Sect. 5.1. The cloud top-k time of all approaches increases with n. But, OPE takes more time than the two other approaches, because it stops deeper in lists, and thus reads more data.

Figure 2 shows the total response time of BuckTop, OPE and TA while varying n, and the other parameters set as default values. Note that the figure are is in logarithmic scale. TA does not need to decrypt any data, so its response time is almost the same as its cloud time. The response time of BuckTop is slightly higher than its cloud top-k time, as in addition to top-k query processing it performs the filtering in the cloud and also needs to decrypt at least k data items. We see that the response time of OPE is much higher than its cloud top-k time. The reason is that OPE returns to the trusted client a lot of false positives, which should be decrypted, and removed from the final result set. But, this is not the case for BuckTop as its filtering algorithm removes almost all the false positives in the cloud (see the results in Sect. 5.5), thus there is no need to decrypt them.

Table 1. False positive elimination by our filtering algorithm over different datasets

Database size (M)	1	2	3	4	5	6
Rate of eliminated false positives	100%	100%	100%	99.99%	99.99%	100%

A: over Uniform dataset

Database size (M)	1	2	3	4	5	6
Rate of eliminated false positives	99.98%	99.99%	99.99%	99.99%	99.99%	99.99%

B: over Real dataset

Database size (M)	1	2	3	4	5	6
Rate of eliminated false positives	99.94%	99.96%	99.97%	99.98%	99.98%	99.98%

C: over Gaussian dataset

5.3 Effect of k

Figure 3 shows the total response times of BuckTop with increasing k, and the other parameters set as default values. We observe that with increasing k the response time increases. The reason is that Bucktop needs to go deeper in the lists to find the top-k results. In addition, increasing k augments the number of data items that the trusted client needs to decrypt (because at least k data items are decrypted by the trusted client).

5.4 Effect of Bucket Size

Figure 4 reports the response time of BuckTop when varying the size of buckets, and the other parameters set as default values. We observe that the response time increases when the bucket size increases. The reason is that the top-k query processing algorithm of Bucktop reads more data in the lists, because the data are read bucket by bucket. In addition, increasing the bucket size increases the number of false positives to be removed by the filtering algorithm, and eventually decrypting the none eliminated false positives in the client side.

5.5 Effect of the Filtering Algorithm

BuckTop's filtering algorithm is used to eliminate/reduce the false positives in the cloud. We study the filtering rate by increasing the size of the dataset. For the uniform synthetic dataset, the results are shown in Table 1A. For datasets with up to three million data items, the filtering method eliminates 100% of the false positives, and the cloud returns to the trusted client only the k data items that are the result of the query. For larger datasets, BuckTop filters up to 99.99% of the false positives. By using the Gaussian dataset, we obtain the results shown in Table 1C. We see that around 99.94% of false positives are eliminated.

Over the real dataset, Table 1B shows the filtering rate. We observe that the filtering algorithm eliminates 99.99% of false positives. Thus, the filtering algorithm is very efficient over all the tested datasets. However, there is a little difference in the filtering rate for different datasets because of the local score distributions. For example, in the Gaussian distribution, the local scores of many data items are very close to each other, thus the filtering rate decreases in this dataset.

6   Related Work

In the literature, there has been some research work to process keyword queries over encrypted data, e.g., [2,17]. For example [2,17] propose matching techniques to search words in encrypted documents. However, the proposed techniques cannot be used to answer top-k queries. There have been also some solutions proposed for secure kNN similarity search, e.g., [3,5,6,14,19]. The problem is to find k points in the search space that are the nearest to a given point. This problem should not be confused with the top-k problem in which the given scoring function plays an important role, such that on the same database and with the same k, if the user changes the scoring function, then the output may change. Thus, the proposed solutions proposed for kNN cannot deal with the top-k problem.

The bucketization technique (i.e., creating buckets) has been used in the literature for answering range queries over encrypted data, e.g., [9,10,16]. For example, in [10], Hore et al. use this technique, and propose optimal solutions for distributing the encrypted data in the buckets in order to guarantee a good performance for range queries.

There have been access pattern attacks against range query processing methods that use the bucketization technique, e.g. [11]. The main idea is to utilize the intersection between the results of the queries and also some background knowledge to guess the bucket boundaries. However, these attacks are not valid for our approach, because there is no range in our queries. In our system, the main plaintext information in the queries is k (i.e., the number of asked top tuples), and this information is not usually useful to violate the privacy of users.

In [12], Kim et al. propose an approach for preserving the privacy of data access patterns during top-k query processing. In [18], Vaidya et al. propose a privacy preserving method for top-k selection from the data shared by individuals in a distributed system. Their objective is to avoid disclosing the data of each node to other nodes. Thus their assumption about the nodes is different from ours, because they can trust the node that stores the data (this is why the data are not encrypted), but in our system we trust no node of the cloud.

Meng et al. [20] propose a solution for processing top-k queries over encrypted data. They assume the existence of two non-colluding nodes in the cloud, one of which can decrypt the data (using the decryption key) and execute a TA-based algorithm. Our assumptions about the cloud are different, as we do not trust any node of the cloud.

7   Conclusion

In this paper, we proposed a novel system, called BuckTop, designed to encrypt sensitive data items, outsource them to a non-trusted cloud, and answer top-k queries. BuckTop has a top-k query processing algorithm that is executed over encrypted data, and returns a set containing the top-k results, without decrypting the data in the cloud. It also comes with a powerful filtering algorithm that eliminates significantly the false positives from the result set.

We validated our system through experimentation over synthetic and real datasets. We compared its response time with OPE over encrypted data, and with the popular TA algorithm over original (plaintext) data. The experimental results show excellent performance gains for BuckTop. They illustrate that the overhead of using BuckTop for top-k processing over encrypted data is very low, because of efficient top-k processing and false positive filtering.

Acknowledgement. The research leading to these results has received funding from the European Union's Horizon 2020 - The EU Framework Programme for Research and Innovation 2014–2020, under grant agreement No. 732051.

References

1. Agrawal, R., Kiernan, J., Srikant, R., Xu, Y.: Order-preserving encryption for numeric data. In: SIGMOD Conference, pp. 563–574 (2004)
2. Chang, Y.-C., Mitzenmacher, M.: Privacy preserving keyword searches on remote encrypted data. In: Ioannidis, J., Keromytis, A., Yung, M. (eds.) ACNS 2005. LNCS, vol. 3531, pp. 442–455. Springer, Heidelberg (2005). https://doi.org/10.1007/11496137_30
3. Choi, S., Ghinita, G., Lim, H.-S., Bertino, E.: Secure kNN query processing in untrusted cloud environments. In: IEEE TKDE, pp. 2818–2831 (2014)
4. Coles, C., Yeoh, J.: Cloud adoption practices and priorities survey report. Technical report, Cloud Security Alliance report, January 2015
5. Ding, X., Liu, P., Jin, H.: Privacy-preserving multi-keyword top-k similarity search over encrypted data. In: IEEE TDSC no. 99, pp. 1–14 (2017)
6. Elmehdwi, Y., Samanthula, B.K., Jiang, W.: Secure k-nearest neighbor query over encrypted data in outsourced environments. In: ICDE Conference (2014)
7. Fagin, R.: Combining fuzzy information from multiple systems. J. Comput. Syst. Sci. **58**(1), 83–99 (1999)
8. Fagin, R., Lotem, A., Naor, M.: Optimal aggregation algorithms for middleware. J. Comput. Syst. Sci. **66**(4), 614–656 (2003)
9. Hore, B., Mehrotra, S., Canim, M., Kantarcioglu, M.: Secure multidimensional range queries over outsourced data. VLDB J. **21**(3), 333–358 (2012)
10. Hore, B., Mehrotra, S., Tsudik, G.: A privacy-preserving index for range queries. In: VLDB Conference, pp. 720–731 (2004)
11. Islam, M.S., Kuzu, M., Kantarcioglu, M.: Inference attack against encrypted range queries on outsourced databases. In: ACM CODASPY, pp. 235–246 (2014)
12. Kim, H.-I., Kim, H.-J., Chang, J.-W.: A privacy-preserving top-k query processing algorithm in the cloud computing. In: Bañares, J.Á., Tserpes, K., Altmann, J. (eds.) GECON 2016. LNCS, vol. 10382, pp. 277–292. Springer, Cham (2017). https://doi.org/10.1007/978-3-319-61920-0_20
13. Li, R., Liu, A.X., Wang, A.L., Bruhadeshwar, B.: Fast range query processing with strong privacy protection for cloud computing. PVLDB **7**(14), 1953–1964 (2014)

14. Liao, X., Li, J.: Privacy-preserving and secure top-k query in two-tier wireless sensor network. In: Global Communications Conference (GLOBECOM), pp. 335–341 (2012)
15. Mahboubi, S., Akbarinia, R., Valduriez, P.: Top-k query processing over outsourced encrypted data. Research report RR-9053, INRIA (2017)
16. Sahin, C., Allard, T., Akbarinia, R., El Abbadi, A., Pacitti, E.: A differentially private index for range query processing in clouds. In: ICDE Conference (2018)
17. Song, D.X., Wagner, D., Perrig, A.: Practical techniques for searches on encrypted data. In: IEEE S&P, pp. 44–55 (2000)
18. Vaidya, J., Clifton, C.: Privacy-preserving top-k queries. In: ICDE Conference, pp. 545–546 (2005)
19. Wong, W.K., Cheung, D.W., Kao, B., Mamoulis, N.: Secure kNN computation on encrypted databases. In: SIGMOD Conference, pp. 139–152 (2009)
20. Zhu, H., Meng, X., Kollios, G.: Top-k query processing on encrypted databases with strong security guarantees. In: ICDE Conference (2018)

R^2-Tree: An Efficient Indexing Scheme for Server-Centric Data Center Networks

Yin Lin, Xinyi Chen, Xiaofeng Gao$^{(\boxtimes)}$, Bin Yao, and Guihai Chen

Shanghai Key Laboratory of Scalable Computing and Systems,
Department of Computer Science and Engineering,
Shanghai Jiao Tong University, Shanghai 200240, China
{ireane,cxinyic}@sjtu.edu.cn, {gao-xf,yaobin,gchen}@cs.sjtu.edu.cn

Abstract. Index plays a very important role in cloud storage systems, which can support efficient querying tasks for data-intensive applications. However, most of existing indexing schemes for data centers focus on one specific topology and cannot be migrated directly to the other networks. In this paper, based on the observation that server-centric data center networks (DCNs) are recursively defined, we propose *pattern vector*, which can formulate the server-centric topologies more generally and design R^2-Tree, a scalable two-layer indexing scheme with a local R-Tree and a global R-Tree to support multi-dimensional query. To show the efficiency of R^2-Tree, we start from a case study for two-dimensional data. We use a layered global index to reduce the query scale by hierarchy and design a method called *Mutex Particle Function* (MPF) to determine the potential indexing range. MPF helps to balance the workload and reduce routing cost greatly. Then, we extend R^2-Tree indexing scheme to handle high-dimensional data query efficiently based on the topology feature. Finally, we demonstrate the superior performance of R^2-Tree in three typical server-centric DCNs on Amazon's EC2 platform and validate its efficiency.

Keywords: Data center network · Cloud storage system
Two-layer index

1 Introduction

Nowadays, cloud storage systems such as Google's GFS [7], Amazon's Dynamo [4], Facebook's Cassandra [2], have been widely used to support data-intensive applications that require PB-scale or even EB-scale data storage across

This work was partly supported by the Program of International S&T Cooperation (2016YFE0100300), the China 973 project (2014CB340303), the National Natural Science Foundation of China (Grant number 61472252, 61672353, 61729202 and U1636210), the Shanghai Science and Technology Fund (Grant number 17510740200), CCF-Tencent Open Research Fund (RAGR20170114), and Guangdong Province Key Laboratory of Popular High Performance Computers of Shenzhen University (SZU-GDPHPCL2017).

© Springer Nature Switzerland AG 2018
S. Hartmann et al. (Eds.): DEXA 2018, LNCS 11029, pp. 232–247, 2018.
https://doi.org/10.1007/978-3-319-98809-2_15

thousands of servers. However, most of the existing indexing schemes for cloud storage systems do not support multi-dimensional query well.

To settle this problem, a load balancing two-layer indexing framework was proposed in [18]. In two-layer indexing scheme, each server will: (1) build indexes in its local layer for the data stored in it, and (2) maintain part of global indexing information which is published by the other servers from their local data.

Based on the two-layer indexing framework, many efforts focus on how to divide the potential indexing range and how to reduce the searching cost. Early researches are mainly focused on Peer-to-Peer (P2P) networks such as RT-CAN [17], while later researches gradually turn to data center networks (DCNs) such as FT-INDEX [6], RT-HCN [12], etc. However, most of researches only focus on one specific network. The design lacks expandability and usually only suits one kind of network. Due to the differences in topology, it is always hard to migrate a specific indexing scheme from one network to another.

In this paper, we first propose a *pattern vector P* to formulate the topologies. Most of the server-centric DCN topologies are recursively defined and a high-level structure is scaled out from several low-level structures by connecting them in a well-defined manner. *Pattern vector* fully exploits the hierarchical feature of the topology by using several parameters to represent the expanding method. The raise of the *pattern vector* makes the migration of the indexing scheme feasible and is the cornerstone of generalization.

Then we introduce a more scalable two-layer indexing scheme for the server-centric DCNs based on P. We design a novel indexing scheme called R^2-Tree where a local R-Tree is used to support query for multi-dimensional local data and a global R-Tree helps to speed up the query for global information. We start from two-dimensional indexing. We reduce the query scale by hierarchy through building global indexes with a layered structure. The hierarchical design prevents repeated query process and achieve better storage efficiency. We also propose a method called *Mutex Particle Function* (MPF) to disperse the indexing range and balance the workload. Furthermore, we extend R^2-Tree to high-dimensional data space. Based on the hierarchy feature of the topology, we assign each level of the topology to be responsible for one dimension of the data. To handle data whose dimension is higher than the levels of the topology, we use *Principal Component Analysis* (PCA) to reduce the dimension. Besides, we design a mapping algorithm to select the nodes in local R-trees as public indexes and publish them on the global R-Trees of corresponding servers.

We evaluate the performance of range and point query for R^2-Tree on Amazon's EC2. We build two-layer indexes on 3 typical server-centric DCNs: **DCell** [10], **Ficonn** [13], **HCN** [11] with both two-dimensional and high-dimensional data and evaluate the query performance. Besides, by comparing the query time with **RT-HCN** [12], we show the technical advancement of our design.

The rest of the paper is organized as follows. The related work will be introduced in Sect. 2. Section 3 introduces the pattern vector to generalize the server-centric architectures. We elaborate the procedure of building two-layer index and the algorithm in Sect. 4 and depict the query processing in Sect. 5. Section 6 exhibits the experiments and the performance of our scheme. Finally, we draw a conclusion of this paper in Sect. 7.

2   Related Work

Data Center Network. Our work aims to construct a scalable, load-balance, and multi-dimensional two-layer indexing on data center networks (DCNs). The underlying topologies of DCN can be roughly separated into two categories. One is the tree-like switch-centric topologies where switches are used for interconnection and routing like the Fat-Tree [1], VL2 [8], Aspen Tree [16], etc. The other one is the server-centric topology, in which the servers are not only used to store the data, but also perform the interconnecting and routing function. Typical server-centric topologies include data centers such as HCN [11], DCell [10], FiConn [13], Dpillar [14], and BCube [9]. Server-centric architectures are mostly recursively defined structures. Our work exploits this hierarchical feature and put forward a pattern vector which can generalize the server-centric topologies.

Two-Layer Indexing. Two-layer indexing [18] maintains two index layers called local layer and global layer to increase parallelism and support efficient query for different data attributes. Given a query, the server will first search its global index to locate the servers which may store the data and then forward the query. The servers which receive the forwarded query will search their local index to retrieve the queried data. Early two-layer index works focus on P2P network, like RT-CAN [17] and the DBMS-like indexes [3]. Subsequently with the rapid development of DCNs, a universal U^2-Tree [15] is proposed for switch-centric DCNs. Apart from that, RT-HCN [12] for HCN and an indexing scheme for multi-dimensional data for BCube [5] are both efficient indexing schemes for server-centric DCNs. Their works are mostly confined to a certain topology. With the generalized pattern vector, we design a highly extendable and flexible indexing scheme which can suit most of the server-centric DCNs.

3   Recursively Defined Data Center

Server-centric DCN topologies have a high degree of scalability, symmetry, and uniformity. Most of the server-centric DCNs are recursively defined, which means that a high-level structure grows from a fixed number of low-level structures recursively. This kind of topologies has a favorable feature to design layered global index. However, due to the diversity of different kinds of topologies, with different number of Network Interface Card (NIC) ports for switches and connection methods, it is hard to migrate a specific indexing scheme from one topology to another. Thus, finding a general pattern for server-centric topologies is of great significance for constructing a scalable indexing scheme. We observe that the scaling out of the topology obeys some certain rules. The ratio of available servers which are actually used for expansion is fixed for every specific

Table 1. Symbol description

Sym.	Description	Sym.	Description
h	Total height of the structure	na_i	Number of servers available to expand
k	Port number of mini-switch	nu_i	Number of servers actually used to expand
α	Expansion factor (≤ 1)	pir_j	potential indexing range of server j
β	Connection method denoter	g_i	Number of ST_{i-1} in ST_i ($g_0 = 1$)
ST_i	A level-i structure	q_i	Position of the meta-block in level-i
mbr	Minimum bounding rectangle	a_i	Position of the server in level-i

topology. In this section, we propose a *pattern vector P* to as a high-level representation to formulate the topologies. For clarity, we summarize the symbols in Table 1. Besides, we also show in Fig. 1 some typical server-centric topologies with the given pattern definition, including **HCN** [11], **DCell** [10], **Ficonn** [13] and **BCube** [5].

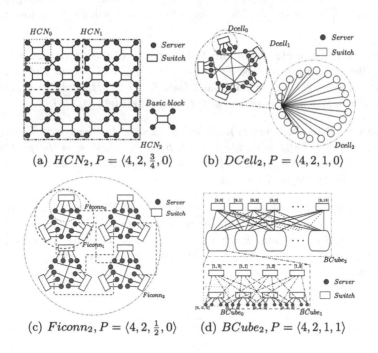

(a) $HCN_2, P = \langle 4, 2, \frac{3}{4}, 0 \rangle$ (b) $DCell_2, P = \langle 4, 2, 1, 0 \rangle$

(c) $Ficonn_2, P = \langle 4, 2, \frac{1}{2}, 0 \rangle$ (d) $BCube_2, P = \langle 4, 2, 1, 1 \rangle$

Fig. 1. Typical server-centric topologies represented by pattern vector P

To formulate the topology completely and concisely, 4 parameters are chosen for *pattern vector*. In the bottom right of Fig. 1(a), we show the basic building block, which contains a mini-switch and 4 servers. The port number of mini-switches which defines the basic recursive unit is denoted as k while the number of levels in the structure which defines the total recursive layers is denoted as h.

Thus, in Fig. 1(a), $k = 4$, $h = 2$. Besides, the recursively scaling out rule for each topology is defined by the expansion factor and the connection method denoter, which are denoted as α and β and are explained in Definitions 1 and 2.

Definition 1 (Expansion factor). *Expansion factor α defines the utilization rate of the servers available for expansion. It can be proved that for every server-centric architecture, α is a constant and different server-centric architectures will have different α, which is given by: $\alpha = \lfloor nu_i/na_i \rfloor$.*

To explain, we use the symbol ST_i to represent the level-i structure. When ST_i scales out to ST_{i+1}, we define na_i as the number of available servers in ST_i that could be used for expansion, while we will use part of them for real expansion, and the total number of those used servers are defined as nu_i. Naturally, $na_i \geq nu_i$. We notice that for each topology, the ratio of servers used for expansion and available servers is surprisingly fixed. Therefore, we can denote a parameter α as $\lfloor nu_i/na_i \rfloor$ to depict the expansion pattern for each topology abstractly, which satisfies $0 < \alpha \leq 1$. For example, in Fig. 1(a), every time when HCN_i grows to HCN_{i+1}, $\alpha = \frac{3}{4}$, since three of four available servers will be used for topology expansion.

Definition 2 (Connection method denotor). *Connection method denotor β defines the connection method of servers, where $\beta = 1$ means the connection type is server-to-server-via-switch, like BCube in Fig. 1(d); and $\beta = 0$ means the connection type is server-to-server-direct, like DCell in Fig. 1(b).*

Definition 3 (Pattern vector). *A server-centric topology can be uniformly represented using a Pattern vector $P = \langle k, h, \alpha, \beta \rangle$, where k is the port number of mini-switches, h is the number of the total level, α is the expansion factor and β represents the connection method.*

To practice, let us first define g_{i+1} as the number of ST_i's in the next recursive expansion ST_{i+1}. Obviously, g_i can be calculated by: $g_i = \alpha \cdot na_{i-1} + 1$. Then take an eye on Fig. 1 again. Each of the subgraph exhibits a topology with $h = 2$. According to their different expansion rules, we can easily calculate the corresponding *pattern vector* values. Actually we can use *pattern vector* to

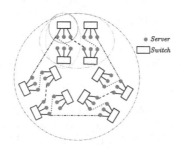

Fig. 2. A new-defined server-centric topology, $P = \langle 3, 3, \frac{1}{3}, 0 \rangle$

construct brand new server-centric topologies, which could provide similar QoS service as other members in the server-centric family. For example in Fig. 2, for a given *Pattern Vector* $P = \langle 3, 3, \frac{1}{3}, 0 \rangle$, we can depict a new server-centric DCN.

4 R^2-Tree Construction

When we use a *pattern vector* to depict any server-centric topologies generally, we can design a more scalable two-layer indexing scheme for efficient query processing requirements. We name this novel design as **R^2-Tree**, as it contains two R-Trees for both local and global indexes. A local R-Tree is an ideal choice for maintaining multi-dimensional data in each server and a global R-Tree helps to speed up the query in the global layer. In this section, we first discuss the hierarchical indexing design for two-dimensional data as an example, and then extend it to multi-dimensional version.

4.1 Meta-block, Meta-server and Representatives

Hierarchical global indexes design can avoid repeated query and achieve better storage efficiency. To build a hierarchical global layer, we divide the two-dimensional indexing space into $h + 1$ levels of meta-blocks, defined as Definition 4.

Definition 4 (Meta-block). *Meta-blocks are a series of abstract blocks which are used to stratify the global indexing range. For a topology with $P = \langle k, h, \alpha, \beta \rangle$, the meta-blocks can be divided into $h + 1$ levels.*

For a recursively defined structure with *pattern vector* $P = \langle k, h, \alpha, \beta \rangle$, we divide the total range in each dimension into g_h parts, where g_h is the number of ST_{h-1} in ST_h, and we can get $g_h{}^2$ meta-blocks on level-$(h$-$1)$. Similarly, we divide the range in each dimension of meta-blocks in the second level into g_{h-1} parts and for each meta-block in second level, we get $g_{h-1}{}^2$ lower level blocks in the next layer. In this way, we can know that in the level-0, there are $\prod_{i=1}^{h} g_i{}^2$ meta-blocks. Thus, the total number of meta-blocks is given by Eq. (1):

$$Total = \sum_{j=1}^{h} \prod_{i=j}^{h} g_i{}^2 + 1 \qquad (1)$$

Each meta-block is assigned an $(h + 1)$-tuple $[q_h, q_{h-1}, \ldots, q_1, q_0]$ in which q_i represents the meta-block's position in level-i. For example in the left part of Fig. 3, the level-0 block at the top left corner is assigned with $[0, 0, 0]$, while the level-1 block at the top left corner is assigned with $[1, 0, 0]$. To simplify the partition and search progress, we merge the $(h + 1)$-tuple of each meta-block as a code ID named *mid*, which can be calculated by Eq. (2).

$$mid_h = \sum_{i=0}^{h} \left(q_i \times \prod_{j=0}^{i} g_j{}^2 \right) \qquad (2)$$

Figure 3 is an example for such range division process. Here in the left sub-graph, the lowest level meta-blocks are coded as $0, 1, \ldots, 143$ and the second level meta-blocks are coded as $144, 153, \ldots, 279$. The highest level meta-block which covers the whole space is coded as 288.

Now we need to assign some representative servers in charge of each meta-block from a server-centric DCN structure.

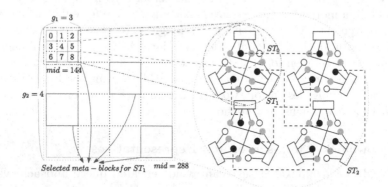

Fig. 3. Mapping meta-blocks to meta-servers

Definition 5 (Meta-server). *For each level-i structure ST_i, we can also denote it using pattern vector as $ST_i = \langle k, i, \alpha, \beta \rangle$, which can be an excellent representative to manage several corresponding meta-blocks, so it is also named as meta-server.*

Respectively, the right part of Fig. 3 shows a *Ficonn$_2$* topology ($P = \langle 4, 2, \frac{1}{2}, 0 \rangle$). ST_2 denotes the meta-server in level-2 while ST_1 is the level-1 meta-server and ST_0 is the level-0 meta-server. Figure 3 also shows a **mapping scheme** to map the meta-blocks to the meta-servers. At level-i, there are g_i ST_i's, $g_i{}^2$ meta-blocks, so we map g_i meta-blocks to each ST_i. For each ST_i, we hope to select meta-blocks sparsely, so we formulate a *Mutex Particle Function (MPF)* to complete this task, motivated by mutex theory in physics. The mapping function will be described in Sect. 4.2.

Figure 3 illustrates this mapping rule thoroughly. The meta-block in the first-layer is mapped to the first-layer meta-server (ST_2). Since ST_2 contains 4 second-layer meta-server (ST_1), the first-layer meta-block contains 4^2 second-layer meta-blocks. Therefore each ST_1 is in charge of 4 second-layer meta-blocks. Similarly, each meta-block which is mapped to the first ST_1 can be divided into 3^2 parts and be mapped to the third-layer meta-server (ST_0) accordingly. After mapping meta-blocks to meta-servers, as meta-servers are just virtual nodes, we should select physical servers as representatives of meta-servers.

Definition 6 (Meta-server representative). *To achieve fast routing process, we select the connecting servers between ST_{i-1}'s as the representatives of ST_i.*

Algorithm 1. Mutex Particle Function (MPF)

Input: A meta-server ST_i
Output: S_i: a set of meta-blocks which are mapped to meta-server ST_i
1 $S_i = \{\emptyset\}$;
2 Select a meta-block in this layer randomly and add it into S_i, and set the centroid of this mapped set as the center of this node;
3 **while** $|S_i| < \prod_{j=i+1}^{h} g_j$ **do**
4   From the set of the non-mapped meta-blocks, select one whose centroid is mostly far away from the centroid of the mapped set. Add this node into the mapped set of this meta sever, and re-calculate the centroid of the mapped set;

In Fig. 3, the grey nodes are the representatives for ST_0 and the black nodes are the representatives for ST_1. Selecting representatives in this method guarantees that the query in the upper layer of the meta-blocks can be forwarded to the lower layer in the least number of hops, and more than one representative to a meta-server guarantees a degree of redundancy.

4.2 Mutex Particle Function

Once the queries appear intensively in a certain area, all the nearby meta-blocks will be searched at a high frequency. Therefore, a carefully designed mapping scheme is needed to balance the request load. We propose *Mutex Particle Function (MPF)* in this subsection. As its name illustrated, we regard the meta-blocks assigned to the same meta-server as the same kind of particles and like mutual exclusion of charges, same kind of particles should be mutually exclusive with each other. That means in two-dimensional space, the distance between the same kind of meta-blocks should be as far as possible. Every time we select a meta-block to a meta-server, we choose the furthest one from the centroid of the meta-blocks which have been chosen. Algorithm 1 describes *MPF* in detail.

4.3 Publishing Local Tree Node

In the process of building R^2-Tree indexes, we first build local R-Tree for every server based on their local data. Then to better locate the servers, information about local data and the corresponding server will be published to global index layer. We first select the nodes to be published from the local R-Trees, which starts from the second layer of local R-Tree to the end layer where all the nodes are leaf nodes. For the layer before the end layer, we select the nodes which have no published ancestors with a certain probability to publish. For the end layer, we publish all the nodes whose ancestors have not been published. In this way, we guarantee the completeness of the publishing scheme. Moreover, we make sure that the nodes in the higher layer have a higher possibility to be published so to reduce the storage pressure in global index layer. After the selection of the local R-Tree node, we find the minimum potential indexing range of a meta-server which covers this selected node exactly. Then, we publish the local R-Tree node to the corresponding representatives in the format of (mbr, ip), where mbr is the minimum bounding rectangle of the local R-Tree node, and ip means the ip address of the server where this node is stored. For each server, it will build a global R-Tree based on all the R-Tree nodes published to it. Global R-Tree can accelerate the speed in searching global indexes and forward the query.

4.4   Multi-dimensional Indexing Extension

The R^2-Tree indexing scheme can also be extended to multi-dimensional space. In our design, multi-dimensional indexing takes advantage of the recursive feature of the topologies to divide the hypercube space and let one level of the structure be in charge of one dimension. In this paper, we will not discuss circumstance where the data dimension is extremely high like image data. This may be solved by LSH-based algorithms, but it is another story from our bottleneck-avoidable two-layer index framework.

Potential Index Range. For a server-centric DCN structure with h levels, we can construct an $(h + 1)$-dimensional indexing space. If the dimension of the data exceeds $h + 1$, methods like principle component analysis (PCA) can be applied to reduce the index dimension. We assign one level of the structure to maintain the global information in one dimension. Since the number of parts in each dimension should be equal to the number of the lower layer structures ST_{i-1} in ST_i which is denoted by g_i, we divide the indexing space in dimension i into g_i parts (k for dimension 0) and every ST_{i-1} in this level will be responsible for one of them. Figure 4 shows the indexing design in detail.

4.5   Potential Indexing Range

As we have mapped several meta-blocks to a meta-server, the potential indexing range of a meta-server is the sum of ranges of those meta-blocks. Taking uniformly distributed data as an example, since there are $\prod_{j=i+1}^{h} g_j^2$ meta-blocks in level-i, the two-dimension boundary $([l_0, u_0], [l_1, u_1])$ can be divided into $\prod_{j=i+1}^{h} g_j$ segments for each dimension in level-i. The range of the highest level meta-block is $pir_h = ([l_0, u_0], [l_1, u_1])$. The range of meta-blocks for each dimension is given by:

$$pir_{i_0} = \left[l_{i_0} + (q_i \bmod g_{i+1}) \times \frac{u_{i_0} - l_{i_0}}{g_{i+1}}, l_{i_0} + (q_i \bmod g_{i+1} + 1) \times \frac{u_{i_0} - l_{i_0}}{g_{i+1}} \right]$$

$$pir_{i_1} = \left[l_{i_1} + (\lfloor q_i \div g_{i+1} \rfloor) \times \frac{u_{i_1} - l_{i_1}}{g_{i+1}}, l_{i_1} + (\lfloor q_i \div g_{i+1} \rfloor + 1) \times \frac{u_{i_1} - l_{i_1}}{g_{i+1}} \right]$$

$$(3)$$

In Eq. (3), the subscript of pir means the level of the meta-block and 0 means the first dimension while 1 means the second dimension. u_i and l_i represent the boundary of the higher level meta-block which just covers it, q_i means the position of meta-block in level-i and i satisfies $0 \le i < h$.

If data is not uniformly distributed, we use the *Piecewise Mapping Function (PMF)* [19] method to balance the skew data. The goal of PMF is partitioning the data evenly into some buckets. We use the cumulative mapping to evenly divide the data into buckets by using hash function.

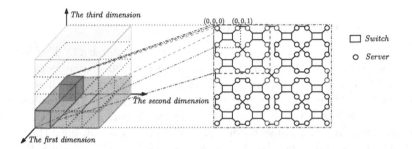

Fig. 4. Potential indexing range of HCN_2 (Color figure online)

In HCN_2, with $P = \langle 4, 2, \frac{3}{4}, 0 \rangle$ which is shown in Fig. 4, the potential indexing range of each server is represented by the purple cuboid. The servers in the level-0 structure will be combined together and ST_0 will manage the potential indexing range represented by the blue long cuboid. The level-1 structure ST_1 consists of 4 ST_0's and will manage the green cuboid consisting of 4 blue cuboids. At the highest level, the data space it manages will be the whole red cuboid.

Suppose the indexing space is bounded by $B = (B_0, B_1, \ldots, B_h)$, and B_i is $[l_i, l_i + w_i]$, $i \in [0, h]$, the potential range of server s is $pir(s)$. Similar to meta-blocks, each meta-server is also assigned an $(h+1)$-tuple $[a_h, a_{h-1}, \ldots, a_1, a_0]$ in which a_i represents the meta-block's position in level-i.

Lemma 1. *For a server s which is represented by tuple $[a_h, a_{h-1}, a_{h-2}, \ldots, a_0]$, its potential indexing range of pir is:*

$$
\begin{aligned}
pir\,(s) &= pir\,([a_h, a_{h-1}, \ldots, a_0]) \\
&= \left(\left[l_0 + a_0 \frac{w_0}{k}, l_0 + (a_0 + 1)\,\frac{w_0}{k} \right], \ldots, \left[l_h + a_h \frac{w_h}{g_h}, l_h + (a_h + 1)\,\frac{w_h}{g_h} \right] \right)
\end{aligned}
\tag{4}
$$

Publishing Scheme. Each server builds its own local R-tree to manage the data stored in it. Meanwhile, every server will select a set of nodes $N_k = \{N_k^1, N_k^2, \ldots, N_k^n\}$ from its local R-tree to publish them into the global index. Similar to the two-dimension situation, the format of the published R-tree node is (mbr, ip). ip records the physical address of server and mbr represents the minimum bounding rectangle of the R-tree node. For each selected R-tree node, we will use center and radius as the criteria for mapping. We set a threshold named R_{max}, to compare with the given radius. Given an R-tree node to be published, we first calculate the center and radius. Then, the node will be published to the server whose potential index range covers the center. If radius is larger than R_{max}, the node will be published to those servers whose potential indexing range intersects with the R-tree node range.

5   Query Processing

5.1   Query in Two-Dimensional Space

Point Query. The point query is processed in two steps: (1) The first step happens among the meta-servers to locate the servers which may possibly store the data. The query point $Q(x_0, x_1)$ will be first forwarded to the nearest level-h level representative which represents the largest meta-block. Then the query will be forwarded to level-$(h\text{-}1)$ representative with corresponding meta-block whose potential indexing range covers Q. The process goes on until the query is forwarded to a level-0 structure. All the representatives which receive the query will search their global R-Trees and forward the query to local servers. (2) In the second step, the servers will search their local R-Trees and return the result. In all, only $(h + 1)$ representatives will be searched in total.

Figure 5 shows a point query example in the global R-Tree on the same topology shown in Fig. 3. Traditionally, we need to perform the query in all servers in the DCN. However, if the hierarchical global indexes are used, we only need to perform query in much fewer servers. For example, for the point query represented by the purple node, the querying process will go through the global index from $Level_2$ to $Level_0$ with 3 representatives, and then the query will be forwarded to the servers who possibly store the result. Therefore, from this case, we can see the effectiveness of this indexing scheme.

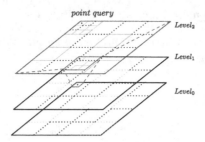

Fig. 5. An example of the point query process in R^2-Tree

Range Query. The range query is similar to point query which is also a two-step processing. Given a range query $R([l_{d_0}, u_{d_0}], [l_{d_1}, u_{d_1}])$, as the same as the processing in point query, we begin query from the largest meta-server to the smallest meta-server which can just cover the range R and then the forwarded servers will search their local R-Trees to find the data. The only difference is that in point query the smallest meta-server must be a physical server.

5.2   Query in High-Dimensional Space

Point Query. The point query is a two-step processing. Given a point query $Q(x_0, x_1, x_2, \ldots, x_d)$, we first create a super-sphere centered at Q with radius R_{max}. We search all the servers whose potential indexing range intersects with the super-sphere. To increase query speed, we forward the query in parallel. After getting the R-tree nodes which cover the point query, we forward the query to the servers which contain these nodes locally.

Range Query. The range query $R([l_{d_0}, u_{d_0}], \ldots, [l_{d_h}, u_{d_h}])$ will be sent to all the servers whose potential indexing range intersects with range query R. These servers will search their global indexes and find the corresponding R-Tree nodes. The query will be forwarded to those local servers. The cost of range query is less than directly broadcasting to all the servers.

6   Experiments

To validate R^2-Tree indexing scheme, we choose three existing server-centric data center network topologies including **DCell** $(P = \langle 4, 2, 1, 0 \rangle)$, **Ficonn** $(P = \langle 4, 2, \frac{1}{2}, 0 \rangle)$, **HCN** $(P = \langle 4, 2, \frac{3}{4}, 0 \rangle)$ to test the performance of our indexing scheme with them on the platform of Amazon's EC2. We implement our R^2-Tree in Python 2.7.9. We use in total 64 instance computers. Each of them has two-core 2.4 GHz Intel Xeon E5-2676v3 processor, 8 GB memory and 8 GB EBS storage. The bandwidth is 100 Mbps. The scale of the DCN topologies ranges from level-0 to level-2. The experiments involve 3 two-dimensional datasets: (1) Uniform_2d which follows uniform distribution, (2) Zipfian_2d which follows zipfian distribution, and (3) Hypsogr which is a real dataset obtained from the R-Tree Portal[1] and one uniform three-dimensional datasets. The detailed information of our experiments is shown in Table 2.

Table 2. Experiment settings

Parameter	Values
DCN topologies	DCell, Ficonn, HCN
Structure level	0, 1, 2
Dimensionality	2, 3
Distribution	Uniform, Zipfian, Real
Uniform datasets	Uniform_2d, Uniform_3d
Skew datasets	Zipfian_2d, Hypsogr
Query method	Point query, range query, centralized point query

Our experiments are conducted as follows. For each DCN topology, we generate 2,000,000 data points for each server. We execute 500 point queries and

---

[1]   http://chorochronos.datastories.org/?q=node/21.

100 range queries and record the total query time as the metric for each dataset. Additionally, to test the effectiveness of the *Mutex Particle Function*, we also perform centralized 500 point queries where all the query are confined to a certain area of the whole data space. By comparing the query time with **RT-HCN** [12], we show the superiority of our global R-Tree design. Besides, by counting the hop number for each point query and the average number of global indexes, we explain a trade-off between the query time and the storage efficiency.

In R^2-Tree, we propose hierarchical global indexes for two-dimensional data and divide the potential indexing range evenly for three-dimensional data. In Fig. 6, we show the point query performance of R^2-Tree in three different datasets. Since it is impossible to manipulate hundreds of thousands of servers in the experiments and a certain number of servers will be representative enough, the server number of **DCell** scales from 4 to 20, while the server number of **Ficonn** scales from 4 to 12 and 12 to 48, and for **HCN**, the server number scales from 4 to 16 and from 16 to 64. The two parallel columns represent the query time for the normal point query and the centralized point query respectively when the server number and the type of dataset are fixed. Based on the result that the query time for the centralized and non-centralized point query is close to each other when the other parameters are fixed, we show that the *Mutex Particle Function* balances the request load effectively.

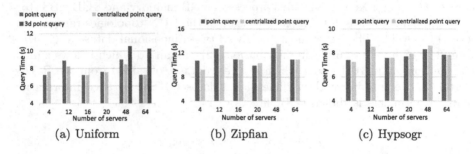

Fig. 6. Point query performance

We observe from Fig. 6 that the query time increases as the DCN structure scales out. By counting the global indexes stored in representatives in different levels, we notice an unbalance of the global information. The representatives in higher level tend to store more global indexes because they have larger potential indexing range. Since most of the chosen-to-published R-Tree nodes are from upper layer, the minimum bounding boxes are larger and will be more likely to be mapped to the meta-blocks which have larger potential indexing range. Nonetheless, in this way, we achieve higher storage efficiency since we do not need to store a lot of global information in each server. Besides, the global R-Tree helps to alleviate this bottleneck to a great extent. Among the three different datasets, we can see that the query time is the shortest for Uniform dataset and longest for Zipfian dataset.

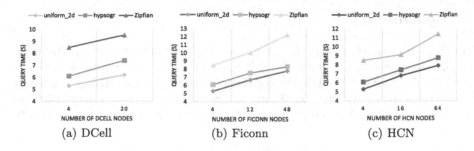

(a) DCell   (b) Ficonn   (c) HCN

Fig. 7. Range query performance

The range query in Fig. 7 also shows a same tendency of query time increase as the structure scales out. From the comparison of query time between different topologies, we find that for the same level number and the same kind of dataset, **DCell** performs the best while **Ficonn** performs the worst. We calculate the number of hops among the servers for a point query to explain the inner reason. In Fig. 8, we can see that the number of hops increases as the structure scales out. For the same level structure, the number of hops for DCell is the least and the hop number for Ficonn is the largest. This can be explained by *expansion factor* α easily. Figure 9 explains the trade-off between query time and storage space clearly. Larger α means that the connection between servers is more compact, and the number of physical hops will reduce and therefore achieve better time efficiency. However, the store efficiency will decrease correspondingly since each server stores more global information in different levels. By Comparing the query hop numbers for 2D and 3D data in Fig. 8, we can see the efficiency for the hierarchical global indexing design. Since the potential indexing range is of different size, we only publish the tree node to the just-cover meta-block. This mechanism avoids the repeated query effectively, and therefore reduce the total query time. Besides, in Fig. 10, we compare the query time of **R^2-Tree** to **RT-HCN** [12]. Global R-Tree accelerates the global query and *PMF* helps to balance the request load. Therefore, R^2-Tree shows superiority over **RT-HCN** [12].

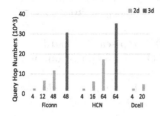

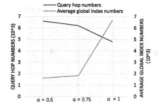

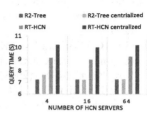

Fig. 8. Hop number   **Fig. 9.** Trade-off   **Fig. 10.** Comparisons

7 Conclusion

In this paper, we propose an indexing scheme named **R²-Tree** for multidimensional query processing which can suit most of the server-centric data center networks. To better formulate the topology of server-centric DCNs, we propose a *pattern vector P* through analyzing the recursively-defined feature of these networks. Based on that, we present a layered mapping method to reduce query scale by hierarchy. To balance the workload, we propose a method called *Mutex Particle Function* to distribute the potential indexing range. We prove theoretically that R^2-Tree can reduce both query cost and storage cost. Besides, we take three typical server-centric DCNs as examples and build indexes on them based on Amazon's EC2 platform, which also validates the efficiency of R^2-Tree.

References

1. Al-Fares, M., Loukissas, A., Vahdat, A.: A scalable, commodity data center network architecture. In: ACM SIGCOMM Computer Communication Review, pp. 63–74 (2008)
2. Beaver, D., Kumar, S., Li, H.C., Sobel, J., Vajgel, P.: Finding a needle in Haystack: Facebook's photo storage. In: OSDI, pp. 47–60 (2010)
3. Chen, G., Vo, H.T., Wu, S., Ooi, B.C., Özsu, M.T.: A framework for supporting DBMS-like indexes in the cloud. Proc. VLDB Endow. **4**(11), 702–713 (2011)
4. Decandia, G., et al.: Dynamo: Amazon's highly available key-value store. In: SOGOPS, pp. 205–220 (2007)
5. Gao, L., Zhang, Y., Gao, X., Chen, G.: Indexing multi-dimensional data in modular data centers. In: Chen, Q., Hameurlain, A., Toumani, F., Wagner, R., Decker, H. (eds.) DEXA 2015. LNCS, vol. 9262, pp. 304–319. Springer, Cham (2015). https://doi.org/10.1007/978-3-319-22852-5_26
6. Gao, X., Li, B., Chen, Z., Yin, M.: FT-INDEX: a distributed indexing scheme for switch-centric cloud storage system. In: ICC, pp. 301–306 (2015)
7. Ghemawat, S., Gobioff, H., Leung, S.T.: The Google file system. In: SOSP, pp. 29–43 (2003)
8. Greenberg, A., et al.: VL2: a scalable and flexible data center network. In: ACM SIGCOMM Computer Communication Review, pp. 51–62 (2009)
9. Guo, C., et al.: BCube: a high performance, server-centric network architecture for modular data centers. ACM SIGCOMM Comput. Commun. Rev. **39**(4), 63–74 (2009)
10. Guo, C., Wu, H., Tan, K., Shi, L., Zhang, Y., Lu, S.: DCell: a scalable and fault-tolerant network structure for data centers. ACM SIGCOMM Comput. Commun. Rev. **38**(4), 75–86 (2008)
11. Guo, D., Chen, T., Li, D., Li, M., Liu, Y., Chen, G.: Expandable and cost-effective network structures for data centers using dual-port servers. IEEE Trans. Comput. **62**(7), 1303–1317 (2013)
12. Hong, Y., Tang, Q., Gao, X., Yao, B., Chen, G., Tang, S.: Efficient R-tree based indexing scheme for server-centric cloud storage system. IEEE Trans. Knowl. Data Eng. **28**(6), 1503–1517 (2016)
13. Li, D., Guo, C., Wu, H., Tan, K.: FiConn: using backup port for server interconnection in data centers. In: INFOCOM, pp. 2276–2285 (2009)

14. Liao, Y., Yin, D., Gao, L.: DPillar: scalable dual-port server interconnection for data center networks. In: ICCCN, pp. 1–6 (2014)
15. Liu, Y., Gao, X., Chen, G.: A universal distributed indexing scheme for data centers with tree-like topologies. In: Chen, Q., Hameurlain, A., Toumani, F., Wagner, R., Decker, H. (eds.) DEXA 2015. LNCS, vol. 9261, pp. 481–496. Springer, Cham (2015). https://doi.org/10.1007/978-3-319-22849-5_33
16. Walraed-Sullivan, M., Vahdat, A., Marzullo, K.: Aspen trees: balancing data center fault tolerance, scalability and cost. In: CoNEXT, pp. 85–96 (2013)
17. Wang, J., Wu, S., Gao, H., Li, J., Ooi, B.C.: Indexing multi-dimensional data in a cloud system. In: SIGMOD, pp. 591–602 (2010)
18. Wu, S., Wu, K.L.: An indexing framework for efficient retrieval on the cloud. IEEE Comput. Soc. Data Eng. Bull. **32**(1), 75–82 (2009)
19. Zhang, R., Qi, J., Stradling, M., Huang, J.: Towards a painless index for spatial objects. ACM Trans. Database Syst. **39**(3), 19 (2014)

Time Series Data

Monitoring Range Motif on Streaming Time-Series

Shinya Kato(✉), Daichi Amagata, Shunya Nishio, and Takahiro Hara

Department of Multimedia Engineering Graduate School of Information Science
and Technology, Osaka University, Yamadaoka 1-5, Suita, Osaka, Japan
kato.shinya@ist.osaka-u.ac.jp

Abstract. Recent IoT-based applications generate time-series in a
streaming fashion, and they often require techniques that enable environmental monitoring and event detection from generated time-series. Discovering a range motif, which is a subsequence that repetitively appears
the most in a time-series, is a promising approach for satisfying such a
requirement. This paper tackles the problem of monitoring a range motif
of a streaming time-series under a count-based sliding-window setting.
Whenever a window slides, a new subsequence is generated and the oldest subsequence is removed. A straightforward solution for monitoring a
range motif is to scan all subsequences in the window while computing
their occurring counts measured by a similarity function. Because the
main bottleneck is similarity computation, this solution is not efficient.
We therefore propose an efficient algorithm, namely SRMM. SRMM is
simple and its time complexity basically depends only on the occurring
counts of the removed and generated subsequences. Our experiments
using four real datasets demonstrate that SRMM scales well and shows
better performance than a baseline.

Keywords: Streaming time-series · Motif monitoring

1 Introduction

Motif discovery is one of the most important tools for analyzing time-series [20].
Given a time-series t, its range motif is a subsequence that appears the most in t,
i.e., a range motif is a frequently occurring subsequence [6,17]. As an example,
in Fig. 1, we illustrate subsequences (red ones) which are repetitively appear
in a streaming time-series of greenhouse gas emission [12], and the left most
red subsequence is the current range motif. (We measure the similarity between
subsequences by z-normalized Euclidean distance, thus the value scale in this
figure is not a problem.) In this paper, we address the problem of monitoring a
range motif (motif in short) of a streaming time-series, because recent IoT-based
applications generate time-series in a streaming fashion [13].

Application Examples. It is not hard to see that this problem has a wide
range of applications. For example, assume that a sensor device measures a sensor value and sends it to a server periodically, which constitutes a streaming

© Springer Nature Switzerland AG 2018
S. Hartmann et al. (Eds.): DEXA 2018, LNCS 11029, pp. 251–266, 2018.
https://doi.org/10.1007/978-3-319-98809-2_16

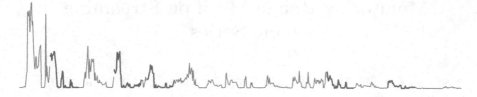

Fig. 1. An example of subsequences (red ones) which are repetitively appear and discovered in a streaming time-series of greenhouse gas emission [12]. We measure the similarity between subsequences by z-normalized Euclidean distance (that corresponds to Pearson correlation), and the current motif is the left most red subsequence. (Color figure online)

time-series. Assume further that a domain expert monitors the time-series, and if its motif changes as time passes, he/she can analyze some underlying phenomenon and form a hypothesis, e.g., sensor values have correlation with not only environmental but also temporal factors. Another example is event detection. Consider that we monitor the current motif and store it every minute. If the current motif is very different from the one obtained at the same time yesterday or we have a significant difference between the current and the previous motifs, it can be expected that there is an anomaly event.

Technical Overview. The above applications require monitoring the current motif in real-time while considering only recent data. We therefore employ a count-based sliding window setting, which considers only the most recent w data, and propose an efficient algorithm, namely SRMM (Streaming Range Motif Monitoring). When a given window slides, a new data is inserted into the window and the oldest data is removed from the window. That is, a new subsequence s_n, which contains the new data, is generated and the oldest one s_e, which contains the oldest data, is removed. A simple approach for updating the current motif, which is used as a baseline algorithm in this paper, is to scan all subsequences while comparing them with s_n and s_e. This can obtain the exact frequency count (the number of other subsequences that are similar to s_n and/or s_e) but incurs an expensive computational cost. SRMM avoids unnecessary computation by focusing on subsequences that can be the motif. The main idea employed in SRMM is to leverage PAA (Piecewise Aggregate Approximation) [7] and kd-tree [2]. This idea brings a technique which upper-bounds the frequency count of s_n with a light-weight cost, and enables pruning the exact frequency count computation. Even if we cannot prune the computation, we do not need to scan all subsequences. Actually, the upper-bounding collects a candidate of subsequences that may be similar to s_n. SRMM therefore needs to compare s_n only with the candidate subsequences.

Contributions. We summarize our contributions below.

– We address, for the first time, the problem of range motif (a subsequence that repetitively appears the most) monitoring on a streaming time-series under a count-based sliding window setting.
– We propose SRMM to efficiently update the current motif when a given window slides. SRMM is simple and efficient, and its time complexity is basically $O(\log(w - l) + m_n + m_e)$, where l is a given subsequence size and m_n and m_e are the upper-bound frequency counts of new and removed subsequences, respectively.
– We conduct experiments using four real datasets, and the results demonstrate that SRMM scales well and the performance of SRMM is better than that of the baseline.

Organization. We provide a preliminary in Sect. 2 and review some related works in Sect. 3. We present SRMM in Sect. 4 and introduce our experimental results in Sect. 5. Finally, Sect. 6 concludes this paper.

2   Preliminary

2.1   Problem Definition

A streaming time-series t is an ordered set of real values, which is described as $t = (t[1], t[2], ...)$, where $t[i]$ is a real value. Because we are interested in an underlying pattern in t, we below define subsequence of t.

Definition 1 (SUBSEQUENCE). *Given t and a length l, a subsequence of t, which starts at p is $s_p = (t[p], t[p + 1], ..., t[p + l - 1])$.*

For ease of presentation, let $s_p[x]$ be the x-th value in s_p. To observe how many similar subsequences s_p have in t (i.e., the occurring count of s_p), we use Pearson correlation, which is a basic function to measure the similarity between time-series [10,15].

Definition 2 (PEARSON CORRELATION). *Given two subsequences s_p and s_q with length l, their Pearson correlation $\rho(s_p, s_q)$ is*

$$\rho(s_p, s_q) = 1 - \frac{\|\hat{s}_p, \hat{s}_q\|^2}{2l}. \tag{1}$$

We have $\rho(s_p, s_q) \in [-1, 1]$. Note that $\|\hat{s}_p, \hat{s}_q\|$ computes the Euclidean distance between $\hat{s}_p$ and $\hat{s}_q$, and

$$\hat{s}_p[i] = \frac{s_p[i] - \mu(s_p)}{\sigma(s_p)},$$

where $\mu(s_p)$ and $\sigma(s_p)$ are the average and the variation of $(s_p[1], s_p[2], ..., s_p[l])$, respectively. Now we see that $\hat{s}_p$ is the z-normalized version of s_p, and Pearson correlation can be converted to the z-normalized Euclidean distance $d(\cdot, \cdot) = \|\cdot, \cdot\|$, i.e., from Eq. (1),

$$d(\hat{s}_p, \hat{s}_q) = \sqrt{2l(1 - \rho(s_p, s_q))}. \tag{2}$$

It is trivial that the time complexity of computing Pearson correlation is $O(l)$. We next define subsequences which are similar to s_p.

Definition 3 (SIMILAR SUBSEQUENCE). *Given s_p, s_q, and a threshold θ, we say that s_q (s_p) is similar to s_p (s_q) if*

$$\rho(s_p, s_q) \geq \theta \Leftrightarrow d(\hat{s}_p, \hat{s}_q) \leq \sqrt{2l(1 - \theta)}. \tag{3}$$

It can be easily seen that s_p and s_{p+1} can be similar to each other, but such a pair is not interesting to obtain a meaningful result. Such overlapping subsequences are denoted by trivial matched subsequences [5,17].

Definition 4 (TRIVIAL MATCH). *Given s_p, its trivial matched subsequences s_q satisfy that $p - l + 1 \leq q \leq p + l - 1$. $\overline{S_p}$ denotes the set of trivial matched subsequences of s_p.*

Now we consider the occurring count of s_p, $score(s_p)$ in other words.

Definition 5 (SCORE). *Given t, l, and θ, the score of a subsequence $s_p \in t$ is defined as:*

$$score(s_p) = |\{s_q \mid s_q \in t, \rho(s_p, s_q) \geq \theta, s_q \notin \overline{S_p}\}|. \tag{4}$$

Here, many applications including the ones in Sect. 1 care only recent data [8,14]. Hence, as with existing works that study streaming time-series [4,9], we employ a count-based sliding window setting, which monitors only the most recent w values. That is, a streaming time-series t in the window is represented as $t = (t[i], t[i+1], ..., t[i+w-1])$ where $t[i+w-1])$ is the newest value, and there are $(w - l + 1)$ subsequences in the window when l is given. When the window slides, we have a new subsequence which consists of the most recent l values. At the same time, the oldest value is removed from the window, so the oldest subsequence expires. We would like to monitor the subsequence of t with the maximum score in this setting. Let S be the set of all subsequences in a given widow with size w, and formally, our problem is:

Definition 6 (RANGE MOTIF MONITORING PROBLEM). *Given t, l, θ, and w, the problem in this paper is to monitor the current range motif s^* that satisfies*

$$s^* = \arg\max_{s \in S} score(s).$$

If the context is clear, range motif is called motif simply.

2.2   Baseline Algorithm

Because this is the first work that tackles this problem, we first provide a naive solution that can monitor the exact result. Section 1 has already introduced the solution, which updates the scores of all subsequences in the window by comparing them with the expired and new subsequences, whenever the window slides. As mentioned earlier, there are $(w - l + 1)$ subsequences in the window and each score computation requires $O(l)$ time. Therefore, the time complexity of this solution is $O((w - l)l)$.

We can intuitively see that, for a subsequence, comparing it with all subsequences incurs redundant computation cost, because the subsequence is interested only in its similar subsequences. To remove such a redundant cost, we propose a technique that efficiently identifies subsequences whose scores need to be updated.

3   Related Work

We introduce existing works that tackle the problem of motif discovery. It is important to note that the term *motif* is sometimes used in different meaning, as claimed in [6]. The first definition of motif is the same as that in this paper. On the other hand, some works, e.g., [10, 14, 15], use motif as the closest subsequence pair in a time-series. In this section, if referred literatures study the problem of discovering the closest subsequence pair, we say that it is pair-motif discovery problem.

3.1   Pair-Motif Discovery Problem

This problem suffers from its quadratic time complexity w.r.t. the number of subsequences, thus it is not trivial to make exact algorithms scale well. Literature [15] first proposed an exact algorithm MK that exploits triangle inequality. MK selects some subsequences as reference points, and utilize them to obtain upper-bound distances when it compares a given subsequence and another one. However, its time complexity is still quadratic. To scale better, [10] proposed Quick-Motif algorithm. Quick-Motif builds an subsequence index in online to reduce the number of subsequence comparisons. Its experiments show that Quick-Motif significantly outperforms MK. Recently, an offline index approach, called Matrix Profile, was proposed in [21, 22]. For all subsequences, this index maintains the distances to other subsequences with the largest similarity. This index makes an online pair-motif discovery algorithm fast [22].

The above studies consider static time-series. The first attempt to monitor the pair-motif is performed in [14]. For each subsequence, the algorithm proposed in [14] maintains its nearest neighbor and reverse nearest neighbor subsequences to deal with the pair-motif update. Literature [8] has optimized a data structure for pair-motif monitoring and the algorithm proposed in [8] outperforms the algorithm of [14].

3.2 Range-Motif Discovery Problem

Patel et al. proposed an approximate algorithm to discover a range motif efficiently [17]. In this algorithm, each subsequence is converted to a string sequence by SAX [11]. Similar to this algorithm, Castro and Azevedo proposed a range motif discovering algorithm [3] that employs iSAX [19]. Both SAX and iSAX approximate a given time-series, thus the discovered motif is not guaranteed to be exact. Some probabilistic algorithms are proposed in [5,20], and again, this approach does not guarantee the correctness. Literature [6] proposed a learning-based motif discovery algorithm. This algorithm requires pre-processing step, thus is hard to be applied in streaming setting. The above literatures consider only a static time-series.

Although [1] considers a streaming time-series, it aims to discover a rare subsequence that has some similar subsequences but with some *very low probability*. The algorithm proposed in [1] also employs approximate approaches (SAX and Bloom filter). [16] also considers a streaming time-series, but this literature considers a distance between subsequences under SAX representation. As can be seen above, the existing works basically consider approximate solutions. In this paper, we provide an *exact* solution for efficient motif monitoring.

4 SRMM: Streaming Range Motif Monitoring

We first note that *the score of each subsequence in the window increases at most one* when the window slides, which can be seen from Definition 5 and the property of count-based sliding window. This observation suggests that the current motif does not change frequently and the score of the new subsequence often does not reach $score(s^*)$.

Let s_n be the new subsequence, and if we can know that $score(s_n) < score(s^*)$ with a *light-weight cost*, we can efficiently monitor the exact motif. To achieve this, we propose a technique that obtains an upper-bound of $score(s_n)$ efficiently and prunes unnecessary exact score computation. We introduce this technique in Sect. 4.1. Recall that the oldest subsequence is removed from the window, which makes the scores of some subsequences decrease by one. This may affect s^*. SRMM can efficiently identify the subsequences whose scores may decrease, which is described in Sect. 4.2. Finally, We elaborate the overall algorithm of SRMM and provide its time complexity in Sect. 4.3.

4.1 Upper-Bounding

First, we obtain an upper-bound of Pearson correlation between s_n and $s \in S$, which corresponds to a lower-bound of the z-normalized distance (see Eq. (2)). We use PAA [7], a dimensionality reduction algorithm, to achieve this. Recall that a subsequence s_p is represented as $(s_p[1], s_p[2], ..., s_p[l])$. This implies that it can be regarded as a point on an l-dimensional space $\mathbb{R}^l$, i.e., a subsequence is an l-dimensional point.

Given a dimensionality $\phi < l$, PAA transforms an l-dimensional point into a ϕ-dimensional point. Let $\hat{s}_p^\phi$ be the transformed $\hat{s}_p$. Each value of $\hat{s}_p^\phi$ is described as

$$\hat{s}_p^\phi[i] = \frac{\phi}{l} \sum_{j=\frac{l}{\phi}i}^{\frac{l}{\phi}(i+1)-1} \hat{s}_p[j].$$

PAA has the following lemma.

Lemma 1 [7]. *Given two subsequences $\hat{s}_p$ and $\hat{s}_q$, we have*

$$\sqrt{\frac{l}{\phi}}dist(\hat{s}_p^\phi, \hat{s}_q^\phi) \leq dist(\hat{s}_p, \hat{s}_q). \tag{5}$$

From PAA, we can obtain a lower-bound of the Euclidean distance between $\hat{s}_p$ and $\hat{s}_q$, i.e., an upper-bound of $\rho(s_p, s_q)$ in $O(\phi)$ time. If $\sqrt{\frac{l}{\phi}}dist(\hat{s}_p^\phi, \hat{s}_q^\phi) > \sqrt{2l(1-\theta)}$, s_q is not similar to s_p (see Definition 3), thus we can safely prune the exact distance computation between $\hat{s}_p$ and $\hat{s}_q$. Given $\hat{s}_n$, an upper-bound of $score(s_n)$ can be obtained if we compute $\sqrt{\frac{l}{\phi}}dist(\hat{s}_n^\phi, \hat{s}_p^\phi)$ for $\forall s_p \in S\backslash\overline{S}_n$. However, this approach is still expensive, incurs $O(\phi(w-l))$ time, and s_n is interested only in s_p such that $\sqrt{\frac{l}{\phi}}dist(\hat{s}_n^\phi, \hat{s}_p^\phi) \leq \sqrt{2l(1-\theta)}$. To obtain such s_p efficiently, we employ a kd-tree [2], which is a binary tree for an arbitrary dimensional space. The behind idea of employing a kd-tree is that kd-tree supports efficient data insertion, deletion, and range query processing.

Assume that all transformed subsequences in the window are indexed by a kd-tree. Now we see that s_p, such that $\sqrt{\frac{l}{\phi}}dist(\hat{s}_n^\phi, \hat{s}_p^\phi) \leq \sqrt{2l(1-\theta)}$, is obtained by a range query where the query point is $\hat{s}_n^\phi$ and the distance threshold is $\sqrt{2\phi(1-\theta)}$. Then we have the following theorem.

Theorem 1. *Assume that we have a new subsequence s_n, a distance threshold $\sqrt{2l(1-\theta)}$, and a kd-tree that maintains all subsequences, except the l most recent ones, which are transformed by PAA. A range query on the kd-tree, where its query point and a distance threshold respectively are $\hat{s}_n^\phi$ and $\sqrt{2\phi(1-\theta)}$, returns S_n^{in} which is a set of transformed subsequences $\hat{s}_p^\phi$ such that $dist(\hat{s}_n^\phi, \hat{s}_p^\phi) \leq \sqrt{2\phi(1-\theta)}$. Let $|S_n^{in}| = m_n$, and we have $m_n \geq score(s_n)$.*

Proof. We want s_p that satisfies $\sqrt{\frac{l}{\phi}}dist(\hat{s}_n^\phi, \hat{s}_p^\phi) \leq \sqrt{2l(1-\theta)}$, which can be seen from Lemma 1. This inequality derives $dist(\hat{s}_n^\phi, \hat{s}_p^\phi) \leq \sqrt{2\phi(1-\theta)}$. Next, the l most recent subsequences can be trivial matched subsequences of s_n, thereby they are not necessary to compute $score(s_n)$. Theorem 1 therefore holds. □

Example 1. Figure 2 illustrates a set of transformed subsequences where $\phi = 2$, i.e., they are 2-dimensional points. To obtain an upper-bound score of s_n, we

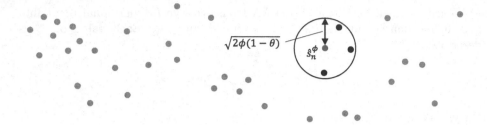

Fig. 2. An example of upper-bounding of $score(s_n)$, where $\phi = 2$. The red point is s_n and $m_n = 3$, since there are three points within the circle centered at $\hat{s}_n^\phi$ with the radius $\sqrt{2\phi(1-\theta)}$. (Color figure online)

set $\sqrt{2\phi(1-\theta)}$ as a distance threshold and execute a range query centered at $\hat{s}_n^\phi$ (the red point). As a query answer, we have three (black) points, which are efficiently retrieved by using a kd-tree, and we have $m_n = 3$.

Theorem 1 provides the following corollary.

Corollary 1. *If $score(s) \geq m_n$ where $s \in S\backslash\{s_n\}$, s_n cannot be the current motif, thus we can safely prune the exact computation of $score(s_n)$.*

Due to Theorem 1, we do not index the l most recent subsequences by a kd-tree. Here, the time complexity of a range query on a kd-tree is $O(\log n + m)$ where n and m are the cardinalities of data in the kd-tree and of data satisfying the distance threshold. The time complexity of the upper-bounding is hence $O(\log(w - l) + m_n)$, and we have $(\log(w - l) + m_n) \ll w$.

4.2 Identifying the Subsequences Whose Scores Can Decrease

When the window slides, the oldest subsequence expires, which makes the scores of some subsequences decrease. One may consider that a range query centered at the expired subsequence can solve this score updates. However, such a duplicate evaluation is not efficient. We overcome this problem by utilizing two lists for each subsequence s_p, similar list SL_p and possible similar list PL_p.

Definition 7 (SIMILAR LIST). *The similar list of s_p, SL_p, is a set of tuples of subsequence identifier q and $\rho(s_p, s_q)$, i.e., $SL_p = \{\langle q, \rho(s_p, s_q)\rangle \mid s_q \in S\backslash S_p, \rho(s_p, s_q) \geq \theta\}$.*

Definition 8 (POSSIBLE SIMILAR LIST). *The possible similar list of s_p, PL_p, is a set of identifiers of subsequences s_q such that $dist(\hat{s}_p^\phi, \hat{s}_q^\phi) \leq \sqrt{2\phi(1-\theta)}$, $s_q \notin S_p$, and $\langle q, \cdot\rangle \notin SL_p$.*

In a nutshell, when we compute an upper-bound score of s_p by a range query, we add q, such that $dist(\hat{s}_p^\phi, \hat{s}_q^\phi) \leq \sqrt{2\phi(1-\theta)}$, into PL_p. We also add p into PL_q. In addition, when we compute $\rho(s_p, s_q)$, we remove q (p) from PL_p (PL_q), and if $\rho(s_p, s_q) \geq \theta$, we update SL_p and SL_q. Now we have two lemmas.

Algorithm 1. SRMM (expiration case)

Input: s_e: the expired subsequence
Output: s^*_{temp}: a temporal motif

1   Delete $\hat{s}^\phi_e$ from kd-tree, $f \leftarrow 0$
2   **for** $\forall p \in SL_e$ **do**
3      $SL_p \leftarrow SL_p \backslash \langle e, \cdot \rangle$
4   **if** $s_p = s^*$ **then**
5         $f \leftarrow 1$

6   **for** $\forall p \in PL_e$ **do**
7      $PL_p \leftarrow PL_p \backslash \{e\}$

8   **if** $s^* = s_e$ **then**
9      $f \leftarrow 1, s^* \leftarrow \varnothing$

10   $s^*_{temp} \leftarrow s^*$
11   **if** $f = 1$ **then**
12   **for** $\forall s_p \in S$ *such that* $|SL_p| + |PL_p| \geq score(s^*_{temp})$ **do**
13         $s^*_{temp} \leftarrow$ Motif-Update(s_p, s^*_{temp})

Lemma 2. $|SL_p| + |PL_p| \geq score(s_p)$.

Lemma 3. *The subsequences s_q, whose scores can decrease due to the expiration of s_e, satisfy that $q \in PL_e$ or $\langle q, \cdot \rangle \in SL_e$.*

Both Lemmas 2 and 3 can be proven by Definitions 7 and 8. Now we see from Lemma 3 that SL_q and PL_q can be updated in $O(1)$ time, so its total update time is $O(|SL_e| + |PL_e|)$.

4.3 Overall Algorithm

We present the detail of SRMM, which exploits the techniques introduced in Sects. 4.1 and 4.2. When the window slides, we first deal with the expired subsequence and obtains a temporal motif s^*_{temp}. After that, we verify whether the new subsequence can be s^*.

Dealing with Expired Subsequence s_e. Algorithm 1 details how SRMM deals with the expired subsequence. Given the expired subsequence s_e, SRMM deletes $\hat{s}^\phi_e$ from the kd-tree, which is done in $O(\log(w - l))$ time, and sets a flag $f = 0$ (line 1). Then, according to Lemma 3, SRMM deletes $\{e\}$ and $\langle e, \cdot \rangle$ from all PL_p and SL_p such that $p \in PL_e$ or $\langle p, \cdot \rangle \in SL_e$ (lines 2–9). Note that if $score(s^*)$ decreases or $s^* = s_e$, we set $f = 1$. Last, if $f = 1$, the current motif can be changed. From Lemma 2, we see the subsequences s_p which can be the motif have to satisfy $|SL_p| + |PL_p| \geq score(s^*_{temp})$. SRMM therefore computes the exact scores of such s_p and obtains a temporal motif s^*_{temp} (line 13), through Motif-Update(s_p, s^*_{temp}), which is introduced later.

We next confirm that the obtained temporal motif is really the current motif or the new subsequence can be the current motif.

Algorithm 2. SRMM (insertion case)

Input: s_n: the new subsequence, s^*_{temp}: a temporal motif
Output: s^*: the current motif

1 Compute $\hat{s}^\phi_n$ by PAA
2 Insert $\hat{s}^\phi_{n-l}$ to kd-tree
3 $SL_n \leftarrow \varnothing$
4 $PL_n \leftarrow$ Range-Search($\hat{s}^\phi_n, \sqrt{2\phi(1-\theta)}$)
5 **for** $\forall p \in PL_n$ **do**
6 $\quad$ **if** $s_p = s^*_{temp}$ **then**
7 $\quad\quad$ Compute $\rho(s_p, s_n)$
8 $\quad\quad$ **if** $\rho(s_p, s_n) \geq \theta$ **then**
9 $\quad\quad\quad$ $SL_p \leftarrow SL_p \cup \langle n, \rho(s_p, s_n)\rangle, \; SL_n \leftarrow SL_n \cup \langle p, \rho(s_p, s_n)\rangle$
10 $\quad\quad$ $PL_n \leftarrow PL_n \backslash \{p\}$
11 $\quad$ **else**
12 $\quad\quad$ $PL_p \leftarrow PL_p \cup \{n\}$
13 $\quad\quad$ **if** $|SL_p| + |PL_p| \geq score(s^*_{temp})$ **then**
14 $\quad\quad\quad$ $s^*_{temp} \leftarrow$ Motif-Update(s_p, s^*_{temp})

15 **if** $|SL_n| + |PL_n| \geq score(s^*_{temp})$ **then**
16 $\quad$ $s^* \leftarrow$ Motif-Update(s_p, s^*_{temp})
17 **else**
18 $\quad$ $s^* = s^*_{temp}$

Dealing with New Subsequence s_n. Algorithm 2 illustrates how SRMM updates the current motif. SRMM first obtains $\hat{s}^\phi_n$ by PAA and inserts $\hat{s}^\phi_{n-l}$ into the kd-tree (lines 1–2). Note that s_{n-l} is the most recent subsequence that does not overlap with s_n. (Recall that our kd-tree does not maintain the l most recent transformed subsequences.) Then SRMM sets $SL_n = \varnothing$ and obtains PL_n by a range query, as explained in Sect. 4.1 (lines 3–4). For $\forall p \in PL_n$, PL_p also needs to be updated. If $s_p = s^*_{temp}$, SRMM computes $\rho(s_p, s_n)$ to obtain $score(s_p)$, and then updates SL_p, SL_n, and PL_n (lines 6–10). On the other hand, if $s_p \neq s^*_{temp}$, PL_p is updated and SRMM checks whether $|SL_p| + |PL_p| \geq score(s^*_{temp})$ or not. In the case where it is true, SRMM executes Motif-Update(s_p, s^*_{temp}) and updates s^*_{temp} if necessary (line 14). Last, if $|SL_n| + |PL_n| \geq score(s^*_{temp})$, SRMM executes Motif-Update(s_n, s^*_{temp}) to verify the current motif (line 15–16). Otherwise, we can guarantee that s^*_{temp} is now s^* (line 18).

Speeding Up Verification. In Motif-Update(s_n, s^*_{temp}), we confirm whether or not $\rho(s_n, s^*_{temp}) \geq \theta$, update their similar and possible similar lists, and replace s^*_{temp} if necessary. We see that updating similar and possible similar lists requires $O(1)$ time, so if we can relieve the confirmation cost, the motif verification cost is reduced. We achieve this by using the following theorem.

Theorem 2. *When s_n, s_p where $p \in PL_n$, s_q where $q \in PL_n \wedge \langle q, \rho(s_p, s_q) \rangle \in SL_p$, and θ are given, we have $\rho(s_n, s_q) \geq \theta$ if $dist(\hat{s}_n, \hat{s}_p) + dist(\hat{s}_p, \hat{s}_q) \leq \sqrt{2l(1-\theta)}$.*

Proof. Recall that $dist(\cdot, \cdot)$ is the z-normalized Euclidean distance. Therefore, from triangle inequality and Eq. (3), Theorem 2 holds. $\qquad\square$

Recall that if $|SL_n| + |PL_n| \geq score(s^*_{temp})$, we need to compute $score(s_n)$. We accelerate this verification, i.e., Motif-Update(s_n, s^*_{temp}) by exploiting Theorem 2. As a reference subsequence, we utilize s_p which is the nearest neighbor to s_n, in the ϕ-dimensional space, among a set of subsequences $s_{p'}$ such that $p' \in PL_n$ and $SL_{p'} \neq \varnothing$. Note that s_p is obtained during Range-Search($\hat{s}_n^\phi, \sqrt{2\phi(1-\theta)}$). First, we compute $dist(\hat{s}_n, \hat{s}_p)$. Then, for $\forall q \in PL_n$, we compute $dist(\hat{s}_n, \hat{s}_p) + dist(\hat{s}_p, \hat{s}_q)$ if $\langle q, \cdot \rangle \in SL_p$. If we have $dist(\hat{s}_n, \hat{s}_p) + dist(\hat{s}_p, \hat{s}_q) \leq \sqrt{2l(1-\theta)}$, we do not need to compute $dist(\hat{s}_n, \hat{s}_q)$. Therefore, we compute $dist(\hat{s}_n, \hat{s}_q)$ only in cases where we have $dist(\hat{s}_n, \hat{s}_p) + dist(\hat{s}_p, \hat{s}_q) > \sqrt{2l(1-\theta)}$ or $\langle q, \cdot \rangle \notin SL_p$.

Time Complexity. As mentioned earlier, inserting/removing a transformed subsequence into/from the kd-tree incurs $O(\log(w-l))$ time. Algorithm 1 requires at least $O(\log(w-l) + m_e)$ time, where $m_e = |SL_e| + |PL_e|$. Also, Algorithm 2 requires at least $O(\log(w-l) + m_n)$ time. Recall that m_n is the cardinality of returned (transformed) subsequences by Range-Search($\hat{s}_n^\phi, \sqrt{2\phi(1-\theta)}$). If we compute the exact score of s_p, $O(l|PL_p|)$ time is required, since we need to scan PL_p and each Pearson correlation computation incurs $O(l)$ time. Let S' be a set of subsequences whose exact scores are computed when the window slides. The total time complexity of SRMM is $O(\log(w-l) + m_e + m_n + \sum_{S'} l|PL_p|)$. It is important to note that $|S'|$ is very small practically. For example, in our experiments, $|S'| \leq 1$ on average. If we consider a polylogarithmic factor, i.e., $\log(w-l)$, can be seen as a constant, the time complexity of SRMM is dependent only on the upper-bound scores of the expired and new subsequences in practice.

5   Experiment

This section introduces our experimental results. We evaluated SRMM and the baseline algorithm introduced in Sect. 2.2. All experiments were conducted on a PC with 3.4 GHz Core i7 CPU and 16 GB RAM, and all the algorithms were implemented in C++.

5.1   Setting

In the following setting, we measured the average update time per a slide of the window.

Datasets. We used four real datasets.

- Google-CPU [18]: this time-series is a merged sequence of CPU usage rate of machines in Google compute cells, and its length is 133,902.
- Google-Memory [18]: this time-series is a merged sequence of memory usage of machines in Google compute cells, and its length is 133,269.
- GreenHouseGas [12]: this is a time-series of green house gas concentrations with length 100,062.
- RefrigerationDevices[1]: this is a sequence of energy consumption of a refrigerator, and its length is 270,000.

Parameters. Table 1 summarizes the parameters used in the experiments and bold values are default values. We set $\phi = \frac{l}{2}$, and when we investigate the impact of a given parameter, the other parameters are fixed.

Table 1. Configuration of parameters

Parameter	Values
Motif length, l	50, **100**, 150, 200
Window-size, w [×1000]	5, **10**, 15, 20
Threshold, θ	0.75, 0.8, 0.85, **0.9**, 0.95

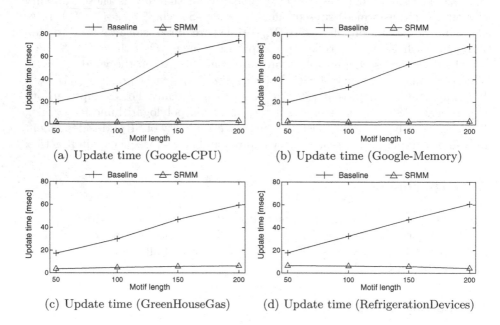

(a) Update time (Google-CPU) (b) Update time (Google-Memory)

(c) Update time (GreenHouseGas) (d) Update time (RefrigerationDevices)

Fig. 3. Impact of l

---

[1] http://timeseriesclassification.com/index.php.

5.2   Result

Varying l. We first investigate the impact of motif length, and Fig. 3 shows the result. We see that the update time of the baseline algorithm linearly increases, as l increases. This is reasonable since its time complexity is $O((w - l)l)$. On the other hand, SRMM is not sensitive to l. As l increases, we need more time to compute Pearson correlation. However, for fixed θ, m_e and m_n decrease as l increases. For a large l, we tend to have a long distance between two subsequences, i.e., their Pearson correlation tends to be low. Hence, it becomes difficult for subsequences to be similar to other ones, which is the reason why m_e and m_n decrease. SRMM therefore has a stable performance even when l varies. This scalability is a good advantage against the baseline, and SRMM is up to 24.5 times faster than the baseline.

Varying w. We next investigate the impact of window size. As can be seen from Fig. 4, we have a very similar result to that in Fig. 3. The time complexity of the baseline is linear to w, so this result is also straightforward. A difference is that the update time of SRMM also increases. As w increases, the score of each subsequence tends to be larger, i.e., m_e and m_n become larger. SRMM therefore needs longer update time when w is large.

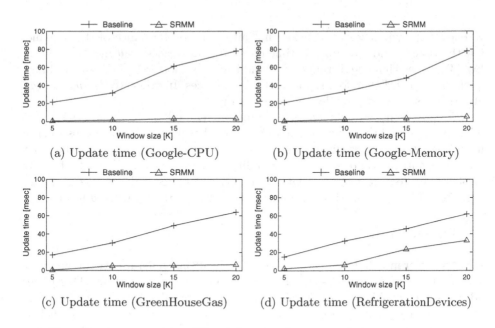

(a) Update time (Google-CPU)   (b) Update time (Google-Memory)

(c) Update time (GreenHouseGas)   (d) Update time (RefrigerationDevices)

Fig. 4. Impact of w

Varying θ. Finally, we report the impact of threshold, and the result is shown in Fig. 5. Because the baseline algorithm scans all subsequences in the window

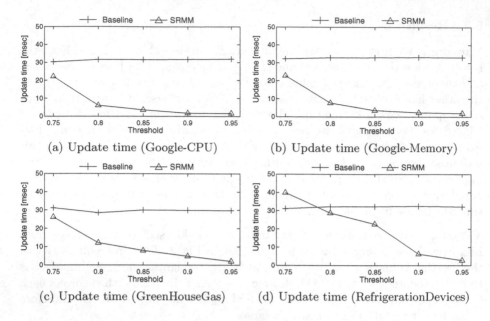

(a) Update time (Google-CPU)   (b) Update time (Google-Memory)

(c) Update time (GreenHouseGas)   (d) Update time (RefrigerationDevices)

Fig. 5. Impact of θ

whenever the window slides, θ does not affect the performance of the baseline. On the other hand, the update time of SRMM decreases as θ increases. From Eq. (3), we see that the distance threshold becomes shorter as θ increases. Range queries in SRMM therefore report less subsequences. In other words, m_e and m_n also decrease, which provides the result in Fig. 5.

We can see that SRMM incurs longer update time than the baseline when $\theta = 0.75$. We observed that there are many similar subsequences for each subsequence in RefrigerationDevices when θ is small. In such cases, we cannot prune the exact score computation and the upper-bounding can be overhead. Note that many applications require a motif that has highly correlated subsequences, and as Figs. 5(a)–(d) show, SRMM can update the motif quite fast when θ is large.

6   Conclusion

Due to the trend that recent IoT-based applications generate streaming time-series, analyzing time-series in real-time becomes more important. This paper addressed the problem of monitoring a range motif (a subsequence which appears repetitively the most in a given time-series), for the first time. As an efficient solution to this problem. we proposed SRMM. This algorithm can avoid unnecessary score computation by exploiting Piecewise Approximate Aggregation and kd-tree. The results of our experiments using four real datasets show the efficiency and scalability of SRMM.

In this paper, we considered an one-dimensional time-series. Recently, a device is becoming to have multiple sensors and can generate a multi-dimensional time-series. As a future work, we plan to address the range motif monitoring of a multi-dimensional streaming time-series.

Acknowledgement. This research is partially supported by JSPS Grant-in-Aid for Scientific Research (A) Grant Number JP26240013, JSPS Grant-in-Aid for Scientific Research (B) Grant Number JP17KT0082, and JSPS Grant-in-Aid for Young Scientists (B) Grant Number JP16K16056.

References

1. Begum, N., Keogh, E.: Rare time series motif discovery from unbounded streams. PVLDB **8**(2), 149–160 (2014)
2. Bentley, J.L.: Multidimensional binary search trees used for associative searching. Commun. ACM **18**(9), 509–517 (1975)
3. Castro, N., Azevedo, P.: Multiresolution motif discovery in time series. In: SDM, pp. 665–676 (2010)
4. Chen, Y., Nascimento, M.A., Ooi, B.C., Tung, A.K.: SpADe: on shape-based pattern detection in streaming time series. In: ICDE, pp. 786–795 (2007)
5. Chiu, B., Keogh, E., Lonardi, S.: Probabilistic discovery of time series motifs. In: KDD, pp. 493–498 (2003)
6. Grabocka, J., Schilling, N., Schmidt-Thieme, L.: Latent time-series motifs. TKDD **11**(1), 6 (2016)
7. Keogh, E., Chakrabarti, K., Pazzani, M., Mehrotra, S.: Dimensionality reduction for fast similarity search in large time series databases. KIS **3**(3), 263–286 (2001)
8. Lam, H.T., Pham, N.D., Calders, T.: Online discovery of top-k similar motifs in time series data. In: SDM, pp. 1004–1015 (2011)
9. Li, Y., Zou, L., Zhang, H., Zhao, D.: Computing longest increasing subsequences over sequential data streams. PVLDB **10**(3), 181–192 (2016)
10. Li, Y., Yiu, M.L., Gong, Z., et al.: Quick-motif: an efficient and scalable framework for exact motif discovery. In: ICDE, pp. 579–590 (2015)
11. Lin, J., Keogh, E., Wei, L., Lonardi, S.: Experiencing sax: a novel symbolic representation of time series. Data Min. Knowl. Disc. **15**(2), 107–144 (2007)
12. Lucas, D., et al.: Designing optimal greenhouse gas observing networks that consider performance and cost. Geosci. Instrum. Methods Data Syst. **4**(1), 121 (2015)
13. Moshtaghi, M., Leckie, C., Bezdek, J.C.: Online clustering of multivariate time-series. In: SDM, pp. 360–368 (2016)
14. Mueen, A., Keogh, E.: Online discovery and maintenance of time series motifs. In: KDD, pp. 1089–1098 (2010)
15. Mueen, A., Keogh, E., Zhu, Q., Cash, S., Westover, B.: Exact discovery of time series motifs. In: SDM, pp. 473–484 (2009)
16. Nguyen, H.L., Ng, W.K., Woon, Y.K.: Closed motifs for streaming time series classification. KIS **41**(1), 101–125 (2014)
17. Patel, P., Keogh, E., Lin, J., Lonardi, S.: Mining motifs in massive time series databases. In: ICDM, pp. 370–377 (2002)
18. Reiss, C., Wilkes, J., Hellerstein, J.L.: Google cluster-usage traces: format+ schema, pp. 1–14. Google Inc., White Paper (2011)
19. Shieh, J., Keogh, E.: i SAX: indexing and mining terabyte sized time series. In: KDD, pp. 623–631 (2008)

20. Yankov, D., Keogh, E., Medina, J., Chiu, B., Zordan, V.: Detecting time series motifs under uniform scaling. In: KDD, pp. 844–853 (2007)
21. Yeh, C.C.M., et al.: Matrix profile I: all pairs similarity joins for time series: a unifying view that includes motifs, discords and shapelets. In: ICDM, pp. 1317–1322 (2016)
22. Zhu, Y., et al.: Matrix profile II: exploiting a novel algorithm and GPUs to break the one hundred million barrier for time series motifs and joins. In: ICDM, pp. 739–748 (2016)

MTSC: An Effective Multiple Time Series Compressing Approach

Ningting Pan[1], Peng Wang[1,2(✉)], Jiaye Wu[1], and Wei Wang[1,2]

[1] School of Computer Science, Fudan University, Shanghai, China
{ntpan17,pengwang5,wujy16,weiwang1}@fudan.edu.cn
[2] Shanghai Key Laboratoray of Data Science, Shanghai, China

Abstract. As the volume of time series data being accumulated is likely to soar, time series compression has become essential in a wide range of sensor-data applications, like Industry 4.0 and Smart grid. Compressing multiple time series simultaneously by exploiting the correlation between time series is more desirable. In this paper, we present MTSC, a novel approach to approximate multiple time series. First, we define a novel representation model, which uses a base series and a *single* value to represent each series. Second, two graph-based algorithms, $MTSC_{mc}$ and $MTSC_{star}$, are proposed to group time series into clusters. $MTSC_{mc}$ can achieve higher compression ratio, while $MTSC_{star}$ is much more efficient by sacrificing the compression ratio slightly. We conduct extensive experiments on real-world datasets, and the results verify that our approach outperforms existing approaches greatly.

1 Introduction

Recent advances in sensing technologies have made possible, both technologically and economically, the deployment of densely distributed sensor networks. In many applications, such as IoT, Smart city and Industry 4.0, thousands or even millions of sensors are deployed to monitor the physical environment. Moreover, more and more applications tend to archive these data over a few years enabling people to do historical comparison and trend analysis [5]. To minimize the overhead of storing, managing and sharing these sensor data, therefore, we must apply smart approximation schemes that significantly reduce the data size without compromising the monitoring and analysis abilities [10]. For many useful data mining tasks, such as analyzing and forecasting resource utilization, anomaly detection, and forensic analysis, the compressed data must guarantee a given maximum (L_∞) decompression error [6].

An individual sensor's measurements can be thought of as a time series. Researchers have proposed many techniques to compress the single time series,

The work is supported by the Ministry of Science and Technology of China, National Key Research and Development Program (No. 2016YFB1000700), National Key Basic Research Program of China (No. 2015CB358800), NSFC (61672163, U1509213), Shanghai Innovation Action Project (No. 16DZ1100200).

S. Hartmann et al. (Eds.): DEXA 2018, LNCS 11029, pp. 267–282, 2018.
https://doi.org/10.1007/978-3-319-98809-2_17

such as DFT, APCA, PLA and DWT [10]. While in many applications, the time series are correlated with each other [6]. For example, the temperature measurements monitored by the closely-located weather stations will fluctuate together. Other examples include, but not limited to, the stock price of the same category and air quality of adjacent regions. Compressing time series individually without considering the correlation will incur much redundant storage.

Inspired from this observation, some works have been proposed to compress multiple sensor series simultaneously [4,6,14]. They collectively approximate multiple series while reducing redundant information. As a pioneer work, SBR [4] groups similar time series into clusters and approximates series of the same cluster with a common base series. However SBR requires similar series to be statically grouped together before running the algorithms, which makes it unsuitable for long time series. Moreover it guarantees the L_2 error bound instead of L_∞, that is, SBR cannot guarantee the error bound in every single time point.

GAMPS is the first work to compress multiple time series guaranteeing the L_∞ error bound. It utilizes a dynamic grouping scheme to group series in different time windows. Within each group of series, it approximates each series based on a common base and a reference series, and compresses both of them with the APCA representation [7]. However the compression quality of GAMPS is inferior to single series compression algorithms, such as APCA, in many cases [10].

In this paper, we propose a new framework to compress multiple time series, named Multiple Time Series Compressing ($MTSC$). Firstly, we define a novel representation model, which uses a base series and a *single* value to represent each series within a cluster. Different from GAMPS, which uses two series to approximate a raw series, our model incurs much less storage cost. The core of our approach is the grouping strategy which groups time series into as few clusters as possible. Two graph-based algorithms, $MTSC_{mc}$ and $MTSC_{star}$, are proposed. $MTSC_{mc}$ can achieve higher compression ratio, while $MTSC_{star}$ is much more efficient by sacrificing the compression ratio slightly. We conduct extensive experiments on multiple real-world datasets, which show that our approach has higher compression ratio than existing approaches in most cases.

The rest of the paper is organized as follows. Preliminary knowledge is introduced in Sect. 2. Section 3 introduces our compression model and theoretical foundation. Sections 4 and 5 describe the $MTSC_{mc}$ and $MTSC_{star}$ algorithms respectively. The experimental results are presented in Sect. 6 and we discuss related work in Sect. 7. Finally, Sect. 8 concludes the paper.

2   Preliminaries

Let $\mathcal{S} = \{S_1, S_2, \cdots, S_N\}$ be a set of N time series with equal length n. S_i is the i-th time series, consisting of a sequence of values at time point from 1 to n, denoted as $S_i = \{s_i(t)|t = 1, 2, \cdots, n\}$. The subsequence of S_i is a continuous subset of the values, denoted as $S_i(l, r) = \{s_i(t), t = l, l+1, \cdots, r\}$.

We produce an approximate representation of $\mathcal{S}$, denoted as Δ. It takes a more concise form, from which, we can reconstruct series of $\mathcal{S}$ within the error bound. Let α_i be the reconstructed series of S_i. In this paper, we utilize L_∞ norm (maximum) error. Formally, the error of our approximation for $\boldsymbol{S}$ is

$$E(\Delta) = \max_{1 \leq i \leq N} \max_{1 \leq t \leq n} |s_i(t) - \alpha_i(t)|$$

which is the maximum difference between the raw series and its representation.

The multiple time series compressing problem is defined as follows. Given a set of series $\boldsymbol{S}$ and an error threshold ε, find the representation Δ such that (1) $E(\Delta) \leq \varepsilon$ and (2) the storage size of Δ is as small as possible. In this case, we say series S_i can be represented by α_i within the maximal error ε ($1 \leq i \leq N$).

2.1   APCA Representation

There exists many approaches to approximating single time series under L_∞ error bound. Based on the experimental results of [10], we know that Adaptive Piecewise Constant Approximation [7] (APCA) outperforms other approaches in most cases. Therefore, we use it to compress the single time series in our approach. Here we introduce it briefly.

Given a series S and an error bound ε, it approximates S by splitting it into k disjoint segments and representing each segment with a single value. Specifically, the form of APCA is $C = \{(c_i, t_i), 1 \leq i \leq k\}$, where t_i is the right endpoint of the i-th segment, and c_i is the representation value of it. The difference between c_i and any value of this segment must be not larger than ϵ.

3   Compression Model and Algorithm Overview

In this section, we present our representation model, and then give the theoretical foundation of our approach.

3.1   Representation Model

First, we give the single-window model, which approximates each series as a whole. Then we extend it to the multi-window model, which splits $\boldsymbol{S}$ into some disjoint windows, and represents each window with the single-window model.

Single-Window Model. Given the set of time series, $\boldsymbol{S} = \{S_1, S_2, \cdots, S_N\}$, the representation model, denoted as $\delta = (\mathcal{C}, \mathcal{B}, \mathcal{O})$, is as follows,

- We dispatch the series in $\boldsymbol{S}$ into disjoint clusters, $\mathcal{C} = \{C_1, C_2, \cdots, C_{|\mathcal{C}|}\}$, each of which contains at least one time series. We use $S_j \in C_i$ to indicate that time series S_j belongs to cluster C_i.
- Each cluster C_i has a corresponding base series, denoted as B_i, which represents the *shape* of all series in cluster C_i. The second parameter of δ, $\mathcal{B} = \{B_1, B_2, \cdots, B_{|\mathcal{C}|}\}$, is the set of base series.

- Each series S_j in C_i can be approximately represented by the combination of the base series B_i and a single value. We call this value as *offset value*, and denote it as o_j. That is, for $S_j \in C_i$, $\alpha_j(t) = B_i(t) + o_j$, such that $|\alpha_j(t) - s_j(t)| \leq \varepsilon$ $(1 \leq t \leq n)$. The third parameter of δ, $\mathcal{O} = \{o_1, o_2 \cdots, o_N\}$, is the set of offset values.

Note that based on the base series, we can represent each series with just a single offset value. Therefore, our goal is to find as few as clusters which can represent all series in $\mathcal{S}$, in order to achieve high compression ratio.

Multi-window Model. The physical environment changes over time, so one series cluster that is optimal at time t may not be optimal in other time. Especially when archiving data over long durations, we expect trends to change. Based on this observation, we extend the single-window model to the multiple one.

Formally, let the window length, denoted as w, be a user-specified threshold. We split the whole time line into $m = \lceil \frac{n}{w} \rceil$ number of disjoint windows, $(W_1, W_2, \cdots, W_m)$. Accordingly, $\mathcal{S}$ is split into m number of windows, denoted as $(\mathcal{S}_1, \mathcal{S}_2, \cdots, \mathcal{S}_m)$. $\mathcal{S}_i$ is composed of subsequences of all series in the i-th window, that is, $\mathcal{S}_i = \{S_j((i-1)*w+1, i*w), 1 \leq j \leq N\}$. To ease the description, we indicate the subsequence of series S_j in the i-th window as S_j^i. That is, $S_j^i = S_j((i-1)*w+1, i*w)$. For each $\mathcal{S}_i$, we can obtain a single-window model, denoted as δ_i, which contains $\mathcal{C}_i$, $\mathcal{B}_i$ and $\mathcal{O}_i$ respectively. The multi-window model is the set of m single-window models, denoted as $\Delta = (\delta_1, \delta_2, \cdots, \delta_m)$.

3.2   Theoretical Foundation

Here we establish a formal theoretical foundation for our approach. As core, we propose a condition under which a set of series can be represented by a base series guaranteeing the L_∞ error bound. We first define the series similarity.

Definition 1 (ε-Similar). *Given two series $X = \{x_i\}$ and $Y = \{y_i\}$ where $1 \leq i \leq n$, we call X and Y are ε-similar if it holds that $\max |x_i - y_i| \leq \varepsilon$.*

Given a set of series $\mathcal{S} = (S_1, S_2, \cdots, S_N)$, where $S_i = \{s_i(t), t = 1, 2, \cdots, n\}$. We construct a base series, $B = \{b(t), t = 1, 2, \cdots, n\}$, as follows. For time point t, let min_t and max_t be the minimum and maximum values of all $s_i(t)$'s $(1 \leq j \leq n)$. We compute $b(t) = \frac{1}{2}(min_t + max_t)$. B has the following property,

Lemma 1. *Given a set of series $\mathcal{S} = (S_1, S_2, \cdots, S_N)$. If any pair of series in $\mathcal{S}$ are 2ε-similar, the base series B can represent all series in $\mathcal{S}$ within the maximum error ε.*

Proof. We just need to prove that for any series S_j $(1 \leq j \leq N)$, it holds that $|s_j(t) - b(t)| \leq \varepsilon$ where $t = 1, 2, \cdots, n$. From the definition of B, we can obtain

$$min_t - \frac{1}{2}(min_t + max_t) \leq s_i(t) - b(t) \leq max_t - \frac{1}{2}(min_t + max_t)$$

After simple transformation, we obtain the following inequality

$$|s_i(t) - b(t)| \leq \frac{1}{2}|max_t - min_t|$$

due to $|max_t - min_t| \leq 2\varepsilon$, So we can get that $|s_i(t) - b(t)| \leq \varepsilon$.   □

The key problem is how to group series into as few clusters as possible, each of which satisfies Lemma 1. In this paper, we propose two graph-based algorithms, $MTSC_{mc}$ and $MTSC_{star}$. We take time series as the vertexes, and the "similarity" of time series as edges to build the graph, and use different techniques to group the series into clusters. $MTSC_{mc}$ can achieve higher compression ratio but is more time consuming. In contrast, $MTSC_{star}$ is much more time efficient while slightly sacrificing the compression ratio. Furthermore, the base series introduced above has the same length of the series. To further improve the compression ratio, we propose a new form of base series with less storage cost.

4   The $MTSC_{mc}$ Algorithm

In this section, we present the first algorithm, $MTSC_{mc}$, which represents $\mathcal{S}$ with the multi-window model. $MTSC_{mc}$ processes $\mathcal{S}_i$ sequentially. In different windows, it groups the series with two alternative strategies. We first introduce the series grouping strategies (Sect. 4.1), and then discuss how to generate the base series for each cluster (Sect. 4.2).

4.1   Series Grouping Strategies

In $MTSC_{mc}$, we solve the series grouping problem with two graph-based approaches, mc-grouping and inc-grouping. Next we introduce them in turn.

Mc-grouping. Assume we group series in window $\mathcal{S}_i = \{S_1^i, S_2^i, \cdots, S_N^i\}$. First of all, we transform all subsequences by removing the *shifting* offset, so that each transformed subsequence has 0 as the mean value. Specifically, suppose the mean value of S_j^i is μ_j^i, we transform each value $s_j(t)$ ($t \in W_i$) into $s_j(t) - \mu_j^i$. We denote the transformed subsequence as $\hat{S}_j^i$ and the new value as $\hat{s}_j(t)$. Then we construct an undirected graph, $G_i = (V_i, E_i)$. V_i contains N number of vertexes, in which vertex v_j corresponds to series S_j. The distance between two vertexes v_j and $v_{j'}$ is the maximal difference of all time points in W_i. That is,

$$D(j, j') = \max_{t \in W_i} |\hat{s}_j(t) - \hat{s}_{j'}(t)|$$

Edge $e(j, j')$ exists in E_i if $D(j, j') \leq 2\varepsilon$. We call graph G_i as 2ε-similar graph.

It is worth noting that in any two windows, say G_i and $G_{i'}$, it always holds that $V_i = V_{i'}$, while E_i and $E_{i'}$ may be different, because two series may be 2ε-similar in some windows, but not in others. After G_i is obtained, we group the series with a *maximum clique* based algorithm. Later, we use series S_j and vertex v_j interchangeably.

Definition 2 (Maximum Clique). *Let G be an undirected graph. A clique refers to a complete subgraph, in which there exists an edge between any pair of vertexes. The maximum clique contains more vertexes than any other cliques.*

The maximum clique problem is a well-known NP-Hard problem. Due to its wide range of applications, many methods are proposed to solve it [8,11]. Here we use the fast deterministic algorithm [11]. The algorithm searches the clique in a certain order, and also uses some pruning strategies to speedup the process.

We use a greedy algorithm to group all series in G_i. Specifically, we first find the maximum clique from G_i, and take all series in it as the first cluster C_1^i. Then we update G_i by deleting the vertexes in C_1^i, as well as edges connecting to at least one vertex in C_1^i. In the second round, we find the maximum clique in the current G_i, and take series in it as C_2^i. This process continues until G_i doesn't contain any edge. In this case, if G_i still contains some vertexes, we take each of them as a cluster, called as *individual cluster*. That is, $\mathcal{C}_i$ is composed of some clusters with multiple series, and some individual clusters.

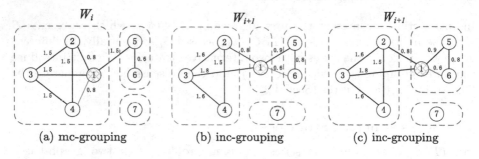

(a) mc-grouping   (b) inc-grouping   (c) inc-grouping

Fig. 1. An example of mc-grouping and inc-grouping

Figure 1(a) shows an example of mc-grouping on G_i, which contains 7 vertexes. Suppose ε is set to 1. Figure 1(a) also shows all edges, each of which is labeled with the distance between two vertexes. It can be seen that $\mathcal{C}_1$ contains two cliques ($C_1^i = \{v_1, v_2, v_3, v_4\}$, $C_2^i = \{v_5, v_6\}$) and one individual cluster $C_3^i = \{v_7\}$.

Inc-grouping. Mc-grouping can achieve high quality clusters, because it always finds the maximum clique. However, it is time consuming due to the high cost of maximum clique mining algorithm. To make it more efficient, we propose another grouping strategy, named inc-grouping. In many applications, it is often that the similarity relationship between series will last for some consecutive windows. In this case, the series clusters of adjacent windows will be similar accordingly. Based on this observation, instead of grouping the series from scratch in each window, inc-grouping strategy inherits the clusters from the previous window, and adjusts them according to the edges of the current window. As a special case, if E_i is exactly same as E_{i-1}, we can directly take $\mathcal{C}_{i-1}$ as $\mathcal{C}_i$.

Now, we introduce the detail of inc-grouping. Suppose we have obtained $\mathcal{C}_{i-1} = \{C_1^{i-1}, C_2^{i-1}, \cdots, C_p^{i-1}\}$, and turn to process $\mathcal{S}_i$. Initially, we compute $\hat{S}_j^i$'s $(1 \leq j \leq N)$ and $G_i = \langle V_i, E_i \rangle$. Then, we construct $\mathcal{C}_i$ as follows. First, we generate a subgraph of G_i, denoted as $G' = \langle V', E' \rangle$, in which, V' has the same vertexes as C_1^{i-1} and $e(j, j') \in E'$ if $v_j \in V'$, $v_{j'} \in V'$ and $e(j, j') \in E_i$.

If G' is a clique in G_i, we directly take it as C_1^i. Otherwise, we transform it into a clique by removing some vertexes. We first select the vertex with the minimal degree, say v, in G' to delete. Here the degree of a vertex is the number of edges connecting to it in G'. After deleting v and all edges connecting to it, we check whether the current G' is a clique. If it is the case, we take current G' as C_1^i, and v as an individual cluster. Otherwise, we repeatedly select the vertex with the minimal degree in G' to delete. We continues this process until G' becomes a clique or it only includes a set of isolated vertexes. In the latter, we take all these vertexes in G' as individual clusters.

Once C_1^i is obtained, we use the same approach to construct C_2^i based on C_2^{i-1}. Again, we obtain a clique which is a shrinking version of C_2^{i-1} and some individual clusters. In the extreme case, all vertexes in C_2^{i-1} will become individual clusters. We iterate this process until all cliques in $\mathcal{C}_{i-1}$ are processed. As the last step, we try to insert individual series into these new cliques.

Figure 1(b) and (c) illustrate the inc-grouping for G_{i+1}. First, we adapts C_1^i to generate C_1^{i+1}. Since $e(v_1, v_4)$ doesn't occur in E_{i+1}, We delete v_1 firstly. The rest vertexes form a clique in G_{i+1}. So either $C_1^{i+1} = \{v_2, v_3, v_4\}$ and v_1 becomes an individual cluster. Next, we process $C_2^i = \{v_5, v_6\}$. Because $e(v_5, v_6) \in E_{i+1}$, C_2^{i+1} is $\{v_5, v_6\}$, as shown in Fig. 1(b). Finally, we check whether v_1 and v_7 can be inserted into C_1^{i+1} or C_2^{i+1}. In this case, v_1 can be added into C_2^{i+1}, since both $e(1,5)$ and $e(1,6)$ exist in E_{i+1}. Figure 1(c) shows the final $\mathcal{C}_{i+1}$.

Put Them Together. Now we introduce how to combine mc-grouping and inc-grouping systematically. Initially, for the first window W_1, we first construct G_1, and then use mc-grouping to obtain $\mathcal{C}_1$. Next, we process $\mathcal{S}_2$. After obtaining G_2, we check how difference between G_1 and G_2. We use the ratio of changed edges to measure the difference. If the difference between G_2 and G_1 doesn't exceed the user-specified threshold, σ, we use inc-grouping to compute $\mathcal{C}_2$. Otherwise, we use mc-grouping. This process continues until all windows are processed.

4.2   Base Series and Offset Value

Once clusters $\mathcal{C}$ in a window is obtained, we need to compute base series for each cluster. Section 3.2 gives a simple format of the base series. However, its length is same as the subsequences. To further reduce the storage cost, we propose a more concise form of base series, which can still guarantees L_∞ error bound. Similarly with the APCA representation, each base series has the form as follows,

$$B = \left(\langle bv_1, br_1 \rangle, \langle bv_2, br_2 \rangle, \cdots, \langle bv_{|B|}, br_{|B|} \rangle \right)$$

where br_i is the right endpoint of the i-th segment and bv_i is a value to represent it. That is, B splits the time window into $|B|$ number of segments, and the i-th segment is $[br_{i-1} + 1, br_i]$. The value of $|B|$ may differ for different clusters.

Given a cluster C, the base series B can be computed by sequentially scanning subsequences in C. To ease the description, we assume cluster C is in window W_1, so the first time point is 1 and the last one is w^1. The first segment, Seg_1, is initialized as $[1, 1]$. We visit all $|C|$ number of values, $\hat{s}_j(1)$'s $(S_j \in C)$, and obtain the minimum and maximum ones in them, denoted as min_1 and max_1 respectively. We use MIN and MAX to represent the minimum and maximum values in the current segment, which are initialized as min_1 and max_1. Next, we visit all values $\hat{s}_j(2)$'s, and obtain min_2 and max_2. If adding time point $t = 2$ into Seg_1 doesn't make $|MAX - MIN| > 2\varepsilon$, we extend segment Seg_1 to $[1, 2]$, and update MAX and MIN if necessary. We sequentially check the next time points until we meet the first time point, say k, adding which into Seg_1 will make $|MAX - MIN| > 2\varepsilon$. In this case, we set $br_1 = k - 1$ and $bv_1 = \frac{MAX+MIN}{2}$. Then we initialize $Seg_2 = [k, k]$ and setting $MAX = max_k$ and $MIN = min_k$. This process continues until time point w is met. The correctness of the base series can be proved by the following lemma.

Lemma 2. *Base series B can represent all series in C within maximal error ε.*

Proof. For the i-th entry of B, $\langle bv_i, br_i \rangle$, $(1 \le i \le |B|)$, we need to prove $|bv_i(t) - s(t)| \le \varepsilon$, where $t \in [br_{i-1} + 1, br_i]$. Let MIN and MAX be the minimum and maximum values in Seg_i, it holds that $bv_i = \frac{MAX+MIN}{2}$ and $|MAX - MIN| \le 2\varepsilon$. For all $t \in [br_{i-1} + 1, br_i]$, it can be inferred that

$$MIN \le min_t \le s(t) \le max_t \le MAX$$

Similar to the proof of Lemma 1, we can get $|bv_i(t) - s(t)| \le \varepsilon$.   □

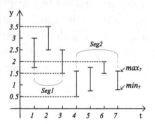

Fig. 2. Base series

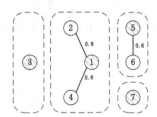

Fig. 3. $MTSC_{star}$

Figure 2 illustrates it with an example. At each time point, we show the value range. For example, at $t = 7$, min_7 and max_7 are 0.7 and 1.5 respectively.

---

[1] Indeed, for window W_i, the first time point is $(i - 1) * w + 1$ and the last one is $i * w$.

$Seg_1 = [1,3]$, because $MAX - MIN = 3.5 - 1.5 \leq 2$. Seg_1 cannot include $t = 4$, because in this case, $MAX - MIN = 3.5 - 0.5 = 3 > 2$. $Seg_2 = [4,7]$, because $MAX - MIN = 2 - 0.5 = 1.5 < 2$.

For any series S_j in cluster C of window W_i, we set the offset value o_j as the mean value μ_j^i. As for the individual clusters, we represent each individual series with APCA, and take it as the base series. In this case, the offset value is 0.

5   The $MTSC_{star}$ Algorithm

In this section, we present the second algorithm $MTSC_{star}$, whose compression quality is slightly lower than that of $MTSC_{mc}$, but has much higher efficiency.

The only difference between $MTSC_{star}$ and $MTSC_{mc}$ is the series grouping strategy. $MTSC_{star}$ still uses the multi-window representation model, and it utilizes the same strategy for all windows. For window $\boldsymbol{S}_i = \{S_j^i, 1 \leq j \leq N\}$, we transform series by removing the shifting offset, and obtain $\hat{\boldsymbol{S}}_i = \{\hat{S}_j^i, 1 \leq j \leq N\}$. Then we compute $G_i = \langle V_i, E_i \rangle$, in which each vertex v_j corresponds to series S_j ($1 \leq j \leq N$). An edge $e(j, j') \in E_i$ if $\hat{S}_j^i$ and $\hat{S}_{j'}^i$ are ε-similar. So the graph is the ε-similar graph.

Different with $MTSC_{mc}$, which groups series by finding cliques, in $MTSC_{star}$, we find star-shape subgraphs. Formally,

Definition 3 (Star-Shape Subgraph). $G = \langle V, E \rangle$ is a star-shape subgraph, if there exists one vertex v in V, so that for any other vertex v' in V, $e(v, v') \in E$.

We can prove that a star-shape subgraph in ε-similar graph is a clique subgraph in 2ε-similar graph with the following lemma.

Lemma 3. Let $G = \langle V, E \rangle$ be the 2ε-similar graph and $G' = \langle V, E' \rangle$ be the ε-similar graph of the same window. Any star-shape subgraph in G' corresponds to a clique in G.

Proof. Suppose SG is a star-shape subgraph of G', and v_a ($\in SG$) connects to all other vertexes in SG. To prove that vertexes of SG can form a clique in G, we only need to prove that any pair of vertexes in SG is 2ε-similar. Based on the definition of v_a, it and any vertex in SG are 2ε-similar. Next we consider any two other vertexes v_b and v_c in SG. It holds that

$$D(a, b) = \max_{t \in W} |\hat{s}_a(t) - \hat{s}_b(t)| \leq \varepsilon \text{ and } D(a, c) = \max_{t \in W} |\hat{s}_a(t) - \hat{s}_c(t)| \leq \varepsilon$$

that means for all time points t's, we have

$$|\hat{s}_a(t) - \hat{s}_b(t)| \leq \varepsilon \text{ and } |\hat{s}_a(t) - \hat{s}_c(t)| \leq \varepsilon$$

So that $|\hat{s}_b(t) - \hat{s}_c(t)| \leq 2\varepsilon$. The distance between v_b and v_c satisfies

$$D(b, c) = \max_{t \in W} |\hat{s}_b(t) - \hat{s}_c(t)| \leq 2\varepsilon$$

So SG will be a clique in 2ε-similar graph G.   □

The advantage of using ε-similar graph is that it is much easier to find star-shape subgraphs than finding cliques. We use a greedy approach to split the graph into a set of star-shape subgraphs (or clusters), and possibly, some individual clusters. Firstly, we select the vertex in G with the highest degree. This vertex and all vertexes connecting to it form the first (and also the maximum) star-shape subgraph in G. Then we update G by removing these vertexes as well as all related edges. Next, we still find the vertex of the highest degree from G, and combine it with all vertexes connecting to it to generate the second star-shape subgraph. This process continues until G doesn't contain any edge. At last, all remainder individual vertexes form a set of individual clusters.

The time complexity of grouping is $O(N^2)$, which is lower than that of generating the graph. So unlike $MTSC_{mc}$ which uses inc-grouping to improve the efficiency, $MTSC_{star}$ deals with all windows with the above grouping strategy. For each cluster, we generate the base series as the same approach as $MTSC_{mc}$.

Figure 3 illustrates the grouping strategy of $MTSC_{star}$ for window W_i. The edges are the subset of edges in Fig. 1(a), that is, it only contains edges for ε-similar vertex pairs ($\varepsilon = 1$). Those edges whose weight is larger than 1 are removed. We first choose vertex v_1 with largest degree 2 and get a cluster $C_1^i = \{v_1, v_2, v_4\}$. Then we construct the second cluster $C_2^i = \{v_5, v_6\}$. The remaining individual vertexes from two individual clusters $C_3^i = \{v_3\}$ and $C_4^i = \{v_7\}$.

6   Experiments

In this section, we evaluate the performance of proposed algorithms by comparing with three approaches, GAMPS, APCA and PLA [9]. GAMPS aims for multiple series, while APCA and PLA are single-series compression approaches that outperform others [10]. For PLA, we use the state-of-the-art algorithm, mixed-PLA [9]. All algorithms are implemented in Java and all experiments are conducted on a 4-core (3.5 GHz) Intel Core i5 desktop with 16 GB memory.

6.1   Datasets

To make fully comparison between algorithms, we use three real-world datasets.

- Gas dataset. It is the Gas Sensor Array Drift Dataset from popular UCI repository, which is collected by 16 chemical sensors used to detect concentrations of 6 kinds of gases [1]. It contains 100 series of length 3,600.
- Google Cluster dataset. It records activities of jobs consisting of many tasks executing on a data center over a seven-hour period [13]. It extracts CPU and memory usage for each task, and contains 2,090 time series of length 74.
- Temperature dataset. It collects the temperature values of 719 climate stations in China [2]. For each station, the temperature is monitored from 1960 to 2012, one value per day. The length of each time series is 19,350.

To make the results on different datasets consistent, we use the relative error threshold ε, which is the fraction of the difference between the maximum and minimum values in the each dataset. The particular parameters of GAMPS are set according to the authors' recommendation. The splitting fraction is set to 0.4ε for base series. GAMPS also splits time series into disjoint windows. The initial window length is set as 100, and the lengths of the next windows are adjusted dynamically according to the fluctuation of series correlation. In $MTSC$ algorithm, the default window length w is set as 100, and the rate of change between two adjacent windows, σ, is set as 0.01.

6.2 Compression Ratio

As traditional time series compression algorithms, we define the *compression ratio* as the ratio between the size of the original dataset and that of the compressed one. Formally, suppose each series value is a 32-bit float number, then the storage cost of the raw time series $\mathcal{S}$ is $32 \times N \times n$.

Our representation model contains three parts, $\mathcal{C}$, $\mathcal{B}$ and $\mathcal{O}$. For the cluster $\mathcal{C}$, each series indicates its cluster ID with a 32-bit integer, so the storage cost of $\mathcal{C}$ is $32 \times N$. The storage cost of $\mathcal{B}$ depends on the number of segments for each base series. For each segment, we use two 32-bit values to store bv and br respectively. Assume the number of segments in B_j^i is $|B_j^i|$, so a base series needs $64 \times |B_j^i|$ bits to store. Each offset value is represented as a 32-bit value, and so the store cost of $\mathcal{O}$ for each window is $32 \times N$. In summary, if we have m number of windows, the total cost of compressed series is $\sum_{i=1}^{m}(64 \times N + 64 \times \sum_{j=1}^{|\mathcal{C}_i|} |B_j^i|)$.

From above, we know the compression ratio mainly depends on two factors, the number of clusters and the storage cost of base series.

6.3 Influence of Error Threshold ε

We test the influence of the error threshold ε on the compression ratio and the runtime. Experiments are conducted on all three datasets. Figure 4 shows the results. The length of series in Cluster dataset is 74, which is less than the default window size (100), so we use the single-window model.

Figure 4(a), (b) and (c) show the results of compression ratio. It can be seen that both $MTSC_{mc}$ and $MTSC_{star}$ have higher compression ratio than APCA, PLA and GAMPS in most cases. When ε becomes larger, the compression ratios of all approaches increase accordingly. However, the increasing is much more obvious in our approaches. Although GAMPS also exploits the correlation between similar series, we can see that its performance is even worse than APCA and PLA. The reason is that GAMPS splits ε into two parts, one for base series and the other for ratio signals. This mechanism makes GAMPS needs more cluster and segments, which causes higher storage cost. Finally, as we analyzed, the compression ratio of $MTSC_{mc}$ is slightly higher than $MTSC_{star}$, due to the maximal clique based approach can use fewer clusters to cover all series.

Figure 4(d), (e) and (f) show the efficiency results. Since APCA and PLA need only one scan to get all segments of each series, they are more efficient and

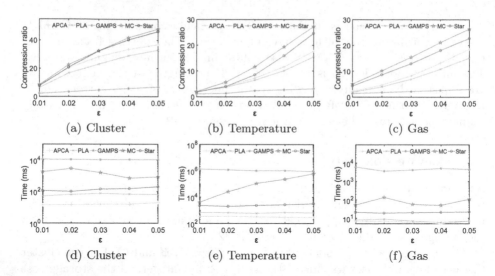

Fig. 4. Compression ratio and time comparison

the runtime doesn't change greatly as ε varies. The running time of $MTSC_{mc}$ demonstrates different trends in three datasets, because it depends on multiple factors, such as number of vertexes and density of the graph. In the Temperature dataset, both the clique size and number of vertexes in cliques become larger as ε increases, which consumes more time searching maximum cliques. Moreover, we find that the searching process in a dense graph is faster than that in a sparse one. The pruning strategy in the maximum clique problem reduce the time to find a clique in the dense graph. When ε exceeds 0.03, the graphs of the Cluster and Gas datasets become very dense, leading to the decrease of the runtime.

Comparing to $MTSC_{mc}$, $MTSC_{star}$ is much more efficient and is more stable as ε increases, because the complexity of series grouping in $MTSC_{star}$ is lower than that of $MTSC_{mc}$ and is less sensitive to the structure of the graph. The running time of GAMPS is highest among all algorithms. It spends most of time to solve the facility location problem which is an NP complete. Though GAMPS uses an approximative algorithm to solve it, it's still not efficient enough.

6.4   The Number of Clusters vs. ε

As shown in Sect. 6.2, the number of clusters has great impact on the compression ratio. Therefore, in this experiment, we investigate the number of clusters in $MTSC_{mc}$, $MTSC_{star}$ and GAMPS. The average number of clusters for all windows is shown in Fig. 5. Moreover, we also show the corresponding compression ratio simultaneously. The numbers of clusters are shown as bars and the corresponding compression ratio as lines.

It can be seen that as ε increases, the number of clusters in both $MTSC_{mc}$ and $MTSC_{star}$ decreases gradually. The reason is that more pairs of series are

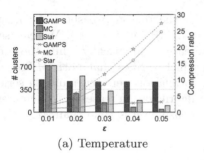
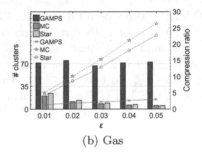

(a) Temperature   (b) Gas

Fig. 5. The number of clusters vs. ε

ε-similar and can be clustered together. In consequence, all series are covered by less clusters. The number of clusters in $MTSC_{mc}$ is larger than that of $MTSC_{star}$, which causes higher compression ratio of $MTSC_{mc}$. In contrast, the number of clusters in GAMPS stays stable in both datasets, which explains why the compression ratio of GAMPS does not increase significantly as ε increases in Fig. 4. Note that when $\varepsilon = 0.01$, although the number of clusters in GAMPS is smaller than that of our algorithms on Temperature dataset, its compression ratio is still lower than ours, because the offset of GAMPS is still a series while it is a single value in our approaches.

6.5 Influence of the Number of Series N

In this experiment, we investigate the influence of the number of series, N, on the performance of our approaches. We randomly extract 100 to 600 number of series from Temperature dataset. The error threshold ε is set to 0.05. Both compression ratio and runtime are compared, and the results are shown in Fig. 6.

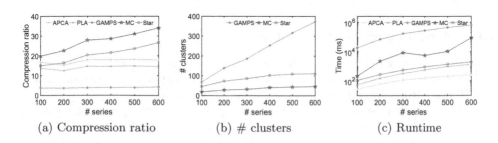

(a) Compression ratio   (b) # clusters   (c) Runtime

Fig. 6. Influence of the number of series N

In Fig. 6(a), as N increases, the compression ratio of our approach increases greatly. Those of APCA and PLA stay stable because they compress each single series individually. The interesting phenomenon is that the compression ratio of GAMPS also doesn't increase. To analyze the reason, we show the number of clusters of both our approaches and GAMPS in Fig. 6(b). We can see that the number of clusters in GAMPS increases dramatically while those of $MTSC_{mc}$ and $MTSC_{star}$ increase slightly, which verifies that both $MTSC_{mc}$ and $MTSC_{star}$ do better in exploiting the correlation between multiple series than GAMPS. In Fig. 6(c), the runtime of all algorithms increases as N increases. Among them, APCA, PLA and $MTSC_{star}$ consume less time than $MTSC_{mc}$ and GAMPS.

6.6   Influence of the Window Length

In both $MTSC_{mc}$ and $MTSC_{star}$, series are split into fixed-length windows. In this experiment, we investigate the impact of window length. We conduct the experiments on Gas dataset and the error threshold ε is set as 0.02.

In Fig. 7, the compression ratio of $MTSC_{mc}$ and $MTSC_{star}$ decreases gradually as w changes from 50 to 250. When w increases, the number of series pairs satisfying 2ε-similar will decrease. In consequence, more clusters are needed to represent all series. On the other hand, the runtime of both $MTSC_{mc}$ and $MTSC_{star}$ decreases, because less windows need to be processed.

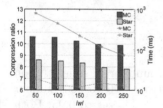   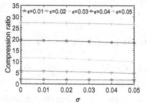   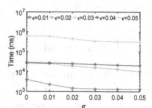

Fig. 7. Influence of w **Fig. 8.** Compression ratio   **Fig. 9.** Runtime

6.7   Mc-grouping vs. Inc-grouping

In this section, we compare the performance of mc-grouping and inc-grouping. Moreover, we also investigate the influence of σ. The experiments are conducted on Temperature dataset. Results are shown in Figs. 8 and 9.

The parameter σ is to measure the change between two graphs of adjacent windows. When σ is set to 0, we use mc-grouping to process all windows, because none of the windows can use clusters of the previous windows. From Figs. 8 and 9, we can see that as σ increases, the compression ratio decreases slightly while the runtime goes down about 30% to 60%. The reason behind is that more windows use the inc-grouping strategy, which is much more efficient than mc-grouping. So, it is a trade-off, larger σ means higher efficiency while lower one means higher compression ratio.

7   Related Work

To reduce the cost of storing large quantities of time series, many compression techniques are proposed [10], which can be divided into two categories, lossless and lossy compression. Most of lossless compression are based on byte stream and have no semantics, such as LZ78 [15]. In an in-memory time series database Gorilla [12] of Facebook, a variable length encoding is used. Time series are compressed by removing the redundant information in the byte-level.

Lossy compression represents time series using well-established approximation models. Moreover the lossy compression is orthogonal to the lossless encoding. There are a lot of work on lossy compression of time series. [10] gives a nice survey about this topic. Most approaches are tailored to the single series, such as Adaptable Piecewise Constant Approximation (APCA) [7], Piecewise Linear Approximation (PLA) [9] and Chebyshev Approximations (CHEB) [3].

On the other hand, some approaches compress multiple time series by exploiting the correlation between series, such as Grouping and AMPlitude Scaling (GAMPS) [6], Self-Based Regression (SBR) [4] and RIDA [14], among which, only GAMPS can guarantee the L_∞ error bound, others are based on L_2 error, which is less desirable than L_∞ in terms of time series compression.

GAMPS [6] groups series and approximates series in each group with base and ratio series together. To deal with the fluctuation of data correlation, it dynamically split series into variable windows and compress subsequence in each window sequentially. Although both series and ratio series of GAMPS can be stored with less cost, the compression ratio may be not satisfactory, GAMPS splits ε into two parts, one for base series and the other for ratio signals. This mechanism makes GAMPS needs more clusters and segments, which causes higher storage cost. Time series clustering is an embedded task in our approach, and there exist many techniques of clustering time series [5]. However, they cannot be applied in our approach due to the different clustering target.

8   Conclusion and Future Work

In this paper, we propose a new framework to compress multiple time series. We first propose a new representation model. Then two graph-based algorithms, $MTSC_{mc}$ and $MTSC_{star}$, are proposed to compress multiple series. Moreover, a concise form of base series is used to further improve the compression quality. Experimental results show that our approach outperforms existing ones greatly.

In the future, we aim to extend the mechanism of fixed-length window to dynamic window lengths, to leverage the data characteristics.

References

1. UCI machine learning repository (2013). http://archive.ics.uci.edu/ml
2. Climatic Data Center. http://data.cma.cn/

3. Cheng, A., Hawkins, S., Nguyen, L., Monaco, C., Seagrave, G.: Data compression using Chebyshev transform. US Patent App. 10/633,447 (2004)
4. Deligiannakis, A., Kotidis, Y., Roussopoulos, N.: Compressing historical information in sensor networks. In: SIGMOD 2004, pp. 527–538 (2004)
5. Esling, P., Agon, C.: Time-series data mining. ACM Comput. Surv. **45**(1), 12:1–12:34 (2012)
6. Gandhi, S., Nath, S., Suri, S., Liu, J.: Gamps: compressing multi sensor data by grouping and amplitude scaling. In: SIGMOD 2009, pp. 771–784 (2009)
7. Guha, S., Koudas, N., Shim, K.: Approximation and streaming algorithms for histogram construction problems. TODS **31**(1), 396–438 (2006)
8. Lu, C., Yu, J.X., Wei, H., Zhang, Y.: Finding the maximum clique in massive graphs. VLDB **10**(11), 1538–1549 (2017)
9. Luo, G., et al.: Piecewise linear approximation of streaming time series data with max-error guarantees. In: ICDE 2015, pp. 173–184 (2015)
10. Nguyen, Q.V.H., Jeung, H., Aberer, K.: An evaluation of model-based approaches to sensor data compression. TKDE **25**(11), 2434–2447 (2013)
11. Östergård, P.R.J.: A fast algorithm for the maximum clique problem. Discrete Appl. Math. **120**(1–3), 197–207 (2002)
12. Pelkonen, T., Franklin, S., Teller, J., Cavallaro, P., Huang, Q., et al.: Gorilla: a fast, scalable, in-memory time series database. VLDB **8**(12), 1816–1827 (2015)
13. Reiss, C., Wilkes, J., Hellerstein, J.L.: Google cluster-usage traces: format + schema. Technical report, Google Inc. (2011)
14. Dang, T., Bulusu, N., Feng, W.: RIDA: a robust information-driven data compression architecture for irregular wireless sensor networks. In: Langendoen, K., Voigt, T. (eds.) EWSN 2007. LNCS, vol. 4373, pp. 133–149. Springer, Heidelberg (2007). https://doi.org/10.1007/978-3-540-69830-2_9
15. Ziv, J., Lempel, A.: Compression of individual sequences via variable-rate coding. IEEE Trans. Inf. Theor. **24**(5), 530–536 (2006)

DancingLines: An Analytical Scheme to Depict Cross-Platform Event Popularity

Tianxiang Gao[1], Weiming Bao[1], Jinning Li[1], Xiaofeng Gao[1(✉)],
Boyuan Kong[2], Yan Tang[3], Guihai Chen[1], and Xuan Li[4]

[1] Shanghai Key Laboratory of Scalable Computing and Systems,
Department of Computer Science and Engineering,
Shanghai Jiao Tong University, Shanghai, China
{gtx9726,wm_bao,lijinning}@sjtu.edu.cn, {gao-xf,gchen}@cs.sjtu.edu.cn
[2] University of California, Berkeley, CA, USA
boyuan_kong@berkeley.edu
[3] Hohai University, Nanjing, China
tangyan@hhu.edu.cn
[4] Baidu, Inc., Beijing, China
xli@baidu.com

Abstract. Nowadays, events usually burst and are propagated online through multiple modern media like social networks and search engines. There exists various research discussing the event dissemination trends on individual medium, while few studies focus on event popularity analysis from a cross-platform perspective. In this paper, we design DANCING-LINES, an innovative scheme that captures and quantitatively analyzes event popularity between pairwise text media. It contains two models: TF-SW, a semantic-aware popularity quantification model, based on an integrated weight coefficient leveraging Word2Vec and TextRank; and ωDTW-CD, a pairwise *event popularity time series* alignment model matching different event phases adapted from Dynamic Time Warping. Experimental results on eighteen real-world datasets from an influential social network and a popular search engine validate the effectiveness and applicability of our scheme. DANCINGLINES is demonstrated to possess broad application potentials for discovering knowledge related to events and different media.

Keywords: Cross-platform analysis · Data mining
Time series alignment

This work has been supported in part by the Program of International S&T Cooperation (2016YFE0100300), the China 973 project (2014CB340303), the National Natural Science Foundation of China (Grant number 61472252, 61672353), the Shanghai Science and Technology Fund (Grant number 17510740200), CCF-Tencent Open Research Fund (RAGR20170114), and Key Technologies R&D Program of China (2017YFC0405805-04).

S. Hartmann et al. (Eds.): DEXA 2018, LNCS 11029, pp. 283–299, 2018.
https://doi.org/10.1007/978-3-319-98809-2_18

1   Introduction

In recent years, the primary media for information propagation have been shifting to online media, such as social networks, search engines, web portals, etc. A vast number of studies have been conducted to analyze the event disseminations comprehensively on single medium [11,12,23]. In fact, an event is less likely to be captured only by single platform, and popular events are usually disseminated on multiple media.

We model the event dissemination trends as *Event Popularity Time Series* (EPTS) at any given temporal resolution. Inspired by the observation that the diversity of the media and their mutual influences cause the EPTSs to be temporally warped, we seek to identify the alignment between pairwise EPTSs to support deeper analysis.

We propose a novel scheme called DANCINGLINES to depict event popularity from pairwise media and quantitatively analyze the popularity trends. DANCINGLINES facilitates cross-platform event popularity analysis with two innovative models, TF-SW (Term Frequency with Semantic Weight) and ωDTW-CD (ωeighted Dynamic Time Warping with Compound Distance).

TF-SW is a semantic-aware popularity quantification model based on Word2Vec [16] and TextRank [15]. The model first discards the words unrelated to certain events; then utilizes semantic and lexical relations to get similarity between words and highlights the semantically related ones with a *contributive words* selection process. Finally based on similarity, TextRank gives us the importance of each word, then the popularity of a certain event. EPTSs generated by TF-SW are able to capture the popularity trend of a specific event at different temporal resolutions.

ωDTW-CD is a pairwise EPTSs alignment model using an extended Dynamic Time Warping method. It generates sequence of matches between temporally warped EPTSs.

Experimental results on eighteen real-world datasets from Baidu, the most popular search engine in China, and Weibo, Chinese version of Twitter, validate the effectiveness and applicability of our models. We demonstrate that TF-SW is in accordance with real trends and sensitive to burst phases, and that ωDTW-CD successfully aligns EPTSs. The model not only gives an excellent performance, but also shows superior robustness. In all, DANCINGLINES has broad application potentials to reveal knowledge of various aspects of cross-platform events and social media.

The rest of this paper is organized as follows. In Sect. 2, related work is discussed. In Sect. 3, we define the problem. In Sect. 4, we introduce the overview of DANCINGLINES. The two models TF-SW and ωDTW-CD are discussed in details respectively in Sects. 5 and 6. Section 7 verifies DANCINGLINES on real-world datasets from Weibo and Baidu. Finally, we conclude the paper in Sect. 8.

2 Related Work

Event Popularity Analysis. Many researches [1,10,19,22] have focused on event evolution analysis for a single medium. The event popularity was evaluated by hourly page view statistics from Wikipedia in [1]. [10] chose the density-based clustering method to group the posts in social text streams into events and tracked the evolution patterns. Breaking news dissemination is studied via network theory based propagation behaviors in [13]. [22] proposed a TF-IDF based approach to analyze event popularity trends. In all, network-based approaches usually have high computational complexity, while frequency-based methods are usually less accurate on reflecting the event popularity.

Cross-Platform Analysis. From a cross-platform perspective, existing researches focus on topic detection, cross-social media user identification, cross-domain information recommendation, etc. [2] selected Twitter, *New York Times* and Flickr to represent multimedia streams, and provided an emerging topic detection method. An attempt, trying to combine Twitter and Wikipedia to do first story detection, was discussed in [18]. [26] proposed an algorithm based on multiple social networks like Twitter, and Facebook to identify anonymous identical users. The relationship between social trends from social network and web trends from search engine are discussed in [5,9]. Recently, a good prediction of social links between users from aligned networks using sparse and low rank matrix is well discussed in [24]. However, few studies have been conducted for popularity analysis from cross-platform perspective.

Dynamic Time Warping. DTW is a well-established method for similarity search between time series. Originating from speech pattern recognition [20], DTW has been effectively implemented in many domains [5]. Recently, remarkable performance on time series classification and clustering by combining KNN classifiers have been achieved in [4,14]. The well-known Derivative DTW is proposed in [8]. Weighted DTW [7] was designed to penalize high phase differences. In [21], the side effect of endpoints which tends to disturb the alignments dramatically in time series is confirmed and an improvement for eliminating such issue is proposed. We are inspired by these related works when designing our own DTW based model for aligning EPTSs.

3 Problem Formulation

3.1 Event Popularity Quantification

We start from dividing the time span T of an event into n periods, which is determined by the time resolution, each stamped with t_i, $T = \langle t_1, \cdots, t_n \rangle$. A record is a set of words preprocessed from datasets, such as a post from social networks or a query from search engines. Then, we use the notation w_k^i to represent, within time interval t_i, the kth word in a record. The notation $R_j^i = \{w_1^i, w_2^i, \cdots, w_{|R_j^i|}^i\}$ is the jth record within time interval t_i. An *event phase*, corresponded to t_i and denoted as E_i, is a finite set of words, and each word is from a related record R_j^i. As a result $E_i = \bigcup_j R_j^i$.

We can now introduce the prototype of our popularity function $pop(\cdot)$. For a given word $w_k^i \in E_i$, the popularity of the word w_k^i is defined as

$$pop(w_k^i) = fre(w_k^i) \cdot weight(w_k^i), \tag{1}$$

where $fre(w_k^i)$ is the word frequency of w_k^i within t_i. The weight function, $weight(w_k^i)$, for a word within t_i, is the kernel we solve in the TF-SW part and is the key to generate event popularity. In this work, we propose a weight function not only utilizing the lexical but also semantic relationships. Details about how to define the weight function is discussed in Sect. 5.

Once we get popularity of word w_k^i within t_i, the popularity of an event phase E_i, $pop(E_i)$, can be generated by summing up all words' popularity,

$$pop(E_i) = \sum_{w_i^k \in E_i} pop(w_k^i). \tag{2}$$

We regard the pair $(t_i, pop(E_i))$ as a point on X-Y plane and get a series of points, formalizing a curve on the plane to reflect the dissemination trend of an event $\mathscr{E}$.

To compare the curves from different media, a further normalization is employed,

$$\overline{pop}(E_i) = \frac{pop(E_i)}{\sum_{1 \le k \le n} pop(E_k)}. \tag{3}$$

After the normalization, the popularity trend of an event on a single medium is represented by a sequence, denoted as $\mathscr{E} = \langle \overline{pop}(E_1), \cdots, \overline{pop}(E_n) \rangle$, which is defined as *Event Popularity Time Series*.

3.2   Time Series Alignment

Two EPTSs generated from two platforms of an event $\mathscr{E}$ are now comparable and can be visualized in a same X-Y plane as Fig. 1, which shows normalized EPTSs of Event *Sinking of a Cruise Ship* generated from Baidu and Weibo.

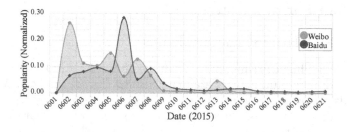

Fig. 1. Normalized EPTSs, *Sinking of a Cruise Ship* (Color figure online)

A Chinese cruise ship called Dongfang Zhi Xing sank into Yangtze River on the night of June 2, 2015 and the following process lasted for about 20 days. X-axis in Fig. 1 represents time and Y-axis indicates the event popularity. If we shifted the orange EPTS, generated from Weibo, to the right for about 4 units, we would notice the blue one approximately overlaps the orange one. This phenomenon indicates a *temporal warp*, which means the trend features are similar, but there exists time differences between EPTSs.

According to Fig. 1, EPTSs are temporally warped. For example, entertainment news tends to be disseminated on social networks and can easily draw extensive attention, but its dissemination on serious media like *Wall Street Journal* is very limited. Another interesting feature is the time differences between EPTSs, the degree of temporal warp, which reveals events' preferences to media. Alignments of EPTSs are quite suitable to reveal such interesting features.

Two temporally-warped EPTSs of an event $\mathscr{E}$ from two media A and B, are denoted as $\mathscr{E}^* = \langle \overline{pop}(E_1^*) \cdots , \overline{pop}(E_n^*) \rangle$, where $\mathscr{E}^*$ represents either $\mathscr{E}^A$ or $\mathscr{E}^B$.

A *match* m_k between E_i^A and E_j^B is defined as $m_k = (i, j)$. Distance between two matched data points is denoted as $dist(m_k)$ or $dist(i, j)$.

There is one problem, *twist*, existing when there are two matches $m_{k_1} = (i_1, j_1)$, $m_{k_2} = (i_2, j_2)$ with $i_1 < i_2$, but $j_1 > j_2$. The reason why there cannot be *twist* is that time sequence and the evolution of events cannot be reversed.

EPTS alignment aims to find a series of twist-free matches $M = \{m_1, \cdots , m_{|M|}\}$ for two $\mathscr{E}^A$ and $\mathscr{E}^B$ that every data point from an EPTS has at least one counterpoint from the other one, and the cumulative distance is the minimum. An intuitive thinking about an optimal alignment is that it should be a feature-to-feature one and differences between aligned EPTSs should be as small as possible. The minimum cumulative distance satisfy these two requirements. The key of alignments is to define a specific, precise, and meaningful distance function $dist(\cdot)$ for our task, which will be fully discussed in Sect. 6.3.

4   Scheme Overview of DANCINGLINES

The overview of DANCINGLINES is illustrated in Fig. 2. We first preprocess the data, then implement the TF-SW and ωDTW-CD models, and finally apply our scheme to real event datasets.

Data Preprocessing is applied on the raw data and has three steps. First of all, in Data-Formatting step, we filter out all irrelevant characters, such as punctuation, hyper links, etc. Secondly, Stopword-Removal step cleans frequently used conjunctions, pronouns and prepositions. Finally, we split every record into words through Word-Segmentation step.

TF-SW is a semantic-aware popularity quantification model based on Word2Vec and TextRank to generate EPTSs at certain temporal resolutions. This model is established by three steps. First of all, a cut-off mechanism is proposed to filter the unrelated words. Secondly, we construct TextRank graph to calculate the relative importance for the remaining words. Finally, a synthesized similarity calculation is defined for the edge weights in TextRank graph. We find

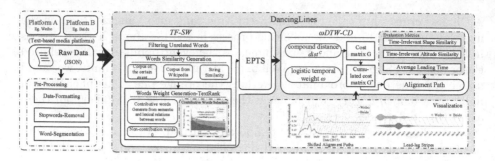

Fig. 2. The overview of DANCINGLINES Scheme

that only the words with both high semantic and lexical relations with other ones truly determine the event popularity. For that, a conception *contributive words* is defined and will be discussed in Sect. 5.

ωDTW-CD is a pairwise EPTSs alignment model derived from DTW. In this model, we innovatively define three distance function for DTW, event phase distance $dist^{\mathcal{E}}(\cdot)$, derivative distance $dist^D(\cdot)$, and Euclidean vertical line distance $dist^L(\cdot)$. Based on these three distance function, a compound distance is generated. A temporal weight coefficient is also introduced into the model for improving the alignment results. We further introduce these in detail in Sect. 6.

5 Semantic-Aware Popularity Quantification Model (TF-SW)

5.1 Filtering Unrelated Words

Since the number of distinct words for an event can be thousands of hundreds and there are tons of them actually not related to the event at all, it is too expensive to take them all into account. We propose a cut-off threshold mechanism to eliminate these unrelated noisy words and significantly reduce the complexity of whole scheme.

In fact, natural language corpus approximately obey the power law distribution and Zipf's Law [17]. Denoting r as the frequency rank of a word in a corpus and f as the corresponded word's frequency, then

$$f = H \cdot r^{-\alpha}, \tag{4}$$

where α and H are feature parameters for a specific corpus.

Since the words with high frequency is the necessary but not sufficient condition for those words to really reflect the actual event trends, an interesting question that where the majority of distribution of r lies is raised. For any power law with exponent $\alpha > 1$, the median is well defined [17]. That is, there is a point $r_{1/2}$ that divides the distribution in half so that half the measured values of r lie above $r_{1/2}$ and half lie below. In our case, r as rank, its minimum is 1, and the point is given by

$$\int_{r_{1/2}}^{\infty} f \; dr = \frac{1}{2} \int_{r_{min}}^{\infty} f \; dr \Rightarrow r_{1/2} = 2^{1/(\alpha-1)} r_{min} = 2^{1/(\alpha-1)}. \qquad (5)$$

Emphasis should be placed on the words that rank ahead of $r_{1/2}$, and the words within the long tail which are occupied by noise should be discarded. Thus cut-off threshold can now be defined as

$$th = H \cdot r_{1/2}^{-\alpha} = \frac{1}{2} \cdot H \cdot 2^{1/(1-\alpha)} \qquad (6)$$

Through this filter, we dramatically reduce the whole complexity of the scheme. For Event *AlphaGO*, the words we need to consider for Baidu reduce from thousands to around 40 and the ones for Weibo reduce to about 350, so the complexity has been reduced by at least 3 orders of magnitude.

5.2 Construction of TextRank Graph

After filtered through threshold, the remaining words are regarded as the representative words that do matter in quantifying the event popularity. However, for the remaining words, the importances are still obscure. They cannot just be naively presented by words' frequency, as a result we introduce TextRank [15] into our scheme.

For our task here, vertex in TextRank algorithm stands for a word that has survived the frequency filter in Sect. 5.1 and we use undirected edges in TextRank instead of directed edges in PageRank, since the relationships between words are bidirectional.

Inspired by the idea of TextRank, we further need to define the weights of edges in the graph described above. We introduce a conception *similarity* between words w_i and w_j, denoted as $sim\,(w_i, w_j)$ for the edges' weights.

However, we notice that there exist some words which passed the first filter but having negative similarity with all the other remaining words, which means these words are semantically far away from the topic of events. This phenomenon, in fact, indicates the existence of paid posters who post a large number of unrelated messages especially on social networks. To address this problem, we focus on the really related words and define a conception *contributive words*, denoted as

$$C_i = \{w_j^i \in E_i \mid \exists w_k^i \in E_i, \; sim(w_k^i, w_k^j) > 0\} \qquad (7)$$

and $\mathscr{C} = \bigcup C_i$. It is worth pointing out that this another filter-like process does not increase any computational complexity and we just do not establish edges when their weights are less than zero, then the non-contributive words will be discarded.

We construct a graph for each event phase E_i, where vertices represent the words and edges refer to their similarity $sim(w_i, w_j)$. We run the TextRank algorithm on the graphs and then get the real importance of each contributive word, $TR(w_i)$. The formula for TextRank is defined as

$$TR(w_i) = \frac{1-\theta}{|\mathscr{C}|} + \theta \cdot \sum_{j \to i} \frac{sim(w_i, w_j)}{\sum\limits_{k \to j} sim(w_k, w_j)} \cdot TR(w_j), \tag{8}$$

where the factor θ, ranging from 0 to 1, is the probability to continue to random surf follow the edges, since the graph cannot be a perfect graph and face potential dead-ends and spider-straps problem in practice. According to [15], θ is usually set to be 0.85. $|\mathscr{C}|$ represents the number of all contributive words, and $j \to i$ refer to words that is adjacent to word w_i.

5.3 Similarity Between Words

In our view, similarity between words are contributed by their semantical and lexical relationships and these two parts will be discussed in this subsection.

First of all, to quantify words' semantic relationships, we adopt Word2Vec [16] to map word w_k to vector $\mathbf{w}_k$. To comprehensively reflect the event characteristics, we integrate two corpora, an event corpus $\mathbb{R}$ from our datasets and a supplementary corpus extracted from Wikipedia with a broad coverage of events (denoted as *Wikipedia Dump*, or $\mathbb{D}$ for short), to train our Word2Vec models. For a word w_k, the corresponding word vectors are $\mathbf{w}_k^{\mathbb{R}}$ and $\mathbf{w}_k^{\mathbb{D}}$ respectively. Both event-specific and general semantic relations between words w_i and w_j are extracted and composed by

$$sem(w_i, w_j) = \beta \cdot \frac{\mathbf{w}_i^{\mathbb{R}} \cdot \mathbf{w}_j^{\mathbb{R}}}{\|\mathbf{w}_i^{\mathbb{R}}\| \cdot \|\mathbf{w}_j^{\mathbb{R}}\|} + (1-\beta) \cdot \frac{\mathbf{w}_i^{\mathbb{D}} \cdot \mathbf{w}_j^{\mathbb{D}}}{\|\mathbf{w}_i^{\mathbb{D}}\| \cdot \|\mathbf{w}_j^{\mathbb{D}}\|}, \tag{9}$$

where β is related to the two corpora and determines which one and to what extent we would like to emphasize.

Secondly, we consider the lexical information and integrate the string similarity so that we can combine the

$$sim(w_i, w_j) = \gamma \cdot sem(w_i, w_j) + (1-\gamma) \cdot str(w_i, w_j), \tag{10}$$

where we introduce a parameter γ to make our model general to different languages. For example, words that look similar are likely to be related in English, while this likelihood is fairly limited for languages like Chinese. We adopt the efficient cosine string similarity as

$$str(w_i, w_j) = \frac{\sum\limits_{c_l \in w_i \cap w_j} num(c_l, w_i) \cdot num(c_l, w_j)}{\sqrt{\sum\limits_{c_l \in w_i} num(c_l, w_i)^2} \cdot \sqrt{\sum\limits_{c_l \in w_j} num(c_l, w_j)^2}}, \tag{11}$$

where $num(c_l, w_i)$ means counts of character c_l in word w_i.

5.4   Definition of Weight Function

Since the sum of vertices' TextRank values for a graph is always 1 regardless of the graph scale, the TextRank value tends to be lower when there are more contributive words within the time interval. Therefore, a compensation factor within each event phase E_i is multiplied to the TextRank values, and the weight function $weight(\cdot)$ for contributive words is finally defined as

$$weight(w_j^i) = \frac{TR(w_j^i)}{|C_i|} \cdot \sum_{w_k^i \in E_i} fre(w_k^i). \tag{12}$$

Recalling that in our scheme, the event popularity $pop(E_i)$ is the sum of popularity of *all* words, for the consistency of Eq. (1), we make the weight function for the non-contributive words identically equal to zero. Then for all words, popularity can be calculated through Eq. (1). For each event phase E_i, according to Eq. (2), we can generate the event popularity within t_i and EPTSs through Eq.(3).

6   Cross-Platform Analysis Model (ωDTW-CD)

6.1   Classic Dynamic Time Warping with Euclidean Distance

We find that, with only the global minimum cost considered, classic DTW with Euclidean distance may provide results suffering from *far-match* and *singularity* problems when aligning pairwise cross-platform EPTSs.

Far-Match Problem. Classic DTW disregards the temporal range, which may lead to *"far-match"* alignments. Since the EPTSs of an event from different platforms keep pace with the event's real-world evolution, alignment of EPTSs' data points that are temporally far away is against the reality. Thus, classic method should be more robust and Euclidean Distance is not ideal enough for EPTS alignment.

Singularity Problem. Classic DTW with Euclidean distance is vulnerable to the *"singularity"* problem elaborated in [8], where a single point in one EPTS is unnecessarily aligned to multiple points in another EPTS. These singular points will generate misleading results for further analysis.

6.2   Event Phase Distance

Recalling Eq. (7) that all the contributive words for an event phase E_i are denoted as C_i and $\mathscr{C}$ is a set of all contributive words for an event $\mathscr{E}$ on single medium, we can utilize the similarity between the contributive word sets C_i to match those event phases. To quantify this similarity, we propose our *event phase distance* measure. Distance between E_i^A and E_j^B is denoted as $dist^{\mathscr{E}}(i, j)$.

Since $\mathscr{C}$ for different platforms are probably not identical, let the general $\mathscr{C}' = \mathscr{C}^A \cup \mathscr{C}^B$. Then, each word list C_i can be intuitively represented as a one-hot vector $\mathbf{z}_i \in \{0,1\}^{|\mathscr{C}'|}$, where each entry of vectors indicates whether corresponding contributive word exists in word list C_i. However, problem arises when calculating the similarity between these very sparse vectors, especially when the event corpus is of a large scale and there are huge amount of data points in EPTSs. To address this problem, we leverage SIMHASH [3], adapted from *locality sensitive hashing* (LSH) [6], to hash the very sparse vectors to small signatures while preserving the similarity among the words.

According to [3], s projection vectors $\mathbf{r}_1, \mathbf{r}_2, \cdots, \mathbf{r}_s$ are selected at random from the $|\mathscr{C}'|$-dimensional Gaussian distribution. A projection vector $\mathbf{r}_l$ is actually a hash function that hashes a one-hot vector $\mathbf{z}_i$ generated from C_i to a scalar -1 or 1. s projection vectors hash the original sparse vector $\mathbf{z}_i$ to a small signature $\mathbf{e}_i$, where $\mathbf{e}_i$ is an s-dimensional vectors with entries equal to -1 or 1. Sparse vectors $\mathbf{z}_i^A$ and $\mathbf{z}_j^B$ can be hashed to $\mathbf{e}_i^A$ and $\mathbf{e}_j^B$ and the distance between these two points can be calculated by

$$dist^{\mathscr{E}}(i,j) = 1 - \frac{\mathbf{e}_i^A \cdot \mathbf{e}_j^B}{\|\mathbf{e}_i^A\| \cdot \|\mathbf{e}_j^B\|}. \tag{13}$$

The dimension of short signatures, s, can be used to tune the accuracy we want to remain versus the low complexity. If we want to dig some subtle information in a high temporal resolution, say half an hour, we should increase s to get more accuracy, while if we just want to have a glimpse of the event, a small s is reasonable.

6.3   The ωDTW-CD Model

To more comprehensively measure the distance between data points from two EPTSs, a ωeighted DTW method with Compound Distance (ωDTW-CD) is proposed to balance temporal alignment and shape-matching. ωDTW-CD tries to synthesize trend characters, Euclidean vertical line distance, and event phase distance all together and this overall distance is measured by compound distance $dist^C(i,j)$,

$$dist(i,j) = dist^C(i,j) + \omega_{i,j}. \tag{14}$$

We regard the difference between estimated derivative of EPTS points, $dist^D(i,j)$, as the trend characters distance. According to [8], $dist^D(i,j)$ generated by

$$dist^D(i,j) = \left| D(E_i^A) - D(E_j^B) \right|, \tag{15}$$

where the estimated derivative $D(x)$ is calculated through

$$D(x) = \frac{x_i - x_{i-1} + \frac{x_{i+1} - x_{i-1}}{2}}{2}. \tag{16}$$

As stated in [8], this estimate is simple but robust to trend characters compared to other estimation methods. The compound distance $dist^C(i,j)$ is generated by

$$dist^C(i,j) = \sqrt[3]{dist^{\mathscr{E}}(i,j) \cdot dist^L(i,j) \cdot dist^D(i,j)}, \tag{17}$$

where $dist^{\mathscr{E}}(i,j)$ is the event phase distance and $dist^{L}(i,j)$ is the Euclidean vertical line distance between data points E_i^A, E_j^B defined as $dist^{L}(i,j) = |E_i^A - E_j^B|$. For the purpose of flexibility [7], we introduce a sigmoid-like temporal weight

$$\omega_{i,j} = \frac{1}{1 + e^{-\eta(|i-j|-\tau)}}. \tag{18}$$

The temporal weight is actually a special cost function for the alignment in our task. It has two parameters, η and τ, to generalize for many other events and languages. Parameter η decides the overall penalty level, which we can tune for different EPTSs. Factor τ is a prior estimated time difference, having the same unit as the temporal resolution we choose, between two platforms based on the natures of different medias.

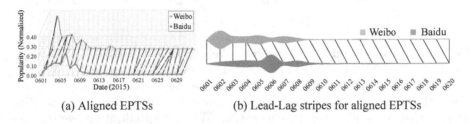

(a) Aligned EPTSs   (b) Lead-Lag stripes for aligned EPTSs

Fig. 3. Visualization of ωDTW-CD, *Sinking of a Cruise Ship*

A visualization is showed in Fig. 3a and it gives a direct way to know how the data points from EPTSs are aligned. The links in the figure represent matches. The lead-lag stripes [25] in Fig. 3b show a more obvious way to know matches. The X-axis represents time and the stripes' vertical width indicates the event popularity in that day. We can find that after the Event *Sinking of a Cruise Ship* happens, the Weibo platform captured and propagated the topic faster than Baidu did in the beginning and then more people started to search on the Baidu for more information so the popularity on Baidu rose.

7   Experiments

7.1   Experiment Setup

Datasets. Our experiments are conducted on eighteen real-world event datasets from Weibo and Baidu, covering nine most popular events that occurred from 2015 to 2016. All the nine events covered in our datasets have provoked intensive discussions and gathered widespread attention. In addition, they are both typical events in distinct categories including disasters, high-tech stories, entertainment news, sports and politics. The detailed information of our datasets is listed in Table 1.

Table 1. Overall information of the datasets

No.	Event name	# of records (k)		Size (MB)	
		Weibo	Baidu	Weibo	Baidu
①	Sinking of a Cruise Ship	308.45	1560.4	320.59	401.48
②	Chinese Stock Market Crash	701.71	420.40	578.77	74.14
③	AlphaGo	838.12	2337.3	654.89	406.83
④	Leonardo DiCaprio, Oscar Best Actor	2569.5	730.82	1788.9	139.52
⑤	Kobe Bryant's Retirement	3655.3	2300.9	2274.8	403.69
⑥	Huo and Lin Went Public with Romance[†]	1535.2	1615.2	1027.1	289.98
⑦	Brexit Referendum	957.16	2160.4	715.51	392.32
⑧	Pokémon Go	936.38	3652.2	695.90	625.87
⑨	The South China Sea Arbitration	7671.0	7815.3	5918.2	1451.9

Implementation and Parameters. We implement CBOW when doing Word2Vec [16]. The parameters involved in TF-SW are set to be $\beta = 0.7$, with $\gamma = 0.02$ considering the nature of Chinese language, that there are many different characters but almost no meaning changes on words. The factor for TextRank is set to be $\theta = 0.85$ by convention. Without specification, we set each time interval to be 1 day. The corresponding parameters for the sigmoid-like temporal weight are set as $\eta = 10, \tau = 2$.

7.2 Verification of TF-SW

To evaluate the effectiveness of TF-SW, we compare the EPTS generated by our model with the EPTSs by other two baselines, naive frequency and TF-IDF [22]. All the EPTSs generated by Naive Frequency and TF-IDF are normalized in the same way as TF-SW through Eq. (3). Based on the three generated EPTSs, we present a thorough discussion and comparison to validate our TF-SW model.

Accuracy. We pick up the peaks in EPTSs and backtrack what exactly happened in reality. An event is always pushed forward by series of "little" events and we call them sub-events, which are reflected as peaks in EPTS figures.

In the Event *Capsizing of a Cruise Ship*, the real-world event evolution involves four key sub-events. On the night of June 1, 2015, the cruise ship sank in a severe thunderstorm. Such a shocking disaster raised tremendous public attention on June 2. On June 5, the ship was hoisted and set upright. A mourning ceremony was held on June 7, and on June 13, total 442 deaths and only 12 survivors were officially confirmed, which marked the end of the rescue work.

The EPTS generated by TF-SW shows four peaks, which is illustrated in Fig. 4. All these peaks are highly consistent with the four key sub-events in real world, while the end of rescue work on June 13 is missed by approaches based on Naive Frequency and TF-IDF. In conclusion, TF-SW model shows the ability to track the development of events precisely.

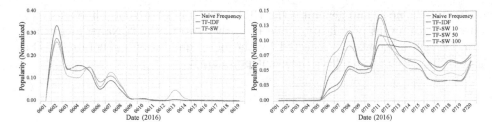

Fig. 4. *Sinking of a Cruise Ship*, Weibo   **Fig. 5.** *Pokémon Go*, Baidu ($th = N.$)

Sensitivity to Burst Phases. Compared with the baselines, our model are more sensitive to the burst phases of an event, as is shown in Fig. 5, especially on data points 07/06, 07/08, and 07/11. The event popularity on these days are larger than those obtained by Naive Frequency and TF-IDF. In another word, the EPTSs generated through TF-SW rises faster, more significant in peaks, and are more sensitive to breaking news which enables the model to capture the burst phases more precisely. From three EPTSs of TF-SW with different th, it is shown that TF-SW is more sensitive to the burst of events with a higher th value, as is shown by the data point 07/06.

An event whose EPTS rises fast at some data points possesses the potential to draw wider attention. It is reasonable for a popularity model not only to depict the current state of event popularity, but also take the potential future trends into consideration. In this way, a quick response to the burst phases of an event is more valuable for real-world applications. This advantage of our model can lead to a powerful technique for first story detection on ongoing events.

Superior Robustness to Noise. To verify whether our model can effectively filter out noisy words, we further implement an experiment on a simulated corpus. We first extract 50K Baidu queries with the highest frequency in the corpus of Event *Kobe's Retirement* and make them as the base data for a 6-day simulated corpus. Then we randomly pick noisy queries from Internet that are not relevant to Event *Kobe's Retirement* at all. The amount of noisy queries is listed in Table 2.

Table 2. Number of noisy records added to each day

Day	1	2	3	4	5	6
# (k)	0.000	1.063	2.235	3.507	4.689	6.026

Since each day's base data are identical, a good model is supposed to filter noisy queries out and generate an EPTS with all identical data points, which form a horizontal line in X-Y plane. EPTSs generated by TF-SW, Naive Frequency and TF-IDF are shown in Fig. 6. It is shown that TF-SW successfully filters out the noise and generates the EPTS which is a horizontal line and captures the real event popularity, while the other two methods Naive Frequency and TF-IDF are obviously effected by the noisy queries and generate EPTSs that cannot accurately reflect the event popularity.

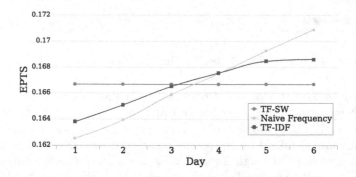

Fig. 6. EPTSs on the simulated corpus

7.3 Verification of ωDTW-CD

To demonstrate the effectiveness of ωDTW-CD, we compare it with seven different DTW extensions listed below.

- *DTW* is the DTW method with Euclidean distance.
- *DDTW* [8] is the Derivative DTW which replaces the Euclidean distance with the difference of estimated derivatives of the data points in EPTSs.
- DTW_{bias} & $DDTW_{bias}$ are the extended DTW and DDTW respectively with a bias towards the diagonal direction.
- *ωDTW* & *ωDDTW* are the temporally weighted DTW and DDTW, where the sigmoid-like temporal weight defined by Eq. (18) is introduced to the cost matrices.
- *DTW-CD* is a simplification of *ωDTW-CD* that implements only $dist^C$ without temporal weight ω.

Singularity. Fig. 7 visualizes the results generated by ωDTW and our proposed model. Classic DTW and DTW_{bias} severely suffer the problem of singularity. Compared with ωDTW, ωDTW-CD presents better and more stable performance when aligning the time series with sharp fluctuations. In general, our model is capable of avoiding the singularity problem by involving the derivative differences.

Far-Match. Considering the fact that the time difference between two aligned sub-event can barely exceed two days, far-match exists in the alignment generated by $DDTW_{bias}$ and DTW-CD in Fig. 8, but not in our results in Fig. 3a. Thus, the sigmoid-like temporal weight introduced to our model helps avoid the far-match problem.

(a) ωDTW ($\eta = 10, m = 2$)

(b) ωDTW-CD ($\eta = 10, m = 2$)

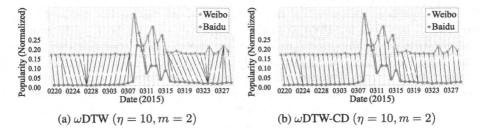

Fig. 7. Alignment results of 2 methods, *AlphaGo*. One data point is categorized as a singular point if it is matched to more than 4 points from the other EPTS.

(a) $\text{DDTW}_{\text{bias}}$ (1.2, 1, 1.2)

(b) DTW-CD

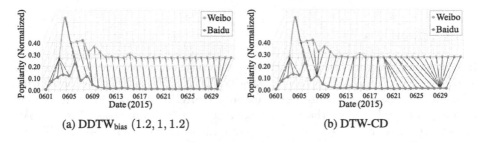

Fig. 8. Alignment results of 2 methods, *Sinking of a Cruise Ship*

Overall Performance. All the comparison results on the eighteen real-world datasets are illustrated in Fig. 9, where each color corresponds to a method, each method are ranked respectively for each event, and methods with higher grades are ranked on the top. Results facing *singularity* or *far-match* are marked by red boxes. The performances are graded under the following criteria. The grades are given to show the relative performances among different methods only regarding one event. The method that does not suffer from *singularity* or *far-match* has higher grades than the one that does. The methods giving same alignment results are further graded considering their complexity.

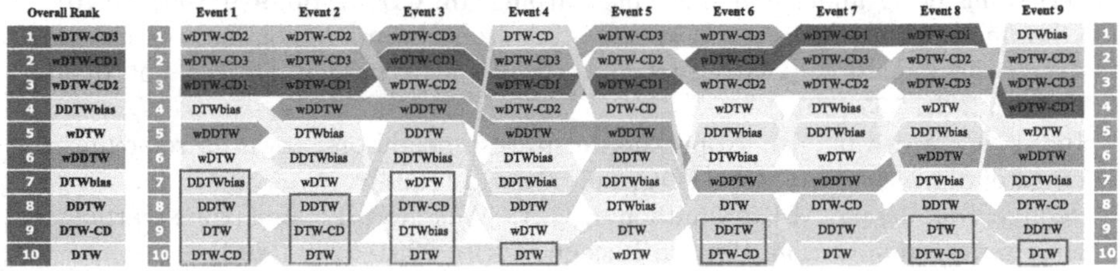

Fig. 9. Ranking visualization of grades for 10 methods on nine real-world events. (Color figure online)

In comparison with existing variants of DTW as well as the reduced version of our method, ωDTW-CD achieves improvements on both performance and robustness on alignment generation and successfully conquers the problem of *singularity* and *far match*. Results shows that the event phase distance, estimated derivative difference, and the sigmoid-like temporal weight simultaneously contribute to the performance enhancement of ωDTW-CD. Moreover, with parameter η and τ, our model is flexible to different temporal resolutions and to events of distinct popularity features. In Fig. 9, ωDTW-CD$_1$ corresponds to $\eta = 5$, $\tau = 3.2$. $\eta = 10$, $\tau = 2$ is for ωDTW-CD$_2$. $\eta = 5$, $\tau = 2.2$ is for ωDTW-CD$_3$. The results show the strong ability of ωDTW-CD to handle specific events.

8   Conclusion

In this paper, we quantify and interpret event popularity between pairwise text media with an innovative scheme, DANCINGLINES. To address the popularity quantification issue, we utilize TextRank and Word2Vec to transform the corpus into a graph and project the words into vectors, which are covered in TF-SW model. To furthermore interpret the temporal warp between two EPTSs, we propose ωDTW-CD to generate alignments of EPTSs. Experimental results on eighteen real-world event datasets from Weibo and Baidu validate the effectiveness and applicability of our scheme.

References

1. Ahn, B., Van Durme, B., Callison-Burch, C.: Wikitopics: what is popular on Wikipedia and why. In: Proceedings of the Workshop on Automatic Summarization for Different Genres, Media, and Languages, pp. 33–40 (2011)
2. Bao, B., Xu, C., Min, W., Hossain, M.S.: Cross-platform emerging topic detection and elaboration from multimedia streams. TOMCCAP **11**(4), 54 (2015)
3. Charikar, M.S.: Similarity estimation techniques from rounding algorithms. In: STOC, pp. 380–388 (2002)
4. Dau, H.A., Begum, N., Keogh, E.: Semi-supervision dramatically improves time series clustering under dynamic time warping. In: CIKM, pp. 999–1008 (2016)
5. Giummolè, F., Orlando, S., Tolomei, G.: A study on microblog and search engine user behaviors: how Twitter trending topics help predict Google hot queries. Human **2**(3), 195 (2013)
6. Indyk, P., Motwani, R.: Approximate nearest neighbors: towards removing the curse of dimensionality. In: STOC, pp. 604–613 (1998)
7. Jeong, Y.S., Jeong, M.K., Omitaomu, O.A.: Weighted dynamic time warping for time series classification. Pattern Recogn. **44**(9), 2231–2240 (2011)
8. Keogh, E.J., Pazzani, M.J.: Derivative dynamic time warping. In: SDM, pp. 1–11 (2001)
9. Kwak, H., Lee, C., Park, H., Moon, S.: What is Twitter, a social network or a news media? In: WWW, pp. 591–600 (2010)
10. Lee, P., Lakshmanan, L.V.S., Milios, E.E.: Keysee: supporting keyword search on evolving events in social streams. In: KDD, pp. 1478–1481 (2013)

11. Li, R., Lei, K.H., Khadiwala, R., Chang, K.: Tedas: a Twitter-based event detection and analysis system. In: ICDE, pp. 1273–1276 (2012)
12. Lin, S., Wang, F., Hu, Q., Yu, P.: Extracting social events for learning better information diffusion models. In: KDD, pp. 365–373 (2013)
13. Liu, N., An, H., Gao, X., Li, H., Hao, X.: Breaking news dissemination in the media via propagation behavior based on complex network theory. Physica A **453**, 44–54 (2016)
14. Maus, V., Câmara, G., Cartaxo, R., Sanchez, A., Ramos, F., Queiroz, G.: A time-weighted dynamic time warping method for land-use and land-cover mapping. J-STARS **9**(8), 3729–3739 (2016)
15. Mihalcea, R., Tarau, P.: TextRank: bringing order into texts. In: EMNLP, pp. 404–411 (2004)
16. Mikolov, T., Sutskever, I., Chen, K., Corrado, G.S., Dean, J.: Distributed representations of words and phrases and their compositionality. In: NIPS, pp. 3111–3119 (2013)
17. Newman, M.: Power laws, pareto distributions and Zipf's law. Contemp. Phys. **46**(5), 323–351 (2005)
18. Osborne, M., Petrovic, S., McCreadie, R., Macdonald, C., Ounis, I.: Bieber no more: first story detection using Twitter and Wikipedia. In: SIGIR 2012 Workshop on Time-Aware Information Access (2012)
19. Rong, Y., Zhu, Q., Cheng, H.: A model-free approach to infer the diffusion network from event cascade. In: CIKM, pp. 1653–1662 (2016)
20. Sakoe, H., Chiba, S.: Dynamic programming algorithm optimization for spoken word recognition. IEEE Trans. Acoust. Speech Signal Process. **26**(1), 43–49 (1978)
21. Silva, D.F., Batista, G.E., Keogh, E.: On the effect of endpoints on dynamic time warping. In: SIGKDD Workshop on Mining Data and Learning from Time Series (2016)
22. Tang, Y., Ma, P., Kong, B., Ji, W., Gao, X., Peng, X.: ESAP: a novel approach for cross-platform event dissemination trend analysis between social network and search engine. In: Cellary, W., Mokbel, M.F., Wang, J., Wang, H., Zhou, R., Zhang, Y. (eds.) WISE 2016. LNCS, vol. 10041, pp. 489–504. Springer, Cham (2016). https://doi.org/10.1007/978-3-319-48740-3_36
23. Wang, J., et al.: Mining multi-aspect reflection of news events in Twitter: discovery, linking and presentation. In: ICDM, pp. 429–438 (2015)
24. Zhang, J., Chen, J., Zhi, S., Chang, Y., Yu, P.S., Han, J.: Link prediction across aligned networks with sparse and low rank matrix estimation. In: ICDE, pp. 971–982 (2017)
25. Zhong, Y., Liu, S., Wang, X., Xiao, J., Song, Y.: Tracking idea flows between social groups. In: AAAI, pp. 1436–1443 (2016)
26. Zhou, X., Liang, X., Zhang, H., Ma, Y.: Cross-platform identification of anonymous identical users in multiple social media networks. TKDE **28**(2), 411–424 (2016)

Social Networks

Community Structure Based Shortest Path Finding for Social Networks

Yale Chai, Chunyao Song$^{(\boxtimes)}$, Peng Nie, Xiaojie Yuan, and Yao Ge

College of Computer and Control Engineering, Nankai University,
38 Tongyan Road, Tianjin 300350, People's Republic of China
{chaiyl,niepeng,geyao}@dbis.nankai.edu.cn,
{chunyao.song,yuanxj}@nankai.edu.cn

Abstract. With the rapid expansion of communication data, research about analyzing social networks has become a hotspot. Finding the shortest path (SP) in social networks can help us to investigate the potential social relationships. However, it is an arduous task, especially on large-scale problems. There have been many previous studies on the SP problem, but very few of them considered the peculiarity of social networks. This paper proposed a community structure based method to accelerate answering the SP problem of social networks during online queries. We devise a two-stage strategy to strike a balance between offline precomputation and online consultations. Our goal is to perform fast and accurate online approximations. Experiments show that our method can instantly return the SP result while satisfying accuracy constraint.

Keywords: Shortest path · Social network · Community structure

1 Introduction

Social network analysis is aimed at quantifying social networks and discovering the latent relationships among social actors, in which social networks can be modeled as a weighted graph $\mathbb{G} = (\mathbb{V}, \mathbb{E})$, where vertices in $\mathbb{V}$ represent social entities (such as individuals or organizations), edges in $\mathbb{E}$ represent relationships between entities. And the closer the two entities are connected, the greater the weight of the edge. Finding the SP in social graphs can help to analyzing social networks, such as information spreading performance and recommendation systems. However, finding the exact SP cannot be adopted for real-world massive networks, especially in online applications where the distance must be provided in a few milliseconds. Thus, this paper focuses on finding a path with a relatively minimum cost in a very short time.

Social networks are often complex and possess some special properties [5]: (*i*) community property, which is also referred to as the small-world property. Connections between the vertices in a community are denser and closer than connections with the rest of the network. (*ii*) scale-free, there can be a large variety of vertices degrees. (*iii*) six degrees of separation, the interval between

© Springer Nature Switzerland AG 2018
S. Hartmann et al. (Eds.): DEXA 2018, LNCS 11029, pp. 303–319, 2018.
https://doi.org/10.1007/978-3-319-98809-2_19

any two social individuals will not exceed six hops. The SP problem has been studied for many years, most are two-stage methods recently [2, 6–17, 21, 22], which provide a tradeoff among space, preprocessing time, querying time, and accuracy. However, rarely are they particularly designed for social networks.

Due to the community property of social networks, we can focus on the connections between communities when searching the SP between two entities. In addition, we distinguish vertices' roles in community. There is a group of people who serve as bridges to connect people inside and outside the community, are denoted as **interface** vertices. For example, in Fig. 1, the number on each edge indicates the edge length, which is the distance between two vertices. If $\{v_1, v_2, v_3, v_6, v_7\}$ want to visit $\{v_9, v_{10}, v_{11}, v_{12}\}$, they must go through interface vertices $\{v_4, v_5, v_8\}$. $\{v_8\}$ is a special class of interface vertices, which belongs to both communities, is denoted as **hub** vertex. Besides, **outlier** vertex $\{v_1\}$ must go through its only neighbor $\{v_2\}$ to access other vertices. In the following, we pay attention to interface vertices which play crucial roles in SP. Chang et al. [1] develop a pSCAN method for scalable structural graph clustering, which distinguishes the different roles of the vertices in the community. However, pSCAN is designed for unweighted graph. Since edge's weight between two vertices can indicate the closeness between two entities, and can reveal more information for social networks, research on weighted graph is more suitable for social networks. Thus this paper develops wSCAN based on pSCAN: we fix the computation method of the structural similarity for every pair of adjacent vertices.

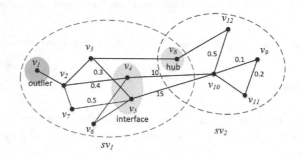

Fig. 1. Different roles of vertices in two adjacent communities

In this paper, we propose a method to find the shortest path based on communities (SPBOC) with two phases: preprocessing and online querying. During the preprocessing step, we construct a sketch of the graph, which is defined as the super graph $\$G$. Specifically, each community in graph G corresponds to a super vertex in $\$G$, and the relationship between communities corresponds to a super edge in $\$G$. At query time, given two vertices $s, t \in G$, we first find the SP between the super vertices that contain s and t respectively. Then the search can narrow down to all the vertices contained by the super vertices on this path.

Our primary contributions are summarized as follows.

1. We propose the concept of super graph in social network, which is based on the result of clustering the original graph, but much smaller scale. In order to cluster weighted social networks, we propose a fast structural clustering method wSCAN. What's more, during preprocessing we: *(i)* compute the shortcuts between all pairs of interface vertices within a super vertex, *(ii)* estimate the distances between adjacent super vertices, and *(iii)* attach labels to each super vertex, so that at query time, we can find out the reachability and the SP between any two super vertices in $O(1)$, and then only focus on the interface vertices of all the super vertices on the SP.
2. We present an approximate SP approach for social networks. This paper draws conclusions from two observations. For two vertices in the same community, SP can be found within the community. For two vertices in different communities, the shortest distance can be estimated by the shortest distance between the communities. By the aid of the pretreatment, the result can be returned in $O(n_{con}logn_{con})$, where $O(n_{con})$ is the size of a single community.
3. We propose three optimizations of which the first one is to reduce the error rate and the next two are to accelerate the query. At query time we: *(i)* expand the SP in $\$G$ to include the neighbors within one hop for each super vertex, *(ii)* deal with oversized and isolated communities after clustering, *(iii)* prune some vertices by predicting the distance towards the target. Pruning can reduce the analysis of many vertices that have little chance to be on the SP.
4. We conduct extensive empirical studies on real social networks and synthetic graphs. Experiments show that SPBOC shows a good mediation between precomputation and online query. It can greatly trim the search vertices range and answer SP queries very effectively in social networks, especially after the optimizations. According to the statistical analysis, our algorithm performs better on datasets with more obvious community nature.

The remainder of this paper is organized as follows. We briefly review related work in Sect. 2. Section 3 introduces some general definitions used in this paper, and discuss some observations and corollaries. We describe our algorithms in Sect. 4 and the optimization techniques in Sect. 5. In Sect. 6, we present our experiments results, and finally reach a conclusion in Sect. 7.

2   Related Work

The traditional Dijkstra algorithm [3] can solve the SP problems in $O(n^2)$, or $O(nlogn+m)$ when using Fibonacci heap. Bidirectional search [4] is an improvement based on Dijkstra, which reduces the time complexity to $O(n^2/8)$ by starting from both the source and the target. These methods do not have any pretreatment, makes it hard to work very well for large-scale social networks.

Afterwards, stimulated by the demands of applications, a lot of impressive algorithms have been proposed. Most of these studies use pre-processing strategies to speed up queries, and they can be roughly divided into three categories: The first one is landmark-based methods [6–9,17,22], they select several vertices as landmarks, which can be used to estimate the distance between any two vertices in the graph. However, the global landmark selection tends to fail to accurately estimate distances between close pairs, and the local landmark selection has a poor scalability because of the extremely large space requirement. Particularly, [22] accelerate queries by using the small-world property of complex networks, however, it is designed only for unweighted graphs.

The second one is label-based methods [10–12,21], which attaches additional information to vertices or edges. Based on the information, the query decides how to prioritize or prune vertices. These kind of methods can be very fast, but they cannot handle billion-scale networks owing to the huge index size.

The third one is hierarchy-based methods [2,13–16], which constructs the hierarchical structure of the graph. Then the SP query can be answered by searching only a small part of the auxiliary graph. According to different application scenarios, it can be further divided into the following three categories: (i) road networks [13,14], which is based on the natural characteristics of the road networks and is not applicable to other networks, (ii) general networks [15,16], which constructing data structure that allows retrieval of a distance estimate for any pair of vertices in $O(1)$. However, the properties of social networks cannot be exploited by common algorithmic techniques, (iii) social networks [2], Gong et. al. in [2] suggests that when the distance between clusters is much longer than the distance between vertices within the cluster, the latter can be ignored. However, [2] is very sensitive to the community property of the datasets, and has to restore the super graph to the original after finding the SP in the super graph, which makes it take a long time to return results on large-scale datasets.

3   Preliminaries

In this section we first list symbols and terms we use in this paper and their corresponding meanings in Table 1, and then present some observations and corollaries. Given a weighted graph $\mathbb{G}$, we transform the weight function $\omega(e)$ into a length function $\ell(e)$ for each edge e, as shown in Table 1. Finding the SP in $\mathbb{G}$ is to find the path with the minimum sum of $\ell(e)$ for all edges on the path. In the following, we refer s, t to be the two particular vertices that we aim to find the SP within $\mathbb{G}$, and let sv_s and sv_t be the communities that contain s, t respectively.

For example, in Fig. 1, $con(sv_1) = \{v_1, v_2, v_3, v_4, v_5, v_6, v_7, v_8\}$, $bel(v_1) = \{sv_1\}$, $con(sv_1, sv_2) = \{(v_8, v_8), (v_4, v_{10}), (v_5, v_{10})\}$, $int(sv_1, sv_2) = \{v_4, v_5, v_8\}$, $int(sv_2, sv_1) = \{v_8, v_{10}\}$, $hub(sv_1, sv_2) = \{v_8\}$, $out(sv_1) = \{v_1\}$.

Observation 1: The shortest distance between two vertices in adjacent communities, is equal to the distance from two vertices to their interface vertices, respectively, plus the distance between interface vertices. For example, in Fig. 1,

Table 1. Notation

Terms, symbols	Meaning
$G = (V, E)$	Original social graph, where V is the set of vertices and E is the set of edges
n	The number of vertices $|V|$
m	The number of edges $|E|$
$(u, v) \in E$	The edge between vertice u and v, where $u, v \in V$
$\omega(e)$	The nonnegative weight function for edge e
$\ell(e)$	The length function for edge e, $\ell(e) = max\{\omega(e_1), ..., \omega(e_m)\} + 1 - \omega(e)$
$SG = (SV, SE)$	Super graph generated based on the clustering result of G, where SV is the set of super vertices and SE is the set of super edges
$\hat{n}$	The number of super vertices $|SV|$
$\hat{m}$	The number of super edges $|SE|$
$con(sv)$	The set of vertices belong to sv, where $sv \in SV$
$bel(v)$	The set of communities that v belongs to, where $v \in V$
$con(sv_1, sv_2)$	The set of edges connect sv_1 and sv_2, where $sv_1, sv_2 \in SV$
$\ell(sv_1, sv_2)$	The length function for super edge (sv_1, sv_2), where $sv_1, sv_2 \in SV$
$int(sv_1, sv_2)$	The set of interface vertices from sv_1 to sv_2. If $(u, v) \in con(sv_1, sv_2)$, $bel(u) = \{sv_1\}$, $bel(v) = \{sv_2\}$, then $u \in int(sv_1, sv_2)$, $v \in int(sv_2, sv_1)$
$hub(sv_1, sv_2)$	The set of intersections of $con(sv_1)$ and $con(sv_2)$
$out(sv)$	The set of vertices $\in con(sv)$, whose degree is 1
n_{con}	The average number of vertices in a single super vertex
n_{int}	The average number of interface vertices in a single super vertex
$p_G(s, t)$	$p_G(s, t) = < s, u_1, u_2, ..., u_\ell, t >$, a path between s and t in G, where $\{u_1, u_2, ..., u_\ell\} \in V$ and $\{(s, u_1), (u_1, u_2), ..., (u_\ell, t)\} \in E$
$P_G(s, t)$	The set of all paths from s to t in G
$d_G(s, t)$	The length of the path with the minimum sum of $\ell(e)$s from s to t in G
$sp_G(s, t)$	A path whose length is equal to $d_G(s, t)$ from s to t
$SP_G(s, t)$	The set of paths whose length is equal to $d_G(s, t)$ from s to t
$p_{SG}(sv_s, sv_t)$	$p_{SG}(sv_s, sv_t) = < sv_s, sv_1, sv_2, ..., sv_\ell, sv_t >$, a path between sv_s and sv_t in SG where $\{sv_1, sv_2, ..., sv_\ell\} \in SV$ and $\{(sv_s, sv_1), (sv_1, sv_2), ..., (sv_\ell, sv_t)\} \in SE$
$d_{SG}(sv_s S, sv_t)$	The length of the path with the minimum sum of $\ell(se)$s from sv_s to sv_t in SG
$sp_{SG}(sv_s, sv_t)$	A path whose length is equal to $d_{SG}(sv_s, sv_t)$ from sv_s to sv_t

$d_G(v_3, v_{12}) = \min\{d_G(v_3, v_4) + d_G(v_4, v_{10}) + d_G(v_{10}, v_{12}), d_G(v_3, v_5) + d_G(v_5, v_{10}) + d_G(v_{10}, v_{12}), d_G(v_3, v_8) + d_G(v_8, v_8) + d_G(v_8, v_{12})\}$. Consequently, $d_G(v_3, v_{12})$ can be indicated as the minimum combination of three phases: $d_G(v_3, int(sv_1, sv_2))$, $d_G(int(sv_1, sv_2), int(sv_2, sv_1))$, and $d_G(int(sv_2, sv_1), v_{12})$. Therefore, we need to focus on interface vertices to find the SP between vertices within adjacent communities.

Observation 2: The lengths of edges within the community are much smaller than the edges between the communities. As we said, connections between the vertices in a community are denser and closer than connections with the rest of the network. In other words, the edges within communities have higher weights and lower lengths than edges between communities. For example, in Fig. 1,

$d_G(v_7, v_9) = d_G(v_7, v_5) + d_G(v_5, v_{10}) + d_G(v_{10}, v_9) = 0.5 + 15 + 0.1 \approx 15$. The distance between communities can be used to represent the whole distance.

Corollary 1. *For two vertices in the same community, the shortest path can be found within the community.*

Proof. According to Observation 2, the distance between communities is much larger than the distance between vertices inside the community, which means the shortest path between two vertices in the same community is unlikely to cross the long distance between communities. Therefore, when it comes to two vertices in the same community, we argue that the search scope can be narrowed down to this community instead of the whole graph.

Corollary 2. *For two vertices in different communities, the shortest distance can be estimated by the shortest distance between the two communities.*

Proof. Let us suppose that $sp_G(s, t) = < s, u, t >$, where $u \in \mathbb{V}$, and u, s, t are in different communities sv_u, sv_s, sv_t respectively. According to Observation 2, $d_G(s, t) = d_G(s, u) + d_G(u, t) \approx d_{SG}(sv_s, sv_u) + d_{SG}(sv_u, sv_t)$. The shortest path between s and t can be estimated by the sum of the distances between the participating communities.

Furthermore, in order to find $sp_G(s, t)$, we firstly need to find $sp_{SG}(sv_s, sv_t)$. Suppose $sp_{SG}(sv_s, sv_t) = < sv_s, sv_1, sv_2, ..., sv_\ell, sv_t >$, where $\{sv_1, sv_2, ..., sv_\ell\} \in \$V$, then the shortest distance between s and t can be estimated as: $d_G(s, t) \approx d_{SG}(sv_s, sv_1) + d_{SG}(sv_1, sv_2) + ... + d_{SG}(sv_\ell, sv_t) = d_{SG}(sv_s, sv_1) + \sum_{i=1}^{\ell-1} d_{SG}(sv_i, sv_{i+1}) + d_{SG}(sv_\ell, sv_t)$. Consequently, we think that $sp_{SG}(S, T)$ can help us find $sp_G(s, t)$.

4   Our Approach

In this section, we will introduce our approach in detail on the basis of previous observations and corollaries. SPBOC is a two-stage strategy which seeks the best balance between scalability (preprocessing time and space) and query performance (query time and precision).

A. Preprocessing Phase
In this phase, we generate the super graph $\$G = (\$V, \$E)$. To be specific, we *(i)* divide the graph into communities using structural clustering method. After clustering, we consider each community as a super vertex, and the connections between two super vertices as a super edge. Besides, *(ii)* for $u, v \in int(sv)$, $sv \in \$V$, we compute the shortcuts between u and v, *(iii)* for $sv_i, sv_j \in \$V$, we estimate $\ell(sv_i, sv_j)$, and *(iv)* for each $sv \in \$V$, attach labels to sv. Next, we will show our implementation methods in detail.

Structural Clustering Method for Weighted Graph: wSCAN

pSCAN [1] is a state-of-the-art graph clustering method, which is based on the idea that vertices in the same community are more structural similar than the rest of the graph. For each vertex v adjacent to u, they compute the structural similarity $\sigma(u, v)$ between u and v in Eq. 1.

$$\sigma(u, v) = \frac{|N[u] \cap N[v]|}{\sqrt{d[u] \cdot d[v]}} \tag{1}$$

where $N[u]$ is the structural neighborhood of u, $N[u] = \{v \in V | (u, v) \in E\}$, and $d[u]$ is the degree of u, $d[u] = |N[u]|$. There shows a weighted graph in Fig. 2, the number on each edge marks its weight. For vertex v_2, $N[v_2] = \{v4, v5, v6\}$, $d[v_2]=3$. $\sigma(v_1, v_2)=3/\sqrt{3*6}=0.71$. Similarly, $\sigma(v_1, v_3) = 0.71$.

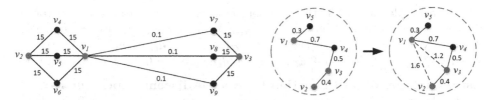

Fig. 2. An example weighted graph   **Fig. 3.** Estimate $d_G(v_1, v_2)$, $d_G(v_1, v_3)$

Apparently, pSCAN does not consider edges' weights when calculating $\sigma(u, v)$. In a weighted graph, for a common neighbor $w \in N[u] \cap N[v]$, the weights between w and u, v are denoted by $\omega(u, w)$, $\omega(v, w)$, respectively. The larger value of $\omega(u, w)$ and $\omega(v, w)$, the higher $\sigma(u, v)$; the less difference value between $\omega(u, w)$ and $\omega(v, w)$, the higher $\sigma(u, v)$. In summary, if there are many common neighbors between u and v, which are closely connect to both u and v, then u, v have a great probability to be in the same community. Hence we propose a new method wSCAN based on pSCAN, in which we modify the formula for calculating the structural similarity between two vertices, as shown in Eq. 2.

$$\sigma(u, v) = \frac{\sum_{w \in N[u] \cap N[v]}((\omega(u, w) + \omega(v, w)) \cdot \phi_w(u, v))}{\partial[u] + \partial[v]} \tag{2}$$

$$\phi_w(u, v) = 1 - |\frac{\omega(u, w) - \omega(v, w)}{\omega(u, w) + \omega(v, w)}| \tag{3}$$

where $\phi_w(u, v)$ evaluates the different between $\omega(u, w)$ and $\omega(v, w)$, as shown in Eq. 3. The closer w is to the middle of two vertices, the larger $\phi_w(u, v)$. $\partial[u]$ is the sum of the weights of edges between u and its neighbors, $\partial[u] = \{\sum \omega(u, v) | v \in N[u]\}$. For each $w \in N[u] \cap N[v]$, $\sigma(u, v)$ takes into account the value of the reciprocity and weight, and is normalized at last. When we use wSCAN to compute the structural similarity in Fig. 2, $\sigma(v_1, v_2)= (30+30+30)/(45+45.3)=0.997$, $\sigma(v_1, v_3)=(0.2+0.2+0.2)/(45+45.3)= 0.007$, obviously, v_1 and v_2 are more likely to be in same community than v_1 and v_3. After clustering, we convert the weight function $\omega(e)$ into a length function $\ell(e)$, to further process the subsequent analysis. Note that a larger $\omega(u, v)$ means a closer connection between u and v, resulting in a less distance between u and v, as indicated by $\ell(e)$ in Table 1.

Estimation for All Pairs of Interface Vertices: Shortcuts

For all $sv \in \$V$, the time complexity to compute the exact SPs between all interface vertices within sv is $O(\hat{n}n_{con}^2 n_{int})$, which is very expensive. Besides, the difference among edges within a community is not much. So given $s, t \in int(S)$, we expand from s, t to neighbors until the intersection, as Eq. 4. Since we do not update the SP based on the newly added shortcuts, we can quickly return the estimation in $O(\hat{n}n_{con}n_{int})$.

$$d_G(s,t) = \begin{cases} d_G(s,u) + d_G(u,v) + d_G(v,t), & u \in N[s], v \in N[t] \\ d_G(s,u) + d_G(u,t), & u \in N[s] \cap N[t] \end{cases} \quad (4)$$

For example, in Fig. 3, the number on each edge indicates the distance between two vertices. $N[v_1] = \{v_4, v_5\}$, $N[v_2] = \{v_3\}$, $N[v_3] = \{v_2, v_4\}$, $N[v_1] \cap N[v_3] = \{v_4\}$, so $d_G(v_1, v_3) = d_G(v_1, v_4) + d_G(v_3, v_4) = 0.7 + 0.5 = 1.2$, $d_G(v_1, v_2) = d_G(v_1, v_4) + d_G(v_3, v_4) + d_G(v_2, v_3) = 0.7 + 0.5 + 0.4 = 1.6$.

Estimation for Length Function Between Adjacent Super Vertices

The length function of super edge directly impacts $sp_{SG}(sv_s, sv_t)$, from where we search $sp_G(s, t)$. A good estimation of $\ell(sv_s, sv_t)$ should reflect the estimated distance between any two vertices in sv_s and sv_t, thereby improve the result's precision. We propose several length functions as below.

- SHORTEST: Let $\ell(sv_s, sv_t) = d_G^*(e) \le d_G(e)$, for edges $e \in con(sv_s, sv_t)$.
- LONGEST: Let $\ell(sv_s, sv_t) = d_G^*(e) \ge d_G(e)$, for edges $e \in con(sv_s, sv_t)$.
- CENTRAL: The above methods do not consider the distance inside the community. Therefore, we think $\ell(sv_s, sv_t)$ can be approximated as the average distance from internal vertices to their interface vertices, respectively, plus the average distance between communities' interface vertices. Furthermore, in order to simplify the process, we select a representative central vertex from each set of interface vertices — landmark. In this paper, we simply use a landmark to replace the interface vertices while calculating. Finally, CENTRAL calculates the length function in Eq. 5:

$$\ell(sv_s, sv_t) = avg(d_G(s, l_{sv_s, sv_t}) + d_G(l_{sv_s, sv_t}, l_{sv_t, sv_s}) + avg(d_G(t, l_{sv_t, sv_s}) \quad (5)$$

$$C_B(u) = \sum_{s,t,u \in V} \frac{\eta_{st}(u)}{\eta_{st}} \quad (6)$$

where $s \in con(sv_s)$ and $s \notin out(sv_s)$, $t \in con(sv_t)$ and $t \notin out(sv_t)$. l_{sv_s, sv_t} is the vertex with the highest betweenness centrality in $int(sv_s, sv_t)$, and has not been chose as a landmark before. The betweenness centrality of the vertex u is defined as $C_B(u)$ [17], where η_{st} denotes the number of SPs from s to t, and $\eta_{st}(u)$ denotes the number of SPs from s to t that u lies on. A higher $C_B(u)$ indicates more SPs pass through u.

Attach to Each Super Vertex: Two Labels

Reachability label $L_{re}(sv)$: Given two vertices s and t, a reachability query asks whether there exists a path between s and t in $\mathbb{G}$. We can judge the reachability between sv_s and sv_t instead, because wSCAN can ensure that vertices are reachable to each other within the community. Therefore, we perform the Breadth-First-Search on $\$G$, and attach $L_{re}(sv_i) = C_i$ to sv_i in the closure C_i. At query time, if there exists $L_{re}(sv_s) = L_{re}(sv_t)$, then s and t can reach each other. For example, in Fig. 4, $L_{re}(sv_1) = L_{re}(sv_2) = \ldots = L_{re}(sv_4) = C_1$, $L_{re}(sv_5) = L_{re}(sv_6) = \ldots = L_{re}(sv_9) = C_2$, so the vertices in sv_1 can reach vertices in $\{sv_1, sv_2, sv_3, sv_4\}$, but cannot reach vertices in $\{sv_5, sv_6, sv_7, sv_8, sv_9\}$.

Shortest path label $L_{sp}(sv)$: According to six degrees of separation, any two vertices can establish a contact within six hops. Thus, for each super vertex $sv \in \$G$, we only calculate the SPs between sv and the neighbors within three hops, then the join of two super vertices can cover the SPs between any pairs of vertices inside them. The SP from sv_s to sv_t is denoted by $L_{sp}(sv_s, sv_t)$. At query time, we can find the SP between any two super vertices in $O(1)$ as Eq. 7. For example, in Fig. 5, $L_{sp}(sv_1, sv_9) = \{9, < sv_1, sv_3, sv_4, sv_9 >\}$, $L_{sp}(sv_5, sv_9)$ $= \{5, < sv_5, sv_6, sv_8, sv_9 >\}$, $L_{sp}(sv_1) \bigcap L_{sp}(sv_5) = \{sv_9\}$, $sp_{SG}(sv_1, sv_5) =$ $sp_{SG}(sv_1, sv_9) + sp_{SG}(sv_5, sv_9) = < sv_1, sv_3, sv_4, sv_9, sv_8, sv_6, sv_5 >$.

$$d_G(sv_s, sv_t) = \min_{sv_i \in L_{sp}(sv_s) \bigcap L_{sp}(sv_t)} \{d_G(sv_s, sv_i) + d_G(sv_t, sv_i)\} \qquad (7)$$

Quick Response to Graph Updates

Social networks update very fast, corresponding to the insertion/deletion of vertices and edges in the social graphs. Instead of performing the preprocessing step all over again, we can quickly adjust the preprocessing results against the update. For insert operation, given a new vertex u and its new edges $\in G$, we: *(i)* let n^i denote the number of vertices whose structure is similar to u in community sv_i. If $n^i \geq \mu$, add u to $contain(sv_i)$, and add u to $int(sv_i, sv_j)$ if u directly connects to a vertex in sv_j; *(ii)* update shortcuts within community sv_i according to u; *(iii)* if $u \in int(sv_i, sv_j)$, and $v = l_{sv_s, sv_t}$, let η_u, η_v denote the number of the shortcuts which u, v lie on, respectively. If $\eta_u > \eta_v$, let u replace v and be the new l_{sv_s, sv_t}; *(iv)* recompute $\ell(sv_i, sv_j)$ according to u.

For the vertex u need to be deleted, there are the following adjustments: *(i)* for each super vertex $sv \in bel(u)$, remove u from $con(sv)$ and $int(sv)$; *(ii)* for each vertex $v \in N[u]$, remove the edge (u, v) from E, remove u from $N[v]$ and check whether the role of v is affected; *(iii)* remove the shortcuts which u lies on; *(iv)* if $u = l_{sv_s, sv_t}$, reselect the landmark and recompute $\ell(sv_i, sv_j)$.

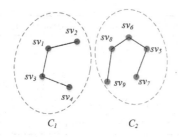

C_1 C_2

Fig. 4. Reachability labels

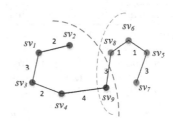

Fig. 5. 3-hops Shortest path labels

B. Online Querying Phase

In Algorithm 1 we describe the online query method. Given two vertices $s, t \in \mathbb{G}$, Set_s and Set_t are the set of super vertices that contain s and t respectively (line 1). We each take one from Set_s and Set_t in turn, are denoted by sv_1 and sv_2 (line 2). There are two situations: *(i)* if s and t are in the same community, then we search $sp_G(s,t)$ within $sv_1(sv_2)$ (line 3–4); *(ii)* if s and t are in different communities, we verify if there exists a path between s and t by using reachability labels, then we add them to the set of candidates if the answer is *true* (line 5–6). Next, we enumerate each pair of candidates $\{sv_1, sv_2\}$ from Set_{con}, and seek $sp_{SG}(sv_s, sv_t)$ with the minimum cost using shortest path labels (line 7–10). Finally, we search $sp_G(s,t)$ based on the vertices in $sp_{SG}(sv_s, sv_t)$ (line 11).

Specifically, for s,t in the same community, we use a modified bidirectional search when finding $sp_G(s,t)$. For each vertex u in the priority queue, we use the minimum sum of $d_G(u, l_i)$ and $d_G(t, l_i)$ as the estimation of $d_G(u,t)$ ($l_i \in L_S =< l_1, l_2, \ldots, l_x >$). Then, instead of ordering vertices by their distance from s, vertices are ordered by their distance from the s plus this estimation. As a result, we can direct the search towards the target and save unnecessary computations.

Algorithm 1. SPBOC

Input: Original graph $\mathbb{G} = (\mathbb{V}, \mathbb{E})$, super graph $\mathbb{SG} = (\mathbb{SV}, \mathbb{SE})$, $s, t \in \mathbb{G}$
Output: $sp_G(s,t)$

1 $Set_s \leftarrow belong(s)$, $Set_t \leftarrow belong(t)$, $Set_{con} \leftarrow \emptyset$;
2 **for** *each $sv_1 \in Set_s$, $sv_2 \in Set_t$* **do**
3   **if** $sv_1 = sv_2$ **then**
4   return $sp_G(s,t) \leftarrow$ use bidirectional Dijkstra algorithm with landmarks;
5   **else if** $L_{re}(sv_1) = L_{re}(sv_2)$ **then**
6           $Set_{con} \leftarrow \{sv_1, sv_2\}$;

7 $minCost \leftarrow \infty$
8 **for** *each $sv_1, sv_2 \in Set_{con}$* **do**
9   **if** $(d_{SG}(sv_1, sv_2) \leftarrow \min\{L_{sp}(sv_1) \bigcap L_{sp}(sv_2)\}) < minCost$ **then**
10          $minCost \leftarrow d_G(sv_1, sv_2)$, $sp_{SG}(sv_s, sv_t) \leftarrow sp_{SG}(sv_1, sv_2)$;

11 return $sp_G(s,t) \leftarrow$ FindShortestPathBetweenCommunities$(s,t,sp_{SG}(sv_s,sv_t))$;

Algorithm 2. FindShortestPathBetweenCommunities$(s,t,sp_{SG}(sv_s,sv_t))$

Input: $s,t,sp_{SG}(sv_s, sv_t)$
Output: $sp_G(s,t)$

1 $SPTillNow \leftarrow$ shortest path from s to $VSet_0$
2 **for** $i=1$; $i<2*(sp_{SG}(sv_s, sv_t).size-2)$; $i++$ **do**
3       $SPTillNow \leftarrow$ CalculateNeighbor$(VSet_i, SPTillNow)$;
4 return $sp_G(s,t) \leftarrow$ Calculate shortest path between $SPTillNow$ and t;

Next, we will focus on the situation when two vertices are in different communities. In order to clarify, we will explain the algorithm in Fig. 6. The vertices in Fig. 6 are all interface vertices except for s and t. According to Observation 1, $d_G(s, t)$ can be indicated as the minimum combination of three phases: (i) the distance from s to interface vertices: $d_G(s, int(sv_s, sv_1))$, (ii) the distance from t to interface vertices: $d_G(int(sv_t, sv_2), t))$, (iii) the distance between interface vertices: $d_G(int(sv_s, sv_1), int(sv_t, sv_2))$. The first two phrases can be transformed to the SP problems within the community. Thus we only focus on the SP between interfaces vertices on $sp_{SG}(sv_1, sv_2)$.

In Fig. 6, $sp_{SG}(sv_s, sv_t) = < sv_s, sv_1, sv_2, sv_t >$. We use $VSet$ to record the collection of all interface vertices sets, such as $VSet_0 = int(sv_s, sv_1) = \{s, v_1\}$, $VSet_1 = int(sv_1, sv_s) = \{v_2, v_3, v_4\}$,....,$VSet_5 = int(sv_t, sv_2) = \{v_{11}\}$. Since we have already estimated the shortcuts between interface vertices within a community, we divide the search processing into parts and progressively calculate the SP from s to the vertices in $VSet_i$ in i-increasing order. For each vertex $v \in VSet_i$, $u \in VSet_{i-1}$, $sp_G(s, v) = \min\{ sp_G(s, u) + sp_G(u, v) \}$. Suppose the number of super vertices on $sp_{SG}(sv_1, sv_2)$ is c, we can get the SP till $VSet_{2c-1}$ in $O(cn_{int})$ for simple sum and compare operations among interface vertices.

Therefore, Algorithm 2 starts from s and calculates the SP till all vertices in $VSet_0$ (line 1). Then, for each interface vertices set $VSet_i \in VSet$, we compute the SP till $VSet_i$, and record it in $SPTillNow$ (line 2–3). Finally, $SPTillNow$ stores the SP from s to interface vertices that sv_t. The problem transforms to a SP problem within the community (line 4). We describe in Algorithm 3 about how to calculate SP till vertices in $VSet_{i+1}$ based on $SPTillNow$.

$SPTillNow$ records the SPs till the vertices in $VSet_{i-1}$ ($i \geq 1$). For each vertex $v \in VSet_i$, we maintain a $minCost$ and a $minPath$ to record the current shortest distance and SP from vertices in $VSet_{i-1}$ to v (line 1–2). If the sum of $d_G(s, u)$ and $d_G(u, v)$ is smaller than $minCost$, then we replace $minCost$ with $d_G(s, u)$ plus $d_G(u, v)$, and also update $minPath$ with the corresponding path (line 3–5). Finally, we add a new SP record about v to $SPNew$ (line 6).

Algorithm 3. CalculateNeighbor($VSet_i, SPTillNow$)

Input: $VSet(i), SPTillNow$
Output: $SPNew$
1 **for** *each* $v \in VSet(i)$ **do**
2      $minCost \leftarrow \infty$, $minPath \leftarrow null$
3   **for** *each* $u \in SPTillNow$ **do**
4   **if** $d_G(s, u) + d_G(u, v) < minCost$ **then**
5             $minCost \leftarrow d_G(s, u) + d_G(u, v); minPath \leftarrow sp_G(s, u) + sp_G(u, v);$
6   add $< v, minPath : minCost >$ to $SPNew$;
7 **return** $SPNew$;

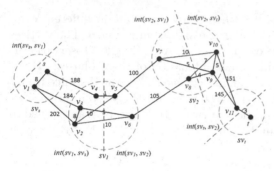

Fig. 6. Finding SP between s and t

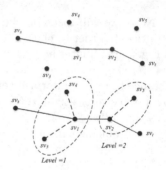

Fig. 7. 1-hop expansion

C. Complexity Analysis

The time complexity of preprocessing is $O(m^{1.5} + \hat{n}n_{con}n_{int} + \hat{n}\hat{m})$. Here, $O(m^{1.5})$ is related to clustering the graph using wSCAN. $O(\hat{n}n_{con}n_{int})$ is for estimating shortcuts within a community. And $O(\hat{n}\hat{m})$ is for computing labels for super vertices. We need extra $O(mn + \hat{m}n_{con}^2/4)$ if using the CENTRAL method to estimate the length function for super edges. The time complexity of online querying is $O(n_{con}logn_{con})$. The space complexity of index is $O(\hat{m} + \hat{n}^2)$ for preserving the super graph and labels.

5   Optimization Techniques

In this section, in order to improve the precision and the speed of querying, we propose the following three optimization techniques.

A. Expand the SP tree

According to six degrees of separation of social networks, the average distance between vertices is usually very small. Thus if we expand the SP in SG to include the neighbors within one hop for each super vertex, we can improve the precision of result. Next, we will explain this process in Fig. 7, where $level(sv_i)$ indicates the number of steps from the source. For example, $level(sv_s)=0$, $level(sv_1)=1$, $level(sv_2)=2$, and $level(sv_3)=3$. For each super vertex $sv_i \in sp_{SG}(sv_s, sv_t)$ except for the both ends, we execute 1-hop expansion and add the neighbors to the same level with sv_i. After that, $level(sv_s)=0$, $level(sv_1)=level(sv_3)=level(sv_4)=1$, $level(sv_2) = level(sv_5)=2$, $level(sv_t)=3$. For all super vertices in the same level, we regard the whole as a new super vertex.

B. Community Size Balancing

The communities after clustering may not be satisfying: some contain only one vertex and some contain too many vertices. Consider two extreme situations: (i) each vertex is a super vertex, (ii) all vertices belong to a very large super vertex. In both cases, our approach is invalid and is equivalent to the traditional Dijkstra. Thus, we need to avoid isolated and oversized communities: (i) for an isolated community sv, where there is only one vertex $v \in sv$, we add v to the neighbor's community whose structure most similar to v, *(ii)* for the oversized community sv, we use re-cluster the vertices in sv, and divide sv into several sub-communities according to the closeness between vertices, so as to avoid excessive number of vertices in each community.

C. Prune during SP Query

We propose an optimization technique to prune some vertices by predicting the distance towards the target, so as to reduce the analysis of many vertices that may not be on the SP and accelerate online query.

Lemma 1. *Given $s,t,u \in G$, sv_s, sv_t, sv_u are the super vertices that contain s, t and u, respectively. Let LD(sv_u, sv_t) and SD(sv_u, sv_t) denote the estimate distance between sv_t and sv_u using LONGEST and SHORTEST. For $u, v \in sv_u$, we prune u if there exists $d_G(s, u) - d_G(s, v) > $ LD(sv_u, sv_t) $-$ SD(sv_u, sv_t).*

Proof. First of all, if $d_G(s, u) + d_G(s, u) > d_G(v, t) + d_G(s, v)$, then u is definitely not on $sp_G(s, t)$. Instead of compute the real distance from u, v to t, we use a simple replacement. There are multiple paths from sv_u to sv_t, if u uses a shortest one and still longer than v use a longest one, then u cannot be on the shortest path. We use SD(sv_u, sv_t)/LD(sv_u, sv_t) to indicate the longest/shortest one, so $d_G(s, u) - d_G(s, v) >$ LD(sv_u, sv_t)$-$SD(sv_u, sv_t) $\geq d_G(v, t) - d_G(u, t) \Leftrightarrow d_G(s, u) - d_G(s, v) >$ LD(sv_u, sv_t) $-$ SD(sv_u, sv_t).

6   Experiment

We try to evaluate the following aspects through experiments: the tradeoff among preprocessing time, querying time, index space and accuracy, and the effect of our optimization methods. We ran all experiments on a computer with an Intel 1.9GHz CPU, 64GB RAM, and Linux OS. We evaluate the performance of algorithms on both real and synthetic graphs as shown in Table 2. First four of them lists the real-world datasets which can be found at the Stanford Network Analysis Platform[1] and DBLP[2]. Enron and DBLP are weighted graphs, others are unweighted graphs. We also evaluate the algorithms on LFR benchmark graphs [18] which can automatically generate undirected weighted graphs. We vary the size of graphs and the clustering coefficient $\bar{c}$ to meet out demands.

Eval-I: Compare wSCAN with pSCAN and SLPA

We compare our wSCAN algorithm with the pSCAN [1] and SLPA [19], and evaluate the communities quality after graph clustering. Modularity [20] of a community network is a measure of how well a community network is divided, denoted by Q. The larger the Q, the better the cluster method. Its ranges is (0,1), and the calculation method for weighted graphs is defined as follows:

$$Q = \frac{1}{2m} \sum_{u,v} (\omega(u,v) - \frac{d[u] \cdot d[v]}{2m}) \delta(u,v) \tag{8}$$

---

[1]   http://snap.stanford.edu/.
[2]   http://dblp.dagstuhl.de/xml/.

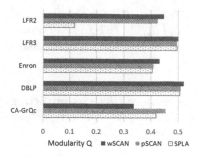

Fig. 8. (Eval-I) Q after clustering

Table 2. Statistics of graphs ($\bar{d}$: average degree, $\bar{c}$: clustering coefficient)

Graph	$|V|$	$|E|$	$\bar{d}$	$\bar{c}$
CA-GrQc	5,242	14,496	6.46	0.530
Enron	33,692	183,831	10.91	0.497
EnAll	265,214	420,045	1.58	0.067
DBLP	1,482,029	10,615,809	7.16	0.561
LFR1	1,000	77,80	15.56	0.752
LFR2	10,000	77,330	15.47	0.169
LFR3	10,000	75,262	15.05	0.754
LFR4	100,000	468,581	9.371	0.745
LFR5	500,000	2,241,850	9.406	0.725

where $\delta(u, v)$ is 1 when vertices u and v belong to the same community, otherwise it equals to 0. The result can be seen in Fig. 8. wSCAN performs better on weighted graph, but not suitable for unweighted graphs such as CA-GrQc.

Eval-II: Evaluate the Effect of Optimization Techniques

In Fig. 9, we evaluate the effect of optimization A by comparing the error rate as Eq. 9, where $\hat{d}_i$ is the estimated shortest distance and d_i is the shortest distance computed by Dijkstra. And in Fig. 10, we evaluate the effect of optimizations B, C by comparing the online processing time. Experiments carry out on four datasets with each N pairs of vertices ($N=500$). In specific, we evaluate the following algorithms:

- SPBOC*: the approach discussed in Sect. 4 (using CENTRAL).
- SPBOC-A: the SPBOC* approach with the optimization technique A.
- SPBOC-B: the SPBOC* approach with the optimization technique B.
- SPBOC-C: the SPBOC* approach with the optimization technique C.
- SPBOC: the SPBOC* approach with all optimization techniques.

$$appr = (\sum_{i=1}^{N} \frac{\hat{d}_i - d_i}{d_i})/N \qquad (9)$$

In Fig. 9, it can be seen that error rate decreases significantly with SPBOC-A because of 1-hop expansion. In Fig. 10, the querying time of SPBOC* is several times larger than SPBOC-B, because that the number of isolated communities and the size of oversized communities are significantly reduced after the adjustment. Besides, the queries can be further accelerated with optimization technique C as a result of pruning useless vertices. The combination of all optimization techniques yields a powerful method — SPBOC, whose processing time is orders of magnitude faster than the approach without optimizations. To sum up, the optimization techniques can improve the query performance.

Eval-III: Compare SPBOC with Other SP Algorithms

In this set of experiments, we evaluate the performance on preprocessing time, querying time, index space as well as the error rate. In particular, SPBOC[1] and

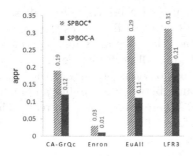

Fig. 9. (Eval-II) Evaluate optimization A

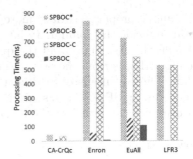

Fig. 10. (Eval-II) Optimizations B, C

	CA-GrQc				Enron				EuAll		
Method	QT(ms)	appr	PT(ms)	DiskSpace	Method	QT(ms)	appr	PT(ms)	DiskSpace	Method	QT(ms)	appr	PT(s)	DiskSpace

Method	QT(ms)	appr	PT(ms)	DiskSpace
SPBOC[1]	2.31	0.120	761	32 KB
SPBOC[2]	2.31	0.119	4,542	32KB
ALT	41.19	0	10,895	1.03 MB
REAL	35.34	0	53,466	1.58 MB
LLS	0.79	0.048	7,078	6.78 MB
SPCD	2.47	0.161	4,229	7 KB

Method	QT(ms)	appr	PT(ms)	DiskSpace
SPBOC[1]	7.11	0.078	174,643	9.3 MB
SPBOC[2]	7.09	0.010	258,747	9.5 MB
ALT	447.13	0	233,890	3.81 MB
REAL	250.63	0	909,832	11.8 MB
LLS	0.69	0.017	222,030	35.9 MB
SPCD	915.07	0.001	3,216	66 KB

Method	QT(ms)	appr	PT(s)	DiskSpace
SPBOC[1]	107.9	0.142	1,827	316 MB
SPBOC[2]	109.4	0.108	98,796	318 MB
ALT	4,161	0	119,879	78.2 MB
REAL	-	-	>2 days	-
LLS	32.4	0.075	77,760	1.58 GB
SPCD	569,005	0.021	11	2.08 MB

	DBLP				LFR1				LFR2		

Method	QT(s)	appr	PT(s)	DiskSpace
SPBOC[1]	1.3	0.120	62,225	99.5 MB
SPBOC[2]	1.3	0.072	564,120	99.5 MB
ALT	-	-	-	-
REAL	-	-	-	-
LLS	0.9	0.085	429,459	8.16 GB
SPCD	1,066	0.046	63	24.04 MB

Method	QT(ms)	appr	PT(ms)	DiskSpace
SPBOC[1]	0.78	0.159	281	15 KB
SPBOC[2]	0.80	0.238	1101	15 KB
ALT	3.84	0	872	242 KB
REAL	2.84	0	3,693	308 KB
LLS	0.32	0.015	1,340	2.88 MB
SPCD	1.32	0.271	334	5 KB

Method	QT(ms)	appr	PT(ms)	DiskSpace
SPBOC[1]	0.14	0.459	654	977 KB
SPBOC[2]	0.11	0.511	84,101	977 KB
ALT	426.14	0	130,540	2.23 MB
REAL	363.45	0	605,233	2.58 MB
LLS	0.97	0.030	187,610	33.8 MB
SPCD	1010.35	0.068	2,242	72 KB

	LFR3				LFR4				LFR5		

Method	QT(ms)	appr	PT(ms)	DiskSpace
SPBOC[1]	0.17	0.209	624	977 KB
SPBOC[2]	0.16	0.311	81,811	977 KB
ALT	302.19	0	132,719	2.23 MB
REAL	282.87	0	543,778	2.58 MB
LLS	0.57	0.030	124,670	33.8 MB
SPCD	2.47	0.068	2,522	72 KB

Method	QT(ms)	appr	PT(s)	DiskSpace
SPBOC[1]	1.01	0.097	7	7.49 MB
SPBOC[2]	1.03	0.198	1,856	7.50 MB
ALT	1008.45	0	3939	21.38 MB
REAL	778.54	0	18,513	25.46 MB
LLS	3.07	0.020	1706	1.1GB
SPCD	21.12	0.019	30	1.02 MB

Method	QT(ms)	appr	PT(s)	DiskSpace
SPBOC[1]	5.36	0.075	39	43.0 MB
SPBOC[2]	5.45	0.105	9,9191	42.9 MB
ALT	-	-	-	-
REAL	-	-	-	-
LLS	7.07	0.005	100,224	6.92 GB
SPCD	120.54	0.009	190	7.13 MB

Fig. 11. (Eval-III) Compare overall performance

SPBOC[2] use SHORTEST and CENTRAL methods in estimating length between super vertices, respectively. And we compare them with two-stage methods: ALT [6], REAL [13], LLS [8] and SPCD [2] by querying SP on four datasets with each N pairs of vertices (N=500). SPCD tries to find $sp_G(s,t)$ among the Top$K$ SPs in \$G. In this paper, we compare the SPCD method with $K = 1$. Among them, ALT, REAL are for exact SP and LLS, SPCD are for approximate SP.

In Fig. 11, QT/PT is short for querying time/preprocessing time. It can be seen that the error rate with SPBOC[1] is lower than SPBOC[2] on synthetic graphs, and has the reverse effect on real social networks. This is because the $\bar{c}$ of these synthetic graphs is very high, and graphs with high $\bar{c}$ can reveal obvious small-world property. However, the real datasets often fail to achieve such strong community property, so it is more suitable to use CENTRAL which takes the distance inside the community into account. The Fig. 12 illustrates the tradeoff between the disk space and the query time on a logarithmic scale. The closer the algorithm is from the origin, the better the overall performance of the algorithm. The advantage of SPBOC in EuAll is not obvious because of the low $\bar{c}$. In general, (i) SPBOC can strike the best balance between scalability and query performance among all methods, (ii) CENTRAL are more suited to real social networks than SHORTEST, (iii) SPBOC performs better on the graphs which show a strong community property than other graphs.

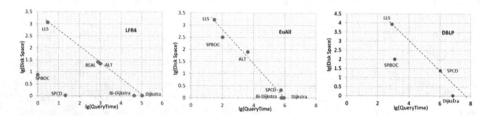

Fig. 12. (Eval-III) Tradeoff between querying time and disk space

7   Conclusion

In this paper, we developed a new SP algorithm for social network based on community structure. We proposed a new structural clustering method for weighted social graph. We made a super graph based on the community structure of the original graph so as to narrow down the scale of searching. To improve the performance of our approach, we further proposed three optimization techniques to improve the query performance. Experiments show that our approach can strike the balance between scalability and query performance, and return an approximate shortest path with allowed accuracy in very short time.

Acknowledgments. This work was supported in part by the National Nature Science Foundation of China under the grants 61702285 and 61772289, the Natural Science Foundation of Tianjin under the grants 17JCQNJC00200, and the Fundamental Research Funds for the Central Universities under the grants 63181317.

References

1. Chang, L., Li, W.: pSCAN: Fast and exact structural graph clustering. ICDE **29**(2), 253–264 (2016)
2. Gong, M., Li, G.: An efficient shortest path approach for social networks based on community structure. CAAI **1**(1), 114–123 (2016)
3. Dijkstra, E.W.: A note on two problems in connexion with graphs. Numer. Math. **1**, 269–271 (1959)
4. Pohl, I.S.: Bi-directional search. Mach. Intell. **6**, 127–140 (1971)

5. Sommer, C.: Shortest-path queries in static networks. ACM Comput. Surv. **46**(4), 1–31 (2014)
6. Goldberg, A.V., Harrelson, C.: Computing the shortest path: A* search meets graph theory. In: 16th SODA, pp. 156–165 (2005)
7. Akiba, T., Sommer, C.: Shortest-path queries for complex networks: exploiting low tree-width outside the core. In: EDBT, pp. 144–155 (2012)
8. Qiao, M., Cheng, H.: Approximate shortest distance computing: a query-dependent local landmark scheme. In: 28th ICDE, pp. 462–473 (2012)
9. Tretyakov, K.: Fast fully dynamic landmark-based estimation of shortest path distances in very large graphs. In: 20th CIKM, pp. 1785–1794 (2012)
10. Cohen, E., Halperin, E.: Reachability and distance queries via 2-hop labels. SIAM J. Comput. **22**, 1338–1355 (2003)
11. Jiang, M.: Hop doubling label indexing for point-to-point distance querying on scale-free networks. PVLDB **7**, 1203–1214 (2014)
12. Akiba, T., Iwata, Y.: Fast exact shortest-path distance queries on large networks by pruned landmark labeling. In: SIGMOD, pp. 349–360 (2013)
13. Goldberg, A.V., Kaplan, H.: Reach for A* shortest path algorithms with preprocessing. In: 9th DIMACS Implementation Challenge, vol. 74, pp. 93–139 (2009)
14. Delling, D., Goldberg, A.V., Werneck, R.F.: Hub label compression. In: Bonifaci, V., Demetrescu, C., Marchetti-Spaccamela, A. (eds.) SEA 2013. LNCS, vol. 7933, pp. 18–29. Springer, Heidelberg (2013). https://doi.org/10.1007/978-3-642-38527-8_4
15. Chechik, S.: Approximate distance oracle with constant query time. arXiv abs/1305.3314 (2013)
16. Chen, W.: A compact routing scheme and approximate distance oracle for power-law graphs. ACM Trans. Algorithms **9**, 349–360 (2012)
17. Potamias, M., Bonchi, F.: Fast shortest path distance estimation in large networks. In: CIKM, pp. 867–876 (2009)
18. Andrea Lancichinetti, A., Fortunato, S.: Benchmarks for testing community detection algorithms on directed and weighted graphs with overlapping communities. Phys. Rev. E **80**, 016118 (2009)
19. Xie, J.: SLPA: uncovering overlapping communities in social networks via a speaker-listener interaction dynamic process. In: ICDMW, pp. 344–349 (2012)
20. Newman, M.E.: Finding and evaluating community structure in networks. Phys. Rev. E **69**(2), 026113 (2004)
21. Fu, A.W.C., Wu, H.: IS-LABEL: an independent-set based labeling scheme for point-to-point distance querying on large graphs. VLDB **6**(6), 457–468 (2013)
22. Hayashi, T., Akiba, T., Kawarabayashi, K.I.: Fully dynamic shortest-path distance query acceleration on massive networks. In: CIKM, pp. 1533–1542 (2016)

On Link Stability Detection for Online Social Networks

Ji Zhang[1(✉)], Xiaohui Tao[1], Leonard Tan[1], Jerry Chun-Wei Lin[2(✉)],
Hongzhou Li[3], and Liang Chang[4]

[1] Faculty of Engineering and Sciences, The University of Southern Queensland,
Toowoomba, Australia
{Ji.Zhang,Xiaohui.Tao,Leonard.Tan}@usq.edu.au
[2] Harbin Institute of Technology Shenzhen Graduate School, Shenzhen, China
jerrylin@ieee.org
[3] School of Life and Environmental Science, Guilin University
of Electronic Technology, Guilin, China
[4] Guangxi Key Laboratory of Trusted Software, Guilin University
of Electronic Technology, Guilin, China

Abstract. Link stability detection has been an important and long-standing problem within the link prediction domain. However, it has often been overlooked as being trivial and has not been adequately dealt with in link prediction. In this paper, we present an innovative method: Multi-Variate Vector Autoregression (MVVA) analysis to determine link stability. Our method adopts link dynamics to establish stability confidence scores within a clique sized model structure observed over a period of 30 days. Our method also improves detection accuracy and representation of stable links through a user-friendly interactive interface. In addition, a good accuracy to performance trade-off in our method is achieved through the use of Random Walk Monte Carlo estimates. Experiments with Facebook datasets reveal that our method performs better than traditional univariate methods for stability identification in online social networks.

Keywords: Link stability · Graph theory · Online social networks
Hamiltonian Monte Carlo (HMC)

1 Introduction

The far reaching social media today contains a rich set of problems that are relationally focused. Some of which include but are not limited to: Exponentially increasing data privacy intrusions on a yearly trend [29]; Rising number of internet suicides from online depression [27,29]; Account poisoning and hacking [26,29]; Terrorism and security breaches [26,29]; Information warfare and cyber attacks [29].

From a structural viewpoint, popular networks like Google, Facebook, Twitter, Youtube, etc. are often used as social and affective means to express

© Springer Nature Switzerland AG 2018
S. Hartmann et al. (Eds.): DEXA 2018, LNCS 11029, pp. 320–335, 2018.
https://doi.org/10.1007/978-3-319-98809-2_20

exchanges and dominance of evolving human ties [26]. This is often done through rich expanses of emotional and sentimental fidelities which fluctuate over topic drifts [26]. Stable links are defined as relations (both benevolent and malevolent) where emotional flux remains relatively high through social evolution [28,29].

Detecting stable links within online social networks is important in many real-life applications. For example, stable links can specifically be applied to analyze and solve interesting problems like detecting a disease outbreak within a community, controlling privacy in networks, detecting fraud and outliers, identifying spam in emails, etc [14]. Identifying stable relations within a social circle as structural pillars of a community is also very important in abating cyber attacks from occuring.

Link stability is a specific problem of link prediction that has been oftentimes overlooked as trivial. Although it shares the same set of domain challenges as link prediction, it does not predict future relations that may occur due to inferences from present observations. Instead, it ranks links shared between actors according to their structural importance to a community by their stability index scores.

There are several major limitations in the study of link stability in literature. First, many existing detection methods use the static node mechanism which fails to consider the intrinsic feature dynamics in the detection process. Additionally, most approaches are tailored to the use of a specific network in question and are not adaptable to more generalized social platforms. Furthermore, stable link identification is a largely unexplored area of research development without a structured framework of approach. This paper will make scientific contributions to enhance the current detection capabilities of stable links to preserve structural integrity within a community and safeguard against detrimental effects of harmful, unstable external social influences.

In this paper, we will present our MVVA (Multi-Variate Vector Autoregression) model for link stability detection, which is developed to encompass the multi-variate feature aspects of links in a single regression model. Its objective function bridges the gap between temporality and stability metrics. The scientific contribution of our work involves the following:

1. Our method bridges the gap between temporality and stability of links in online social networks. As an improvement to conventional static node and neighbor link occurrence methods, our approach is able to handle dynamic link features efficiently in the "prediction" process;
2. An innovative Hamiltonian Monte Carlo estimator is developed to help the MVVA model scale up to increasing dimensionality as the data volume grows arbitrarily large;
3. Experiment results show that the MVVA is able to offer a good modeling of the ground truth growth distribution of stable links within a Facebook clique with a good accuracy performance.

The rest of the paper is organized as follows. Section 2 presents a brief outlook and overview of related work and literature reviews. Section 3 elaborates on the implemented methodologies and theoretical frameworks. Section 4 presents the results and discusses the analysis of the graphs and figures. Section 5 summarizes and concludes with a short indication of the future direction for the research work on link stability within the domain of structural integrity of OSNs and SISs.

2   Related Work

Social Network Analysis (SNA) has a long history based on key foundational principles of similarity. It has long been postulated that similar relationships between actors contain crucial information about social structure integrity [13]. The paradigm of link dynamics and their impact on structure is a question most social models struggle with solving. Furthermore, this has recently been made more complicated with the emergence of Heterogeneous Networks (HNs) and Social Internetworking Scenarios (SIS). In this section, we briefly review the state-of-the-art techniques and approaches of research done in two major areas of stable community and stable link detection.

2.1   Stable Community Detection

A community is intuitively recognized by strong internal bonds and weak external connections. The measure of strength in connectivity is usually represented by quantity over quality of connections within a group. These measures therefore, represent relational densities of varying scales. Thus, most clearly defined communities are often characterized by dense intra-community relationships and sparse intercommunity links at node edges [6, 16]. However, similar classical techniques suffer from several drawbacks because the detected community structure will not remain stable over time [17]. Detection of stable communities requires the identification of stable links to serve as core structures of influence upon which a group of actors establishes online relations around [7].

In [23], a proposed framework to detect stable communities was developed. This was achieved by enriching the structure with mutual relationship estimations of observed links. In their study, link reciprocity estimation of backward edges and link stability scores were first established. The focus was given to detecting the presence of mutual links by preserving the original strength of backward edges, which scales better with longer time observable windows. Stable communities are then discovered using the enriched graphical representation containing link stability information. This was done through a correlation of persistence probability (repeated time existence/occurrence) of each community and its local topology.

In [4], Charkraborty et al. studies how results from community detection algorithms change when vertex orderings stay invariant. By stabilizing the ranking of vertices, they show that the variation of community detection results can be significantly reduced. Using the node invariance technique, they define constant communities as regions over which the structure remains constant over different perturbations and community detection algorithms over time.

2.2 Stable Link Detection

In [24], the authors suggest an activity-based approach to establish the strength (stability) of a social link. In contrast to friendship structures, their approach centers around a common disregarded aspect of activity networks. They argue that over time, social links can grow stronger(stable) or weaker(unstable) as a measure of social transaction activities. The study involves an observation of the evolutionary nature of link activities on Facebook. Their findings indicate that link prediction tasks relying on link occurrences as baseline metrics of measurements are inaccurate. As their results show, links in an activity network tend to fluctuate rapidly over time. Furthermore, the authors explain that decaying strength(stability) of ties correlate to decreasing social activity as the social network ages.

The study in [25] presents an overview of how links and their corresponding structures are being perceived from common link mining tasks. Such tasks include object ranking, group detection, collective classification, link prediction and subgraph discovery. The authors argue that these techniques address the discovery of patterns and collections of Independent Identically Distributed (I.I.D.) instances. Their methods are focused around finding patterns in data by exploiting and explicitly modeling time-aware links among data instances. In addition, their paper contribution presents some of the more common research pathways into applications which are emerging from the fast-growing field of link mining like [22].

In summary, detecting stable links is an important aspect of many inference and prediction tasks which online applications use all the time [1,3]. Community detection and link prediction are concerned with identifying correlated distributions from a social scene [19]. These distributions can then be used as measures for decision support and recommendation systems [20].

3 Our Method

In this section, we detail our method for detecting stable links. The core of our model is developed from a regressional technique and was later refined to integrate with a stochastic approach for the cross-validation of accuracy and performance within a small Facebook clique.

3.1 Multi-variate Vector Autoregression

The time series regression technique was chosen as the main approach to compute the stability index of links within a network. For small-scale datasets, vector regression methods (VAR) offer a very simple yet elegant means of analysis. Time series regressions are very simple and direct approaches. They are most often used in two forms to solve problems from a topological perspective. The first of these are the reduced (primary) form used in forecasting while the second is the structural (extended) form used in structural analysis.

In our work, we have adopted the structural framework as one of the core methods of approach towards identifying stability in links. Structural regressions have the ability to benchmark relational behavior against known dynamic models in the social scene. It can also be used to investigate the response to disruptive surprises. Such social disruptions often occur as shocks from world events (e.g. The Brexit from the E.U., etc.).

A multiple linear regression model essentially extends the single regression model by considering multiple independent variable relationships to estimate the state of a dependent variable. MVVA extends this principle further by correlating the multi-linear regression relationships through time. Given a series of past dependent observables Y_τ, one can predict the unobserved dependent variable at the current time Y_t from the following mathematical formula:

$$Y_t = B_0 + \sum_{n=0,\tau=0}^{m,t-n} (G_n Y_\tau + \varepsilon_\tau) \tag{1}$$

where B_0 is the array of residual constants and ε_τ is the error vector with zero variance co-variance.

Under the MVVA model which we have proposed, the six chosen variables of our study have been identified to be pivotal contributors of link stability. These identifications were studied from correlations, scatter plots and simple regressions between independent and dependent observables. It allows useful interpretation of observed relational behaviors which can be used for a variety of other tasks as well.

The stability matrix at time t is calculated from the predicted contributions of the six independent variables used in our study. We define the Stability index from Node Feature Similarity as $N(S)_t$, Cumulative Frequency as $F(Q)_t$, Sentiment as $I(S)_t$, Trust as $R(S)_t$, Betweenness as $B(S)_t$ and Transactions as $W(S)_t$. Thus, the stability contribution matrix S_t of all the six features is given as: $S_t = [N(S)_t, F(Q)_t, I(S)_t, R(S)_t, B(S)_t, W(S)_t]^T$.

From a structural perspective, the model we have developed follows the following mathematical formulation:

$$AS_t = \beta_0 + \sum_{\tau=1}^{p} (\beta_\tau S_{t-\tau}) + U_t \tag{2}$$

where A is the restricted correlation matrix between the endogenous variables (dynamic feature stability contributions) identified through its past variations. β_0 and β_τ are structural parameters estimated through the method of Ordinary Least Squares (OLS). Hence, $\beta_\tau = A * G_\tau$. Finally, U_t are the time-independent disruptions caused by unsettling world events. This is derived from the (linear) system of equations as:

$a_{11}N(S)_t + a_{12}F(Q)_t + a_{13}I(S)_t + a_{14}R(S)_t + a_{15}B(S)_t + a_{16}W(S)_t =$
$\beta_{10} + \beta_{11}N(S)_t + \beta_{12}F(Q)_t + \beta_{13}I(S)_t + \beta_{14}R(S)_t + \beta_{15}B(S)_t + \beta_{16}W(S)_t + U_{N(S)_t}$

$a_{21}N(S)_t + a_{22}F(Q)_t + a_{23}I(S)_t + a_{24}R(S)_t + a_{25}B(S)_t + a_{26}W(S)_t =$
$\beta_{20} + \beta_{21}N(S)_t + \beta_{22}F(Q)_t + \beta_{23}I(S)_t + \beta_{24}R(S)_t + \beta_{25}B(S)_t + \beta_{26}W(S)_t + U_{F(Q)_t}$

.
.
.

$a_{61}N(S)_t + a_{62}F(Q)_t + a_{63}I(S)_t + a_{64}R(S)_t + a_{65}B(S)_t + a_{66}W(S)_t =$
$\beta_{60} + \beta_{61}N(S)_t + \beta_{62}F(Q)_t + \beta_{63}I(S)_t + \beta_{64}R(S)_t + \beta_{65}B(S)_t + \beta_{66}W(S)_t + U_{W(S)_t}$

In its primary form,

$$S_t = C_t + \sum_{\tau=1}^{m,t-n} G_\tau S_\tau + \varepsilon_t \qquad (3)$$

where, $C_t = A^{-1} * \beta_0$, $G_\tau = A^{-1} * \beta_\tau$ and the residual errors $\varepsilon_t = A^{-1} * U_t$.

The number of independence restrictions imposed on the correlation matrix A is simply the difference between the unknown and known elements obtained from the variance co-variance matrix of the errors, $E(\varepsilon_t \varepsilon_t') = \Sigma_\varepsilon$. For the symmetric matrix of our model, $A = A^T$, which is $\frac{n^2-n}{2}$.

We define the feature rate coupling ratio w_t as the weighted impulse responses due to the structural disruptions on the endogenous feature observables. Each dynamic link feature response includes the effect of specific disruptions on one or more of the variables in the social system - at first occurrence t, and in subsequent time frames, $t+1$, $t+2$, etc.

The feature rate coupling ratio is thus given as:

$$\sum_{\tau=1}^{n} w_{U_\tau} = \sum_{\tau=1}^{n} (\dot{w}_{U_{\tau-1}} * [F_{U_\tau} - F_{U_{\tau-1}}]) \qquad (4)$$

where $\dot{w}_{U_{\tau-1}}$ is the first derivative response lag, which measures the momentum vector of social activity and F_{U_τ} and $F_{U_{\tau-1}}$ are endogenous feature observable vectors at current and lag time frames respectively.

Then, we can express our structural autoregressive model in a vector sum of social disruptions as:

$$S_t^i = \mu + \sum_{i=0}^{k} w_{t,i} S_{t,i} \qquad (5)$$

where S_t^i is the stability matrix (with each feature element in i indicating how stable link is). $w_{t,i}$ is the feature rate coupling ratio at time t and $S_{t,i}$ is the stability contribution; both across i endogenous feature observables. Finally, μ is the impulse residual constant.

The MVVA model is not without its drawbacks. The complexity of the OLS problem involving a Cholesky decomposition of matrix M is at least $O(C^2N)$, where N is the sample data size and C is the total number of features. By direct inference, MVVA entropies to the squared growth in network complexity. Furthermore, two additional problems may arise as complexity of the social network grows; i.e. overfitting and multi-collinearity.

To overcome the above problems, we explore the Hamiltonian Monte Carlo (HMC) as an important extension to address the limitations of MVVA from a stochastic perspective for link stability detection. Since the social network we obtain from the repositories of common crawl contains missing links and partial information, stochastic estimations are used to measure the accuracy and reliability of our experimental MVVA results [12]. Additionally, HMC models are powerful samplers of potential energy distributions and its partial derivatives - which are representative of online social structures [29]. This means that overfitting and multi-collinearity will be tackled through high acceptance ratios [29]. Furthermore, the complexity per transition is O(GN). Where G is the gradient cost of the exact model which scales linearly with data and N is the number of steps [5].

3.2   Hamiltonian Monte Carlo

The condition that full form adaptive MCMC methods satisfy is:

$$\sum_x T(x' \leftarrow x)P(x) = P(x^i) \tag{6}$$

For a good sample x from the distribution P(x). x' is the next step-wise sample from x. The Hamiltonian Monte Carlo extends the sampling efficiency of posteriors made by MCMC, through the use of Hamiltonian dynamics [8]. As an energy-based method, it is postulated that the sum total of all energies within a closed link-dynamics based system is conserved [10].

Hence, for every feature identified in the belief state graph G, its stability index score can be correlated to vector positional (static, potential) energy function $e^{H(G)}$ for any combinational variant of the graph $g \in G$ [15]. The Hamiltonian dynamics recognizes that a single form of energy cannot exist alone because it has to be conserved. Therefore, wherever potentials are the effects, the kinetics are the casuals [8]. By introducing another variable which isn't our main information of interest, we are able to conserve this "relational energy" within the closed social belief system [11]. This can be identified as the transitional tensor (moving, kinetic) energy function $e^{-v^T v/2}$ between the different features and their states, such that this joint distribution is given as:

$$P(x, v) \propto e^{-E(x)}e^{-v^T v/2} = e^{-H(x,v)} \tag{7}$$

where $P(x, v)$ is the conditional state transition probability between energy vectors x and v.

Firstly, the Leapfrog integration $L(\epsilon, M)$ is performed M times with an arbitrarily chosen step size ϵ. This means that $L(\zeta)$ is the final resulting state from M steps from the HMC dynamics with predefined step size ϵ. The next state transition step is given as:

$$\zeta^{(t,1)} = \sum_{n=1}^{k} L^n \zeta^{(t,0)} \text{ with probability } \pi_L^n(\zeta^{(t,0)}) \tag{8}$$

It is probabilistically defined as a Markov transition on its own [5]. The state transition momentum vector resulting from the secondary added accountable term for kinetic energy is then further corrupted by Gaussian noise so that there are uncertainties during the transition of the states [9]. This is important because the non-deterministic nature of the momentum during transitions allow for proposals from current states onto new and further displaced states.

The randomization operator $R(\beta)$ mixes Gaussian noise determined by $\beta \in [0, 1]$ into the velocity vector given as:

$$R(\beta) = x, v' \tag{9}$$

$$v' = v\sqrt{1 - \beta} + n\beta \tag{10}$$

where n is drawn from a normal distribution:

$$n \sim N(0, I) \tag{11}$$

The transition probabilities are then chosen as:

$$\pi_{L^a}(\zeta) = min \left\{ \begin{array}{l} \pi_{L^b}(\zeta), \\ \sum_{b \leq a} \frac{p(FL^a(\zeta))}{p(\zeta)}(1 - \sum_{b \leq a} \pi_{L^b}(FL^a(\zeta))) \end{array} \right\}$$

Which satisfies the reversibility of the Markov Chain fixed positional transitional vector.

4   Experimental Results

In this section, we present the setup and results of our experimental evaluations on both MVVA and HMC algorithms.

4.1   Experimental Setup

The dataset chosen for this study was crawled from Facebook and obtained from the repository of the Common Crawl (August 2016). It includes the following relational features between any two arbitrary nodes: The Cumulative Frequency of the type of wall posts, the sentiment of the content in context of the post (Neutral, Positive, Somewhat Positive, Mildly Positive, Negative, Somewhat Negative, Mildly Negative), the Node-betweenness Feature Similarity (Roles and

Proximity metrics), the Trust Reciprocity Index (Similar in quantization to Sentiment Index) and the number of posts at defined quantized Unix time sample space as a measure of link virility. In this study, the Node Feature Similarity Index is used as a performance benchmark against multivariate analysis.

The experiments were conducted on our Multi-Variate Vector Auto-Regression Model on undirected small world topologies with a clique size of 20–100 nodes. A subset of nodes (<10) was first chosen for this study as the defining seed community. Then, this chosen community was allowed to grow to a maximum size of 1019 nodes by adjoining nodes to establish new relationships.

The links in the network are tagged based on their Stability Index (SI) scores. Stable links are labeled with SI scores higher than or equal to 80, while Neutral links are labeled with SI scores in the range of [50, 80), the slightly unstable links are labeled with SI scores in the range of [30, 50) and the unstable links are labeled with SI scores in the range of [0, 30) respectively. The new and existing links which are SI score labeled (satisfying their respective threshold conditions) were then subsequently evaluated for their Aggregated Link Stability Index over time at every sample (whenever social transactions were captured by the crawler across posts) based on the variate features discussed above. The aggregated link stability index is calculated as:

$$AG_t = \sum_{i,E=0}^{k,m} S_t^{i,E} \qquad (12)$$

where AG_t is the aggregated link stability index of the topograph at time instant t and $S_t^{i,E}$ is the feature i stability index of edge E in the network.

The prediction error is given simply as:

$$e_t = |Y_t - F_t| \qquad (13)$$

where e_t is the Aggregated Stability Index Prediction error, Y_t is the observed Aggregated Stability Index at time t - this is given by the HMC Stability State Index values after a 100 times iteration over the samples of the 5 multivariates. F_t is the predicted Stability Index based on both the MVVA model and the univariate (Similarity Index) regression model.

The scaled error across the two different datasets is given by the equation as:

$$\varepsilon = \frac{e_t}{\frac{1}{n-1} \sum_{i=2}^{n} |Y_i - Y_{i-1}|} \qquad (14)$$

where ε_t is the absolute scale free error of the predicted data set F_t against the observed dataset Y_t. n is the number of sampled forecasts. The Mean Absolute Scaled Error (MASE) of a distribution plot Q is given as:

$$MASE(Q) = \sum_{t=0}^{k} \frac{\varepsilon_t}{k} \qquad (15)$$

4.2   Link Stability Evaluation for MVVA

In our experiment, the link growth comparison is conducted between the univariate node based Common Neighbor (CN) feature and our univariate features (which includes Similarity Index as well). As seen in Fig. 2, by considering the dynamics of the relational features within those established links in the multivariate time regression process we have proposed, our Multivariate Link Stability Index outperformed the CN-Node Similarity based Stability Index by over twice the score of the traditional metric used in the link prediction process with an AUC of 0.87 by comparison to the latter's AUC of 0.46; which is a tremendous improvement in terms of efficacy. The number of labeled links of the fully evolved topograph at the end of 30 days and the calculated aggregated stability index are given in Tables 1 and 2. Figure 1 shows the growth distribution of the stability index scores of links within a Facebook clique for a period of 30 days. The experiment was done using the MVVA autoregressive algorithm for both univariate and multivariate modes of calculation, of the dataset acquired from Facebook. It measures the aggregate stability scores accumulated within the clique against the time - which has been normalized to fit into the scale window of the plot.

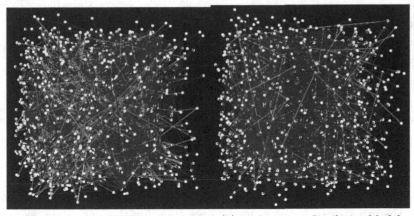

(a) Univariate Similarity Model   (b) Multivariate Similarity Model

Fig. 1. Topograph of univariate similarity based stability index (left) and Topograph of the multivariate stability index (right)

4.3   MVVA Accuracy Evaluation

Based on Fig. 2, Tables 1 and 2, it can be seen that stable link detection accuracy using our model has been vastly improved by 78.29% with 3184 links being detected as stable in the univariate analysis; and only a similar 694 links detected in the multivariate analysis. Furthermore, with only 694 links identified as stable in the Multivariate Analysis, the aggregated scores of the topology are 2.34 times higher than the Univariate Analysis; suggesting a noticable improvement in

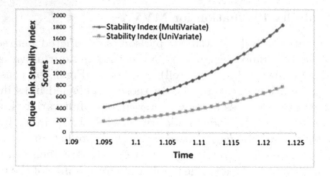

Fig. 2. Link stability index comparison over time: both scales (Stability Index Scores and Unix Time) have been normalized to fit into the plot frame window.

Table 1. Tabulation of the number of identified labeled links for both univariate and multivariate regression analysis

Type of link	Univariate	Multivariate	Score range
Stable	3184	694	>80
Neutral	5257	7782	50-79
Somewhat stable	35	0	30-49
Unstable	0	0	0-29

Table 2. 30-day normalized aggregated stability index

Multivariate	1835
Univariate	783

terms of efficacy - making our model far more reliable than traditional univariate methods throughout the prediction process.

4.4 Prediction Error Evaluation

The prediction error results can be summarized in Table 3. As can be seen from Fig. 4, the error score index ε_t grows over time for the univariate regression analysis, whereas the error score index ε_t of the MVVA model which we proposed decreases over time. Additionally, as can be seen from Table 3, the MASE score for the MVVA model improves both the In-Sample and Out-Sample prediction accuracy of the underlying stability index distribution for the Facebook clique over the 30-day time frame by 8.3 times more than the MASE score for the conventional univariate regression model.

4.5 HMC Results and Evaluation

Figure 3 shows good (small) autocorrelations between the training data of features in most sets, although there are some sets which present spurious/biased information where a Gaussian distributed and noise-corrupted momentum sampled model could not correlate well to with respect to log distributions of its

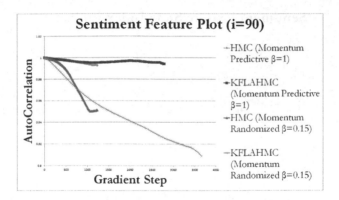

Fig. 3. Graph of sentiment autocorrelation against the number of gradient iterations for predictive ($\beta = 1$) and randomized ($\beta = 0.15$) momentum vectors of HMC for 10 burn in data sets of the similarity feature from the Facebook wall posts.

Table 3. Tabulation of Mean Squared Errors (MASE) of both multivariate and univariate analysis at the end of the 30-day clique evolution period.

	MVVA		Univariate	
	In	Out	In	Out
MASE	0.074268	0.0944732	0.616677	0.572323

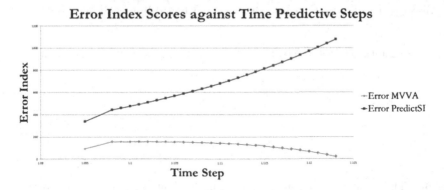

Fig. 4. Error score ϵ_t comparison over time between MVVA and the univariate regression models.

momenta and positional gradients. However, it can be seen that from more burn in data samples and more randomized (corrupted by noise - $\beta = 0.1$) momenta sampling behavior, the performance of the gradient autocorrelation improves during the learning phase of our HMC implementation.

Figure 5 is a posterior sample of Sentiment index scores. The horizontal axis reflects the normalized time which has elapsed during the process and is also

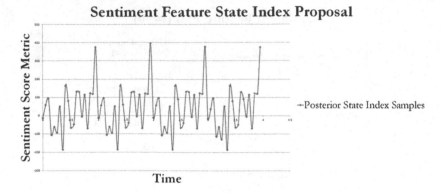

Fig. 5. Plot of posterior sentiment feature state samples.

directly proportional to the number of iterations progressed through this window (as displayed on the graphs).

Figures 6, 7, 8 and 9 show progressively how the random walk proposed distribution converges towards the actual distribution of the Stability Index data set from a fixed point condition (the very first initial feature belief state at t = 0) being held constant. Figure 9 is the Monte Carlo approximation for the actual 30-day aggregated stability index distribution repeated over Hamiltonian dynamics for 100 cycles. It shows a good convergence towards our MVVA model; which reflects very closely to the actual growth of aggregated stability index over time - as opposed to univariate (similarity feature) based link stability prediction.

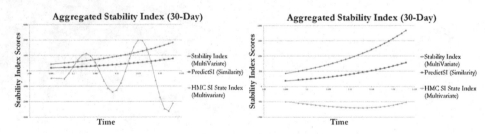

Fig. 6. Link stability index comparison over time with HMC iterated over 10 times for posterior states of the 5 multivariates (Time Delta, Frequency, Similarity, Sentiment, Trust).

Fig. 7. Link stability index comparison over time with HMC iterated over 50 times for posterior states of the 5 multivariates (Time Delta, Frequency, Similarity, Sentiment, Trust).

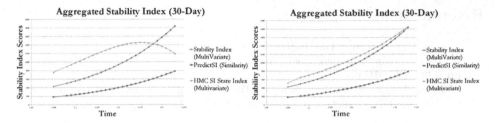

Fig. 8. Link stability index comparison over time with HMC iterated over 80 times for posterior states of the 5 multivariates (Time Delta, Frequency, Similarity, Sentiment, Trust).

Fig. 9. Link stability index comparison over time with HMC iterated over 100 times for posterior states of the 5 multivariates (Time Delta, Frequency, Similarity, Sentiment, Trust).

5 Conclusion

In conclusion, the Multivariate model (MVVA) which we have proposed for the detection and identification of stable links works well and is far more superior to univariate models or models which consider only static node based features and link temporality. Our system has been tested on a small Facebook clique which was evolving. This dynamic growth can now be better understood and comprehended through the existence of stable links as other seed clusters form around it. However, the tighter, more stringent constraints of a small world model used in this study should not be overlooked. In larger hyper-graphical models, where boundaries fall apart due to sheer volume distributions of scattered data, a larger scope of stochastic lemmas surrounding both high complexities and large volumes of social features have to be re-discovered [21].

Some advantages of our methods and experimentation include a strongly connected network with a firm belief structure and sufficient access to new information being made readily available during the data mining process. However, in larger dimensional frameworks where the constraints of such structure break down and data is made even wider and more sparse, deep learning knowledge discovery methods like Monte Carlo estimates and the DNNs are powerful variations which can be used for online social prediction and inference tasks [18].

Acknowledgment. This research was partially supported by Guangxi Key Laboratory of Trusted Software (No. kx201615), Shenzhen Technical Project (JCYJ20170307151733005 and KQJSCX20170726103424709), Capacity Building Project for Young University Staff in Guangxi Province, Department of Education of Guangxi Province (No. ky2016YB149).

References

1. Ozcan, A., Oguducu, S.G.: Multivariate temporal link prediction in evolving social networks. In: International Conference on Information Systems 2015 (ICIS-2015), pp. 113–118 (2015)
2. Mengshoel, O.J., Desai, R., Chen, A., Tran, B.: Will we connect again? machine learning for link prediction in mobile social networks. In: Eleventh Workshop on Mining and Learning with Graphs. Chicargo, Illinois 2013, pp. 1–6 (2013)
3. Hyndman, R.J., Koehler, A.B.: Another look at measures of forecast accuracy. Int. J. Forecast. **22**(4), 679–688 (2006)
4. Chakraborty, T., Srinivasan, S., Ganguly, N., Bhowmick, S., Mukherjee, A.: Constant communities in complex networks (2013). arXiv preprint arXiv:1302.5794
5. Sohl-Dickstein, J., Mudigonda, M., DeWeese, M.R.: Hamiltonian Monte Carlo without detailed balance. In: Proceedings of the 31st International Conference on Machine Learning (JMLR), vol. 32 (2014)
6. Farasat, A., Nikolaev, A., Srihari, S.N., Blair, R.H.: Probabilistic graphical models in modern social network analysis. Soc. Netw. Anal. Min. **5**(1), 62 (2015)
7. Viswanath, B., Mislove, A., Cha, M., Gummadi, K.P.: On the evolution of user interaction in facebook. In: Proceedings of the 2nd ACM Workshop on Online Social Networks, pp. 37–42. ACM (2009)
8. Girolami, M., Calderhead, B., Chin, S.A.: Riemannian manifold Hamiltonian Monte Carlo. Arxiv preprint, 6 July 2009
9. Meyer, H., Simma, H., Sommer, R., Della Morte, M., Witzel, O., Wolff, U., Alpha Collaboration: Exploring the HMC trajectory-length dependence of autocorrelation times in lattice QCD. Comput. Phys. Commun. **176**(2), 91–97 (2007)
10. Read, J., Martino, L., Luengo, D.: Efficient monte carlo methods for multi-dimensional learning with classifier chains. Pattern Recogn. **47**(3), 1535–1546 (2014)
11. Pakman, A., Paninski, L.: Auxiliary-variable exact Hamiltonian Monte Carlo samplers for binary distributions. In: Advances in Neural Information Processing Systems, pp. 2490–2498 (2013)
12. Hoffman, M.D., Gelman, A.: The No-U-turn sampler: adaptively setting path lengths in hamiltonian monte carlo. J. Mach. Learn. Res. **15**, 1351–1381 (2014)
13. Rodriguez, A.: Modeling the dynamics of social networks using Bayesian hierarchical blockmodels. Stat. Anal. Data Min. **5**(3), 218–234 (2012)
14. Hunter, D.R., Krivitsky, P.N., Schweinberger, M.: Computational statistical methods for social network models. J. Comput. Graph. Stat. **21**(4), 856–882 (2012)
15. Nightingale, G., Boogert, N.J., Laland, K.N., Hoppitt, W.: Quantifying diffusion in social networks: a Bayesian approach. In: Animal Social Networks, pp. 38–52. Oxford University Press, Oxford (2014)
16. Fan, Y, Shelton. C.R.: Learning continuous-time social network dynamics. In: Proceedings of the Twenty-Fifth Conference on Uncertainty in Artificial Intelligence, 18 Jun 2009, pp. 161–168. AUAI Press (2009)
17. Yang, T., Chi, Y., Zhu, S., Gong, Y., Jin, R.: Detecting communities and their evolutions in dynamic social networks - a Bayesian approach. Mach. Learn. **82**(2), 157–189 (2011)
18. Mossel, E., Sly, A., Tamuz, O.: Asymptotic learning on bayesian social networks. Probab. Theor. Relat. Fields **158**(1–2), 127–157 (2014)
19. Gale, D., Kariv, S.: Bayesian learning in social networks. Games Econ. Behav. **45**(2), 329–346 (2003)

20. Needham, C.J., Bradford, J.R., Bulpitt, A.J., Westhead, D.R.: A primer on learning in Bayesian networks for computational biology. PLoS Comput. Biol. **3**(8), e129 (2007)
21. Gardella, C., Marre, O., Mora, T.: A tractable method for describing complex couplings between neurons and population rate. In: eNeuro, 1 July 2016, vol. 3, no. 4 (2016). ENEURO-0160
22. Getoor, L., Diehl, C.P.: Link mining: a random graph models approach. J. Soc. Struct. **7**(2), 3–12 (2005). 2002 Apr survey. ACM SIGKDD Explorations Newsletter
23. Nguyen, N.P., Alim, M.A., Dinh, T.N., Thai, M.T.: A method to detect communities with stability in social networks. Soc. Netw. Anal. Min. **4**(1), 1–15 (2014)
24. Liu, F., Liu, B., Sun, C., Liu, M., Wang, X.: Deep belief network-based approaches for link prediction in signed social networks. Entropy **17**(4), 2140–2169 (2015). Multidisciplinary Digital Publishing Institute
25. Wang, P., Xu, B., Wu, Y., Zhou, X.: Link prediction in social networks: the state-of-the-art. Sci. China Inf. Sci. **58**(1), 1–38 (2015)
26. Zhou, X., Tao, X., Rahman, M.M., Zhang, J.: Coupling topic modelling in opinion mining for social media analysis. In: Proceedings of the International Conference on Web Intelligence, pp. 533–540. ACM (2017)
27. Tao, X., Zhou, X., Zhang, J., Yong, J.: Sentiment analysis for depression detection on social networks. In: Li, J., Li, X., Wang, S., Li, J., Sheng, Q.Z. (eds.) ADMA 2016. LNCS (LNAI), vol. 10086, pp. 807–810. Springer, Cham (2016). https://doi. org/10.1007/978-3-319-49586-6_59
28. Zhang, J., Tan, L., Tao, X., Zheng, X., Luo, Y., Lin, J.C.-W.: SLIND: Identifying stable links in online social networks. In: Pei, J., Manolopoulos, Y., Sadiq, S., Li, J. (eds.) DASFAA 2018. LNCS, vol. 10828, pp. 813–816. Springer, Cham (2018). https://doi.org/10.1007/978-3-319-91458-9_54
29. Zhang, J., Tao, X., Tan, L.: On relational learning and discovery: a survey. Int. J. Mach. Learn. Cybern. **2**(2), 88–114 (2018)

EPOC: A Survival Perspective Early Pattern Detection Model for Outbreak Cascades

Chaoqi Yang, Qitian Wu, Xiaofeng Gao$^{(\boxtimes)}$, and Guihai Chen

Shanghai Key Laboratory of Scalable Computing and Systems,
Department of Computer Science and Engineering, Shanghai Jiao Tong University,
Shanghai 200240, People's Republic of China
ycqsjtu@gmail.com, echo740@sjtu.edu.cn, {gao-xf,gchen}@cs.sjtu.edu.cn

Abstract. The past few decades have witnessed the booming of social networks, which leads to a lot of researches exploring information dissemination. However, owing to the insufficient information exposed before the outbreak of the cascade, many previous works fail to fully catch its characteristics, and thus usually model the burst process in a rough manner. In this paper, we employ survival theory and design a novel survival perspective Early Pattern detection model for Outbreak Cascades (in abbreviation, EPOC), which utilizes information both from the static nature and its later diffusion process. To classify the cascades, we employ two Gaussian distributions to get the optimal boundary and also provide rigorous proof to testify its rationality. Then by utilizing both the survival boundary and hazard ceiling, we can precisely detect early pattern of outbreak cascades at very early stage. Experiment results demonstrate that under three practical and special metrics, our model outperforms the state-of-the-art baselines in this early-stage task.

Keywords: Early-stage detection · Outbreak cascade
Survival theory · Cox's model · Social networks

1 Introduction

The rapid development of modern technology has changed the lifestyles to a large extent compared to a few years ago. Every day millions of people express ideas and interact with friends through online platforms like Twitter and Weibo. On these platforms, registered users are able to *tweet* short messages (e.g., up to 140 characters in Twitter), and others who are interested in it will give likes, comments, or more commonly, *retweets*. Such retweeting would potentially disseminate and further spread information to a large number of users, which forms a *cascade* [1]. While the cascade grows larger and get more individuals involved, a sudden *burst* will definitely arrive, which we call a *spike*. As a matter of fact, detecting and predicting the burst pattern of a cascade, especially at early stage,

© Springer Nature Switzerland AG 2018
S. Hartmann et al. (Eds.): DEXA 2018, LNCS 11029, pp. 336–351, 2018.
https://doi.org/10.1007/978-3-319-98809-2_21

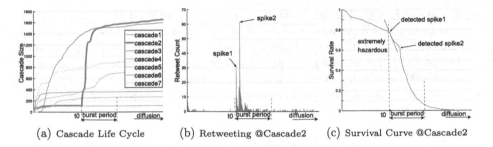

(a) Cascade Life Cycle   (b) Retweeting @Cascade2   (c) Survival Curve @Cascade2

Fig. 1. Samples of cascade diffusion on Twitter

attract lots of attention in various domains: meme tracking [2], stock bubble diagnosis [3], and sales prediction [4], etc.

However, to fully understand the burst pattern of cascades ahead of time will meet three major challenges. **First** and foremost, due to the deficiency of available information and its disorder nature at early stage [5], one can hardly catch distinguishing signs on whether a cascade will break out. The **second** challenge stems from the significantly distinct life span of different cascades [6], which makes it tough to extract typical features. Worse still, this distinctiveness makes it hard for researchers to set suitable observation time, owing to the variety of life spans. The **third** challenge is that the burst pattern of cascades usually follows a quick *rise and fall* law [7], which lasts a few minutes but causes magnificent influence. In this situation, the correlations between the history and the near future can be hardly characterized by traditional models.

Shown in Fig. 1(a), we plot the diffusion process of seven real-world cascades from Twitter. We can see that @Cascade2 shares almost the same pattern with @Cascade1 before it outbreaks at time t_0, which means that it is hard for us to catch the distinguishing signs using the early information. As the second challenge states, @Cascade1~7 represent different life span at early stage. While @Cascade6 ends its diffusion, @Cascade3 is just about to start propagation, and it still enlarges even at the end of observation. The third challenge can be vividly described in Fig. 1(b), where we focus on @Cascade2 and plot how it is retweeted. Figure 1(b) shows that @Cascade2 experiences a mild propagation when it appears, but after time t_0, it goes through two large retweeting spikes (sudden falls in survival curve ploted in Fig. 1(c)), and the final amount of retweeting explodes to about 1600 during the burst period. These three core challenges motivate us to design a model that can handle this quick rise and fall pattern, characterize different cascades uniformly, and detect the burst pattern as early as possible.

Motivated by the study of death in biological organisms, in this paper, we regard the diffusion of cascades as the growing process of biological organisms. Since Cox's model is widely used to characterize the life span of biological organisms, here we adopt Cox's model with the knowledge of cascades, transforming the burst detection task into diagnosis of cascade life table, and then we build a survival perspective Early Pattern detection model for Outbreak Cascades, in abbreviation, EPOC. Though previous work [8] has also tried Cox's model, their work is mainly based on unsubstantiated observations as well as only taking one feature into consideration, which does not address the above challenges at all.

In our EPOC, to consider the influential factors from different perspectives, we harness three features from each cascade (retweet sequence, follower number sequence, and original timestamp) to capture the effectiveness of temporal information [9], the influence of involved users [10], and the dynamics of user activity [11]. Then, to study the distinctiveness of cascades' life span, we train an effective Cox's model and employ two Gaussian distributions to fit the survival probability of viral and non-viral cascades at different time point respectively, and obtaining a survival boundary between the viral and the non-viral, which is further proven to be well-defined theoretically. Finally, as the static and dynamic nature of cascade diffusion are both important indicators of cascade virality, we jointly consider survival probability and hazard rate, which considerably enhances our model's performance in handling the quick rise and fall pattern. We then employ three special metrics (K-coverage, Cost, Time ahead) to compare EPOC with two basic machine learning methods (LR, SVR) and three powerful baselines published in recent literatures (PreWhether [12], SEISMIC [10], SansNet [8]) on two large real-world datasets: Twitter and Weibo. Experiment results show that EPOC outperforms these five methods in burst pattern detection at very early stage.

Our main contributions are summarized as:

- We adopt survival theory and establish a powerful burst detection model EPOC for cascade diffusion, which can handle the quick rise-and-fall pattern as well as the significantly distinct life span of cascades at the early stage.
- We utilize both static and dynamic information from cascades, obtain a dimidiate boundary with two Gaussian distributions, and then novelly use the burst pattern to help predict the popularity of an online content.
- We adopt three special metrics and conduct extensive experiments on two large real-world data sets (Twitter and Weibo). The results show that EPOC gives the best performance comparing with five state-of-the-art approaches.

The remainder of the paper is organized as follows. Some common notions of survival theory and the basic Cox's model are introduced in Sect. 2. The design of our proposed model EPOC is specified in Sect. 3. We evaluate and analyze our model on Twitter and Weibo in Sect. 4. We review several related works in Sect. 5. Finally, we conclude our work and highlight the possible future perspectives in Sect. 6.

2   Survival Analysis and Cox's Model

In this section, we give some definitions about survival theory in social networks. Initially, when a user shares the content with her set of friends, several of these friends share it with their respective sets of friends, and a *cascade* of resharing can develop [13]. Once the size of this cascade grows above a certain *threshold*

ρ, we call it goes *viral*, and otherwise *non-viral*. To quantitively describe these statues of cascade diffusion, we introduce *survival function* and *hazard function* respectively in Definitions 1 and 2.

Definition 1. *(Survival Function): let $S(t) \in (0,1)$ denote the survival probability of cascade subject to time t, i.e., at time t, cascade has the probability of $S(t)$ to be non-viral, where $S(t)$ is naturally monotonic decreasing with time t.*

Definition 2. *(Hazard Function): let $h(t) \in (0,\infty)$ denote the hazard rate of cascade at time t on the condition that it survives until time t ,i.e., h(t) is the negative derivative of survival probability $-\frac{dS(t)}{dt}$ to the survival function $S(t)$, specifically given by the following formula,*

$$h(t) = -\frac{dS(t)}{dt} \cdot \frac{1}{S(t)}. \tag{1}$$

Since Cox's survival model was proposed [14], it has been widespread used in the analysis of time-to-event data with censoring and covariates [15]. In this work, we use Cox's proportional hazard model with time-dependent covariates (also called Cox-extended model) to characterize the association between early information and the cascade statues (viral or non-viral).

Basic Model: For cascades $i = 1, 2, \cdots, n$, they share the same baseline hazard function denoted as $h_0(t)$, and $\boldsymbol{X}_i(t) = \{x_1^{(i)}, x_2^{(i)}, \cdots, x_m^{(i)}\}$ denotes the feature vector of the i_{th} cascade, where $h_0(t)$ does not depend on each $\boldsymbol{X}_i(t)$ but only on t. $\boldsymbol{\beta} = \{\beta_1, \beta_2, \cdots, \beta_m\}$ is the parameter vector of our hazard model. We specify the hazard function of i_{th} cascade as follows,

$$h_i(t) = h_0(t) \cdot \exp\left(\boldsymbol{\beta}^T \boldsymbol{X}_i(t)\right). \tag{2}$$

Because the model is proportional, i.e., given i_{th} and j_{th} cascade, the relative hazard rate $\lambda_{i,j}$ can be concretely given by,

$$\lambda_{i,j} = \frac{h_i(t)}{h_j(t)} = \frac{h_0(t) \cdot \exp\left(\boldsymbol{\beta}^T \boldsymbol{X}_i(t)\right)}{h_0(t) \cdot \exp\left(\boldsymbol{\beta}^T \boldsymbol{X}_j(t)\right)} = \frac{\exp\left(\boldsymbol{\beta}^T \boldsymbol{X}_i(t)\right)}{\exp\left(\boldsymbol{\beta}^T \boldsymbol{X}_j(t)\right)} \tag{3}$$

where $\boldsymbol{\beta}$ is the parameter vector, $\boldsymbol{X}_i(t)$ and $\boldsymbol{X}_j(t)$ are respectively the feature vectors of i_{th} and j_{th} cascade. From Eq. (3), it is easy to conclude that the baseline hazard does not play any role in relative hazard rate $\lambda_{i,j}$, i.e., the model is also a semi-parametric approach. Therefore, instead of considering the absolute hazard function, we only care about the relative hazard rate of cascades, which only concerns parameter vector $\boldsymbol{\beta}$. Then we use Maximum Likelihood Estimation to get parameter vector $\boldsymbol{\beta}$. We denote i_{th} cascade time-to-event as t_i, and assume that $0 < t_1 < t_2 < \cdots < t_n$. The Cox's partial likelihood is given by,

$$L(\boldsymbol{\beta}) = \prod_{i=1}^{n} \left(\frac{h_i(t_i)}{\sum_{j=i}^{n} h_j(t_i)}\right)^{\delta_i} = \prod_{i=1}^{n} \left(\frac{exp\left(\boldsymbol{\beta}^T X_i(t_i)\right)}{\sum_{j=i}^{n} exp\left(\boldsymbol{\beta}^T X_j(t_i)\right)}\right)^{\delta_i}, \tag{4}$$

where δ_i means whether the data from i_{th} cascade is censored, i.e., if the event happens to i_{th} cascade, then δ_i equals to 1, and otherwise 0. Then the log-partial likelihood of parameter vector β can be calculated as,

$$\log L(\beta) = \sum_{i=1}^{n} \delta_i \left[\beta^T X_i(t_i) - \log \left(\sum_{j=i}^{n} \exp \left(\beta^T X_j(t_i) \right) \right) \right], \qquad (5)$$

maximizing the log-partial likelihood by solving equation $\frac{\mathrm{d} \log L(\beta)}{\mathrm{d}\beta} = 0$, then we can get the numerical estimation of parameter vector β using Newton method.

3   EPOC: Detecting Early Pattern of Outbreak Cascades

Based on the basic model stated previously, in this section, we combine the Cox's model with our knowledge of cascades, and make it suitable to handle the task of detecting the early pattern of outbreak cascades. Here we regard cascades as complex dynamic objects that pass through successive stages as they grow. During this process of growth, the survival probability and the hazard rate of cascades will change dynamically. The high survival probability and low hazard rate suggest that cascades are unlikely to be viral in the future, while the low survival probability as well as high hazard rate imply the opposite. In this sense, we introduce the *survival boundary* and the *hazard ceiling* to help accomplish this challenging task at very early stage.

Feature Selection: As is stated previously, the effectiveness of temporal information, the influence of involved users, and the dynamics of user activity are all powerful indicators of the cascade statues. Therefore, in this experiment, we utilize three features accordingly: *timestamp of each retweet, number of followers* of every user involved in the cascade, and *timestamp of the first tweet*.

3.1   Survival Boundary: A Static Perspective

To detect the early pattern of outbreak cascades, firstly, we characterize the survival functions of all cascades. Shown in Fig. 2(a), the red lines represent the survival functions of viral cascades, and the blue lines show the non-virals'. Then we are supposed to divide the estimated survival functions of all cascades into two classes (viral and non-viral). In other word, we need to find a *survival boundary*. As is illustrated in Fig. 2(b), the red dashed line separates the two categories of blue (non-viral cascades) and red (viral cascades).

Previous works [16] have demonstrated that at a fixed observing time t, the distribution of survival probability of different cascades obeys Gaussian distribution. Based on this knowledge, we employ two random variables: f_v^t (for viral cascades) and f_n^t (for non-viral cascades) subject to time t, which satisfy the Gaussian. Formally, we specify this assumption in Definition 3.

Definition 3. *For any Given time t, we have $f_v^t \sim \mathcal{N}(\mu_v^t, \sigma_v^t)$ and $f_n^t \sim \mathcal{N}(\mu_n^t, \sigma_n^t)$, where μ_v^t, σ_v^t and μ_n^t, σ_n^t are the parameters of Gaussian distribution for viral and non-viral cascades subject to time t.*

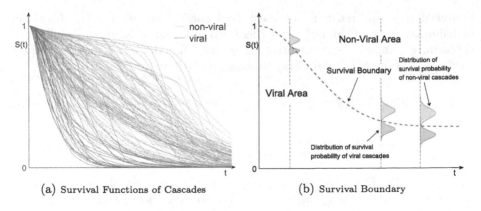

(a) Survival Functions of Cascades   (b) Survival Boundary

Fig. 2. Survival functions and survival boundary (Color figure online)

Based on Definition 3, for a given time t, the survival probability of viral and non-viral cascades can be respectively characterized as f_v^t and f_n^t. Therefore, the task to find the optimal survival boundary is to give the suitable separation between two Gaussian distributions.

Definition 4. *(Survival Boundary): for any given time t, assume the survival boundary to be $S^*(t)$, which is given by the following formula,*

$$\int_{-\infty}^{S^*(t)} \frac{1}{\sqrt{2\pi}\sigma_v^t} \exp\left(-\frac{(x-\mu_v^t)^2}{2\sigma_v^{t\,2}}\right) dx = \int_{S^*(t)}^{+\infty} \frac{1}{\sqrt{2\pi}\sigma_n^t} \exp\left(-\frac{(x-\mu_n^t)^2}{2\sigma_n^{t\,2}}\right) dx.$$

(6)

Then the optimal survival boundary can be calculated as $S^*(t) = \frac{\mu_v^t \sigma_n^t + \mu_n^t \sigma_v^t}{\sigma_v^t + \sigma_n^t}$.

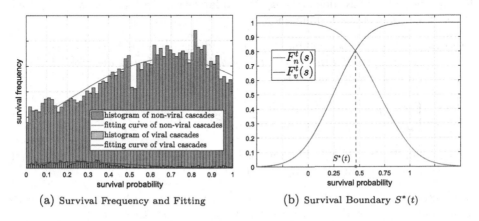

(a) Survival Frequency and Fitting   (b) Survival Boundary $S^*(t)$

Fig. 3. Survival frequency and survival boundary at time t (Color figure online)

As is shown in Fig. 3(a), given time t, we plot the frequency histograms of survival probabilities of both viral and non-viral cascades (blue bars represent non-viral ones, and red bars represent viral ones). Then we use two Gaussian distribution curves f_v^t and f_n^t to fit these two histograms. Next, to simplify our problem, we employ the cumulative distribution function of f_v^t and f_n^t, respectively denoted as $F_v^t(s)$ and $F_n^t(s)$, specifically we have,

$$F_v^t(s) = P(S < s) = \int_{-\infty}^{s} \frac{1}{\sqrt{2\pi}\sigma_v^t} \exp\left(-\frac{(x - \mu_v^t)^2}{2\sigma_v^{t\,2}}\right) \, dx, \tag{7a}$$

$$F_n^t(s) = P(S > s) = \int_{s}^{+\infty} \frac{1}{\sqrt{2\pi}\sigma_n^t} \exp\left(-\frac{(x - \mu_n^t)^2}{2\sigma_n^{t\,2}}\right) \, dx. \tag{7b}$$

Finally, we plot $F_v^t(s)$ and $F_n^t(s)$ in Fig. 3(b), and the x-coordinate of the only intersection $S^*(t)$ is the optimal survival boundary subject to time t.

3.2 Well-Definedness of Survival Boundary

In order to make the problem more complete and rigorous, in this subsection, we mainly discuss the monotonicity of the survival boundary, which is given in Definition 4, i.e., we will prove that the optimal survival boundary is itself a survival function.

In fact, during the observation period, we conclude three solid facts. First of all, the survival probabilities of both viral and non-viral cascades are naturally monotonic decreasing with time t, so the average survival probabilities of both cascades are also monotonic decreasing. Besides, non-viral cascades intuitively possess a higher survival probability, thus the average survival probability for non-viral cascades μ_n^t is reasonably larger than that of viral ones μ_v^t. Further more, real-word data shows that the survival probability range of non-viral cascades appears to be more dynamic and uncertain, which means its relative fluctuation of standard deviation σ_n^t is also larger than σ_v^t. Formally, we specify these three conclusions in Lemma 1.

Lemma 1. *For any given time t, μ_v^t, σ_v^t and μ_n^t, σ_n^t respectively represent the average survival probability and its standard deviation of viral and non-viral cascades. Given time $t' > t$, we have*

$$\begin{cases} \mu_v^t \geq \mu_v^{t'} \\ \mu_n^t \geq \mu_n^{t'} \end{cases}, \quad \mu_n^t \geq \mu_v^t, \quad \frac{\sigma_n^{t'} - \sigma_n^t}{\sigma_n^t} \geq \frac{\sigma_v^{t'} - \sigma_v^t}{\sigma_v^t}, \quad \forall \ 0 < t < t'. \tag{8}$$

Based on Definition 4 and Lemma 1, we given detailed proof that the optimal survival boundary is itself a survival function.

Theorem 1. *The optimal survival boundary $S^*(t)$ is monotonic decreasing with time t, i.e., $S^*(t)$ is also a survival function. Formally, we have*

$$S^*(t) \geq S^*(t'), \qquad \forall \ 0 < t < t', \tag{9}$$

Proof. For $\forall\, 0 < t < t'$, we have

$$
\begin{aligned}
S^*(t) - S^*(t') &= \frac{\mu_n^t \sigma_v^t + \mu_v^t \sigma_n^t}{\sigma_n^t + \sigma_v^t} - \frac{\mu_n^{t'} \sigma_v^{t'} + \mu_v^{t'} \sigma_n^{t'}}{\sigma_n^{t'} + \sigma_v^{t'}} \\
&= \frac{(\mu_n^t - \mu_v^{t'})\sigma_v^t \sigma_n^{t'} + (\mu_v^t - \mu_n^{t'})\sigma_n^t \sigma_v^{t'} + (\mu_v^t - \mu_v^{t'})\sigma_n^t \sigma_n^{t'} + (\mu_n^t - \mu_n^{t'})\sigma_v^t \sigma_v^{t'}}{(\sigma_n^t + \sigma_v^t)(\sigma_n^{t'} + \sigma_v^{t'})} \\
&\geq \frac{(\mu_v^t - \mu_v^{t'})\sigma_v^t \sigma_n^{t'} + (\mu_n^t - \mu_n^{t'})\sigma_n^t \sigma_v^{t'} + (\mu_v^t - \mu_v^{t'})\sigma_n^t \sigma_n^{t'} + (\mu_n^t - \mu_n^{t'})\sigma_v^t \sigma_v^{t'}}{(\sigma_n^t + \sigma_v^t)(\sigma_n^{t'} + \sigma_v^{t'})} \\
&\geq 0,
\end{aligned}
\tag{10}
$$

according to Lemma 1. We can easily conclude that $S^*(t) \geq S^*(t')$.

3.3 Hazard Ceiling: A Dynamic Perspective

As is defined in Definition 2, hazard function is specifically denoted as $h(t) = -\frac{dS(t)}{dt} \cdot \frac{1}{S(t)}$, we can easily monitor the hazard function $h(t)$ of a cascade when given its survival function $S(t)$.

To detect the early pattern of outbreak cascades, many previous works usually ignore the underlying arrival process of retweets, instead, they only consider the relationship between the static size of cascade and a predefined threshold [6,17], then determine whether the cascade is suffering a burst period. However, before the static size of a cascade accumulates to a certain threshold, its burst pattern can be exactly uncovered from dynamic information, such as the hazard function $h(t)$ in this problem. Intuitively, we conclude that if at a certain time t_0, the hazard function $h(t)$ of a cascade suddenly rises above a *hazard ceiling* α, in other word, $h(t_0) > \alpha$, we deem that the burst period of this cascade begins.

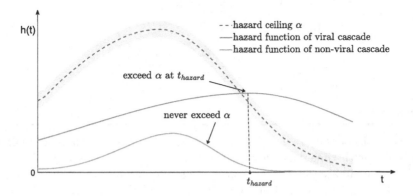

Fig. 4. Hazard functions and hazard ceiling (Color figure online)

However, instead of utilizing a fix threshold, we employ the baseline hazard function with a 5% hazard-tolerant interval as hazard ceiling (illustrated in

Fig. 4), since intuitively the characteristics of cascades may vary a lot during the diffusion process. In Fig. 4, the hazard ceiling is drawn in red dash line with a grey hazard-tolerant interval, and the red solid line and blue solid line respectively denote the hazard functions of a viral cascade and a non-viral cascade. We can clearly conclude that the blue line never exceeds hazard ceiling α, and the red line exceeds α and its hazard-tolerant interval at t_{hazard}. Therefore, we deem that at t_{hazard}, this cascade goes viral and starts to burst.

3.4   Incorporation of Two Techniques

In this subsection, we conclude our method and integrate survival boundary and hazard ceiling. The whole process of EPOC is shown in Algorithm 1.

Algorithm 1. Algorithm of EPOC

Input: training data D, test data D', threshold ρ, hazard ceiling α.
Output: status vector V, detect time T.

1 Set labels for each cascade from D using threshold ρ ;
2 Train a Cox's model C with time-dependent data D ;
3 Initialize survival function set as S ;
4 **foreach** d *in* D **do**
5 $\quad$ estimate the survival function $S_d(t)$ of d using C ;
6 $\quad$ add $S_d(t)$ to S;

7 Train an optimal survival boundary S^* with S ;
8 **foreach** d' *in* D' **do**
9 $\quad$ estimate the survival function $S_{d'}(t)$ and hazard function $h_{d'}(t)$ of d' ;
10 $\quad$ **if** $S_{d'}(t)$ *firstly falls down below* $S^*(t)$ *at time* t_0 **then**
11 $\quad\quad$ add 1 to S ;
12 $\quad\quad$ **if** $h_{d'}(t)$ *firstly rises up above* α *at time* t_1 **then**
13 $\quad\quad\quad$ add $min\{t_0, t_1\}$ to T ;
14 $\quad\quad$ **else**
15 $\quad\quad\quad$ add t_0 to T ;
16 $\quad$ **else**
17 $\quad\quad$ add 0 to S ;
18 $\quad\quad$ add *none* to T ;

19 **return** S *and* T.

In Algorithm 1, *Line1*∼*Line3* is the initialization, and especially we train the Cox's model with time-dependent features in *Line2*. Then the optimal survival boundary is estimated in *Line4*∼*Line7*, after that, we detect the burst pattern between *Line8* and *Line18* using both survival probability and hazard rate.

4   Experiments

In this section, we conduct comprehensive experiments to verify our model in early pattern detection of outbreak cascades. Firstly, we describe the data sets

(Twitter and Weibo) and five comparative state-of-the-art baselines in detail. Then we conduct our experiments as well as providing corresponding analysis.

4.1 Data Sets

We implement our model EPOC on two large real-world data sets: *Twitter* and *Weibo*. Twitter is one of the most famous social platforms in the world with annually 0.5 billion users. We densely crawl the tweets that contains hashtags with Twitter search API. In our experiments, a cascade is considered to consist of all tweets with the same hashtag. Another large dataset Weibo is from an online resource[1]. However, different from Twitter, due to the sparsity of hashtags in Weibo, a cascade is defined by the diffusion of a single microblog. More detailed information of two data sets can be found in Table 1.

Table 1. Data sets information

Data set	# of cascades	Type	Range	Year	Size (GB)
Twitter	166,076	Hashtag	Aug.13th–Sep.10th	2017	3.827
Weibo	300,000	Microblog	Sept.28th–Oct.29th	2012	1.426

4.2 Experiment Setting

For our model implementation, we need to specify some settings. Because large cascades are rare [13], in this paper, we set threshold for viral and non-viral cascades to be 95 percentile in both Twitter and Weibo, where a larger size will be regarded as viral cascade, and otherwise non-viral. As cascades are formed by large resharing activities and can potentially reach a large number of people [13], we only consider the cascades with a tweet count larger than 50 in Twitter and filter out the remains. As for Weibo, the out line is set to be 80.

In the outset of our experiments, we randomly divide each data set into two parts, 80% of the cascades is employed as training data, and the remaining one-fifth as test data. As for the hazard ceiling, in this paper, we use the baseline hazard function as ceiling and set 5% as the hazard-tolerant interval.

4.3 Baselines

From previous literatures, we select a variety of approaches from different perspectives to compare our EPOC: traditional machine learning methods, Bayesian methods, survival methods, and time series methods.

- *Linear Regression (LR)*: Linear regression is a simple and feasible way to characterize the relationship between variables and final result. In this paper, we divide the observation time into twelve time periods, then implement LR with L1 regularization based on different time periods, utilizing the observed information to predict whether or when a cascade goes viral.

---

[1] arnetminer.org/Influencelocality.

- *Support Vector Regression (SVR)*: As is widely used in various areas, SVR is a powerful regression model. We use SVR with Gaussian kernel as a baseline to predict whether a cascade will go viral or even burst in the near future. More detailed implementation of SVR is similar to linear regression.
- *PreWhether* [12]: From a Beyassian perspective, PreWhether is one of the pioneers in social content prediction, which utilizes three temporal features (sum, velocity, and acceleration) to infer the content ultimate popularity. In our experiments, we also use the same time period manner to implement PreWhether.
- *SEISMIC* [10]: SEISMIC is a point process based time series model, which takes individual's influence into consideration. Since the model itself is designed to predict the popularity of single tweets in social networks, we extend it to suit our goals of cascades' burst pattern detection.
- *SansNet* [8]: SansNet is a network-agnostic approach proposed in recent literature, which also regards the burst detection task as a judgement of viral and non-viral. This method shows its detection performance using only the time series information of a cascade.

4.4 Burst Pattern Detection

Burst or Not: To detect the early pattern of outbreak cascades, we primarily divide this problem into two steps. Firstly, we detect whether a cascade will outbreak based on the observed information. Since large cascades are arguably more striking [13], in this classification task, we employ two special metrics: k-coverage and Cost. k-coverage mainly focuses on those cascades with a very large size. Specifically, it is calculated by $\frac{n}{k}$, ($k \geq n$), where k is the number of the largest cascades being concentrated on, and n denotes the number of cascades we successfully detect from the top-k viral cascades. Here in this work, n equals 50. Cost (more precisely called sensitive cost) is a targeted metric, which is selected to handle the problem of unequal-cost. If a viral cascade (like a rumor [1]) is classified to be non-viral, it will cost a lot when this cascade gets larger and causes a big trouble. On the contrary, if we misclassify a non-viral cascade, it only costs some additional labor. Cost is specified in Eq. (11),

$$Cost = \frac{FNR \times p \times Cost_{FN} + FPR \times (1 - p) \times Cost_{FP}}{p \times Cost_{FN} + (1 - p) \times Cost_{FP}}, \qquad (11)$$

where FNR is the false negative rate, FPR is the false positive rate, p is the proportion of viral cascades in all cascades, $Cost_{FN}$ and $Cost_{FP}$ are entries in cost matrix. We also specify the cost matrix in Table 2.

Performance Analysis. The results of burst detection are aggregated in Table 3 and the underlined numbers show the best results. One can see that in general, our EPOC performs relatively better than five baselines in terms of both k-coverage and Cost. LR also shows great performance in k-coverage on Weibo, and it works much better than SVR and SEISMIC, which means that the L1 regularization comes into effect. As a probabilistic model, PreWhether gives a

Table 2. Unequal-cost matrix

Real class	Detected class	
	Viral	Non-viral
Viral	$Cost_{TP} = 0$	$Cost_{FN} = 5$
Non-viral	$Cost_{FP} = 1$	$Cost_{TN} = 0$

slightly poor detection result due to the assumption that all the features are independent. Though less effective than EPOC, SansNet outperforms all the other baselines in this classification task, since SansNet only employs one feature from cascades. However, it is plausible to note that SansNet gives stable k-coverage and Cost results in both Twitter and Weibo, which indicates that survival perspective models are suitable in this scenario.

Table 3. Result of burst detection on Twitter

		LR	SVR	PreWheter	SEISMIC	SansNet	EPOC
Twitter	k-coverage	0.7781	0.5969	0.7490	0.5188	0.8275	0.8471
	Cost	0.1032	0.0998	0.0956	0.1677	0.0776	0.0701
Weibo	k-coverage	0.6805	0.4918	0.6512	0.4589	0.7720	0.7784
	Cost	0.0951	0.1229	0.1271	0.1581	0.0961	0.0881

Change of Observation Periods. To explore the connection between observing period and the performance of methods, we conduct experiments on Twitter with six time periods from 0.5 to 3 h and organize the results in Fig. 5. Intuitively, the performances of EPOC and five baselines improve gradually as the observing period increases. We can clearly see that EPOC performs the best with a pretty high k-coverage at about 87% and a pretty low cost at around 0.068. Besides, it is worth noticing that SEISMIC is far behind other approaches no matter in k-coverage or in Cost, which suggests that time series model depends on a relatively longer observing period, and can not do a good job the burst detection task at early stage.

Time Ahead (Similar to *EPA* from [8]**):** Further, we try to figure out how early we can detect the outbreak cascades with EPOC. As [13] states, it is a pathological task to estimate the final size of a cascade if only given a short initial portion, since almost all cascades are small. Besides, comparing with getting the final size of a cascade, it is more meaningful and practical to detect how early a cascade will break out. Therefore, in this experiment of Twitter and Weibo, we only probe into the early pattern of outbreak cascades, and mainly focus on *absolute time ahead*, which is the interval between the predicted burst time $t_{predict}$ and the actual burst time t_{actual}. Specifically during the experiments, if $t_{actual} \geq t_{predict}$, we record as $t_{actual} - t_{predict}$, and otherwise, 0. Also, we consider the *relative time ahead*, which is given by $\frac{t_{actual} - t_{predict}}{t_{actual}}$ or 0.

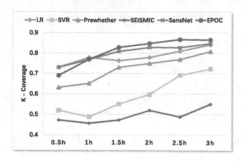 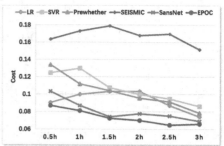

Fig. 5. k-Coverage and cost under different observing periods on Twitter

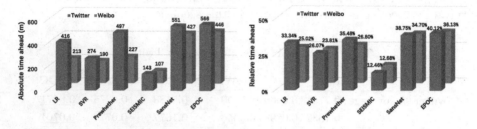

Fig. 6. Absolute and relative time ahead on Twitter and Weibo

Performance Analysis. Figure 6 illustrates the corresponding experiment results on Twitter and Weibo. We conclude that all the methods have a similar rank in terms of absolute time ahead and relative time ahead. SansNet and our EPOC steadily keep a leading role in this regression task at about 38.75% and 40.12% respectively ahead of the actual burst time in Twitter. PreWhether and LR work mildly, and they can successfully predict the occurrence of burst, when the diffusion process of cascades only goes on about two thirds. Though SVR possesses much better performance than the poorest SEISMIC, it falls behind comparing with other baselines, which suggests that the notion of support vector may not be applicable in this problem.

5   Related Work

In recent years, social networks have successfully attracted researchers' attention, and plenty of achievements have been made in the past few decades, especially when it comes to the study of information cascades, including the prediction of cascade size, how the cascade grows and disseminates, etc.

5.1   Information Cascade and Social Networks

The study of information cascades has been going for a long time, and it is of great use in many applications, such as meme tracking [2], stock bubble diagnosis [3], and sales prediction [4]. The literature concerning cascade in social networks can be divided into three categories. The first category lays on user level prediction. One of the pioneers is Iwata et al. [18], they propose a Bayesian inference model with stochastic EM algorithm, trying to discover the latent influence among online users. [19] also utilizes user-related features to help social event detection. Additionally, some other researchers also analyze the topology, since structural feature is said to be one of the predictors of cascade size [13]. PageRank of retweeting graph is taken into consideration [20], while [21] utilizes the number of directed followers as one of the important infectors. Another significant category is temporal features. Many experimental results, such as [9,10], reveal that temporal features are the most effective type of indicators. To depict the connection between early cascade and its final state, both [5,12] propose Bayesian networks with temporal information. Other temporal information, like mean time and maximum time interval, has also been considered [9].

5.2 Outbreak Detection and Modeling

Burst or outbreak, defined as "a brief period of intensive activity followed by long period of nothingness" [6], is a common phenomenon during the diffusion of social content, which is worthy of studying and may bring benefits to modern society. Existing works probing into cascades mainly focus on prediction of its future popularity [5,12,20] or final aggregate size [10,13]. However, how to detect the burst pattern of large cascade in early stage remains an intriguing problem. Recently, based on the transformation of time window, Wang et al. [6] proposes a classification model to predict the burst time of cascade. Unfortunately, their approach acquires laborious feature extraction, and the traditional classifiers they used can hardly take the best use of the features. [17] implements a logistic model, which considers all the nodes as cascade sensors. Just as bad, when the number of nodes in networks turns to be billions, the implementation of this method will be particularly difficult.

In this work, adopting survival theory, we can exactly overcome these drawbacks from the perspective of cascade dynamics. Other researchers also employ survival models to understand the burst of cascades. SansNet is proposed in [8], predicting whether and when a cascade goes viral. This approach utilizes only the size of cascades as feature, making it weak to apply to multiply cases, since the features of an author [22] and the inherent network [13] are sometimes more important than features from cascade itself [22]. Another drawback of this approach is that the survival curve cannot totally reveal the status of cascades.

6 Conclusion and Perspectives

In social networks, detecting whether and when a cascade will outbreak is a non-trivial but beneficial task. In this paper, we novelly employ survival theory, proposing a survival model EPOC to detect the early pattern of outbreak cascades. We extract both dynamic and static features from cascades and utilize

Gaussian distributions to characterize their survival probabilities, then accompanied with hazard rate, we successfully detect the burst pattern of cascades at very early stage. Extensive experiment shows that our EPOC outperforms five state-of-the-art methods in this practical task.

As future work, firstly we will mainly concentrate on how to choose a better standard baseline for hazard ceiling, and more experiment observation might be made. Then, we will consider more influential and relevant features or try another suitable survival theory based model. Finally, we hope that our work will pave ways to richer and deeper understanding of cascades.

Acknowledgements. This work is supported by the Program of International S&T Cooperation (2016YFE0100300), the China 973 project (2014CB340303), the National Natural Science Foundation of China (61472252, 61672353), the Shanghai Science and Technology Fund (17510740200), and CCF-Tencent Open Research Fund (RAGR20170114).

References

1. Adrien, F., Lada, A., Dean, E., Justin C.: Rumor cascades. In: ICWSM (2014)
2. Bai, j., Li, L., Lu, L., Yang, Y., Zeng, D.: Real-time prediction of meme burst. In: IEEE ISI (2017)
3. Jiang, Z., Zhou, W., Didier, S., Ryan, W., Ken, B., Peter, C.: Bubble diagnosis and prediction of the 2005–2007 and 2008–2009 Chinese stock market bubbles. J. Econ. Behav. Organ. **74**, 149–162 (2010)
4. Daniel, G., Ramanathan, V. Ravi, K., Jasmine, N., Andrew, T.: The predictive power of online chatter. In: SIGKDD (2005)
5. Ma, X., Gao, X., Chen, G.: BEEP: a Bayesian perspective early stage event prediction model for online social networks. In: ICDM (2017)
6. Wang, S., Yan, Z., Hu, X., Philip, S., Li, Z.: Burst time prediction in cascades. In: AAAI (2015)
7. Matsubara, Y., Sakurai, Y., Prakash, B., Li, L., Faloutsos C.: Rise and fall patterns of information diffusion: model and implications. In: SIGKDD (2012)
8. Subbian, K., Prakash, B., Adamic, L.: Detecting large reshare cascades in social networks. In: WWW (2017)
9. Gao, S., Ma, J., Chen, Z.: Effective and effortless features for popularity prediction in microblogging network. In: WWW (2014)
10. Zhao, Q., Erdogdu, M., He, H., Rajaraman, A., Leskovec, J.: SEISMIC: a self-exciting point process model for predicting tweet popularity. In: SIGKDD (2015)
11. Gao, S., Ma, J., Chen, Z.: Modeling and predicting retweeting dynamics on microblogging platforms. In: WSDM (2015)
12. Liu, W., Deng, Z, Gong, X., Jiang, F., Tsang, I.: Effectively predicting whether and when a topic will become prevalent in a social network. In: AAAI (2015)
13. Cheng, J., Adamic, L., Dow, P., Kleinberg, J., Leskovec, J.: Can cascades be predicted? In: WWW (2014)
14. Cox, R.: Regression models and life-tables. In: Kotz, S., Johnson, N.L. (eds.) Breakthroughs in Statistics, pp. 527–541. Springer, New York (1992). https://doi.org/10.1007/978-1-4612-4380-9_37
15. Aalen, O., Borgan, O., Gjessing, H.: Survival and Event History Analysis. Springer, Heidelberg (2008). https://doi.org/10.1007/978-0-387-68560-1

16. Anderson, J.R., Bernstein, L., Pike, M.C.: Approximate confidence intervals for probabilities of survival and quantiles in life-table analysis. Int. Biom. Soc. JSTOR **38**(2), 407–416 (1982)
17. Cui, P., Jin, S., Yu, L., Wang, F., Zhu, W., Yang, S.: Cascading outbreak prediction in networks: a data-driven approach. In: SIGKDD (2013)
18. Iwata, T., Shah, A., Ghahramani, Z.: Discovering latent influence in online social activities via shared cascade poisson processes. In: SIGKDD (2013)
19. Mansour, E., Tekli, G., Arnould, P., Chbeir, R., Cardinale, Y.: F-SED: feature-centric social event detection. In: Benslimane, D., Damiani, E., Grosky, W.I., Hameurlain, A., Sheth, A., Wagner, R.R. (eds.) DEXA 2017. LNCS, vol. 10439, pp. 409–426. Springer, Cham (2017). https://doi.org/10.1007/978-3-319-64471-4_33
20. Hong, L., Dan, O., Davison, B.: Predicting popular messages in Twitter. In: WWW (2011)
21. Feng, Z., Li, Y., Jin, L., Feng, L.: A cluster-based epidemic model for retweeting trend prediction on micro-blog. In: Chen, Q., Hameurlain, A., Toumani, F., Wagner, R., Decker, H. (eds.) DEXA 2015. LNCS, vol. 9261, pp. 558–573. Springer, Cham (2015). https://doi.org/10.1007/978-3-319-22849-5_39
22. Petrovic, S., Osborne, M., Lavrenko, V.: RT to Win! Predicting message propagation in Twitter. In: ICWSM (2011)

Temporal and Spatial Databases

Analyzing Temporal Keyword Queries for Interactive Search over Temporal Databases

Qiao Gao[1(✉)], Mong Li Lee[1], Tok Wang Ling[1], Gillian Dobbie[2],
and Zhong Zeng[3]

[1] National University of Singapore, Singapore, Singapore
{gaoqiao,leeml,lingtw}@comp.nus.edu.sg
[2] University of Auckland, Auckland, New Zealand
g.dobbie@auckland.ac.nz
[3] Data Center Technology Lab, Huawei, Hangzhou, China
zengzhong4@huawei.com

Abstract. Querying temporal relational databases is a challenge for
non-expert database users, since it requires users to understand the
semantics of the database and apply temporal joins as well as temporal conditions correctly in SQL statements. Traditional keyword search
approaches are not directly applicable to temporal relational databases
since they treat time-related keywords as tuple values and do not
consider the temporal joins between relations, which leads to missing
answers, incorrect answers and missing query interpretations. In this
work, we extend keyword queries to allow the temporal predicates, and
design a schema graph approach based on the Object-Relationship-
Attribute (ORA) semantics. This approach enables us to identify temporal attributes of objects/relationships and infer the target temporal
data of temporal predicates, thus improving the completeness and correctness of temporal keyword search and capturing the various possible
interpretations of temporal keyword queries. We also propose a two-level
ranking scheme for the different interpretations of a temporal query, and
develop a prototype system to support interactive keyword search.

1 Introduction

Temporal relational databases enable users to keep track of the changes of data
and associate a time period to the temporal data to indicate its valid time period
in the real world. Then users can retrieve information by specifying the time
period (e.g. find patients who have fever in 2015), or the temporal relationship
between the time periods of temporal data (e.g. find patients who have cough
and fever *on the same day*). While such queries can be written precisely in
SQL statements, it is a challenge for non-expert database users to write the
statements correctly since it requires users to understand the temporal database
schema well, associate the temporal conditions to the appropriate temporal data,
and apply temporal joins between multiple relations.

© Springer Nature Switzerland AG 2018
S. Hartmann et al. (Eds.): DEXA 2018, LNCS 11029, pp. 355–371, 2018.
https://doi.org/10.1007/978-3-319-98809-2_22

Keyword queries over relational databases free users from writing complicated SQL statements and has become a popular search paradigm. However, introducing temporal periods in keyword queries may lead to the problems of (a) missing answers, (b) missing interpretations and (c) incorrect answers if the temporal periods are not handled properly, as we will elaborate.

Missing Answers. This issue arises because traditional keyword search engines treat time-related keywords as tuple values. Figure 1 shows a hospital database that records the temperature and symptoms of patients, salary of doctors, and the dates that patients consult doctors. Suppose we issue a keyword query {Patient cough 2015-05-10} to find patients who have cough on 2015-05-10. Traditional keyword search engine will retrieve patient $p1$ since tuple t_{31}:<$p1$, cough, 2015-05-10,2015-05-13> in relation PatientSymptom matches the DATE keyword "2015-05-10". Patient $p2$ is not returned as an answer even though tuple t_{34}:<$p2$, cough, 2015-05-07,2015-05-11> indicates that $p2$ has a cough on 2015-05-10. This is because $p2$ does not have a tuple matching "2015-05-10" in PatientSymptom.

The work in [9] first adapts relational keyword search to temporal relational database by allowing keywords to be constrained by time periods, and temporal predicates such as BEFORE and OVERLAP between keywords. As such, their method will check if "2015-05-10" is contained within the time period of patients' symptom and retrieve both patients $p1$ and $p2$.

Patient

	Pid	Pname	Gender
t_{11}	p1	Smith	Male
t_{12}	p2	Green	Male
t_{13}	p3	Alice	Female

PatientTemperature

	Pid	Temperature	Temperature_Date
t_{21}	p1	36.7	2015-05-10
t_{22}	p1	39.2	2015-05-11
t_{23}	p1	36.3	2015-06-04
t_{24}	p2	36.7	2015-05-07
t_{25}	p2	38.8	2015-07-13
t_{26}	p3	37.2	2015-10-21

PatientSymptom

	Pid	Symptom	Symptom_Start	Symptom_End
t_{31}	p1	cough	2015-05-10	2015-05-13
t_{32}	p1	fever	2015-05-11	2015-05-13
t_{33}	p1	cough	2015-06-03	2015-06-07
t_{34}	p2	cough	2015-05-07	2015-05-11
t_{35}	p2	fever	2015-07-13	2015-07-15
t_{36}	p3	headache	2015-10-19	2015-10-23

Clinic

	Cid	Cname
t_{41}	c1	Internal Medicine
t_{42}	c2	Cardiology

Doctor

	Did	Dname	Doctor_Start	Doctor_End	Cid
t_{51}	d1	Smith	2000-01-01	2016-12-31	c1
t_{52}	d2	George	2005-01-01	now	c2
t_{53}	d3	John	2010-01-01	now	c2

DoctorSalary

	Did	Salary	Salary_Start	Salary_End
t_{61}	d1	8,000	2000-01-01	2004-12-31
t_{62}	d1	10,000	2005-01-01	2012-12-31
t_{63}	d1	12,000	2013-01-01	2016-12-31
t_{64}	d2	8,000	2005-01-01	Now
t_{65}	d2	10,000	2010-01-01	Now

Consult

	Pid	Did	Consult_Date
t_{71}	p1	d1	2015-05-12
t_{72}	p1	d2	2015-05-13
t_{73}	p1	d1	2015-05-15
t_{74}	p2	d1	2015-05-12
t_{75}	p2	d2	2015-07-13
t_{76}	p3	d1	2015-10-21

Fig. 1. Example hospital database.

Missing Interpretations. This issue arises because the work in [9] assume that a *time condition* (temporal predicates and time periods) is always associated with the nearest keyword in the query. This may miss other possible interpretations and their answers to the query. For example, the keyword query {Patient Doctor DURING [2015-01-01,2015-01-31]} has two possible interpretations depending on the user search intention:

- find patients who consult doctor during January 2015,
- find patients who consult doctor who work in hospital during January 2015.

By assuming that the time condition "DURING [2015-01-01,2015-01-31]" is associated with the nearest keyword "Doctor" that matches the relation name Doctor with a valid time period [*Doctor_Start*, *Doctor_End*] indicating the work period of doctor in the hospital, the work in [9] will only return answers for the second interpretation, and miss answers for the first interpretation which is more likely the user search intention.

Incorrect Answers. This issue arises when the time periods in a join operation are not handled correctly, in other words, there is no support for temporal join. Consider the query {Patient temperature fever DURING [2015-05-01,2015-05-31]} to find the temperature of patients who had a fever during May 2015. This requires a temporal join (joining two records if their keys are equal and their time periods intersect [5]) of the relations PatientSymptom and PatientTemperature. The expected result is 39.2, obtained by joining tuples t_{22} and t_{32}, which gives the temperature of patient p_1 who had a fever during May 2015. The work in [9] only applies the time condition to the nearest keyword "fever" without considering the intersection of time periods during the join operation. Then tuples t_{21} and t_{23} are also joined with tuple t_{32}, adding temperatures 36.7 and 36.3 to the results, which are incorrect because they are not associated with the fever that p_1 had in May 2015.

In this work, we generalize the syntax for temporal keyword queries to include basic keywords and temporal keywords. We design a semantic approach to process complex temporal keyword queries involving temporal joins, taking into consideration the various ways a time condition can be applied. We use an Object-Relationship-Mixed (ORM) schema graph to capture the semantics of objects, relationships and attributes in the temporal databases. With this, we can generate a set of initial query patterns to capture the interpretations of the basic keywords of a query. Then we infer the target time period of the temporal predicate and generate temporal constraints to capture the different interpretations of temporal keywords including an interpretation involving temporal join. We propose a two-level ranking scheme for the different interpretations of a temporal query, and develop a prototype system to support interactive keyword search over a temporal database. Finally, a set of SQL statements is generated from the user-selected query patterns with the temporal constraints translated into temporal joins or select conditions correctly. Experiments on two datasets show the effectiveness of our proposed approach to handle complex temporal keyword queries and retrieve relevant results.

2   Related Work

Methods for keyword search over temporal databases [9,13] can be extended from existing relational keyword search methods which can be broadly classified into data graph [3,6,8,10,16] and schema graph [2,7,11,12,14,15] approaches.

The former models a database as a graph where each node represents a tuple and each edge represents a foreign key-key reference, and an answer to a keyword query is a minimal connected subgraph (Steiner tree) containing all the keywords. The latter models a database as a graph where each node represents a relation and each edge represents a foreign key-key constraint, and a keyword query is translated into a set of SQL statements. All these works do not distinguish the Object-Relationship-Attribute (ORA) semantics in the database, which leads to incomplete and meaningless results. They also do not handle time-related keywords properly and do not support temporal joins between relations, which leads to missing answers and missing interpretations as we have highlighted.

[9] extends keyword queries with temporal predicates and focuses on keyword query efficiency utilizing a data graph approach. However, this work applies the temporal predicate to the nearest keyword in the query and does not consider temporal joins between relations, which leads to missing interpretations and incorrect answers. [13] extends the solution in [8] to improve the efficiency of keyword query over temporal graphs. This work does not handle queries with implicit time period (see Sect. 4), and also suffers from missing interpretations. Futher, without considering the ORA semantics, both works [9,13] also have the problem of missing answers and returning incomplete and meaningless results.

The works in [17,18] distinguish the ORA semantics and extend keyword queries with meta-data to reduce the ambiguity of keyword queries, and retrieve user intended information and meaningful results. Our work builds upon these works and focuses on identifying the temporal relations in a temporal database and infers the target temporal period of the temporal predicate in the database.

3   Preliminaries

Temporal databases support transaction time and valid time. Here, we focus on *valid time* which can be a closed time period or a time point. Besides augmenting keyword queries with temporal predicates and time periods, users can explicitly indicate their search intention with metadata keywords that match relation/attribute names to reduce the ambiguity of queries.

Definition 1. *A temporal keyword query $Q = \{k_1 \cdots k_n\}$ is a sequence of basic and temporal keywords with syntax constraints.*

*A **basic keyword** is*

- *a data-content keyword that matches a tuple value, or*
- *a metadata keyword that matches a relation name or an attribute name.*

*A **temporal keyword** is*

- *a time period expressed as a closed time period $[s, e]$ or time point $[s]$, or*
- *a temporal predicate such as AFTER, DURING [1].*

*The **syntax constraints** are*

- *the first keyword k_1 and the last keyword k_n cannot be a temporal predicate,*
- *time periods must be adjacent to a temporal predicate,*

– *for a temporal predicate k_i, previous keyword k_{i-1} and next keyword k_{i+1} cannot be temporal predicates, and k_{i-1} and k_{i+1} cannot both be time periods.*

Basic keywords specify *what* information users care about, while temporal keywords provide *time condition* on the information. Temporal predicates are based on [1] and Table 1 gives their mathematical meanings. Syntax constraints imposed on the keywords ensure meaningful temporal keyword queries, e.g., it does not make sense to have a temporal predicate *AFTER* as the first keyword of a query, and it is meaningless to have a temporal predicate with two time operands.

Table 1. Mathematical meaning of temporal predicates

Temporal predicate	Meaning	Temporal predicate	Meaning
$[s_1, e_1]$ BEFORE $[s_2, e_2]$	$e_1 < s_2$	$[s_1, e_1]$ AFTER	$s_1 > e_2$
$[s_1, e_1]$ MEETS $[s_2, e_2]$	$e_1 = s_2$	$[s_1, e_1]$ MET_BY $[s_2, e_2]$	$s_1 = e_2$
$[s_1, e_1]$ DURING $[s_2, e_2]$	$s_1 > s_2 \wedge e_1 < e_2$	$[s_1, e_1]$ CONTAINS $[s_2, e_2]$	$s_1 < s_2 \wedge e_1 > e_2$
$[s_1, e_1]$ STARTS $[s_2, e_2]$	$s_1 = s_2 \wedge e_1 < e_2$	$[s_1, e_1]$ STARTED_BY $[s_2, e_2]$	$s_1 = s_2 \wedge e_1 > e_2$
$[s_1, e_1]$ FINISHES $[s_2, e_2]$	$s_1 > s_2 \wedge e_1 = e_2$	$[s_1, e_1]$ FINISHED_BY $[s_2, e_2]$	$s_1 < s_2 \wedge e_1 = e_2$
$[s_1, e_1]$ EQUAL $[s_2, e_2]$	$s_1 = s_2 \wedge e_1 = e_2$	$[s_1, e_1]$ INTERSECT $[s_2, e_2]$	$s_1 \leqslant e_2 \wedge e_1 \geqslant s_2$
$[s_1, e_1]$ OVERLAPS $[s_2, e_2]$	$s_1 < s_2 \wedge s_2 < e_1 < e_2$	$[s_1, e_1]$ OVERLAPPED_BY $[s_2, e_2]$	$e_1 > e_2 \wedge s_2 < s_1 < e_2$

A database can be represented using an Object-Relationship-Mixed (ORM) schema graph $G = (V, E)$. Each node $u \in V$ is an object/relationship/mixed node comprising of an object/relationship/mixed relation and its component relations. An object (or relationship) relation captures the single-valued attributes of objects (or relationships). Multivalued attributes are captured in component relations. A mixed relation contains information of both objects and many-to-one relationships. Two nodes u and v are connected by an undirected edge $(u, v) \in E$ if there exists a foreign key-key constraint from the relations in u to those in v. Figure 2 shows the ORM schema graph for the database in Fig. 1. Note that an ORM node can have multiple relations, e.g., node **Patient** contains object relation *Patient* and component relations *PatientSymptom* and *PatientTemperature*.

Fig. 2. ORM schema graph of Fig. 1

Based on the ORM schema graph, we can generate a set of query patterns to capture the possible interpretations of the query basic keywords. Details of pattern generation process are in [17]. We illustrate the key ideas with an example.

Example 1 (Query Patterns). Consider the query {Smith cough} which contains basic keywords *Smith* and *cough*. The keyword *Smith* matches some tuple value in relation Patient, while keyword *cough* matches some tuple value in component relation PatientSymptom (see Fig. 1). These relations are mapped to the Patient node in the ORM schema graph in Fig. 2. Based on the matches, we generate the query pattern in Fig. 3(a) which shows an annotated Patient object node.

Another interpretation which finds patients who have a cough and consult doctor Smith is shown in Fig. 3(b). This is because the keyword *Smith* also matches tuple values in the Doctor relation.   □

(a) Query pattern P_1 (b) Query pattern P_2

Fig. 3. Query patterns for query {Smith cough}

4   Temporal Query Interpretations

A keyword query that has only basic keywords can be interpreted using the traditional keyword search. However, in temporal databases, we have another interpretation involving temporal join.

Recall that a query pattern P has a set of object/relationship/mixed nodes. We identify the set of temporal relations S with respect to P that will be involved in a temporal join. A relation R is a *temporal relation* if it has a time period $R[A.Start, A.End]$ or a time point $R[A.Date]$. Here, we also represent a time point $R[A.Date]$ as a time period $R[A.Date, A.Date]$.

For each node $u \in P$, we add the temporal relation $R \in u$ to S if R is the object/relationship/mixed relation of u, or if R is matched by some query keywords. If $|S| > 1$, then P has two interpretations. The first interpretation does not consider the temporal aspect of relations in P, i.e., no temporal join or *null* temporal constraint. The second interpretation involves a temporal join between all the temporal relations $R_1, R_2, \cdots, R_m$ in S, indicated by a temporal constraint that restricts the temporal objects, relationships and attributes in P to the same time periods:

$R_1[A_1.Start, A_1.End]$ INTERSECT $R_2[A_2.Start, A_2.End]$ INTERSECT
$\cdots$ $R_m[A_m.Start, A_m.End]$

In other words, we can generate a set of temporal constraints for each query pattern. One query pattern with one temporal constraint forms one complete interpretation of a keyword query.

Example 2 (Temporal constraints). Figure 4 shows a query pattern P_3 for the query {Patient cough Doctor}. Keyword *Doctor* matches the name of the temporal relation *Doctor* in Doctor node, while keyword *cough* matches some tuple values in the temporal relation *PatientSymptom* in **Patient** node. The set of temporal relations $\mathcal{S} = \{Doctor, Consult, PatientSymptom\}$. Table 2 shows the temporal constraints generated to interpret P_3. One interpretation has a *null* temporal constraint TC_{11} and finds patients who had a cough and consulted a doctor without any consideration of time. Another interpretation has a temporal constraint TC_{12} and finds patients who consulted a doctor when they had a cough, which requires temporal joins of the relations in $\mathcal{S}$.   □

Fig. 4. Query pattern P_3

Table 2. Temporal constraints for {Patient cough Doctor} w.r.t. P_3 in Fig. 4

TC_{11}	*null*
TC_{12}	Doctor[Doctor_Start,Doctor_End] INTERSECT Consult[Consult_Start,Consult_End] INTERSECT PatientSymptom[Symptom_Start,Symptom_End]

On the other hand, when a query has temporal keywords, there is always some temporal predicate TP and the time period may be explicit or implicit.

Queries with Explicit Time Period. Consider the query {Patient cough Doctor DURING [2015-01-01,2015-12-31]} which has a temporal predicate $DURING$ with an explicit time period [2015-01-01,2015-12-31] forming a *time condition*. A query pattern for this query is shown in Fig. 4, which can be generated without considering the temporal keywords. We can apply the time condition "$DURING$ [2015-01-01,2015-12-31]" to the underlying temporal relations associated with this query pattern in several ways, leading to different interpretations of the query. Table 3 shows all possible interpretations of the time conditions in the form of temporal constraints. Some example interpretations include:

1. (TC_{23}) Apply time condition to temporal relation *Consult* to find patients who had a cough and consulted a doctor during this period.
2. (TC_{24}) Apply time condition to temporal relation *PatientSymptom* to find patients who had a cough during this period and consulted a doctor.

The above interpretations assume the traditional join between the relations that matches the basic query keywords. An additional interpretation is obtained when we apply the time condition after performing a temporal join of the relations. This will find patients who had a cough (during this period) and they consulted a doctor (during this period) who worked in a clinic during this period (TC_{26}).

All the interpretations without temporal join can be obtained by applying the time condition to each temporal relation in a query pattern P. Note that these include temporal component relations in P which are not matched by query keywords, e.g., TC_{22} and TC_{25} in Table 3. The interpretation involving temporal join is obtained by identifying the set of temporal relations S in P that are involved in the temporal join and applying the time condition to restrict the temporal objects, relationships and attributes in P to the same time periods.

Table 3. Temporal constraints for query {Patient cough Doctor DURING [2015-01-01, 2015-12-31]} w.r.t query pattern P_3 in Fig. 4.

TC_{21}	Doctor[Doctor_Start,Doctor_End] DURING [2015-01-01,2015-12-31]
TC_{22}	DoctorSalary[Salary_Start,Salary_End] DURING [2015-01-01,2015-12-31]
TC_{23}	Consult[Consult_Start,Consult_End] DURING [2015-01-01,2015-12-31]
TC_{24}	PatientSymptom[Symptom_Start,Symptom_End] DURING [2015-01-01,2015-12-31]
TC_{25}	PatientTemperature[Temperature_Start,Temperature_End] DURING [2015-01-01,2015-12-31]
TC_{26}	(Doctor[Doctor_Start,Doctor_End] INTERSECT Consult[Consult_Start,Consult_End] INTERSECT PatientSymptom[Symptom_Start,Symptom_End]) DURING [2015-01-01,2015-12-31]

Queries with Implicit Time Period. Consider the query {Patient Doctor AFTER cough} which has a temporal predicate AFTER with no explicit time period. The keyword *cough* matches the temporal relation *PatientSymptom*, and the time period for this query is derived from the tuples that match the keyword *cough*. A query pattern for this query is the same as P_3 in Fig. 4, since these two queries have the same set of basic keywords. Depending on where we apply the time condition, AFTER cough, to the underlying temporal relations associated with this query pattern, we have a number of interpretations, including:

1. (TC_{31}) Apply the time condition to temporal relation *Doctor* to find patients who consulted a doctor who worked in a clinic after the patient had a cough.
2. (TC_{33}) Apply the time condition to temporal relation *Consult* to find patients who consulted a doctor after the patient had a cough.

Note that since a patient could consult doctor several times after s/he had a cough, we may have a set of time periods to consider for the time condition AFTER cough. Here we take the time period with the earliest start time, i.e., the nearest consultation after a patient has cough. Again, these interpretations assume the traditional join between the relations that match the basic keywords in the query. We have an additional interpretation when we apply the time condition after performing a temporal join of the relations (TC_{35}). Table 4 shows the temporal constraints obtained. Since the temporal relation *PatientSymptom* (matched by keyword *cough*) is already in the time condition and there is no other keywords matches this relation, we will not apply the time condition to this relation and not include it in the temporal join.

Table 4. Temporal constraints for query {Patient Doctor AFTER cough} w.r.t. query pattern P_3 in Fig. 4.

TC_{31}	Doctor[Doctor_Start,Doctor_End] AFTER PatientSymptom[Symptom_Start,Symptom_End]
TC_{32}	DoctorSalary[Salary_Start,Salary_End] AFTER PatientSymptom[Symptom_Start,Symptom_End]
TC_{33}	Consult[Consult_Start,Consult_End] AFTER PatientSymptom[Symptom_Start,Symptom_End]
TC_{34}	PatientTemperature[Temperature_Start,Temperature_End] AFTER PatientSymptom[Symptom_Start,Symptom_End]
TC_{35}	(Doctor[Doctor_Start,Doctor_End] INTERSECT Consult[Consult_Start,Consult_End]) AFTER PatientSymptom[Symptom_Start,Symptom_End]

Details of the temporal constraints generation is given in [4]. A special case occurs when the keywords before and after a temporal predicate matches the same relation, e.g., query {Patient Doctor fever AFTER cough} has both keywords fever and cough matching the same temporal relation *PatientSymptom*. Figure 5 shows the corresponding query pattern. We have one interpretation where we apply the temporal predicate to the temporal relation *PatientSymptom* to find patients who consulted a doctor and had a fever after a cough (TC_{41}), and another interpretation where we apply the temporal predicate after performing a temporal join of the relations (TC_{42}). Table 5 shows the constraints obtained.

Fig. 5. Query pattern for {Patient Doctor fever AFTER cough}.

Table 5. Temporal constraints for query {Patient Doctor fever AFTER cough} w.r.t. query pattern in Fig. 5.

TC_{41}	PatientSymptom$_1$[Symptom_Start,Symptom_End] AFTER PatientSymptom$_2$[Symptom_Start,Symptom_End]
TC_{42}	(Doctor[Doctor_Start,Doctor_End] INTERSECT Consult[Consult_Start,Consult_End] INTERSECT PatientSymptom$_1$[Symptom_Start,Symptom_End]) AFTER PatientSymptom$_2$[Symptom_Start,Symptom_End]

5   Ranking Temporal Query Interpretations

We have discussed how a temporal keyword query can have multiple query patterns, and each pattern can have multiple temporal constraints depending on how the temporal predicate is applied to the underlying temporal relations.

In this section, we describe a two-level ranking mechanism where the first level ranks query patterns without considering the temporal constraints, and the second level ranks the temporal constraints within each query pattern.

For the first level ranking, we adopt the approach in [18]. This work identifies the *target* and *value condition* nodes in a query pattern P. A target node specifies the search target of the query, typically the node that matches the first query keyword, while a value condition node is annotated with the attribute value conditions. Query patterns are ranked based on their number of object/mixed nodes and the average number of object/mixed nodes between the target and value condition nodes. Patterns with fewer object/mixed nodes and a smaller average number of object/mixed nodes between target and value condition nodes are ranked higher. Equation (1) gives the scoring function for this first level ranking.

$$score_1(P) = \frac{1}{N * \sum_{v \in V} \frac{count(u, v, P)}{|V|}} \tag{1}$$

where u is the target node, V is the set of value condition nodes, $count(u, v, P)$ is the total number of object/mixed nodes in the path connecting two nodes u and v in P, and N is the number of object and mixed nodes in P.

The query {Smith cough} has two query patterns P_1 and P_2 (see Fig. 3), and P_1 is ranked higher than P_2. The Patient node in P_1 is both a value condition node and a target node, with $count(Patient, Patient, P_1) = 1$ and $score_1(P_1) = \frac{1}{1*1} = 1$. For pattern P_2, Doctor and Patient nodes are value condition nodes, and Doctor node is the target node since the first keyword *Smith* matches doctor's name. We have $count(Doctor, Patient, P_2) = 2$ and $score_1(P_2) = \frac{1}{2*\frac{2+1}{2}} = \frac{1}{3}$.

For the second level ranking, we compute a score for each temporal constraint TC of a query pattern P. The temporal constraint with temporal join is ranked the highest since it involves all the temporal relations related to the query. Note that there is at most one temporal constraint with temporal join with respect to one query pattern. For the temporal constraints without temporal join, we first identify the *time condition* node in the query pattern with respect to this constraint. A time condition node contains the temporal relation that the time condition is applied to. There is only one time condition node for each temporal constraint without temporal join. Here, we count the number of object/mixed nodes between target node and time condition node in the query pattern, and rank temporal constraint with smaller number of object/mixed nodes between target node and time condition node higher. Equation (2) gives the ranking function:

$$score_2(TC, P) = \begin{cases} 2 & \text{if TC has temporal join} \\ \frac{1}{count(u, w, P)} & \text{otherwise} \end{cases} \tag{2}$$

where $u \in P$ is the target node, $w \in P$ is the time condition node w.r.t temporal constraint TC. The maximum score for a temporal constraint without temporal join is 1. Temporal constraint with temporal join has a score of 2 so that it is always ranked highest among all constraints.

Note that when the query only contains basic keywords, there are at most two temporal constraints generated (recall Example 2). In this case, we rank the temporal constraint with temporal join first, followed by the null constraint.

Example 3 (Second-Level Ranking). Consider query {Patient cough Doctor DURING [2015-01-01,2015-12-31]} and its temporal constraints in Table 3 w.r.t. the query pattern P_3 in Fig. 4. TC_{26} has a score of 2 since it involves a temporal join. TC_{21} to TC_{25} have no temporal join, and we compute their scores by counting the number of object/mixed nodes between target node Patient and the time condition node for each constraint. Both TC_{21} and TC_{22} have a score of $\frac{1}{2}$ since the time condition nodes is *Doctor* and $count(Patient, Doctor, P_3) = 2$. TC_{23} has a score of 1 since the time condition node is node *Consult* and $count(Patient, Consult, P_3) = 1$. TC_{24} and TC_{25} have a score of 1 since $count(Patient, Patient, P_3) = 1$. □

6   Generating SQL Statements

Finally, we generate a set of SQL statements based on the query patterns and their temporal constraints to retrieve results from the database.

We first consider the query pattern and generate the SELECT, FROM and WHERE clause according to [17]. The SELECT clause includes the attributes of the target node and the FROM clause includes the relations of every node in P. The WHERE clause joins the relations in the FROM clause based on the foreign key-key constraints and translates attribute value condition such as $A = value$ into a selection condition "$contains(R_u.A, value)$". The SQL statement for the query pattern in Fig. 4 for the query {Patient cough Doctor DURING [2015-01-01,2015-12-31]} is as follows. Note that the FROM clause includes relation *PatientSymptom* since it is matched by keyword *cough*.

```
1    SELECT P.*
2    FROM Doctor D, Consult C, Patient P, PatientSymptom PS
3    WHERE D.Did=C.Did AND C.Pid=P.Pid AND P.Pid=PS.Pid
4    AND contains(PS.Symptom,"cough")
```

Next, we consider the temporal constraints of the query pattern. For each temporal constraint of the form of "$R[A.Start, A.End]\ TP\ [s, e]$" where $[s, e]$ is an explicit time period, we translate the temporal predicate TP into a set of comparison operators between $[A.Start, A.End]$ and $[s, c]$ based on Table 1. For example, we translate TC_{24} in Table 3 to the following conditions in the WHERE clause:

"PS.Symptom_Start>2015-01-01 AND PS.Symptom_End<2015-12-31"

For each temporal constraint involving temporal joins, e.g., TC_{26} in Table 3, we first translate the temporal predicate INTERSECT into a set of comparison operator between its adjacent time periods according to Table 1. Then we apply the temporal predicate with time period, like "$TP\ [s, e]$", to one of time periods involved in the temporal join. For example, TC_{26} in Table 3 is translated into conditions as follows, in which lines 5-6 indicate temporal joins and line 7 indicates the temporal predicate with time period.

5   *D.Doctor_Start $\leqslant$ C.Consult_Date AND D.Doctor_End $\geqslant$ C.Consult_Date AND*
6   *C.Consult_Date $\leqslant$ PS.Symptom_End AND C.Consult_Date $\geqslant$ PS.Symptom_Start AND*
7   *PS.PatientSymptom_Start $>$ '2015-01-01' AND PS.PatientSymptom_End $<$ '2015-12-31'*

7   PowerQ^T System Prototype

Given the inherent ambiguity of keyword queries, we propose to generate various interpretations of the query based on all possible matching of basic keywords and apply the temporal predicate to the different temporal relations. However, it is difficult for users to find the correct interpretation of their query. As such, we design a prototype system called PowerQ^T to allow interactive keyword search over a temporal database. PowerQ^T also includes our two-level ranking mechanism to rank the generated query interpretations, which facilitate users to choose the interpretation that best captures their search intention.

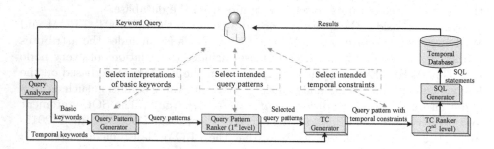

Fig. 6. Architecture of PowerQ^T

Figure 6 shows the main components of PowerQ^T. Given a keyword query Q, the Query Analyzer distinguishes the basic keywords and temporal keywords in Q. Each basic keyword may have different interpretations as they may have different matches, e.g. keyword *Smith* could be a patient's name or a doctor's name. We allow users to choose the intended interpretations of each basic keyword. Then the Query Pattern Generator generates a set of query patterns based on the selected interpretations of each basic keyword. This reduces the number of query patterns generated. The Query Pattern Ranker uses the first level ranking scheme to rank the generated query patterns for the user to choose. For each selected query pattern, the Temporal Constraint (TC) Generator analyzes the temporal relations and the temporal keywords to generate a set of temporal constraints that depict how the time condition is handled. The Temporal Constraint (TC) Ranker uses the second level ranking scheme to rank the temporal constraints within each query pattern for the user to choose. Finally, we generate SQL statements to retrieve the answers to Q. Note that the answers are grouped by the query interpretations.

This interactive process allows users to consider the interpretations of the basic keywords and temporal keywords separately, and users will not be overwhelmed by too many interpretations.

8 Evaluation

We evaluate the expressive ability of our proposed approach (PowerQ^T) and compare it with the method in [9] (ATQ) which does not consider multiple temporal relations involved in the query and support temporal join. We use the following datasets in our evaluation.

1. *Basketball dataset*[1]. It contains information about NBA players, teams and coaches from 1946 to 2009. We modify the schema to create time period attributes (*from* and *to*) based on the original time point attribute (*year*) to make it a temporal database.
2. *Employee dataset*[2]. It contains the job histories of employees, as well as the department where the employees have worked from 1985 to 2003.

Table 6 shows the schema of these two datasets. A temporal relation is indicated by a superscript T. The DATE type attributes are in *italics*.

Table 6. Dataset schemas

Basketball	Employee
Team(<u>tid</u>, location, name)	Department(<u>deptno</u>, dname)
Coach(<u>cid</u>, name)	Employee(<u>empno</u>, ename, gender)
Player^T(<u>pid</u>, name, position, weight, college,	EmployeeTitle^T(<u>empno</u>, *from*, title, *to*)
first_season, last_season)	EmployeeSalary^T(<u>empno</u>, *from*, salary, *to*)
PlayerSeason^T(<u>pid</u>, *year*, game, point)	Workfor^T(<u>empno</u>, *from*, deptno, *to*)
TeamSeason^T(<u>tid</u>, *year*, won, lost)	Manage^T(<u>deptno</u>, *from*, empno, *to*)
PlayFor^T(<u>pid</u>, tid, *from*, *to*)	
CoachFor^T(<u>cid</u>, *from*, tid, *to*)	

Table 7 shows the 3 types of queries we designed for each dataset: (a) queries without time constraint, (b) queries with explicit time period, and (c) queries with implicit time period. We evaluate whether PowerQ^T and ATQ are able to retrieve the correct answers with respect to the user search intention.

---

[1] https://github.com/briandk/2009-nba-data/.
[2] https://dev.mysql.com/doc/employee/en/.

Table 7. Queries for Basketball (B) and employee (E) datasets

(a) Type I Queries (no time constraint)

#	Query	Keywords	PowerQ^T	ATQ
B_1	Find the position of player "Michael Jordan"	position "Michael Jordan"	✓	✓
B_2	Find the point record of players when they were coached by "Pat Riley"	player point coach "Pat Riley"	✓	✗
B_3	Find how many times the teams won when they were coached by "Phil Jackson"	team won coach "Phil Jackson"	✓	✗
E_1	Find the employee named "Mark" and has been an engineer	employee Mark engineer	✓	✓
E_2	Find the departments where employee "Mark" working as an engineer	department Mark engineer	✓	✗
E_3	Find the department where employees "Mark" and "Leon" are colleagues	department Mark Leon	✓	✗

(b) Type II Queries (with explicit time period)

#	Query	Keywords	PowerQ^T	ATQ
B_4	Find the players who played in team Lakers from 2000 to 2002	player team Lakers OVERLAPS [2000,2002]	✓	✓
B_5	Find the coaches of player "Magic Johnson" from 1990 to 2000	coach "Magic Johnson" OVERLAPS [1990,2000]	✓	✗
B_6	Find the point record of players when they was coached by "Phil Jackson" from 1989 to 2003	point coach "Phil Jackson" OVERLAPS [1989,2003]	✓	✗
E_4	Find the salary from 2008 to 2009 for employee Jackson	salary OVERLAPS [2008,2009] employee Jackson	✓	✓
E_5	Find the salary from 2008 to 2009 for employees who worked in department finance at that time	salary OVERLAPS [2008,2009] department finance	✓	✗
E_6	Find the managers of employee "Danny" in 2005	employee manage Danny OVERLAPS [2005,2005]	✓	✗

(c) Type III Queries (with implicit time period)

#	Query	Keywords	PowerQ^T	ATQ
B_7	Find the number of win times of a team before coach "Phil Jackson" joined in	team won BEFORE coach "Phil Jackson"	✓	✓
B_8	Find the points of players before they coached by "Pat Riley"	player points BEFORE coach "Pat Riley"	✓	✓
B_9	Find the player who played for team Cavaliers then moved to team Suns	player Cavaliers MEETS Suns	✓	✗
E_7	Find the employees who had been an engineer before becoming a technique leader	employee engineer BEFORE "technique leader"	✓	✓
E_8	Find the employees's tittle before they who manage department development	employee tittle BEFORE manage development	✓	✓
E_9	Find the employees who worked in department finance then move to department marketing	employee finance MEETS marketing	✓	✗

Type I Queries. These queries do not contain any time constraint, i.e., no explicit temporal predicate or time period (see Table 7(a)). Queries B_1 and E_1 do not involve temporal join, and both PowerQ^T and ATQ retrieve the correct results by matching the query keywords to the database tuples.

Queries $B_2 \sim B_3$ and $E_2 \sim E_3$ involve temporal join and only PowerQ^T could retrieve the correct results. Take for example query B_2. PowerQ^T retrieves the correct results by applying temporal join over the temporal relations *PlayerSeason*, *PlayFor* and *CoachFor* which ensures that only the point history of players who were coached by "Pat Riley" are retrieved. However, ATQ uses the standard join over these temporal relations and also returns the players' point history when they were coached by other coaches.

Type II Queries. These are queries with explicit time period (see Table 7(b)). Queries B_4 and E_4 involves only one temporal relation, and both PowerQ^T and ATQ retrieve the correct results by applying the time period to this relation. However, queries $B_5 \sim B_6$ and $E_5 \sim E_6$ involve multiple temporal relations, and only PowerQ^T retrieves the correct results for them. This is because ATQ does not apply temporal join between relations.

Take for example query B_5. PowerQ^T retrieves the correct results by carrying out a temporal join over the temporal relations *PlayFor* and *CoachFor*, and applying the time condition "OVERLAPS [1990, 2000]" to the result of the temporal join. This ensures that we find the coaches for "Magic Johnson" from 1990 to 2000. In contrast, ATQ associates the time period separately to the relations *PlayFor* and *Coachfor*, and returns incorrect results, e.g., "Randy Pfund" is not a correct result since he coached the team "Los Angeles Lakers" from 1992 to 1993, while "Magic Johnson" played for this team only on 1990 and 1995, indicating that Randy did not coach "Magic Johnson" from 1990 to 2000.

Type III Queries. These are queries with implicit time period (see Table 7(c)). Both PowerQ^T and ATQ could retrieve correct results for queries $B_7 \sim B_8$ and $E_7 \sim E_8$ since the target relations of the temporal predicate are easily found by matching the adjacent keywords.

However, for queries B_9 and E_9, only PowerQ^T could retrieve the correct results, and no answers are returned by ATQ. This is because ATQ is unable to interpret the temporal predicate in these queries since the keywords adjacent to the temporal predicate match non-temporal relations. In contrast, PowerQ^T interprets the temporal predicate over the query pattern generated by matching the basic keywords, which finds the temporal relationship relations as the operands of the temporal predicate correctly.

Take for example query B_9. The keywords "Cavaliers" and "Suns" match the relation *Team* which is not a temporal relation. PowerQ^T is able to identify the temporal relation *PlayFor* involved in the generated query pattern as the target relation of temporal predicate MEETS. Thus it is able to retrieve the players who played for team "Cavaliers" then playing for team "Suns".

In summary, we have shown that PowerQ^T is able to retrieve the correct answers for all given queries in each dataset, while ATQ is able to return correct results for some of the queries. There are two reasons why PowerQ^T performs better than ATQ. First, PowerQ^T handles the basic keywords and temporal keywords separately, which enable us to identify temporal relations involved in a keyword query which is not explicitly specified by the users, e.g., queries N_9 and E_9. Second, by analyzing the temporal relations involved in a query pattern, PowerQ^T is able to handle keyword queries that require temporal join between relations, which is not considered in ATQ, e.g., queries N_5 and E_5. Besides these two reasons, there is another advantage of PowerQ^T over ATQ. PowerQ^T

helps users to reduce the multiple interpretations of one keyword query into some interpretations which match their search intention based on the interactive search and the two-level ranking mechanism. However, ATQ returns the results of all possible interpretations of one keyword query, which requires additional work on the user's part to filter out the results.

9   Conclusion

In this work, we have studied the problem of evaluating keyword query with temporal keywords (temporal predicate and time period) over temporal relational databases. Existing works do not consider temporal join and the multiple interpretations of temporal keywords, which leads missing answers, missing query interpretations, and incorrect answers. We addressed these problems by considering the Object-Relationship-Attribute semantics of the database to identify the temporal attributes of objects/relationships and infer the target temporal data of temporal predicates. After generating an initial set of query patterns, we can infer the target time period of the temporal predicate and generate temporal constraints to capture the different interpretations of a temporal keyword query. We have also developed a two-level ranking scheme and a prototype system to support interactive keyword search. Evaluation of queries over two datasets demonstrate the expressiveness and effectiveness of the proposed approach.

References

1. Allen, J.F.: Maintaining knowledge about temporal intervals. CACM **26**, 832–843 (1983)
2. de Oliveira, P., da Silva, A., de Moura, E.: Ranking candidate networks of relations to improve keyword search over relational databases. In: ICDE (2015)
3. Ding, B., Yu, J.X., Wang, S., Qin, L., Zhang, X., Lin, X.: Finding top-k min-cost connected trees in databases. In: ICDE (2007)
4. Gao, Q., Lee, M.L., Ling, T.W., Dobbie, G., Zeng, Z.: Analyzing temporal keyword queries for interactive search over temporal databases. Technical report TRA3/18. National University of Singapore (2018)
5. Gunadhi, H., Segev, A.: Query processing algorithms for temporal intersection joins. In: ICDE (1991)
6. Hristidis, V., Hwang, H., Papakonstantinou, Y.: Authority-based keyword search in databases. ACM TODS **33**(1), 1:1–1:40 (2008)
7. Hristidis, V., Papakonstantinou, Y.: DISCOVER: keyword search in relational databases. In: VLDB (2002)
8. Hulgeri, A., Nakhe, C.: Keyword searching and browsing in databases using BANKS. In: ICDE (2002)
9. Jia, X., Hsu, W., Lee, M.L.: Target-oriented keyword search over temporal databases. In: Hartmann, S., Ma, H. (eds.) DEXA 2016. LNCS, vol. 9827, pp. 3–19. Springer, Cham (2016). https://doi.org/10.1007/978-3-319-44403-1_1
10. Kacholia, V., Pandit, S., Chakrabarti, S.: Bidirectional expansion for keyword search on graph databases. In: VLDB (2005)

11. Kargar, M., An, A., Cercone, N., Godfrey, P., Szlichta, J., Yu, X.: Meaningful keyword search in relational databases with large and complex schema. In: ICDE (2015)
12. Liu, F., Yu, C., Meng, W., Chowdhury, A.: Effective keyword search in relational databases. In: ACM SIGMOD (2006)
13. Liu, Z., Wang, C., Chen, Y.: Keyword search on temporal graphs. TKDE **29**(8), 1667–1680 (2017)
14. Luo, Y., Lin, X., Wang, W., Zhou, X.: SPARK: top-k keyword query in relational databases. In: ACM SIGMOD (2007)
15. Qin, L., Yu, J.X., Chang, L.: Keyword search in databases: the power of RDBMS. In: ACM SIGMOD (2009)
16. Yu, X., Shi, H.: CI-Rank: ranking keyword search results based on collective importance. In: ICDE (2012)
17. Zeng, Z., Bao, Z., Le, T.N., Lee, M.L., Ling. T.W.: ExpressQ: identifying keyword context and search target in relational keyword queries. In: ACM CIKM (2014)
18. Zeng, Z., Bao, Z., Lee, M.L., Ling, T.W.: A semantic approach to keyword search over relational databases. In: ER (2013)

Implicit Representation of Bigranular Rules for Multigranular Data

Stephen J. Hegner[1(✉)] and M. Andrea Rodríguez[2]

[1] DBMS Research of New Hampshire, PO Box 2153, New London, NH 03257, USA
dbmsnh@gmx.com

[2] Millennium Institute for Foundational Research on Data, Departamento Ingeniería Informática y Ciencias de la Computación, Universidad de Concepción, Edmundo Larenas 219, 4070409 Concepción, Chile
andrea@udec.cl

Abstract. Domains for spatial and temporal data are often multigranular in nature, possessing a natural order structure defined by spatial inclusion and time-interval inclusion, respectively. This order structure induces lattice-like (partial) operations, such as join, which in turn lead to *join rules*, in which a single domain element (granule) is asserted to be equal to, or contained in, the join of a set of such granules. In general, the efficient representation of such *join rules* is a difficult problem. However, there is a very effective representation in the case that the rule is *bigranular*; i.e., all of the joined elements belong to the same granularity, and, in addition, complete information about the (non)disjointness of all granules involved is known. The details of that representation form the focus of the paper.

1 Introduction

In a multigranular attribute, the domain elements are related by order-like and even lattice-like operations, leading to a much richer family of integrity constraints than is found in the traditional monogranular setting. The ideas are best illustrated via example. Let $R_{\mathsf{sumb}}\langle A_{\mathsf{Plc}}, A_{\mathsf{Tim}}, B_{\mathsf{Bth}}\rangle$ be the schema in which the spatial attribute A_{Plc} identifies certain geographical areas of Chile, the temporal attribute A_{Tim} identifies intervals of time, and the thematic attribute B_{Bth} has numerical values representing the number of births. A tuple of the form $\langle p, t, b \rangle$ denotes that in the region defined by p, for the time interval defined by t, the number of births was b. An example instance for this schema is shown in Fig. 1. Think of the two tables of that figure to be part of a single relation; the division is for expository reasons, as well as to conserve space. In that instance, for domain elements (called *granules*) of A_{Plc}, the suffix _*prv* identifies the name as that of a province, _*rgn* identifies a region, _*cmn* identifies a county, while _*urb* identifies a metropolitan area. For A_{Tim}, *Y2017Qx* denotes quarter *x* of year 2017, while *Y2017* represents the entire year. Such a multigranular schema and instance may arise, for example, when data of varying granularities of space and

© Springer Nature Switzerland AG 2018
S. Hartmann et al. (Eds.): DEXA 2018, LNCS 11029, pp. 372–389, 2018.
https://doi.org/10.1007/978-3-319-98809-2_23

A_{Plc}	A_{Tim}	B_{Bth}
Los_Lagos_rgn	Y2017Q1	b_1
Osorno_prv	Y2017Q1	b_2
Llanquihue_prv	Y2017Q1	b_3
Chiloé_prv	Y2017Q1	b_4
Palena_prv	Y2017Q1	b_5
Puerto_Montt_cmn	Y2017Q1	b_6
Puerto_Varas_cmn	Y2017Q1	b_7
Gran_Puerto_Montt_urb	Y2017Q1	b_8

A_{Plc}	A_{Tim}	B_{Bth}
BíoBío_rgn	Y2017	b_1'
BíoBío_rgn	Y2017Q1	b_2'
BíoBío_rgn	Y2017Q2	b_3'
BíoBío_rgn	Y2017Q3	b_4'
BíoBío_rgn	Y2017Q4	b_5'

Fig. 1. Multigranular relational instance

time are integrated, into a single schema, with respect to the same *thematic attribute* (here B_{Bth}).

It is clear that the ordinary functional dependency (FD) $\{A_{\mathsf{Plc}}, A_{\mathsf{Tim}}\} \to B_{\mathsf{Bth}}$ is expected to hold. However, there are also several other natural dependencies, induced by the structure of the multigranular domains. Each of the four listed provinces is contained in the region Los Lagos, expressed formally as *Osorno_prv* $\sqsubseteq$ *Los_Lagos_rgn*, *Llanquihue_prv* $\sqsubseteq$ *Los_Lagos_rgn*, *Chiloé_prv* $\sqsubseteq$ *Los_Lagos_rgn*, and *Palena_prv* $\sqsubseteq$ *Los_Lagos_rgn*. Similarly, both counties, as well as the metropolitan area of Gran Puerto Montt, are contained in the province Llanquihue; *Puerto_Montt_cmn* $\sqsubseteq$ *Llanquihue_prv*, *Puerto_Varas_cmn* $\sqsubseteq$ *Llanquihue_prv*, and *Gran_Puerto_Montt_urb* $\sqsubseteq$ *Llanquihue_prv*. For the temporal domain, each of the quarters of 2017 is contained in the entire year: *Y2017Qx* $\sqsubseteq$ *Y2017* for $x \in \{1, 2, 3, 4\}$. Since the number of births is monotonic with respect to region size and time-interval size, these conditions in turn lead to the constraints $b_i \leq b_1$ for $i \in \{2, 3, 4, 5\}$, $b_i \leq b_3$ for $i \in \{6, 7, 8\}$, and $b_i' \leq b_1'$ for $i \in \{2, 3, 4, 5\}$.

More is true, however. The region Los Lagos is composed exactly of the four provinces listed, without any overlap, written as the *disjoint-join equality rule* (r-LLr) below.

$$Los\_Lagos\_rgn = \left\lfloor \bot \right\rfloor \{Osorno\_prv, Llanquihue\_prv, Chiloé\_prv, Palena\_prv\}$$

$$\text{(r-LLr)}$$

Specifically, the symbol $\bigsqcup$ means that the four provinces cover the region completely, while the embedded $\bot$ means that the join is *disjoint*; that is, that the regions do not overlap. This leads to the *spatial aggregation constraint* $\sum_{i=2}^{5} b_i = b_1$. Additionally, the metropolitan area of Gran Puerto Montt lies entirely within the combined areas of the counties Puerto Montt and Puerto Varas, leading to the *disjoint-join subsumption rule* (r-Llp) shown below, and consequently the spatial aggregation constraint $b_8 \leq b_6 + b_7$.

$$Gran\_Puerto\_Montt\_urb \sqsubseteq \left\lfloor \bot \right\rfloor \{Puerto\_Montt\_cmn, Puerto\_Varas\_cmn\} \quad \text{(r-Llp)}$$

Such aggregation constraints arise in the same fashion for temporal multi-granular attributes, such as A_{Tim}. For example, the disjoint-join equality rule (r-YQ2017) shown below holds, leading to the *temporal aggregation constraint* $\sum_{i=2}^{5} b_i' = b_1'$.

$$Y2017 = \boxed{}^{\perp} \{Y2017Q1, Y2017Q2, Y2017Q3, Y2017Q4\} \qquad \text{(r-YQ2017)}$$

Aggregation constraints arising from join rules, as illustrated by the examples above, are instances of *TMCDs* or *thematic multigranular comparison dependencies*, which are developed in detail in [8], including a notion of *tolerance* which replaces absolute equality with an approximate one (to account for differences arising from rounding and measurement errors). In order to enforce such TMCDs, it is first of all essential to know which ones hold. This, in turn, requires a means to determine which disjoint-join rules hold. Although a formal semantics and inference mechanism for such rules is developed in [8], it is quite resource expensive to enforce all TMCDs by identifying the associated join rules via direct inference. The focus of this paper is the development of a compact and efficient representation for certain types of join rules which occur frequently in practice.

Key to these results are the observation that the granules of a multigranular attribute may be partitioned naturally into so-called *granularities* (hence the term *multigranular*) of *disjoint* members, as illustrated in Fig. 2 for both space and time. Arrows of the form $G_1 \prec G_2$ represent the basic refinement order of granularities, in the sense that for every granule g_1 of granularity G_1 there is a granule g_2 of granularity G_2 with $g_1 \sqsubseteq g_2$. Inline, this typically written $G_1 \leq G_2$. Thus, every county is contained in a (unique) province, every province is contained in a (unique) region, and every region is contained in Chile. Similarly, every metropolitan area is contained in a region, (although not necessarily in a single province.)

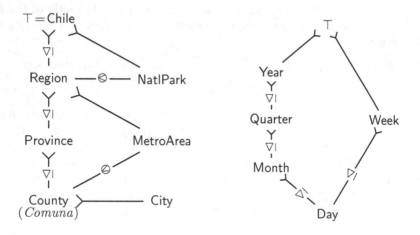

Fig. 2. Granularity hierarchies for Chile and for time

In support of the representation of rules, there are two additional binary relations on granularities which are of fundamental importance, *equality join order*, denoted $\trianglelefteq$, and *subsumption join order*, denoted $\otimes$. $G_1 \trianglelefteq G_2$ holds just in case every granule g_2 of granularity G_2 is the (necessarily disjoint) join of some granules of granularity G_1; i.e., if $g_2 = \biguplus S$ holds for some finite set S of granules of G_2. As can be seen in Fig. 2, with the symbol $\trianglelefteq$ embedded in a line indicating that this relation holds between the granularities which it connects, this condition characterizes many practical situations. As a concrete example, Province $\trianglelefteq$ Region, with (r-LLp) a specific instance of a join rule arising from it. Similarly, for the time hierarchy, (r-YQ2017) is a specific instance of a rule arising from Quarter $\trianglelefteq$ Year.

The main result of this paper regarding $\trianglelefteq$ may be summarized as follows. Let $\mathsf{NRel}_{\langle G_1, G_2 \rangle}$ denote the relation which identifies pairs $\langle g_1, g_2 \rangle$ of granules from $\langle G_1, G_2 \rangle$ (i.e., with g_1 of granularity G_1 and g_2 of granularity G_2) which are not disjoint. Then, it must be the case that $S = \{g_2 \mid \langle g_1, g_2 \rangle \in \mathsf{NRel}_{\langle G_1, G_2 \rangle}\}$; in other words, S must be exactly the set of all granules of g_2 which are not disjoint from g_1. As a specific example, to identify those provinces which lie in *Los_Lagos_rgn*, it is only necessary to retrieve $\{g \mid \langle \textit{Los\_Lagos\_rgn}, g \rangle \in \mathsf{NRel}_{\langle \mathsf{Region}, \mathsf{Province} \rangle}\}$; no complex inference procedure is necessary. In assessing this solution, it must be remembered that knowledge about granules, including subsumption, disjointness, and join, is specified via statements. There is the possibility that a given assertion is *unresolvable*; i.e., it is not possible to establish that it is true or it is false. (See Summary 2.7 for details.) What is remarkable about this result is that no such unresolvability can occur for $\langle G_1, G_2 \rangle$ disjointness. For $G_1 \trianglelefteq G_2$ to hold, it must be the case that for any pair $\langle g_1, g_2 \rangle$ of granules of $\langle G_1, G_2 \rangle$, it is the case that the disjointness of $\langle g_1, g_2 \rangle$ is resolvable.

This idea applies also, subject to an additional condition, when subsumption replaces equality. $G_1 \otimes G_2$ holds just in case every granule of G_1 is subsumed by the join of some granules in G_2; i.e., if $g_2 \sqsubseteq \biguplus S$ holds for some finite set S of granules of G_2. This is illustrated in particular by rule (r-Llp), as an instance of County $\otimes$ MetroArea. Of course, $G_1 \trianglelefteq G_2$ always implies $G_1 \otimes G_2$, but this example shows that the converse need not hold. The additional condition which must be imposed is that the join be *resolved minimal*, meaning that if any element is removed from the join set, the assertion becomes resolvably false. In other words, both *Gran_Puerto_Montt_urb* $\not\sqsubseteq$ *Puerto_Montt_cmn* and *Gran_Puerto_Montt_urb* $\not\sqsubseteq$ *Puerto_Varas_cmn* must follow from the rules. In this case, to determine the counties in which *Gran_Puerto_Montt_urb* lies, it is only necessary to retrieve $\{g \mid \langle \textit{Gran\_Puerto\_Montt\_urb}, g \rangle \in \mathsf{NRel}_{\langle \mathsf{County}, \mathsf{MetroArea} \rangle}\}$.

To clarify the terminology, a join rule $g = \biguplus S$ is *bigranular* if every granule in S is of the same granularity G_2. (Since granules of the same granularity are disjoint, it must be the case that the granularity G_1 of g is different from that of the members of S, hence the term bigranular.) Thus, any rule arising from the application of a condition of the form $G_1 \trianglelefteq G_2$ or $G_1 \otimes G_2$ is necessarily bigranular.

The representations developed above are termed *implicit,* since a rule of the form $g = \bigsqcup S$ or $g \sqsubseteq \bigsqcup S$ is represented by a way to recover S from the appropriate $\mathsf{NRel}_{\langle -, - \rangle}$. In the remainder of this paper, the details of how and why this method of representing of join rules works are developed.

The paper is organized as follows. Section 2 provides necessary details of the multigranular framework developed in [8]. Section 3 develops the general ideas of minimality for join rules, while Sect. 4 contains the main results of the paper on the representation of bigranular join rules. Finally, Sect. 5 contains conclusions and further directions.

2   Multigranular Attributes and Their Semantics

The results of this paper are based upon the formal model of multigranular attributes, as developed in [8]. It is thus appropriate to begin with a summary of that framework. Although [7] covers similar material, it is of a preliminary nature, so the reader is always referred to [8] for clarification of details. For terminology and notation regarding logic, consult [11], while for issues surrounding order structures, including posets, see [3]. For basic concepts surrounding the relational model, see [9].

Notation 2.1 (Special mathematical notation). $X_1 \subsetneq X_2$ (resp. $X_1 \subseteq_f X_2$) denotes that X_1 is a proper (resp. finite) subset of X_2. The cardinality of the set X is denoted $\mathsf{Card}(X)$.

Overview 2.2 (Constrained granulated attribute schemata). In the ordinary relational model with SQL used for data definition, several attributes may use the same data type. For example, two distinct attributes may be declared to be of the same type `VARCHAR(10)`. Similarly, in the multigranular model, several distinct attributes may be declared to be of the same type. Such a type is called a *constrained granulated attribute schema,* or *CGAS,* and is a triple $\mathfrak{S} = (\mathsf{Glty}\langle\mathfrak{S}\rangle, \mathsf{GrAsgn}\langle\mathfrak{S}\rangle, \mathsf{Constr}^{\pm}\langle\mathfrak{S}\rangle)$ in which $\mathsf{Glty}\langle\mathfrak{S}\rangle$ is a poset of *granularities* and $\mathsf{GrAsgn}\langle\mathfrak{S}\rangle$ is a *granule assignment,* both elaborated in Summary 2.3 below, while $\mathsf{Constr}^{\pm}\langle\mathfrak{S}\rangle$ is a unified set of constraints, elaborated in Summary 2.5 below.

Summary 2.3 (Granularities and granules). A *granularity poset* for the CGAS $\mathfrak{S}$ is an upper-bounded poset $\mathsf{Glty}\langle\mathfrak{S}\rangle = (\mathsf{Glty}\langle\mathfrak{S}\rangle, \leq_{\mathsf{Glty}\langle\mathfrak{S}\rangle}, \top_{\mathsf{Glty}\langle\mathfrak{S}\rangle})$; that is, it is poset with a greatest element $\top_{\mathsf{Glty}\langle\mathfrak{S}\rangle}$. The two diagrams of Fig. 2 represent the specific granularity posets for $\mathfrak{S}$ replaced by $\mathfrak{C}$ and $\mathfrak{T}$, respectively, with $G_1 \leq_{\mathsf{Glty}\langle\mathfrak{C}\rangle} G_2$ (resp. $G_1 \leq_{\mathsf{Glty}\langle\mathfrak{T}\rangle} G_2$) iff there is an arrow of the form $G_1 \dashleftarrow G_2$ in the associated diagram. In that which follows, $\mathfrak{S}$ will be used to represent a general CGAS, while $\mathfrak{C}$ (for Chile) and $\mathfrak{T}$ (for time) will be used to represent, respectively, the spatial and the temporal schema whose granularities are depicted in Fig. 2.

A *granule assignment* $\mathsf{GrAsgn}\langle\mathfrak{S}\rangle = (\mathsf{Gnle}\langle\mathfrak{S}\rangle, \varPi_{\mathsf{Gnle}}\langle\mathfrak{S}\rangle)$ for $\mathfrak{S}$ extends the idea of a domain assignment for an ordinary relational attribute, in the sense that it assigns (with one exception) every granule to a granularity. $\mathbf{Gnle}\langle\mathfrak{S}\rangle =$ (Granules$\langle\mathfrak{S}\rangle, \sqsubseteq_\mathfrak{S}, \top_\mathfrak{S}, \bot_\mathfrak{S}$) is the (bounded) *granule preorder*, while $\Pi_{\mathsf{Gnle}}\langle\mathfrak{S}\rangle =$ {Granules$\langle\mathfrak{S}|G\rangle \mid G \in \mathsf{Glty}\langle\mathfrak{S}\rangle$} is a partition of Granules$_\chi\langle\mathfrak{S}\rangle =$ Granules$\langle\mathfrak{S}\rangle \setminus$ {$\bot_\mathfrak{S}$} that identifies which granules are assigned to which granularities. The bottom granule $\bot_\mathfrak{S}$ (the least element of the preorder $\mathbf{Gnle}\langle\mathfrak{S}\rangle$) is not a member of Granules$\langle\mathfrak{S}|G\rangle$ for any granularity G, while the top granule $\top_\mathfrak{S}$ (the greatest element of the preorder $\mathbf{Gnle}\langle\mathfrak{S}\rangle$) lies in Granules$\langle\mathfrak{S}|\top_{\mathsf{Glty}\langle\mathfrak{S}\rangle}\rangle$.

The orders of granularities and granules are closely related. Specifically, for granularities G_1 and G_2, $G_1 \leq_{\mathsf{Glty}\langle\mathfrak{S}\rangle} G_2$ iff for every $g_1 \in$ Granules$\langle\mathfrak{S}|G_1\rangle$, there is a $g_2 \in$ Granules$\langle\mathfrak{S}|G_2\rangle$ with the property that $g_1 \sqsubseteq_\mathfrak{S} g_2$. Since $\mathbf{Gnle}\langle\mathfrak{S}\rangle$ is only a preorder, distinct granules may be equivalent, in the sense that $g_1 \sqsubseteq_\mathfrak{S} g_2 \sqsubseteq_\mathfrak{S} g_1$. Write $[g_1]_{\mathsf{Gnle}\langle\mathfrak{S}\rangle}$ to denote the equivalence class of g_1; thus, with g_1, g_2 as above, $g_2 \in [g_1]_{\mathsf{Gnle}\langle\mathfrak{S}\rangle}$ and $[g_1]_{\mathsf{Gnle}\langle\mathfrak{S}\rangle} = [g_2]_{\mathsf{Gnle}\langle\mathfrak{S}\rangle}$. To avoid problems, the special notation $g_1 \overset{\text{id}}{=} g_2$ will be used to mean that g_1 and g_2 are the same granule, with the meaning of $g_1 = g_2$ deferred until Summary 2.5, when semantics are discussed. With this in mind, further conditions may be stated. First of all, the top granularity $\top_{\mathsf{Glty}\langle\mathfrak{S}\rangle}$ is the only one which may contain equivalent but not identical granules. It contains the top granule $\top_\mathfrak{S}$ (the greatest element of the poset $\mathbf{Gnle}\langle\mathfrak{S}\rangle$), as well as any granule equivalent to it. For example, in the CGAS $\mathfrak{C}$, $[\top_\mathfrak{C}]_{\mathsf{Gnle}\langle\mathfrak{C}\rangle} = [\mathsf{Chile}]_{\mathsf{Gnle}\langle\mathfrak{C}\rangle}$ (see Fig. 2). Otherwise, non-identical granules of the same granularity may not be equivalent, and they furthermore must have the bottom granule as GLB (greatest lower bound). More precisely, if g_1 and g_2 are of the same non-$\top_{\mathsf{Glty}\langle\mathfrak{S}\rangle}$ granularity, and $g_1 \overset{\text{id}}{\neq} g_2$, then both $([g_1]_{\mathsf{Gnle}\langle\mathfrak{S}\rangle} \neq [g_2]_{\mathsf{Gnle}\langle\mathfrak{S}\rangle})$ and $(\mathsf{GLB}_{\mathsf{Gnle}\langle\mathfrak{S}\rangle}\langle\{g_1, g_2\}\rangle = \bot_\mathfrak{S})$ hold.

Summary 2.4 (Semantics of granules). A *granule structure* $\sigma = \sigma = $ (Dom$\langle\sigma\rangle$, GnletoDom$_\sigma$) for the granule assignment GrAsgn$\langle\mathfrak{S}\rangle$ provides set-based semantics. Dom$\langle\sigma\rangle$ is a (not necessarily finite) set, called the *domain* of σ, and GnletoDom$_\sigma$: Granules$\langle\mathfrak{S}\rangle \rightarrow 2^{\mathsf{Dom}\langle\sigma\rangle}$ is a function which assigns to each granule a subset of the domain. In this assignment, granule subsumption translates to set inclusion ($g_1 \sqsubseteq_\mathfrak{S} g_2$ implies GnletoDom$_\sigma(g_1) \subseteq$ GnletoDom$_\sigma(g_2)$), granule disjointness translates to empty intersection (if g_1 and g_2 are of the same granularity with $g_1 \overset{\text{id}}{\neq} g_2$, then GnletoDom$_\sigma(g_1) \cap$ GnletoDom$_\sigma(g_2) = \emptyset$); equivalent granules have identical semantics ((GnletoDom$_\sigma(g_1) = $ GnletoDom$_\sigma(g_2)$) $\Leftrightarrow$ $[g_1]_{\mathsf{Gnle}\langle\mathfrak{S}\rangle} = [g_2]_{\mathsf{Gnle}\langle\mathfrak{S}\rangle}$); and the bottom granule maps to the empty set (GnletoDom$_\mathfrak{S}(\bot_\mathfrak{S}) = \emptyset$).

As already mentioned in Sect. 1, for a spatial attribute such as $\mathfrak{C}$, a natural granular structure might be σ_{Chile}, the subset of the real plane $\mathbb{R} \times \mathbb{R}$ representing Chile, with GnletoDom$_{\sigma_{\mathsf{Chile}}}(g)$ exactly the geographic region corresponding to granule g. While such a structure is mathematically correct, it involves an enormous amount of detail, much more than is necessary in many cases. It is for this reason that the semantics of a multigranular attribute is modelled not by a single granular structure, but rather by any such structure which satisfies the constraint, or rules, of the schema, as defined in Summary 2.5 below. For a more complete explanation, see [8, Sect. 3.6].

Summary 2.5 (Rules). In [8, Sect. 3], general constraints for GGASs and their semantics are developed extensively. In this paper, only those constraint types which are used in the theory developed here are sketched.

The *primitive basic rules* over the CGAS $\mathfrak{S}$, denoted, $\mathsf{PrBaRules}\langle\mathfrak{S}\rangle$ are of the following two forms.

(pjrule-i) A *subsumption join rule* is of the form $(g \sqsubseteq_\mathfrak{S} \bigsqcup_\mathfrak{S} S)$ for $\{g\} \cup S \subseteq \mathsf{Granules}_\chi\langle\mathfrak{S}\rangle$. The *elemental subsumption rule* $(g_1 \sqsubseteq_\mathfrak{S} g_2)$, with $g_1, g_2 \in \mathsf{Granules}_\chi\langle\mathfrak{S}\rangle$, is shorthand for $(g_1 \sqsubseteq_\mathfrak{S} \bigsqcup_\mathfrak{S} \{g_2\})$.

(psrule-ii) A *basic disjointness rule* is of the form $(\bigsqcap_\mathfrak{S} \{g_1, g_2\} = \bot_\mathfrak{S})$ for $g_1, g_2 \in \mathsf{Granules}_\chi\langle\mathfrak{S}\rangle$ and $[g_1]_\mathfrak{S} \neq [g_2]_\mathfrak{S}$.

Extending the notion of semantics of Summary 2.4 to $\mathsf{PrBaRules}\langle\mathfrak{S}\rangle$, a granule structure σ for $\mathfrak{S}$ is a *model* of the subsumption rule $(g \sqsubseteq_\mathfrak{S} \bigsqcup_\mathfrak{S} S)$ if $\mathsf{GnletoDom}_\sigma(g) \subseteq \bigcup_{s \in S} \mathsf{GnletoDom}_\sigma(s)$, while σ is *model* of the basic disjointness rule $(\bigsqcap_\mathfrak{S} \{g_1, g_2\} = \bot_\mathfrak{S})$ if $\mathsf{GnletoDom}_\sigma(g_1) \cap \mathsf{GnletoDom}_\sigma(g_2) = \emptyset$. For $\Phi \subseteq \mathsf{PrBaRules}\langle\mathfrak{S}\rangle$, $\mathsf{Models}_\mathfrak{S}\langle\Phi\rangle$ denotes the collection of all models of Φ.

For any CGAS $\mathfrak{S}$, the *built-in rules* $\mathsf{BuiltInRules}\langle\mathfrak{S}\rangle$ are those which are satisfied by every granular structure σ for $\mathfrak{S}$. These include the subsumption rule $(g_1 \sqsubseteq_\mathfrak{S} g_2)$ whenever $g_1 \sqsubseteq_\mathfrak{S} g_2$ holds,[1] as well as $\bigsqcap_\mathfrak{S} \{g_1, g_2\} = \bot_\mathfrak{S}$ whenever $g_1 \stackrel{\mathrm{id}}{\neq} g_2$ are of the same granularity.

A *complex rule* is a conjunction of primitive basic rules. Write $\mathsf{Conjuncts}\langle\varphi\rangle$ to denote the set of conjuncts of the complex rule φ. Thus, if $\varphi = \varphi_1 \wedge \varphi_2 \wedge \ldots \wedge \varphi_k$, then $\mathsf{Conjuncts}\langle\varphi\rangle = \{\varphi_1, \varphi_2, \ldots, \varphi_k\}$. The most important kind of complex rules are the *complex join rules*:

(cjrule-i) An *equality join rule* is of the form $(g = \bigsqcup_\mathfrak{S} S)$, for $\{g\} \cup S \subseteq \mathsf{Granules}_\chi\langle\mathfrak{S}\rangle$. Its definition in terms of primitive basic rules is

$$\mathsf{Conjuncts}_\mathfrak{S}\langle(g = \bigsqcup_\mathfrak{S} S)\rangle = \{(g \sqsubseteq_\mathfrak{S} \bigsqcup_\mathfrak{S} S)\} \cup \{(g_i \sqsubseteq_\mathfrak{S} g) \mid g_i \in S\}.$$

(cjrule-ii) A *disjoint-join subsumption rule*, written as $(g \sqsubseteq_\mathfrak{S} \biguplus_\mathfrak{S} S)$ for $\{g\} \cup S \subseteq \mathsf{Granules}_\chi\langle\mathfrak{S}\rangle$, is defined in terms of primitive basic join rules as $\mathsf{Conjuncts}_\mathfrak{S}\langle(g \sqsubseteq_\mathfrak{S} \biguplus_\mathfrak{S} S)\rangle =$

$$\mathsf{Conjuncts}\langle(g \sqsubseteq_\mathfrak{S} \bigsqcup_\mathfrak{S} S)\rangle \cup \{(\bigsqcap_\mathfrak{S} \{g_1, g_2\} = \bot_\mathfrak{S}) \mid g_i, g_j \in S \text{ and } g_i \stackrel{\mathrm{id}}{\neq} g_2\}.$$

(cjrule-iii) A *disjoint-join equality rule*, written as $(g = \biguplus_\mathfrak{S} S)$ for $\{g\} \cup S \subseteq \mathsf{Granules}_\chi\langle\mathfrak{S}\rangle$ is defined in terms of primitive basic join rules as

$$\mathsf{Conjuncts}_\mathfrak{S}\langle(g = \biguplus_\mathfrak{S} S)\rangle = \mathsf{Conjuncts}_\mathfrak{S}\langle(g = \bigsqcup_\mathfrak{S} S)\rangle \cup \mathsf{Conjuncts}_\mathfrak{S}\langle(g \sqsubseteq_\mathfrak{S} \biguplus_\mathfrak{S} S)\rangle.$$

For convenience, a complex rule will be represented by its set of conjuncts. Thus, every complex rule is a regarded as a finite nonempty set of primitive basic rules.

---

[1] $\sqsubseteq_\mathfrak{S}$ is the granule preorder defined in the granule assignment $\mathsf{GrAsgn}\langle\mathfrak{S}\rangle$ (see Summary 2.3) while $\sqsubseteq_\mathfrak{S}$ is the general subsumption relation used to define rules. For $g_1, g_2 \in \mathsf{Granules}\langle\mathfrak{S}\rangle$, it is always the case that $g_1 \sqsubseteq_\mathfrak{S} g_2$ implies $(g_1 \sqsubseteq_\mathfrak{S} g_2)$. The converse is not required to hold, although in practice it usually does.

For simplicity, the example rules in Sect. 1 were presented without qualifying subscripts on the operators. Using the notation for specific granular attributes introduced in Summary 2.3, for example, rule (r-Llp) should be written more properly as $Gran\_Puerto\_Montt\_urb \sqsubseteq_{\mathfrak{C}} \bigsqcup_{\mathfrak{C}}^{\perp}$ $\{Puerto\_Montt\_cmn, Puerto\_Varas\_cmn\}$. It is assumed that the reader will add these qualifying symbols, as necessary.

Summary 2.6 (Negation of rules). It is also necessary to work with negations of primitive basic rules over the CGAS $\mathfrak{S}$; the most important example is negation of disjointness; for $g_1, g_2 \in \mathsf{Granules}_{\chi}\langle\mathfrak{S}\rangle$, write $(\bigsqcap_{\mathfrak{S}}\{g_1, g_2\} \neq \perp_{\mathfrak{S}})$ to mean $\neg(\bigsqcap_{\mathfrak{S}}\{g_1, g_2\} = \perp_{\mathfrak{S}})$. Similarly, $(g_1 \not\sqsubseteq_{\mathfrak{S}} g_2)$ means $\neg(g_1 \sqsubseteq_{\mathfrak{S}} g_2)$ and $(g_1 \not\sqsubseteq_{\mathfrak{S}} S)$ means $\neg(g_1 \sqsubseteq_{\mathfrak{S}} S)$. The set of all negations of primitive basic rules is denoted $\mathsf{NegPrBaRules}\langle\mathfrak{S}\rangle$. The granule structure σ is a model of $\psi = \neg\varphi \in \mathsf{NegPrBaRules}\langle\mathfrak{S}\rangle$, iff it is not a model of φ; i.e., $\mathsf{Models}_{\mathfrak{S}}\langle\psi\rangle$ is the collection of all granule structures which do not lie in $\mathsf{Models}_{\mathfrak{S}}\langle\varphi\rangle$.

For $\Phi, \Phi' \subseteq \mathsf{PrBaRules}\langle\mathfrak{S}\rangle$, define $\mathsf{Not}\langle\Phi\rangle = \{(\neg\varphi) \mid \varphi \in \Phi\}$. Thus, $\mathsf{NegPrBaRules}\langle\mathfrak{S}\rangle = \mathsf{Not}\langle\mathsf{PrBaRules}\langle\mathfrak{S}\rangle\rangle$.

Finally, it is convenient to combine positive and negated rules into one set. Define $\mathsf{AllPrBaRules}\langle\mathfrak{S}\rangle = \mathsf{PrBaRules}\langle\mathfrak{S}\rangle \cup \mathsf{NegPrBaRules}\langle\mathfrak{S}\rangle$. For $\Phi \subseteq \mathsf{AllPrBaRules}\langle\mathfrak{S}\rangle$, $\mathsf{Models}_{\mathfrak{S}}\langle\Phi\rangle = \bigcap\{\mathsf{Models}_{\mathfrak{S}}\langle\varphi\rangle \mid \varphi \in \Phi\}$.

Summary 2.7 (Satisfiability and Resolvability). Continuing with $\mathfrak{S}$ a CGAS, for $\varphi \in \mathsf{AllPrBaRules}\langle\mathfrak{S}\rangle$ and $\Phi \subseteq \mathsf{AllPrBaRules}\langle\mathfrak{S}\rangle$, define semantic entailment $\Phi \models_{\mathfrak{S}} \varphi$ to mean that $\mathsf{Models}_{\mathfrak{S}}\langle\Phi\rangle \subseteq \mathsf{Models}_{\mathfrak{S}}\langle\varphi\rangle$, and for $\Phi' \subseteq \mathsf{AllPrBaRules}\langle\mathfrak{S}\rangle$, $\Phi \models_{\mathfrak{S}} \Phi'$ to mean that $\mathsf{Models}_{\mathfrak{S}}\langle\Phi\rangle \subseteq \mathsf{Models}_{\mathfrak{S}}\langle\Phi'\rangle$. In other words, Φ imposes stronger constraints than does Φ'. φ (resp. Φ) is *satisfiable* (or *consistent*) if it has a model; i.e., $\mathsf{Models}_{\mathfrak{S}}\langle\varphi\rangle \neq \emptyset$ (resp. $\mathsf{Models}_{\mathfrak{S}}\langle\Phi\rangle \neq \emptyset$).

Let $\Phi \subseteq \mathsf{AllPrBaRules}\langle\mathfrak{S}\rangle$ and $\varphi \in \mathsf{PrBaRules}\langle\mathfrak{S}\rangle$. Say that φ is *resolvable* from Φ, written $\Phi \models\!\!\models_{\mathfrak{S}} \varphi$, if one of $\Phi \models_{\mathfrak{S}} \varphi$ or else $\Phi \models_{\mathfrak{S}} \neg\varphi$ holds. In other words, the truth value of φ is determined by Φ; either φ is true in every model of Φ, or else φ is false in every model of φ.

The set $\mathsf{PrBaRules}\langle\mathfrak{S}\rangle$ has the property of admitting *Armstrong models* [6], in the precise sense that for any consistent $\Phi \subseteq \mathsf{PrBaRules}\langle\mathfrak{S}\rangle$, there is a model which satisfies only those members of Φ. This means that members of $\mathsf{NegPrBaRules}\langle\mathfrak{S}\rangle$ whose negations are not entailed by Φ may be added to Φ in any combination while retaining satisfiability. See [8, Sects. 3.15–3.20] for details.

Finally, $\mathsf{Constr}^{\pm}\langle\mathfrak{S}\rangle \subseteq \mathsf{AllPrBaRules}\langle\mathfrak{S}\rangle$ is a consistent set of rules, representing the set of constraints of $\mathfrak{S}$, as first identified in Overview 2.2. In [8] this set is represented as a pair $\langle\mathsf{Constr}(\mathfrak{S}), \mathsf{cwa}\langle\mathfrak{S}\rangle\rangle$, with $\mathsf{Constr}(\mathfrak{S})$ the positive constraints and $\mathsf{cwa}\langle\mathfrak{S}\rangle$ those to be negated; $\mathsf{Constr}^{\pm}\langle\mathfrak{S}\rangle = \mathsf{Constr}(\mathfrak{S}) \cup \mathsf{Not}\langle\mathsf{cwa}\langle\mathfrak{S}\rangle\rangle$ provides the equivalence of notation.

3 Minimality of Join Rules

Roughly, a join rule is minimal if removing any of the joined granules results in a rule which is no longer a consequence of the constraints. In this section, this idea of minimality is developed formally.

Context 3.1 (CGAS). Unless stated specifically to the contrary, for the remainder of this paper, let $\mathfrak{S} = (\mathsf{Glty}\langle\mathfrak{S}\rangle, \mathsf{GrAsgn}\langle\mathfrak{S}\rangle, \mathsf{Constr}^{\pm}\langle\mathfrak{S}\rangle)$ denote an arbitrary CGAS.

Notation 3.2 (Components of join rules). There are four variants of join rule over $\mathfrak{S}$, identified in (pjrule-i) and (cjrule-i)–(cjrule-iii) of Summary 2.5, collectively denoted $\mathsf{JRules}\langle\mathfrak{S}\rangle$. A join rule over $\mathfrak{S}$ is thus a statement of the form $(g \circledast \lfloor ? \rfloor S)$ with $\circledast \in \{=, \sqsubseteq_{\mathfrak{S}}\}$, and $\lfloor ? \rfloor \in \{\bigsqcup_{\mathfrak{S}}, \biguplus_{\mathfrak{S}}\}$, for $g \in \mathsf{Granules}_{\chi}\langle\mathfrak{S}\rangle$, and $S \subseteq \mathsf{Granules}_{\chi}\langle\mathfrak{S}\rangle$ nonempty. Using terminology borrowed from logic, g is called the *head* of the rule while S is called the *body*, denoted by $\mathsf{Head}\langle\varphi\rangle$ and $\mathsf{Body}\langle\varphi\rangle$, respectively, for $\varphi \in \mathsf{JRules}\langle\mathfrak{S}\rangle$. In addition, $\mathsf{CompOp}\langle\varphi\rangle \in \{=, \sqsubseteq_{\mathfrak{S}}\}$ denotes the operator of the rule, and $\mathsf{JoinOp}\langle\varphi\rangle \in \{\bigsqcup_{\mathfrak{S}}, \biguplus_{\mathfrak{S}}\}$ denotes the join operation of the rule. In other words, $\mathsf{CompOp}\langle\varphi\rangle$ is just $\circledast$ and $\mathsf{JoinOp}\langle\varphi\rangle$ is just $\lfloor ? \rfloor_{\mathfrak{S}}$, as defined above. The new notation is introduced in order to be able to parameterize these items in terms of the underlying rule φ. Thus, φ may be written, somewhat cryptically, as $(\mathsf{Head}\langle\varphi\rangle\ \mathsf{CompOp}\langle\varphi\rangle\ \mathsf{JoinOp}\langle\varphi\rangle\ \mathsf{Body}\langle\varphi\rangle)$.

Definition 3.3 (Primitive reduction and minimality of join rules). The *primitive reduction* of $\varphi \in \mathsf{JRules}\langle\mathfrak{S}\rangle$ by $Z \subseteq \mathsf{Body}\langle\varphi\rangle$, denoted $\mathsf{PrReduct}\langle\varphi : Z\rangle$, is obtained by removing the members of Z from $\mathsf{Body}\langle\varphi\rangle$, and by replacing, if necessary, equality with subsumption as the comparison operator. Formally, $\mathsf{PrReduct}\langle\varphi : Z\rangle$ is the rule $\varphi' \in \mathsf{JRules}\langle\mathfrak{S}\rangle$ with $\mathsf{Body}\langle\varphi'\rangle = \mathsf{Body}\langle\varphi\rangle \setminus Z$ and $\mathsf{JoinOp}\langle\varphi\rangle = \bigsqcup_{\mathfrak{S}}$, while $\mathsf{Head}\langle\varphi'\rangle$ and $\mathsf{CompOp}\langle\varphi'\rangle$, remain unchanged from φ. If $\mathsf{Body}\langle\varphi'\rangle$ is a proper subset of $\mathsf{Body}\langle\varphi\rangle$; i.e., $\mathsf{Body}\langle\varphi'\rangle \subsetneq \mathsf{Head}\langle\varphi\rangle$, then φ' is called a *proper primitive reduction* of φ. For example, letting φ be the rule (r-LLr) of Sect. 1, with $Z = \{\textit{Osorno\_prv}, \textit{Chiloé\_prv}\}$,

$$\mathsf{PrReduct}\langle\varphi : Z\rangle = (\textit{Los\_Lagos\_rgn} \sqsubseteq_{\mathfrak{c}} \bigsqcup_{\mathfrak{c}} \{\textit{Llanquihue\_prv}, \textit{Palena\_prv}\}).$$

$\varphi \in \mathsf{JRules}\langle\mathfrak{S}\rangle$ is *minimal* (for $\mathfrak{S}$) if for no proper primitive reduction φ' of φ is it the case that $\mathsf{Constr}^{\pm}\langle\mathfrak{S}\rangle \models_{\mathfrak{S}} \varphi'$. More formally, φ is minimal if for no nonempty $Z \subseteq \mathsf{Body}\langle\varphi\rangle$ is it the case that $\mathsf{Constr}^{\pm}\langle\mathfrak{S}\rangle \models_{\mathfrak{S}} \mathsf{PrReduct}\langle\varphi : Z\rangle$. In other words, if any nonempty subset of the body is removed, the resulting rule is no longer a consequence of $\mathsf{Constr}^{\pm}\langle\mathfrak{S}\rangle$. φ is *resolved minimal* (for $\mathfrak{S}$) if for every nonempty $Z \subseteq \mathsf{Body}\langle\varphi\rangle$ it is the case that $\mathsf{Constr}^{\pm}\langle\mathfrak{S}\rangle \models_{\mathfrak{S}} \neg\mathsf{PrReduct}_{\mathfrak{S}}\langle\varphi : Z\rangle$. Put another way, if any element of the body is removed, and the comparison operator is replaced by subsumption, the rule becomes false. If φ is minimal but not resolved minimal, then it is called *unresolved minimal*. Both forms of minimality may be characterized by the removal of single elements from the body. Define the *primitive reduction set* of φ, denoted $\mathsf{RedSet}\langle\varphi\rangle$, to be

$$\{\mathsf{PrReduct}_{\mathfrak{S}}\langle\varphi : \{h\}\rangle \mid h \in \mathsf{Body}\langle\varphi\rangle\}\ \text{if}\ \mathsf{Card}(\mathsf{Body}\langle\varphi\rangle) \geq 2,$$

and to be $\emptyset$ otherwise. For example, letting φ again be (r-LLr),

$$\mathsf{RedSet}\langle\varphi\rangle = \{(\mathit{Los\_Lagos\_rgn} \sqsubseteq_{\mathfrak{e}} \textstyle\bigsqcup_{\mathfrak{e}}\{\mathit{Osorno\_prv}, \mathit{Llanquihue\_prv}, \mathit{Chilo\acute{e}\_prv}\}),$$
$$(\mathit{Los\_Lagos\_rgn} \sqsubseteq_{\mathfrak{e}} \textstyle\bigsqcup_{\mathfrak{e}}\{\mathit{Osorno\_prv}, \mathit{Llanquihue\_prv}, \mathit{Palena\_prv}\}),$$
$$(\mathit{Los\_Lagos\_rgn} \sqsubseteq_{\mathfrak{e}} \textstyle\bigsqcup_{\mathfrak{e}}\{\mathit{Osorno\_prv}, \mathit{Chilo\acute{e}\_prv}, \mathit{Palena\_prv}\}),$$
$$(\mathit{Los\_Lagos\_rgn} \sqsubseteq_{\mathfrak{e}} \textstyle\bigsqcup_{\mathfrak{e}}\{\mathit{Llanquihue\_prv}, \mathit{Chilo\acute{e}\ prv}, \mathit{Palena\_prv}\})\}.$$

For φ to be minimal, no element of $\mathsf{RedSet}\langle\varphi\rangle$ may be implied by the constraints, while to be resolved minimal, the negation of every such element must be so implied. This is formalized by the following, whose proof is immediate.

Observation 3.4 (Removing single elements suffices). Let $\varphi \in \mathsf{JRules}\langle\mathfrak{S}\rangle$ with $\mathsf{Constr}^{\pm}\langle\mathfrak{S}\rangle \models_{\mathfrak{S}} \varphi$.

(a) φ is minimal iff for no $\psi \in \mathsf{RedSet}\langle\varphi\rangle$ does $\mathsf{Constr}^{\pm}\langle\mathfrak{S}\rangle \models_{\mathfrak{S}} \psi$ hold.
(b) φ is resolved minimal iff $\mathsf{Constr}^{\pm}\langle\mathfrak{S}\rangle \models_{\mathfrak{S}} \mathsf{Not}\langle\mathsf{RedSet}\langle\varphi\rangle\rangle$.

Proposition 3.5 (Disjoint equality join implies resolved minimality). *A disjoint equality join rule φ for which $\mathsf{Constr}^{\pm}\langle\mathfrak{S}\rangle \models_{\mathfrak{S}} \varphi$ is resolved minimal.*

Proof. Writing φ as $(g = \bigsqcup_{\mathfrak{S}} S)$, according to Summary 2.5, it has the representation $\mathsf{Conjuncts}_{\mathfrak{S}}\langle\varphi\rangle =$
$(g \sqsubseteq_{\mathfrak{S}} \bigsqcup_{\mathfrak{S}} S) \cup \{(s \sqsubseteq_{\mathfrak{S}} g) \mid s \in S\} \cup \{(\bigsqcap_{\mathfrak{S}}\{s,s'\} = \bot_{\mathfrak{S}}) \mid s,s' \in S \text{ and} s \overset{\mathsf{id}}{\neq} s'\}$ in terms of primitive basic rules. Now, let $\sigma \in \mathsf{Models}_{\mathfrak{S}}\langle\mathsf{Constr}^{\pm}\langle\mathfrak{S}\rangle\rangle$ and choose any $s \in S$. Since $\sigma(s) \neq \emptyset$, $\sigma(s) \cap \sigma(s') = \emptyset$ for all $s' \in S \setminus \{s\}$, and $\sigma(g) = \bigcup\{\sigma(s'') \mid s'' \in S\}$, it follows that $\sigma(g) \subsetneq \bigcup\{s'' \in S \mid s'' \overset{\mathsf{id}}{\neq} s\}$. Since σ is an arbitrary model of $\mathsf{Constr}^{\pm}\langle\mathfrak{S}\rangle$, it follows that $\mathsf{Constr}^{\pm}\langle\mathfrak{S}\rangle \models_{\mathfrak{S}} \neg(g \sqsubseteq_{\mathfrak{S}} S \setminus \{s\}) = \neg\mathsf{PrReduct}_{\mathfrak{S}}\langle\varphi : \{s\}\rangle$. Finally, since s is arbitrary, the proof follows from Observation 3.4(b). $\qquad\square$

Discussion 3.6 (Subsumption join and minimal rules). In view of Proposition 3.5, (r-LLr) is automatically resolved minimal. This is clear, since if any of the provinces are removed from the body, the subsumption will fail. However, this idea does not extend to subsumption join. For example, any metropolitan area of Chile lies within the join of all counties; e.g.,

$$(\mathit{Gran\_Puerto\_Montt\_urb} \sqsubseteq_{\mathfrak{e}} \bigsqcup_{\mathfrak{e}} \mathsf{Granules}\langle\mathfrak{C}|\mathit{County}\rangle).$$

This rule is not even unresolved minimal; there are only two counties with which Gran Puerto Montt is not disjoint. Thus, resolved minimality must be asserted explicitly for a rule such as (r-Llp) of Sect. 1.

Definition 3.7 (Resolved-minimal join rules). For any $\varphi \in \mathsf{JRules}\langle\mathfrak{S}\rangle$, define $\mathsf{RMinSet}\langle\varphi\rangle = \mathsf{Not}\langle\mathsf{RedSet}\langle\varphi\rangle\rangle$, and define the *resolved minimization* of φ to be $\mathsf{ResMin}\langle\varphi\rangle = \mathsf{Conjuncts}_{\mathfrak{S}}\langle\varphi\rangle \cup \mathsf{RMinSet}\langle\varphi\rangle$. In light of Observation 3.4(b), $\mathsf{RMinSet}\langle\varphi\rangle$ consists of exactly those constraints necessary to make φ a resolved minimal join rule. For φ set to (r-Llp) of Sect. 1,

$$\mathsf{ResMin}\langle\varphi\rangle = \{\neg(\mathit{Gran\_Puerto\_Montt\_urb} \sqsubseteq_{\mathfrak{e}} \mathit{Puerto\_Montt\_cmn}),$$
$$\neg(\mathit{Gran\_Puerto\_Montt\_urb} \sqsubseteq_{\mathfrak{e}} \mathit{Puerto\_Varas\_cmn})\}$$

Just as the basic join symbol $\bigsqcup_{\mathfrak{S}}$ is embellished with $\perp$ to yield $\biguplus_{\mathfrak{S}}$ to indicate disjoint join, it is also useful to embellish the symbol to indicate resolved minimal joins. More precisely, for any type of join rule φ identified in Notation 3.2, replacing $\bigsqcup_{\mathfrak{S}}$ by $\overset{\text{rmin}}{\bigsqcup}_{\mathfrak{S}}$, or $\biguplus_{\mathfrak{S}}$ by $\overset{\text{rmin}}{\biguplus}_{\mathfrak{S}}$, denotes its resolved minimization. For this paper, the concrete case of interest is the *resolved-minimal disjoint subsumption join rule* $(g \sqsubseteq_{\mathfrak{S}} \overset{\text{rmin}}{\biguplus}_{\mathfrak{S}} S)$, shorthand for $\mathsf{Conjuncts}_{\mathfrak{S}}\langle(g \sqsubseteq_{\mathfrak{S}} \biguplus_{\mathfrak{S}} S)\rangle \cup \mathsf{RMinSet}\langle(g \sqsubseteq_{\mathfrak{S}} \biguplus_{\mathfrak{S}} S)\rangle$. Formally, the *resolved-minimal disjoint equality join rule* $(g = \overset{\text{rmin}}{\biguplus}_{\mathfrak{S}} S)$, shorthand for $\mathsf{Conjuncts}_{\mathfrak{S}}\langle(g = \biguplus_{\mathfrak{S}} S)\rangle \cup \mathsf{RMinSet}\langle(g = \biguplus_{\mathfrak{S}} S)\rangle$, is also used, but in view of Proposition 3.5, every disjoint equality join rule is resolved minimal, so the property is redundant. The set of all rules which are of one of these resolved forms is called the *resolved minimal join rules*, denoted $\mathsf{RMJRules}\langle\mathfrak{S}\rangle$. $\varphi \in \mathsf{RMJRules}\langle\mathfrak{S}\rangle$ has $\mathsf{JoinOp}\langle\varphi\rangle \in \{\overset{\text{rmin}}{\bigsqcup}_{\mathfrak{S}}, \overset{\text{rmin}}{\biguplus}_{\mathfrak{S}}\}$ but is otherwise syntactically identical to a rule in $\mathsf{JRules}\langle\mathfrak{S}\rangle$. As a concrete example, to express that it is resolved minimal, (r-Llp) may be rewritten as

$$Gran\_Puerto\_Montt\_urb \sqsubseteq_{\mathcal{C}} \overset{\text{rmin}}{\biguplus}_{\mathcal{C}} \{Puerto\_Montt\_cmn, Puerto\_Varas\_cmn\} \quad (\text{r-Llp}')$$

4 Bigranular Join Rules and Their Representation

In this section, the main results of the paper, on the implicit representation of multigranular join rules, are developed.

Definition 4.1 (Granularity pairs). A *granularity pair* over $\mathfrak{S}$ is an ordered pair $\langle G_1, G_2 \rangle \in \mathsf{Glty}\langle\mathfrak{S}\rangle \times \mathsf{Glty}\langle\mathfrak{S}\rangle$ with $G_1 \neq G_2$.

Context 4.2 (Granularity names and granularity pairs). For the remainder of this section, unless stated specifically to the contrary, let $G_1, G_2, G_3 \in \mathsf{Glty}\langle\mathfrak{S}\rangle$. In particular, $\langle G_1, G_2 \rangle$ and $\langle G_2, G_3 \rangle$ are granularity pairs.

Definition 4.3 (Join-order properties of granularity pairs). The notions of equality-join order and subsumption-join order, introduced informally in Sect. 1, are formalized as follows.

(ej-ord) $\langle G_1, G_2 \rangle$ has the *equality-join order property*, written $G_1 \trianglelefteq_{\mathfrak{S}} G_2$, if

$$(\forall g_2 \in \mathsf{Granules}\langle\mathfrak{S}|G_2\rangle)(\exists S \subseteq_f \mathsf{Granules}\langle\mathfrak{S}|G_1\rangle)$$
$$(\mathsf{Constr}^{\pm}\langle\mathfrak{S}\rangle \models_{\mathfrak{S}} (g_2 = \bigsqcup_{\mathfrak{S}} S)).$$

(sj-ord) $\langle G_1, G_2 \rangle$ has the *subsumption-join order property*, written $G_1 \otimes_{\mathfrak{S}} G_2$, if
$$(\forall g_2 \in \mathsf{Granules}\langle\mathfrak{S}|G_2\rangle)(\exists S \subseteq_f \mathsf{Granules}\langle\mathfrak{S}|G_1\rangle)$$
$$(\mathsf{Constr}^{\pm}\langle\mathfrak{S}\rangle \models_{\mathfrak{S}} (g_2 \sqsubseteq_{\mathfrak{S}} \overset{\text{rmin}}{\bigsqcup}_{\mathfrak{S}} S)).$$

While the join in these rules is not explicitly disjoint, in applications to bigranular rules (Definition 4.6), it will always be disjoint (Proposition 4.7).

Observation 4.4 (Equality join implies subsumption join). *If* $G_1 \trianglelefteq_{\mathfrak{S}} G_2$ *holds, then so too does* $G_1 \ominus_{\mathfrak{S}} G_2$.

Proof. Equality is a special case of subsumption, and equality join is always minimal (Proposition 3.5).   □

Definition 4.5 (Biresolvability and equiresolvability). In order to characterize these order properties in terms of simpler ones, several new notions are essential. Local resolvability (for disjointness, subsumption, or both) characterizes resolvability at a fixed $g_2 \in \mathsf{Granules}\langle \mathfrak{S}|G_2\rangle$, while full resolvability characterizes the corresponding property for all such g_2. Formally, given $g_2 \in \mathsf{Granules}\langle \mathfrak{S}|G_2\rangle$, the pair $\langle G_1, G_2\rangle$ is *locally disjointness resolvable* (resp. *locally subsumption resolvable*) at g_2 if for every $g_1 \in \mathsf{Granules}\langle \mathfrak{S}|G_1\rangle$, $\mathsf{Constr}^{\pm}\langle \mathfrak{S}\rangle \models_{\mathfrak{S}}$ $(\bigsqcap_{\mathfrak{S}}\{g_1, g_2\} = \bot_{\mathfrak{S}})$ (resp. $\mathsf{Constr}^{\pm}\langle \mathfrak{S}\rangle \models_{\mathfrak{S}} (g_1 \sqsubseteq_{\mathfrak{S}} g_2)$). If $\langle G_1, G_2\rangle$ is locally disjointness resolvable (resp. locally subsumption resolvable) for every $g_2 \in \mathsf{Granules}\langle \mathfrak{S}|G_2\rangle$, then it is called *fully disjointness resolvable* (resp. *fully subsumption resolvable*). Call $\langle G_1, G_2\rangle$ *locally biresolvable* at g_2 (resp. *fully biresolvable*) if it is both locally disjointness resolvable and locally subsumption resolvable at g_2 (resp. both fully disjointness resolvable and fully subsumption resolvable).

The pair $\langle G_1, G_2\rangle$ is equiresolvable if subsumption and nondisjointness resolve equivalently. More formally, $\langle G_1, G_2\rangle$ is *equiresolvable* at g_2 if, for every $g_1 \in \mathsf{Granules}\langle \mathfrak{S}|G_1\rangle$, $\mathsf{Constr}^{\pm}\langle \mathfrak{S}\rangle \models_{\mathfrak{S}} (g_1 \sqsubseteq_{\mathfrak{S}} g_2)$ holds iff $\mathsf{Constr}^{\pm}\langle \mathfrak{S}\rangle \models_{\mathfrak{S}}$ $(\bigsqcap_{\mathfrak{S}}\{g_1, g_2\} \neq \bot_{\mathfrak{S}})$ holds; and $\mathsf{Constr}^{\pm}\langle \mathfrak{S}\rangle \models_{\mathfrak{S}} (g_1 \not\sqsubseteq_{\mathfrak{S}} g_2)$ holds iff $\mathsf{Constr}^{\pm}\langle \mathfrak{S}\rangle \models_{\mathfrak{S}} (\bigsqcap_{\mathfrak{S}}\{g_1, g_2\} = \bot_{\mathfrak{S}})$ holds. Call $\langle G_1, G_2\rangle$ *fully equiresolvable* if it is equiresolvable at each $g_2 \in \mathsf{Granules}\langle \mathfrak{S}|G_2\rangle$.

Definition 4.6 (Bigranular join rules). A join rule φ is of type $\langle G_1, G_2\rangle$ if $\mathsf{Head}\langle \varphi\rangle \in \mathsf{Granules}\langle \mathfrak{S}|G_1\rangle$ and $\mathsf{Body}\langle \varphi\rangle \subseteq \mathsf{Granules}\langle \mathfrak{S}|G_2\rangle$. Such a rule is also called *bigranular*.

Proposition 4.7 (Bigranular implies disjoint). *If a join rule φ is bigranular, then it is disjoint; i.e.,* $\mathsf{JoinOp}\langle \varphi\rangle \in \{\bigsqcup_{\mathfrak{S}}, \overset{\mathrm{rmin}}{\bigsqcup}_{\mathfrak{S}}\}$.

Proof. Distinct granules of the same granularity are disjoint; in particular, the granules of $\mathsf{Body}\langle \varphi\rangle$ have that property.   □

The main characterization result for resolved minimality, in its most general form, is presented next.

Proposition 4.8 (Characterization of resolved minimality). *Let φ be a minimal join rule of type $\langle G_1, G_2\rangle$ with the property that $\mathsf{Constr}^{\pm}\langle \mathfrak{S}\rangle \models_{\mathfrak{S}} \varphi$. The following three conditions are then equivalent.*

(a) $\langle G_1, G_2\rangle$ is locally disjointness resolvable at $\mathsf{Head}\langle \varphi\rangle$.
(b) φ is resolved minimal.
(c) $\mathsf{Body}\langle \varphi\rangle =$
$$\{g_1 \in \mathsf{Granules}\langle \mathfrak{S}|G_1\rangle \mid \mathsf{Constr}^{\pm}\langle \mathfrak{S}\rangle \models_{\mathfrak{S}} (\bigsqcap_{\mathfrak{S}}\{g_1, \mathsf{Head}\langle \varphi\rangle\} \neq \bot_{\mathfrak{S}})\}.$$

Proof. (a) $\Rightarrow$ (c): Regardless of whether or not (a) holds,

$$\{g_1 \in \mathsf{Granules}\langle\mathfrak{S}|G_1\rangle \mid \mathsf{Constr}^{\pm}\langle\mathfrak{S}\rangle \models_{\mathfrak{S}} (\textstyle\bigsqcap_{\mathfrak{S}}\{g_1, \mathsf{Head}\langle\varphi\rangle\} \neq \bot_{\mathfrak{S}})\} \subseteq \mathsf{Body}\langle\varphi\rangle,$$

since distinct elements of $\mathsf{Granules}\langle\mathfrak{S}|G_1\rangle$ must be disjoint. If (a) holds, then every $g_1' \in \mathsf{Granules}\langle\mathfrak{S}|G_1\rangle \setminus$
$$\{g_1 \in \mathsf{Granules}\langle\mathfrak{S}|G_1\rangle \mid \mathsf{Constr}^{\pm}\langle\mathfrak{S}\rangle \models_{\mathfrak{S}} (\textstyle\bigsqcap_{\mathfrak{S}}\{g_1, \mathsf{Head}\langle\varphi\rangle\} \neq \bot_{\mathfrak{S}})\}$$
must have the property that $\mathsf{Constr}^{\pm}\langle\mathfrak{S}\rangle \models_{\mathfrak{S}} (\bigsqcap_{\mathfrak{S}}\{g_1', \mathsf{Head}\langle\varphi\rangle\} = \bot_{\mathfrak{S}})$, by the very definition of local disjoint resolvability. Clearly, such a granule is not needed in $\mathsf{Body}\langle\varphi\rangle$. Hence (c) holds.

(c) $\Rightarrow$ (b): Assume that (c) holds. For any $g_1' \in \mathsf{Body}\langle\varphi\rangle$, it is clear that $\mathsf{Constr}^{\pm}\langle\mathfrak{S}\rangle \models_{\mathfrak{S}} \neg\mathsf{PrReduct}\langle\varphi : \{g_1'\}\rangle$, since there is no way that $(\mathsf{Head}\langle\varphi\rangle \sqsubseteq_{\mathfrak{S}} \mathsf{Body}\langle\varphi\rangle \setminus \{g_1'\})$ can hold, owing to the disjointness of distinct granules of G_1. Hence φ is resolved minimal.

(b) $\Rightarrow$ (a): Assume that φ is resolved minimal. Then for any $g_1' \in \mathsf{Body}\langle\varphi\rangle$, $\mathsf{Constr}^{\pm}\langle\mathfrak{S}\rangle \models \neg(\mathsf{PrReduct}\langle\varphi : \{g_1'\}\rangle)$. Since distinct granules of G_1 are disjoint, this implies that $\mathsf{Constr}^{\pm}\langle\mathfrak{S}\rangle \models_{\mathfrak{S}} (\bigsqcap_{\mathfrak{S}}\{g_1', \mathsf{Head}\langle\varphi\rangle\} \neq \bot_{\mathfrak{S}})$. On the other hand, let $g_1'' \in \mathsf{Granules}\langle\mathfrak{S}|G_1\rangle \setminus \mathsf{Body}\langle\varphi\rangle$. If $\mathsf{Constr}^{\pm}\langle\mathfrak{S}\rangle \not\models_{\mathfrak{S}} (\bigsqcap_{\mathfrak{S}}\{g_1'', \mathsf{Head}\langle\varphi\rangle\} = \bot_{\mathfrak{S}})$, then there must be a model σ of $\mathsf{Constr}^{\pm}\langle\mathfrak{S}\rangle$ for which $\sigma \in \mathsf{Models}_{\mathfrak{S}}\langle(\bigsqcap_{\mathfrak{S}}\{g_1'', \mathsf{Head}\langle\varphi\rangle\} \neq \bot_{\mathfrak{S}})\rangle$ also. In that case, owing to the disjointness of distinct granules of G_1, it would necessarily be the case that $g_1'' \in \mathsf{Body}\langle\varphi\rangle$, a contradiction. Hence it must be the case that $\mathsf{Constr}^{\pm}\langle\mathfrak{S}\rangle \models_{\mathfrak{S}} (\bigsqcap_{\mathfrak{S}}\{g_1'', \mathsf{Head}\langle\varphi\rangle\} = \bot_{\mathfrak{S}})$, and so $\langle G_1, G_2\rangle$ is locally disjointness resolvable at $\mathsf{Head}\langle\varphi\rangle$, as required. $\qquad\square$

The above result provides in particular a succinct characterization of the subsumption join order $\otimes$ in terms of subsumption join rules. Notice that, in contrast to the case for $\trianglelefteq$, resolved minimality must be asserted explicitly.

Theorem 4.9 (Characterization of subsumption join order). *Let $\langle G_1, G_2\rangle$ be a granularity pair. The following conditions are equivalent.*

(a) $G_1 \otimes_{\mathfrak{S}} G_2$.
(b) For each $g_2 \in \mathsf{Granules}\langle\mathfrak{S}|G_2\rangle$,

$$g_2 \sqsubseteq_{\mathfrak{S}} \overset{\mathsf{rmin}}{\bigsqcup}_{\mathfrak{S}}\{g_1 \in \mathsf{Granules}\langle\mathfrak{S}|G_1\rangle \mid \mathsf{Constr}^{\pm}\langle\mathfrak{S}\rangle \models_{\mathfrak{S}} (\textstyle\bigsqcap_{\mathfrak{S}}\{g_1, g_2\} \neq \bot_{\mathfrak{S}})\},$$

and this is the only possibility for a resolved minimal rule φ with

$$\mathsf{Head}\langle\varphi\rangle = g_2 \text{ and } \mathsf{Body}\langle\varphi\rangle \subseteq \mathsf{Granules}\langle\mathfrak{S}|G_1\rangle.$$

Furthermore, if either (a) *or* (b) *holds, then* $\langle G_1, G_2\rangle$ *is both fully biresolvable and fully equiresolvable.*

Proof. Follows directly from Proposition 4.8 using Definition 4.3(sj-ord). $\qquad\square$

For the special case of equality join, the results of Proposition 4.8 may be refined as follows, establishing resolved minimality, local biresolvability and equiresolvability, as well as characterization of the body in terms of both subsumption and nondisjointness.

Proposition 4.10 (Resolved minimality for equality join). *Let φ be an equality-join rule of type $\langle G_1, G_2 \rangle$ with the property that $\mathsf{Constr}^{\pm}\langle \mathfrak{S} \rangle \models_{\mathfrak{S}} \varphi$. The following properties then hold.*

(a) φ is resolved minimal.
(b) $\langle G_1, G_2 \rangle$ is locally biresolvable as well as locally equiresolvable at $\mathsf{Head}\langle \varphi \rangle$.
(c) $\mathsf{Body}\langle \varphi \rangle = \{g_1 \in \mathsf{Granules}\langle \mathfrak{S}|G_1 \rangle \mid \mathsf{Constr}^{\pm}\langle \mathfrak{S} \rangle \models_{\mathfrak{S}} (g_1 \sqsubseteq_{\mathfrak{S}} \mathsf{Head}\langle \varphi \rangle)\}$
$= \{g_1 \in \mathsf{Granules}\langle \mathfrak{S}|G_1 \rangle \mid \mathsf{Constr}^{\pm}\langle \mathfrak{S} \rangle \models_{\mathfrak{S}} (\bigsqcap_{\mathfrak{S}} \{g_1, \mathsf{Head}\langle \varphi \rangle\} \neq \bot_{\mathfrak{S}})\}$.

Proof. Part (a) follows immediately from Proposition 4.7, Proposition 3.5, and Proposition 4.8(b), whereupon the equality of the first and third expressions of (c) follows from Proposition 4.8(c). To complete the proof, it suffices to note that, by the very definition of disjoint-join equality rule (Summary 2.5(cjrule-iii)), $(g \sqsubseteq_{\mathfrak{S}} \mathsf{Head}\langle \varphi \rangle)$ for every $g \in \mathsf{Body}\langle \varphi \rangle$. Since granules of G_1 are pairwise disjoint, and since $\mathsf{Head}\langle \varphi \rangle = \bigsqcup_{\mathfrak{S}} \mathsf{Body}\langle \varphi \rangle$, is follows that no granule $g \in \mathsf{Granules}\langle \mathfrak{S}|G_1 \rangle \setminus \mathsf{Body}\langle \varphi \rangle$ can have the property that $(g \sqsubseteq_{\mathfrak{S}} \mathsf{Head}\langle \varphi \rangle)$. Hence, the remaining equality of (c) holds, from which (b) then follows directly. $\square$

A characterization of equality join order $\trianglelefteq$, similar to that of Theorem 4.9 but expanded to include subsumption, may now be established.

Theorem 4.11 (Characterization of equality-join order). *Let $\langle G_1, G_2 \rangle$ be a granularity pair. The following conditions are equivalent.*

(a) $G_1 \trianglelefteq_{\mathfrak{S}} G_2$.
(b) For each $g_2 \in \mathsf{Granules}\langle \mathfrak{S}|G_2 \rangle$,

$$g_2 = \overset{\mathrm{min}}{\biguplus}_{\mathfrak{S}} \{g_1 \in \mathsf{Granules}\langle \mathfrak{S}|G_1 \rangle \mid \mathsf{Constr}^{\pm}\langle \mathfrak{S} \rangle \models_{\mathfrak{S}} (g_1 \sqsubseteq_{\mathfrak{S}} g_2)\}$$
$$= \overset{\mathrm{min}}{\biguplus}_{\mathfrak{S}} \{g_1 \in \mathsf{Granules}\langle \mathfrak{S}|G_1 \rangle \mid \mathsf{Constr}^{\pm}\langle \mathfrak{S} \rangle \models_{\mathfrak{S}} (\bigsqcap_{\mathfrak{S}} \{g_1, g_2\} \neq \bot_{\mathfrak{S}})\},$$

and this is the only possibility for a minimal rule φ with

$$\mathsf{Head}\langle \varphi \rangle = g_2 \; and \; \mathsf{Body}\langle \varphi \rangle \subseteq \mathsf{Granules}\langle \mathfrak{S}|G_1 \rangle.$$

Furthermore, if either (a) or (b) holds, then $\langle G_1, G_2 \rangle$ is both fully biresolvable and fully equiresolvable.

Proof. Follows directly from Proposition 4.10 using Definition 4.3(ej-ord). $\square$

Discussion 4.12 (Consequences of the characterizations). The main thrust of the results developed so far in this section is that even though there may be many granule structures which are models for the constraints associated with $G_1 \otimes_{\mathfrak{S}} G_2$ and $G_1 \trianglelefteq_{\mathfrak{S}} G_2$, all of these models agree on which granules of G_1 are and are not disjoint from granules of G_2. Furthermore, this disjointness information is sufficient to recover completely the join rules. This information is represented via the relation nondisjointness relation $\mathsf{NRel}_{\mathfrak{S}:\langle -,- \rangle}$, as introduced in

Sect. 1. The corresponding relation $\mathsf{SRel}_{\mathfrak{S}:\langle -,- \rangle}$ for subsumption is similarly used, as its special properties will prove to be useful in the representation of rules associated with $\trianglelefteq_{\mathfrak{S}}$. The formalization of these ideas are found in Definition 4.13 and Theorem 4.14 below.

Definition 4.13 (The fundamental relations of a granularity pair).
Define the *nondisjointness relation* for $\langle G_1, G_2 \rangle$ as
$$\mathsf{NRel}_{\mathfrak{S}:\langle G_1,G_2 \rangle} = \{\langle g_1, g_2 \rangle \in \mathsf{Granules}\langle \mathfrak{S}|G_1 \rangle \times \mathsf{Granules}\langle \mathfrak{S}|G_2 \rangle \mid$$
$$\mathsf{Constr}^{\pm}\langle \mathfrak{S} \rangle \models_{\mathfrak{S}} (\textstyle\bigsqcap_{\mathfrak{S}} \{g_1, g_2\} \neq \bot_{\mathfrak{S}})\}.$$
Similarly, define the *subsumption relation* for $\langle G_1, G_2 \rangle$ as
$$\mathsf{SRel}_{\mathfrak{S}:\langle G_1,G_2 \rangle} = \{\langle g_1, g_2 \rangle \in \mathsf{Granules}\langle \mathfrak{S}|G_1 \rangle \times \mathsf{Granules}\langle \mathfrak{S}|G_2 \rangle \mid$$
$$\mathsf{Constr}^{\pm}\langle \mathfrak{S} \rangle \models_{\mathfrak{S}} (g_1 \sqsubseteq_{\mathfrak{S}} g_2)\}.$$
Note that if $\langle G_1, G_2 \rangle$ is fully equiresolvable (Definition 4.5), in particular if $G_1 \trianglelefteq_{\mathfrak{S}} G_2$ (Theorem 4.11), then $\mathsf{NRel}_{\mathfrak{S}:\langle G_1,G_2 \rangle} = \mathsf{SRel}_{\mathfrak{S}:\langle G_1,G_2 \rangle}$.
The main theorem for implicit representation is the following.

Theorem 4.14 (Representation of bigranular join rules using fundamental relations)

(a) If $G_1 \otimes_{\mathfrak{S}} G_2$ holds, then for every $g_2 \in \mathsf{Granules}\langle \mathfrak{S}|G_2 \rangle$ and every $S \subseteq_f \mathsf{Granules}\langle \mathfrak{S}|G_1 \rangle$,

$$\mathsf{Constr}^{\pm}\langle \mathfrak{S} \rangle \models_{\mathfrak{S}} (g_2 \sqsubseteq_{\mathfrak{S}} \textstyle\bigsqcup_{\mathfrak{S}} S) \, iff \, \{g_1 \mid \langle g_1, g_2 \rangle \in \mathsf{NRel}_{\mathfrak{S}:\langle G_1,G_2 \rangle}\} \subseteq S.$$

In particular,

$$\mathsf{Constr}^{\pm}\langle \mathfrak{S} \rangle \models_{\mathfrak{S}} (g_2 \sqsubseteq_{\mathfrak{S}} \textstyle\overset{\text{min}}{\biguplus}_{\mathfrak{S}} S) \, iff \, S = \{g_1 \mid \langle g_1, g_2 \rangle \in \mathsf{NRel}_{\mathfrak{S}:\langle G_1,G_2 \rangle}\}.$$

(b) If $G_1 \trianglelefteq_{\mathfrak{S}} G_2$ holds, then for every $g_2 \in \mathsf{Granules}\langle \mathfrak{S}|G_2 \rangle$ and every $S \subseteq_f \mathsf{Granules}\langle \mathfrak{S}|G_1 \rangle$, $\mathsf{Constr}^{\pm}\langle \mathfrak{S} \rangle \models_{\mathfrak{S}} (g_2 = \textstyle\bigsqcup_{\mathfrak{S}} S)$ iff

$$S = \{g_1 \mid \langle g_1, g_2 \rangle \in \mathsf{NRel}_{\mathfrak{S}:\langle G_1,G_2 \rangle}\} = \{g_1 \mid \langle g_1, g_2 \rangle \in \mathsf{SRel}_{\mathfrak{S}:\langle G_1,G_2 \rangle}\}.$$

Proof. The proof follows immediately from Theorems 4.9 and 4.11. □

Discussion 4.15 (Equality-join order is transitive). It is easy to see that the equality-join order relation is transitive. More precisely, if $G_1 \trianglelefteq_{\mathfrak{S}} G_2$ and $G_2 \trianglelefteq_{\mathfrak{S}} G_3$ both hold, then so too does $G_1 \trianglelefteq_{\mathfrak{S}} G_3$. This follows immediately from the first equality of Theorem 4.11(b) and the fact that the subsumption relation $\sqsubseteq_{\mathfrak{S}}$ is transitive. To illustrate the utility of this observation via example, referring to the hierarchy to the left in Fig. 2, since both Province $\trianglelefteq_{\mathfrak{C}}$ Region and County $\trianglelefteq_{\mathfrak{C}}$ Province, it is also the case that County $\trianglelefteq_{\mathfrak{C}}$ Region, and, furthermore,

$$\mathsf{SRel}_{\mathfrak{C}:\langle \mathsf{County},\mathsf{Region} \rangle} = \mathsf{SRel}_{\mathfrak{C}:\langle \mathsf{County},\mathsf{Province} \rangle} \circ \mathsf{SRel}_{\mathfrak{C}:\langle \mathsf{Province},\mathsf{Region} \rangle},$$

with ∘ denoting relational composition. Thus, it is not necessary to represent all pair of the form $G_i \trianglelefteq_{\mathfrak{S}} G_j$, but rather only a base set, from which the others may be obtained via transitivity. In both diagrams of Fig. 2, the edges labelled with $\trianglelefteq$ identify such base sets.

This transitivity property is not shared by the subsumption-join order relation $\otimes_{\mathfrak{S}}$, as is easily verified by example.

Discussion 4.16 (Implementation of bigranular constraints via implicit representation). A PostgreSQL-based system, providing multigranular features, is under development at the University of Concepción. Called MGDB, it is based upon the theory of [8], employing further the ideas elaborated in this paper. MGDB supports neither detailed spatial models (based upon regions in $\mathbb{R}^2$) nor the detailed spatial operations described in [4]. Rather, it is a relational extension which supports multigranular attributes. A main feature is support for basic spatial relationships, such as nondisjointness, subsumption, and join, without the need for an elaborate $\mathbb{R}^2$ model. A second feature is that spatial and temporal attributes are both recaptured using the same underlying formalism.

Currently, MGDB is implemented via additional relations on top of an ordinary relational schema. Thus, each multigranular attribute $\mathfrak{S}$ is represented as an ordinary attribute, together with additional relations which recapture its special properties. In particular, for each such attribute and each granularity pair $\langle G_1, G_2 \rangle$, the relations $\mathsf{NRel}_{\mathfrak{S}:\langle G_1,G_2\rangle}$ and $\mathsf{SRel}_{\mathfrak{S}:\langle G_1,G_2\rangle}$ are stored, either fundamentally or as views (see below for more detail), to the extent that the associated information is known. In addition, there is a special ternary relation $\mathsf{GrPrProp}_{\mathfrak{S}}$, with a tuple of this relation of the form $\langle G_1, G_2, c \rangle$, with c a code which identifies the relationship between the granularities G_1 and G_2. The code may represent combinations of $G_1 \leq_{\mathfrak{S}} G_2$, $G_1 \trianglelefteq_{\mathfrak{S}} G_2$, and $G_1 \otimes_{\mathfrak{S}} G_2$, as well as other relationships not covered in this paper. Given a granule $g_2 \in \mathsf{Granules}\langle \mathfrak{S}|G_2\rangle$, and a request to determine which granules of G_1 are related to it via a join rule which is a consequence of a bigranular property, it is only necessary to look in $\mathsf{GrPrProp}_{\mathfrak{S}}$ to determine the type of join rule (e.g., equality or subsumption), and then to determine the body via a lookup, in $\mathsf{NRel}_{\mathfrak{S}:\langle G_1,G_2\rangle}$, which granules of G_1 form the body of that rule. Since the rules are recovered via retrieval of the appropriate tuples in these relations, and not directly as formulas, the representation is termed *implicit*.

For economy, some of the relations of the form $\mathsf{DRel}_{\mathfrak{S}:\langle G_1,G_2\rangle}$ and $\mathsf{SRel}_{\mathfrak{S}:\langle G_1,G_2\rangle}$ are implemented as views. For example, if either of $G_1 \leq_{\mathfrak{S}} G_2$ or $G_1 \trianglelefteq_{\mathfrak{S}} G_2$ holds, then $\mathsf{DRel}_{\mathfrak{S}:\langle G_1,G_2\rangle}$ and $\mathsf{SRel}_{\mathfrak{S}:\langle G_1,G_2\rangle}$ are the same relation, so only one need be stored explicitly. Likewise, $\mathsf{SRel}_{\mathfrak{S}:\langle G_1,G_3\rangle} = \mathsf{SRel}_{\mathfrak{S}:\langle G_1,G_2\rangle} \circ \mathsf{SRel}_{\mathfrak{S}:\langle G_2,G_3\rangle}$ if either of $G_1 \leq_{\mathfrak{S}} G_2 \leq_{\mathfrak{S}} G_3$ or $G_1 \trianglelefteq_{\mathfrak{S}} G_2 \trianglelefteq_{\mathfrak{S}} G_3$ holds, so $\mathsf{SRel}_{\mathfrak{S}:\langle G_1,G_3\rangle}$ may then be represented as a view defined by relational join. This means that relationships such as equality join, as sketched in Discussion 4.15, require virtually no additional storage for representation. While a tuple of the form $\langle G_1, G_3, c \rangle$ must be present in $\mathsf{GrPrProp}_{\mathfrak{S}}$, no additional space is required to represent $\mathsf{SRel}_{\mathfrak{S}:\langle G_1,G_3\rangle}$ or $\mathsf{NRel}_{\mathfrak{S}:\langle G_1,G_3\rangle}$.

A substantial superset of the hierarchies shown in Fig. 2, including electoral as well as administrative subdivisions of Chile in the spatial case, forms the core of the test database. All such data are obtained from publicly available sources. This spatial hierarchy is very rich in granularity pairs related by $\trianglelefteq_{\mathfrak{e}}$ and $\otimes_{\mathfrak{e}}$. Time intervals, as illustrated in the rightmost hierarchy of Fig. 2, form part of the test database as well. The system will be discussed in more detail in a future paper.

Discussion 4.17 (Relationship to other work). An extensive literature comparison for the general multigranular framework used in this paper may be found in [8, Sect. 6]. Only literature relevant to the topics of this paper which are not developed in [8] are noted here. A fairly extensive presentation of granular relationships may be found in [1], including in particular the equality join relation $\trianglelefteq$, there called *groups into*, as well as the combination of ordinary granularity order $\leq$ and equality join $\trianglelefteq$, there called *partitions*. It does not cover the subsumption join relation $\ominus$. Although [1] is specifically about the time domain, many of the concepts presented there apply equally well to spatial and other domains. This is reinforced not only by the work of this paper, but also by papers such as [2,10], which apply the concepts of [1] to the spatial domain. In addition, [12] provides a development of the equality-join operator $\trianglelefteq$ for the spatial domain, there denoted $\models$. Reference [5] provides further insights into the multigranular framework within the context of time granularity.

5   Conclusions and Further Directions

A method for representing bigranular join rules implicitly in a multigranular relational DBMS has been developed. As such rules occur frequently in practice, the technique promises to prove central to an implementation. Indeed, they have already been used in an early implementation of the system MGDB.

There are two main avenues for future work. First, the main reason that the techniques of this paper were developed is that direct implementation of join rules proved too inefficient in practice. While most rules are bigranular, there are often some which are not. One topic of future work is to find a way to integrate the methods of this paper with representation of non-bigranular rules, in a way which preserves the efficacy of the implementation. A second and very major topic is to extend MGDB with its own query language and interface. Currently, MGDB is a testbed for ideas, but to be useful as a stand-alone system, it must be augmented to have its own query language and interface, so that the implementation of the multigranular features is transparent to the user.

Acknowledgment. The work of M. Andrea Rodríguez, as well as three visits of Stephen J. Hegner to Concepción, during which many of the ideas reported here were developed, were funded in part by Fondecyt-Conicyt grant number 1170497.

References

1. Bettini, C., Dyreson, C.E., Evans, W.S., Snodgrass, R.T., Wang, X.S.: A glossary of time granularity concepts. In: Etzion, O., Jajodia, S., Sripada, S. (eds.) Temporal Databases: Research and Practice. LNCS, vol. 1399, pp. 406–413. Springer, Heidelberg (1998). https://doi.org/10.1007/BFb0053711
2. Camossi, E., Bertolotto, M., Bertino, E.: A multigranular object-oriented framework supporting spatio-temporal granularity conversions. Int. J. Geogr. Inf. Sci. **20**(5), 511–534 (2006)

3. Davey, B.A., Priestly, H.A.: Introduction to Lattices and Order, 2nd edn. Cambridge University Press, Cambridge (2002)
4. Egenhofer, M.J.: Deriving the composition of binary topological relations. J. Vis. Lang. Comput. **5**(2), 133–149 (1994)
5. Euzenat, J., Montanari, A.: Time granularity. In: Fisher, M., Gabbay, D.M., Vila, L. (eds.) Handbook of Temporal Reasoning in Artificial Intelligence, vol. 1, pp. 59–118. Elsevier, New York (2005)
6. Fagin, R.: Horn clauses and database dependencies. J. Assoc. Comp. Mach. **29**(4), 952–985 (1982)
7. Hegner, S.J., Rodríguez, M.A.: Integration integrity for multigranular data. In: Pokorný, J., Ivanović, M., Thalheim, B., Šaloun, P. (eds.) ADBIS 2016. LNCS, vol. 9809, pp. 226–242. Springer, Cham (2016). https://doi.org/10.1007/978-3-319-44039-2_16
8. Hegner, S.J., Rodríguez, M.A.: A model for multigranular data and its integrity. Informatica Lith. Acad. Sci. **28**, 45–78 (2017)
9. Kifer, M., Bernstein, A., Lewis, P.M.: Database Systems: An Application-Oriented Approach, 2nd edn. Addison-Wesley, Boston (2006)
10. Mach, M.A., Owoc, M.L.: Knowledge granularity and representation of knowledge: towards knowledge grid. In: Shi, Z., Vadera, S., Aamodt, A., Leake, D. (eds.) IIP 2010. IAICT, vol. 340, pp. 251–258. Springer, Heidelberg (2010). https://doi.org/10.1007/978-3-642-16327-2_31
11. Monk, J.D.: Mathematical Logic. Springer, New York (1976). https://doi.org/10.1007/978-1-4684-9452-5
12. Wang, S., Liu, D.: Spatio-temporal database with multi-granularities. In: Li, Q., Wang, G., Feng, L. (eds.) WAIM 2004. LNCS, vol. 3129, pp. 137–146. Springer, Heidelberg (2004). https://doi.org/10.1007/978-3-540-27772-9_15

QDR-Tree: An Efficient Index Scheme for Complex Spatial Keyword Query

Xinshi Zang, Peiwen Hao, Xiaofeng Gao$^{(\boxtimes)}$, Bin Yao, and Guihai Chen

Shanghai Key Laboratory of Scalable Computing and Systems,
Department of Computer Science and Engineering, Shanghai Jiao Tong University,
Shanghai 200240, China
{fei125,williamhao}@sjtu.edu.cn, {gao-xf,yaobin,gchen}@cs.sjtu.edu.cn

Abstract. With the popularity of mobile devices and the development of geo-positioning technology, location-based services (LBS) attract much attention and top-k spatial keyword queries become increasingly complex.It is common to see that clients issue a query to find a restaurant serving pizza and steak, low in price and noise level particularly.However, most of prior works focused only on the spatial keyword while ignoring these independent numerical attributes.

In this paper we demonstrate, for the first time, the *Attributes-Aware Spatial Keyword Query* (ASKQ), and devise a two-layer hybrid index structure called *Quad-cluster Dual-filtering R-Tree* (QDR-Tree). In the keyword cluster layer, a Quad-Cluster Tree (QC-Tree) is built based on the hierarchical clustering algorithm using kernel k-means to classify keywords.In the spatial layer, for each leaf node of the QC-Tree, we attach a Dual-Filtering R-Tree (DR-Tree) with two filtering algorithms, namely, keyword bitmap-based and attributes skyline-based filtering. Accordingly, efficient query processing algorithms are proposed.

Through theoretical analysis, we have verified the optimization both in processing time and space consumption. Finally, massive experiments with real-data demonstrate the efficiency and effectiveness of QDR-Tree.

Keywords: Top-k spatial keyword query · Skyline algorithm
Keyword cluster · Location-based service

1 Introduction

With the growing popularity of mobile devices and the advance in geo-positioning technology, location-based services (LBS) are widely used and spatial keyword

This work was partly supported by the Program of International S&T Cooperation (2016YFE0100300), the China 973 project (2014CB340303), the National Natural Science Foundation of China (Grant number 61472252, 61672353, 61729202 and U1636210), the Shanghai Science and Technology Fund (Grant number 17510740200), CCF-Tencent Open Research Fund (RAGR20170114), and Guangdong Province Key Laboratory of Popular High Performance Computers of Shenzhen University (SZU-GDPHPCL2017).

© Springer Nature Switzerland AG 2018
S. Hartmann et al. (Eds.): DEXA 2018, LNCS 11029, pp. 390–404, 2018.
https://doi.org/10.1007/978-3-319-98809-2_24

query becomes increasingly complex. Clients may have special requests on numerical attributes, such as price, in addition to the location and keywords.

Example 1. Consider some spatial objects in Fig. 1(a), where dots represent spatial objects such as restaurants, whose keywords and three numerical attributes are listed in Fig. 1(b). Dots with the same color own similar keywords, e.g., red dots share keywords about food. The triangle represents a user issuing a query to find a nearest restaurant serving pizza and steak with low level in price, noise, and congestion. At a first glance, o_8 seems to be the best choice for the close range, while o_1 surpasses o_8 in the numerical attributes obviously. This common situation shows that such complex queries deserve careful treatment.

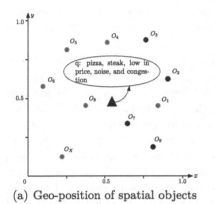

Spatial objects	Keywords	Numerical attributes		
		price	noise	congestion
O_1	pizza, steak	0.3	0.3	0.4
O_2	hospital, dentist	0.7	0.5	0.3
O_3	cinema, theater	0.3	0.5	0.6
O_4	hamburger, pizza	0.2	0.4	0.2
O_5	pizza, steak	0.6	0.3	0.3
O_6	sandwich, bread	0.7	0.2	0.2
O_7	youth, hotel	0.5	0.3	0.4
O_8	pizza, steak	0.5	0.5	0.5
O_9	market, mall	0.2	0.4	0.8

(a) Geo-position of spatial objects   (b) Keywords and attributes table

Fig. 1. A set of spatial objects and a query (Color figure online)

Extensive efforts have been made to support spatial keyword query. However, prior works [7,9,15] mainly focused on the keywords of spatial objects but neglected or failed to distinguish independent numerical attributes. Recently, Sasaki [16] schemed out SKY R-Tree which incorporates R-tree with skyline algorithm to deal with the numerical attributes. However, it does not work well for multi keywords, which reduces their usage for various applications. Liu [10] proposed a hybrid index structure called Inverted R-tree with Synopses tree (IRS), which can search many different types of numerical attributes simultaneously. However, the IRS-based search algorithm requires providing exact ranges of attributes which is a heavy and unnecessary burden to the users. What's more, the exact match in in attributes can also lead to few or no query results to be returned.

Correspondingly, in this paper, we named and studied, for the first time, the *attributes-aware spatial keyword query* (ASKQ). This complex query needs to take location proximity, keywords' similarity, and the value of numerical attributes into consideration, that is respectively, the Euclidean spatial distance, the relevance of different keywords, and the integrated attributes of users' preference. Obviously the ASKQ has wide apps in the real world.

Tackling with the ASKQ in Example 1, common search algorithms [7,9,15] ignoring numerical attributes may retrieve finally o_1, o_5, o_8 indiscriminately, and SKY R-Tree-based algorithm may return o_4 as one of results, and IRS-Tree-based algorithm may retrieve no objects when the query predicate is set as *"price < 0.3 & noise < 0.3 & congestion < 0.4"*. Apparently, none of these algorithms can satisfy the users' need. These gaps motivate us to investigate new approaches that can deal with the ASKQ efficiently.

In this paper, we propose a novel two-layer index structure called *Quad-cluster Dual-filter R-Tree* (QDR-Tree) with query processing algorithms. In the first layer we deal with keyword specifically. Considering numbers of keywords share the similar semanteme and clients tend to query objects of the same class, we cluster and store the keywords in a Quad-Cluster Tree (QC-Tree) by hierarchical clustering algorithm using kernel k-means clustering [6]. With keyword relaxation operation and Cut-line theorem to avoid redundance, QC-Tree can balance search time and space cost well.

In the second layer we deal with spatial objects with numerical attributes. At each leaf node of the first layer, a Dual-filter R-Tree (DR-Tree) is attached according to two filtering algorithms, namely, keyword bitmap-based filtering and attributes skyline-based filtering, which effectively reduce the false positives.

Moreover, we also propose a novel method to measure the relevance of one spatial object with the query keywords. We measure the similarity of different keywords from both textual and semantic aspects. For the latter one, the *term vectors* that are obtained by word2vec [12] are applied to represent every keywords, and therefore, the similarity can be quantified. Note that both queries and spatial objects usually own several keywords, a bitmap of keywords is used to measure the relevance between two lists of keywords lightly and efficiently.

Table 1 compares the current index with QDR-Tree in three aspects. Apparently, QDR-Tree outperform existing methods in tackling with the ASKQ, and can achieve great improvements in query processing time and space consumption. This will be demonstrated in both theoretical and experimental analysis. Massive experiments with real-data also confirm the efficiency of QDR-Tree.

Table 1. Comparisons among current indexes and QDR-tree

Index	From	Location proximity	Muti-keywords	Fuzzy attributes
IR-Tree	TKDE (2011) [9]	✓	✓	✗
IL-Quadtree	ICDE (2013) [18]	✓	✓	✗
SKY R-Tree	DASFAA (2014) [16]	✓	✗	✓
IRS-Tree	TKDE (2015) [10]	✓	✓	✗
QDR-Tree	DEXA (2018)	✓	✓	✓

To sum up, the main contributions of this paper are summarized as follows:

- We formulate the attributes-aware spatial keyword query, which takes spatial proximity, keywords' similarity and numerical attributes into consideration.

- We design a novel hybrid index structure, i.e., QDR-Tree which incorporates Quad-Cluster Tree with Dual-filtering R-Trees and accordingly propose the query processing algorithm to tackle the ASKQ.
- We propose a novel method to measure the relevance of one spatial objects with query keywords based on word2vec and bitmap of keyword.
- We conduct an empirical study that demonstrates the efficiency of our algorithms and index structures for processing the ASKQ on real-world datasets.

The rest of the paper is organized as follows. Section 2 reviews the related works. Section 3 formulates the problem of ASKQ. Section 4 presents the QDR-Tree. Section 5 introduces the query processing algorithm based on the QDR-Tree. Three baseline algorithms are proposed in Sect. 6 and considerable experimental results are reported. Finally, Sect. 7 concludes the paper.

2 Related Work

Existing works concerning the ASKQ include spatial keyword search, keyword relevance measurement, and the skyline operator.

Spatial Keyword Search. There are many studies on spatial keyword search recently [7,17,18]. Most of them focus on integrating inverted index and R-tree to support spatial keyword search. For example, IR2-tree [7] combines R-trees with signature files. It preserves objects spatial proximity, which is the key to solve spatial queries efficiently, and can filter a considerable portion of the objects that do not contain all the query keywords. Thus it significantly reduces the number of objects to be examined. SI-index [18] overcomes IR2-trees' drawbacks and outperform IR2-tree in query response time significantly. [17] proposes inverted linear quadtree, which is carefully designed to exploit both spatial and keyword-based pruning techniques to effectively reduce the search space.

Keyword Relevance Measurement. The traditional measurement on keyword relevance includes textual and semantic relevance. The textual relevance can be computed using an information retrieval model [2,4,5]. They are all TF-IDF variants essentially sharing the same fundamental principles. The semantic relevance is measured by many methods. [13,14] apply the Latent Dirichlet Allocation (LDA) model to calculate the topic distance of keywords. Gao [3] proposed an efficient disk-based metric access method which achieves excellent performance in the measurement of keywords' similarity.

The Skyline Operator. The skyline operator deals with the optimization problem of selecting multi-dimension points. A skyline query returns a set of points that are not dominated by any other points, called a skyline. It is said that a point o_i dominates another point o_j if o_i is no worse than o_j in all dimensions of attributes and is better than o_j at least in one dimension. Borzsonyi et al. [1] first introduced the skyline operator into relational database systems and introduced three algorithms. Geng et al. [11] propose a method which combines the spatial information with non-spatial information to obtain skyline results. Lee [8] et al. focused on two methods about multi-dimensional subspace skyline computation and developed orthogonal optimization principles.

3   Problem Statement

Given an geo-object dataset O in which each object o is denoted as a tuple $\langle \lambda,$ K, A$\rangle$, where $o.\lambda$ is a location descriptor which we assume is at a two dimensional geographical space and is composed of latitude and longitude, $o.K$ is the set of keywords, and $o.A$ represents the set of numerical attributes. Without loss of generality, we assume the attributes $o.a_i$ in $o.A$ are numeric attributes and normalize each $o.a_i \in [0,1]$. We assume that smaller values of these numerical attributes, e.g., price and noise, are preferable. As for other numerical attributes' values which are better if higher, such as the rating and health score, we convert them decreasingly as $o.a_i = 1 - o.a_i$. The query q is represented as a tuple $\langle \lambda, K, W \rangle$, where $q.\lambda$ and $q.K$ represent the location of the user and the required keywords respectively, and $q.W$ represents the set of weight for different numerical attributes and user's different preference on these attributes. $\forall q.w_i \in q.W, q.w_i \geq 0$ $(i = 1, \ldots, |q.W|)$ and $\sum_{i=1}^{|q.W|} q.w_i = 1$. The reason for assigning weight to each attribute instead of qualifying exact range of attributes is to prepare for the fuzzy query on numerical attributes. In order to elaborate the QDR-Tree , we firstly define the keyword distance and the keyword cluster as follows.

Definition 1 (Keyword Distance). *Given two keywords k_1, k_2, their keyword distance, denoted as $d(k_1, k_2)$, includes both textual distance and semantic distance. The textual similarity between two keywords is denoted as $d_t(k_1, k_2)$ which is measured by the Edit Distance. The semantic distance between two keywords denoted as d_s is measured by the Euclidean distance of the term vector generated by word2vec. With a parameter $\delta (\in [0,1])$ controlling their relative weights, Eq. (1) describes the formulation of $d(k_1, k_2)$.*

$$d(k_1, k_2) = \delta d_t(k_1, k_2) + (1 - \delta) d_s(k_1, k_2) \tag{1}$$

Definition 2 (Keyword Cluster). *A keyword cluster (C_i) is formed by similar keywords. The cluster diameter is defined as the maximum keyword distance within the cluster. One keyword can be allocated into the cluster if the diameter after adding it does not exceed the threshold τ, i.e. $\forall k_i, k_j \in C_i, d(k_i, k_j) < \tau$. Each cluster has a center object denoted as $C_i.cen$. All the keyword clusters (C_i) make up the set of keyword clusters $(\mathbb{C})$.*

Definition 3 (Attributes-Aware Spatial Keyword Query). *Given a geo-object set O and the attributes-aware spatial keyword query q, the result includes a set of $Top_\kappa(q),$[1] $Top_\kappa(q) \subset O$, $|Top_\kappa(q)| = \kappa$ and $\forall o_i, o_j : o_i \in Top_\kappa(q), o_j \in O - Top_\kappa(q)$, it holds that $score(q, o_i) \leq score(q, o_j)$.*

As for the evaluation function, $score(q, o)$ in Definition 3, it is composed of three aspects, including the location proximity, the keywords similarity, and the value of numerical attributes, and will be discussed at large in the Sect. 5.

---

[1] Hereafter, Top-k is denoted as Top-κ to avoid confusion with the k-means algorithm.

4   QDR-Tree

In this section, we introduce a new hybrid index structure QDR-Tree, which is a new indexing framework for efficiently processing the ASKQ. The QDR-Tree can be divided into two layers, the keyword cluster layer and the spatial layer where the QDR-Tree can be split up into two sub-trees, named as Quad-Cluster Tree (QC-Tree) and Dual-filtering R-tree (DR-Tree) respectively.

4.1   Keyword Cluster Layer

The keyword cluster layer deals with keyword search with both textual and semantic similarities. Neither appending an R-Tree to each keyword with a huge space redundancy, nor just clustering all keywords into k groups with a high false positive ratio during query search, QC-Tree smartly splits keyword set into hierarchical levels and link them by a Quad-Tree.

To improve the searching efficiency, we propose a new hierarchical quad clustering algorithm based on the *kernel k-means* [6]. Compared with the traditional k-means clustering, kernel k-means will have better clustering effect even the samples do not obey the normal distribution and is more suitable to cluster the keywords. Moreover, different from the common clustering, hierarchical clustering can form a meaningful relationship between different clusters, which is helpful to allocate a new sample and decrease the cost of misallocation. After the clustering process finishs, a quad-cluster tree (QC-Tree) is used to arrange all of these clusters, which is the core composition of the keyword cluster layer. In Algorithm 1, the critical part is applying the kernel k-means to each keyword cluster per level, with k fixed as 4. Furthermore, when the diameter of the keyword cluster is smaller than the $\tau_{cluster}$, the duplication operation is executed, which is presented in Algorithm 2 and will be discussed later.

Algorithm 1. Hierarchical quad clustering algorithm

Input: keyword set K, cluster number k
Output: Quad-Cluster Tree: T_{qc}
1   T_{qc}.add(K)
2   Insert K into a priority queue U /* insterst as a set */
3   **while** $U \neq \emptyset$ **do**
4 | $S \leftarrow U$.Pop()   /* pop the whole set */
5 | $\{S_1,S_2,S_3,S_4\} \leftarrow$ **Kernel$k$Means** (k, S) /* k=4 by default */
6 | **foreach** $S_i \in \{S_1,S_2,S_3,S_4\}$ **do**
7 | | **if** $S_i.diameter < \tau_{cluster}$ **then**
8 | | | Duplication (S_1,S_2,S_3,S_4)
9 | | **else**
10 | | | insert S_i in to U
11 | | |_ T_{qc}.add(S_i)   /* S_i are children of S */

Algorithm 2. Duplication

Input: Four keyword sets: S_1, S_2, S_3, S_4
Output: Duplicated keyword sets: S_1', S_2', S_3', S_4'

1 **for** $\forall k_i \in S_1 \bigcup S_2 \bigcup S_3 \bigcup S_4$ **do**
2 | **if** $\sigma(d(k_i, S_j.cen)) < \tau_{dup}$ **then**   /* Variance */
3 | └ $S_j' \leftarrow k_i \bigcup S_j$, if $k_i \notin S_j$ with $j \in \{1, 2, 3, 4\}$
4 $\{S_1, S_2, S_3, S_4\} \leftarrow \{S_1', S_2', S_3', S_4'\}$

Figure 2(a) illustrates the hierarchical clustering in Algorithm 1, where each dot represents a keyword and different aggregation of these dots presents different keyword clusters. The dots marked in different color are the centroid of these clusters, and moreover, same color denotes their clusters stay in the same level.

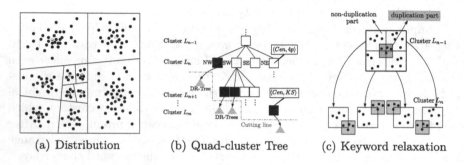

(a) Distribution   (b) Quad-cluster Tree   (c) Keyword relaxation

Fig. 2. Overview of the keyword cluster layer

Notice that, the main target of QC-Tree is to improve the pruning effect of keywords while making the future query keyword set located in only one keyword cluster. As is shown in both Algorithm 1 and Fig. 2(a), with the cluster level growing, the cluster will be more centralized and compact. That means the possibility of one query being allocated to different clusters increases layer by layer. It is necessary to decide an optimal $\tau_{cluster}$ to terminate the hierarchical cluster proceeding, if not, there would only be a single keyword in each cluster finally. The basic structure of QC-Tree is displayed in Fig. 2(b), where each internal node keeps the centroid keyword (cen) and four pointers (4p) to its four descendants nodes, and each leaf node will keep the keyword set in this cluster and the pointer to a new DR-Tree. Additionally, a cut-line is drawn to emphasize the shift of index structure, which is mainly dependent on the value of $\tau_{cluster}$.

As is analyzed above, the leaf cluster is where a query would most likely be scattered into different clusters. We will take a keyword-relaxation operation by duplicating some keywords among the four clusters sharing the same parent node. In Fig. 2(c), for a keyword cluster, its keywords are grouped into four sub-clusters and the duplication operation need to be executed. The dots in the shadow represent the keywords that will be duplicated and allocated to all of these four sub-clusters because they are closed to all of the sub-clusters. Here, we introduce another threshold (τ_{dup}) to decide whether to execute the duplication operation. Although this keyword-relaxation operator will cause redundancy of keywords and extra space consumption, it will largely improve the time efficiency, which will also be demonstrated in the experimental verification.

4.2 Spatial Layer

Under each keyword cluster in the bottom of QC-Tree, we build a DR-Tree based on dual-filtering technique to organize the spatial objects in this cluster.

In Fig. 3, a basic structure of DR-Tree is shown in the spatial layer. Each internal node N records a two-element tuple: $\langle SP, KB \rangle$. The first element SP stands for the skyline points of the numerical attributes of all objects in the subtree rooted at the node. The second element is a bitmap of the keywords included in this cluster, which uses 1 and 0 to denote the existence of keywords.

Keyword Bitmap Filter Algorithm: In the DR-Tree, each node just records the keyword bitmap, and then the specific keywords list is kept only in the leaf keyword cluster. Then, the keyword relevance can be calculated just by Bitwise AND within the pair of bitmaps, which can decrease the storage consumption and increase the query efficiency.

Because bitwise AND within bitmaps need an exact keywords matching, in order to support similar keyword matching, we also implement the relaxation in each query process. In Fig. 3, as is highlighted in blue, the bitmap of query keywords performs a search-relaxation by switching some 0-bits to 1-bits based on the keyword similarity The search-relaxation algorithm will be proposed in Algorithm 4 in Sect. 4.2.

Multidimensional Subspace Skyline Filter Algorithm: In order to satisfy the needs of user's intention on multiple attributes, a filter called Multidimensional Subspace Skyline Filter, which is inspired by [1,8], is employed to amortize the query false positive and the cost of computation. We use the Evaluate() algorithm proposed in [8] to gain the multidimensional skyline points efficiently, and then let every QC-Tree node record the skyline points of its descendants. Furthermore, in order to reduce the complexity of recording multidimensional skyline points, we will take the point-compression operation by merging the closed skyline points in the attributes space. We calculate the cosine distance between skyline points' attributes to measure the similarity, and then merge these closed points when cosine distance is larger than a threshold.

5 QDR-Based Query Algorithm

In this section, we will introduce the ASKQ processing algorithms based on QDR-Tree. The process includes finding the Leaf Cluster, making search-relaxation and searching in the DR-Tree.

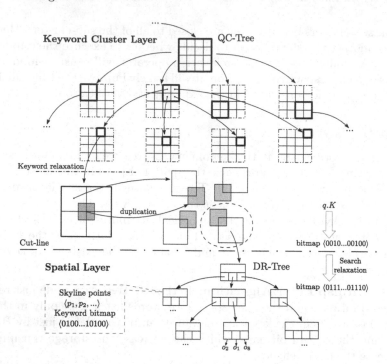

Fig. 3. Structure of QDR-Tree

Find the leaf cluster. The leaf keyword cluster that is best-matched with q can be obtained by iteratively comparing q with the four sub-clusters in each cluster level. If the combination of keywords in the query is typical and can be allocated into the same cluster, only one keyword cluster will be found. Otherwise, more than one keyword cluster may be returned.

Search-Relaxation. As is stated in Sect. 4.2, by means of executing search-relaxation, bitmap-based filter can support similar keyword matching. In Algorithm 4, a bitmap of relaxed query keyword is obtained by switching 0-bit to 1-bit if their keyword distance is under a threshold. By adopting a rational threshold, we can make a good trade-off between time cost and space occupation.

Algorithm 3. FindLeafCluster

Input: q, QC-Tree T_{qc}

Output: the leaf cluster: LC

1 $LC \leftarrow \emptyset$

2 **foreach** $k \in q.K$ **do**

3 | $lc \leftarrow T_{qc}.root$

4 | **while** lc *is not leaf cluster* **do**

5 | $\lfloor$ $ls \leftarrow lc.sub_i$, with $d(k, lc.sub_i.cen)$ is minimum among 4 $lc.subs$

6   $\lfloor$ $LC \leftarrow LC \cup lc$

Algorithm 4. Search relaxation
___

Input: bitmap of query keyword: bmq, bitmap of keyword cluster: bmc
Output: bitmap of relaxed query keyword: bmr
1 **for** $i \leftarrow 1$ to $/bmq/$ **do**
2   **if** $bmq[i] = 1$ **then**
3         $bmr[i] \leftarrow 1$
4   **for** $j \leftarrow 1$ to $/bmc/$ **do**
5   **if** $d(k_i, k_j) < \tau$ **then**
6               $bmr[j] \leftarrow 1$

Algorithm 5 illustrates the query processing mechanism over QDR-Tree. Given a query q, the object retrieval is carried out firstly by traversing the QC-Tree to locate the best-matched keyword cluster. Secondly, after executing search-relaxation, it will traverse the DR-Tree in the ascending order of the scores and keep a minimum heap for the scores. Notice that, if more than one keyword cluster is located, it will traverse all of them. At last the Top-κ results can be returned.

The ranking score of an object o for ASKQ is calculated by Eq. (2). Here, $\alpha, \beta \in [0,1]$ are parameters indicating the relative importance of these three factors. $\psi(q,o)$ is the Euclidian distance between q and o. The D_s^{max} is the maximal spatial distance that the client will accept. $\phi(q,o)$ which represents the keyword relevance between q and o is determined by the result of Bitwise AND between their keyword bitmaps. The smaller the score, the higher the relevance.

$$score(q,o) = \alpha\beta \times \frac{\psi(q,o)}{D_s^{max}} + (1-\beta) \times \frac{1}{\phi(q,o)} + (1-\alpha)\beta \times \sum_{i=1}^{|q.W|} q.w_i \times o.a_i \quad (2)$$

What is more, the score for non-leaf node N can also been measured to represent the optimal score of its descendant nodes, which is defined as Eq. (3)

$$score(q,N) = \alpha\beta \times \frac{\min \psi(q, N.MBR)}{D_s^{max}} + (1-\beta) \times \frac{1}{\phi(q,N)}$$
$$+ (1-\alpha)\beta \times \min_{\forall p \in N.sp} \sum_{i=1}^{|q.W|} q.w_i \times p.a_i \quad (3)$$

where the $\min \psi(q, N.MBR)$ represents the minimum Euclidian distance between the N's MBR and the $\phi(q,N)$ is can also be calculated by the bitmap of keywords kept in this node. We can prove that $Top_\kappa(q)$ is an exact result by the Theorem 1. If the score of the internal node dose not satisfy the ASKQ, there is no need to search its descendant nodes. Hence, the final Top-κ objects will have the least κ scores.

Theorem 1. *The score of an internal node N is the best score of its descendant object o to the query q.*

Proof. the score factors in location proximity, keyword relevance and non-spatial attributes' value. First, the MBR of the N encloses all of its descendant objects, then $\forall o_i \in$ descendant objects of N, $\min \psi(q, N.MBR) \leq \psi(q, o_i)$. Second, the keyword bitmap includes all of the keywords existing in the descendant objects of N. Obviously, $\phi(q, N) \geq \phi(q, o)$. Finally, the skyline points dominate or are equal to all of descendent objects concerning the value of attributes, i.e., $\min_{\forall p \in N.SP} \sum_{i=1}^{|q.W|} q.w_i \times p.a_i \leq \sum_{i=1}^{|q.W|} q.w_i \times o.a_i$. All these inequalities contribute to that $score(q, N) \leq score(q.o)$. ☐

Algorithm 5. QDR-Search algorithm

Input: a query q, Top$_\kappa$ results κ, and a QDR-Tree T_{qdr}
Output: Top$_\kappa(q)$

1 $LC = $ FindLeafCluster (q, T_{qc});
2 **for** $i \leftarrow 1$ to $|LC|$ **do**
3   q.bitmap $\leftarrow$ SearchRelaxation $(q$.bitmap, $LC[i]$.bitmap$)$
4   Minheap.insert($LC[i]$.root, 0)
5   **while** *Minheap.size()* $\neq 0$ **do**
6          $N \leftarrow$ Minheap.first()
7   **if** N *is an object* **then**
8   Top$_\kappa(q)$.insert(N)
9   **if** *Top$_\kappa(q).size()$* $\geq k$ **then**
10   **break**
11   **else**
12   **for** $n_i \in N.entry$ **do**
13   **if** *Number of objects with smaller score than score(q, n_i) in Minheap* $< (\kappa - Top_\kappa(q).size())$ **then**
14   Minheap.insert(n_i, score(q, n_i))

---

6 Experiment Study

6.1 Baseline Algorithm

In this section, we propose three baseline algorithms which are based on the three existing indexes listed in Table 1, including IR-Tree [9], SKY R-Tree [16] and IRS-Tree [10]. As is discussed in Sect. 1, none of these existing indexes can be qualified for the ASKQ due to different drawbacks. The specific algorithm designs will be respectively explained in detail as follows.

Because the IR-Tree pays no attention on the value of numerical attributes, all spatial objects containing the query keywords and numerical attributes will be extracted. After that they will be ranked by the comprehensive value of numercial attributes. Eventually, the top-κ spatial objects are just the result of the ASKQ.

Different from the IR-Tree, the SKY R-Tree fails to support multi-keywords query because one SKY R-Tree can only arrange one keyword and its corresponding spatial objects, such as restaurant. In order to deal with the ASKQ, all of the SKY R-Trees containing the query keywords will be searched and merged to obtain the final top-κ results.

The last baseline algorithm is proposed based on the IRS-Tree which is originally intended to address the GLPQ. Unlike ASKQ, the GLPQ requires specific range of attributes to leverage the IRS-Tree. To copy with the ASKQ, we will firstly set some different suitable ranges of each attributes as the input, which insures that enough spatial objects can be returned. Afterwards, we will further to select top-κ objects from the results in the first stage. Apparently, in our experiments, the IRS-Tree will not make much sense anymore.

Notice that, all of these three baseline algorithms cannot solve the ASKQ directly at a time and need subsequent elimination of redundancy, which determines their inefficiency in the ASKQ.

In the experiment section, we conduct extensive experiments on both real and synthetic datasets to evaluate the performance of our proposed algorithms.

6.2 Experiment Setup

The real dataset is crawled from the famous location-based service platform, Foursquare. After information cleaning, the dataset has about 1M objects consisting of geographical location, the keyword list written in English, and the normalized value of attributes. Each spatial object contains the keywords such as steak, pizza, coffee, etc. and four numerical attributes, including price, environment, service and rating.

In the synthetic dataset, each object is composed of coordinates, various keyword, and multi-dimensional numerical attributes. The size of the synthetic dataset varies in the experiments. The coordinates are randomly generated in $(0, 10000.0)$, and the average number of keywords per object is decided by a parameter r which denotes the ratio of the number of object's keywords to the cluster's. Without loss of generality, the values of each numercial attribute are randomly and independently generated, following a normal distribution.

We compare the query cost of proposed algorithms with different datasets respectively. The experimental settings are given in Table 2. The default values are used unless otherwise specified. All algorithms are implemented in Python and run with Intel core i7 6700HQ CPU at 2.60 GHz and 16 GB memory.

6.3 Performance Evaluation

In this section, we campare different baseline algorithms proposed in Sect. 6.1 with our framework. We evaluate the processing time and disk I/O of all the proposed methods by varying the parameters in Table 2 and investigate their effects. In the first part we study the experimental results on the real dataset.

Table 2. Default value of parameters

Parameter	Default value	Descriptions
κ	10	Top-κ query
$|o.A|$	4	No. of attributes' dimension
δ	0.5	Weight factor of Eq. (1)
α	0.5	Weight factor of Eq. (2)
β	0.67	Weight factor of Eq. (2)
$\tau_{cluster}$	0.3	Threshold of quad clustering
τ_{dup}	0.05	Threshold of duplication
$|O|$	1M	Number of objects
M	25	Maximum number of DR-tree entries

Index Construction Cost: We first evaluate the construction costs of various methods. The cost of an index is measured by its construction time and space budget. The costs of various methods are shown in Fig. 4, where IRS refers to the baseline structure IRS-Tree, the SKY-R, IR refer to SKY R-Tree, IR-Tree respectively, and the QDR represents our design. We can see that, SKY R-Tree and IR-Tree exceed both in time consumption and space cost, because they are short of attention of either attributes or multi-keywords. Moreover, since we employ bitmap and skyline points to measure numercial attributes, QDR is more lightweight than IRS.

Effect of κ**:** We investigate the effect of κ on the processing time and disk I/O of the proposed algorithms by randomly generate 100 queries. Here, considering that the SKY R-Tree and IR-Tree do not take into account either attributes or keywords, we add a filter operation after their query process. For example, the SKY R-Tree returns Top-κ results of each keyword and merges them in the second stage. Obviously, this redundancy of result is the main reason of the high time cost. As shown in Fig. 5, with the increase of κ, IRS and QDR have the same smoothly increasing trend on query time and disk I/O. QDR exceeds in query time cost with different parameters. It indicates that we can effectively receive the Top-κ results from one branch to another.

Effect of $|O|$**:** Parameter $|O|$ denotes the number of objects in the QDR-Tree, We increase the number of objects in the synthetic dataset from 0.1 to 5 M. It can be seen from Fig. 7(a) and (b), that, SKY-R and IR have more obvious increases in query time cost and disk I/O when the data size increases, which can be explained because of their larger redundancy along with the larger dataset. On the other hand, the QDR is more stable and surpasses another three indexes.

Effect of $|o.A|$**:** Parameter $|o.A|$ denotes the number of attributes the object o covers. As shown in Fig. 7(c) and (d), the query time and Disk I/O of the IR-Tree based on synopses tree has distinct increase, because it fails to consider the attributes, while another three frameworks are more stable, which is mainly because of either the skyline filter algorithm or the synopses tree.

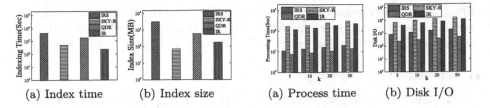

(a) Index time   (b) Index size   (a) Process time   (b) Disk I/O

Fig. 4. Construction cost   **Fig. 5.** Effect of κ

Effect of $\tau_{cluster}$ & τ_{dup}: $\tau_{cluster}$ and τ_{dup} are the crucial parameters in our QDR-Tree, which are analyzed theoretically in Sect. 4. The experimental results also verify their effect on the Processing Time and Index Size. Figure 6(a) and (c) show that both $\tau_{cluster}$ and τ_{dup} have an optimal value to minimize the processing time. Smaller or larger value will both increase the processing time because of keyword scattring or redundancy. In Fig. 6(b) and (d), index size decreases as the $\tau_{cluster}$ becomes larger and reach saturation at some point, while it increases along with the τ_{dup} because of the increase of keyword redundancy.

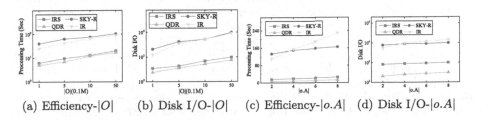

(a) Efficiency-$|O|$ (b) Disk I/O-$|O|$ (c) Efficiency-$|o.A|$ (d) Disk I/O-$|o.A|$

Fig. 6. Synthetic dataset

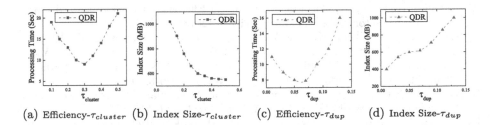

(a) Efficiency-$\tau_{cluster}$ (b) Index Size-$\tau_{cluster}$ (c) Efficiency-τ_{dup} (d) Index Size-τ_{dup}

Fig. 7. Effect of $\tau_{cluster}$ & τ_{dup}

7 Conclusion

In this paper, we formulated the attributes-aware spatial keyword query (ASKQ) and proposed a novel index structure call Quad-cluster Dual-filtering R-Tree (QDR). QDR-Tree is a two-layer hybrid index based on two index structures and two searching algorithms. We also proposed a novel method to measure the relevance of spatial objects with query keywords, which applies keyword-bitmap and search-relaxation to achieve exact and similar keyword match. Moreover, by employing the keyword-relaxation, we greatly improve the time efficiency at the sacrifice of a little space consumption. Finally, massive experiments with real datasets demonstrate the efficiency and effectiveness of QDR-Tree.

References

1. Borzsonyi, S., Stocker, K., Kossmann, D.: The skyline operator. In: IEEE International Conference on 2002 Data Engineering (ICDE), pp. 421–430 (2002)
2. Cao, X., Cong, G., Jensen, C.S.: Retrieving top-k prestige-based relevant spatial web objects. Int. Conf. Very Large Data Bases (VLDB) **3**, 373–384 (2010)
3. Chen, L., Gao, Y., Li, X., Jensen, C.S., Chen, G.: Efficient metric indexing for similarity search. IEEE Trans. Knowl. Data Eng. (TKDE) **29**(3), 556–571 (2017)
4. Cong, G., Jensen, C., Wu, D.: Efficient retrieval of the top-k most relevant spatial web objects. Int. Conf. Very Large Data Bases (VLDB) **2**(1), 337–348 (2009)
5. Cong, G., Jensen, C.S., Wu, D.: Efficient retrieval of the top-k most relevant spatial web objects. Int. Conf. Very Large Data Bases (VLDB) **2**(1), 337–348 (2009)
6. Dhillon, I.S., Guan, Y., Kulis, B.: Kernel k-means: spectral clustering and normalized cuts. In: ACM SIGKDD International Conference on Knowledge Discovery and Data Mining (KDD), pp. 551–556 (2004)
7. Felipe, I.D., Hristidis, V., Rishe, N.: Keyword search on spatial databases. In: IEEE International Conference on Data Engineering (ICDE), pp. 656–665 (2008)
8. Lee, J., Hwang, S.: Toward efficient multidimensional subspace skyline computation. Int. Conf. Very Large Data Bases (VLDB) **23**(1), 129–145 (2014)
9. Li, Z., Lee, K.C., Zheng, B., Lee, W.C., Lee, D., Wang, X.: Ir-tree: an efficient index for geographic document search. IEEE Trans. Knowl. Data Eng. (TKDE) **23**(4), 585–599 (2011)
10. Liu, X., Chen, L., Wan, C.: LINQ: a framework for location-aware indexing and query processing. IEEE Trans. Knowl. Data Eng. (TKDE) **27**(5), 1288–1300 (2015)
11. Ma, G., Arefin, M.S., Morimoto, Y.: A spatial skyline query for a group of users having different positions. In: Third International Conference on Networking and Computing, pp. 137–142 (2012)
12. Mikolov, T., Chen, K., Corrado, G., Dean, J.: Efficient estimation of word representations in vector space. Computer Science (2013)
13. Qian, Z., Xu, J., Zheng, K., Sun, W., Li, Z., Guo, H.: On efficient spatial keyword querying with semantics. In: Navathe, S.B., Wu, W., Shekhar, S., Du, X., Wang, X.S., Xiong, H. (eds.) DASFAA 2016. LNCS, vol. 9643, pp. 149–164. Springer, Cham (2016). https://doi.org/10.1007/978-3-319-32049-6_10
14. Qian, Z., Xu, J., Zheng, K., Zhao, P., Zhou, X.: Semantic-aware top-k spatial keyword queries. World Wide Web (WWW), pp. 1–22 (2017)
15. Ray, S., Nickerson, B.G.: Dynamically ranked top-k spatial keyword search. In: ACM International Conference on Management of Data (SIGMOD), pp. 6–18 (2016)
16. Sasaki, Y., Lee, W.-C., Hara, T., Nishio, S.: Sky R-tree: an index structure for distance-based top-k query. In: Bhowmick, S.S., Dyreson, C.E., Jensen, C.S., Lee, M.L., Muliantara, A., Thalheim, B. (eds.) DASFAA 2014. LNCS, vol. 8421, pp. 220–235. Springer, Cham (2014). https://doi.org/10.1007/978-3-319-05810-8_15
17. Tao, Y., Sheng, C.: Fast nearest neighbor search with keywords. IEEE Trans. Knowl. Data Eng. (TKDE) **26**(4), 878–888 (2014)
18. Zhang, C., Zhang, Y., Zhang, W., Lin, X.: Inverted linear quadtree: efficient top k spatial keyword search. In: IEEE International Conference on Data Engineering (ICDE), pp. 901–912 (2013)

Graph Data and Road Networks

Approximating Diversified Top-k Graph Pattern Matching

Xin Wang[1]([⊠]) and Huayi Zhan[2]

[1] Southwest Jiaotong University, Chengdu, China
xinwang@swjtu.cn
[2] ChangHong Inc., Mianyang, China
huayi.zhan@changhong.com

Abstract. Graph pattern matching has been increasingly used in *e.g.,* social network analysis. As the matching semantic is typically defined in terms of subgraph isomorphism, several problems are raised: (1) matching computation is often very expensive, due to the intractability of the problem, (2) the semantic is often too strict to identify meaningful matches, and (3) there may exist excessive matches which makes inspection very difficult. On the other hand, users are often interested in diversified top-k matches, rather than entire match set, since result diversification has been proven effective in improving users' satisfaction, and top-k matches not only eases result understanding but also can save the cost of matching computation. Motivated by these, this paper investigates approximating diversified top-k graph pattern matching. (1) We extend traditional notion of subgraph isomorphism by allowing edge to path mapping, and define matching based on the revised notion. With the extension, more meaningful matches could be captured. (2) We propose two functions for ranking matches: a relevance function $w(\cdot)$ based on tightness of connectivity, and a distance function $d(\cdot)$ measuring match diversity. Based on relevance and distance functions, we propose diversification function $F(\cdot)$, and formalize the *diversified top-k graph pattern matching* problem using $F(\cdot)$. (3) Despite hardness of the problem, we provide two approximation algorithms with *performance guarantees*, and one of them even preserves *early termination property*. (4) Using real-life and synthetic data, we experimentally verify that our approximation algorithms are effective, and outperform traditional matching algorithms.

1 Introduction

Graph pattern matching has being widely used in social data analysis [4,18], among other things. A number of algorithms have been developed for graph pattern matching that, given a pattern graph Q and a data graph G, compute $M(Q,G)$, the set of matches of Q in G (*e.g.,* [7,15]). As social graphs are typically very large, with millions of nodes and billions of edges, several challenges to social data analysis with graph pattern matching are brought out.

© Springer Nature Switzerland AG 2018
S. Hartmann et al. (Eds.): DEXA 2018, LNCS 11029, pp. 407–423, 2018.
https://doi.org/10.1007/978-3-319-98809-2_25

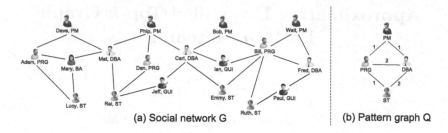

(b) Pattern graph Q

Fig. 1. Querying social network

(1) Traditionally, graph pattern matching is defined in terms of subgraph isomorphism [13]. The matching semantic only allows edge to edge mapping, which is often too strict to identify important and meaningful matches.

(2) The matching algorithms often return an excessive number of results. Indeed, $M(Q, G)$ may contain exponentially many subgraphs of G when matching is defined by subgraph isomorphism. It is a daunting task for the users to inspect such a large match set $M(Q, G)$ and find what they are searching for.

(3) The sheer size of social graphs makes matching computation costly: it is NP-complete to decide whether a match exists (cf. [16]), not to mention identifying the complete match set.

These highlight the need for approximating *diversified top-k graph pattern matching*: given Q, G and k, it is to find top-k matches of Q in $M(Q, G)$, such that the quality of the k-element match set has provable bounds. The benefits of identifying diversified top-k matches with quality bounds are twofold: (1) users only need to check k matches of Q rather than a large match set $M(Q, G)$; (2) if we have an algorithm for computing top-k matches with *the early termination property*, *i.e.*, it finds top-k matches of Q *without* computing the entire match set $M(Q, G)$, we do not have to pay the price of full-fledged graph pattern matching.

Example 1. A fraction of a social network is given as graph G in Fig. 1(a). Each node in G denotes a person, with attributes such as *job title*, *e.g.*, project manager (PM), database administrator (DBA), programmer (PRG), business analyst (BA), user interface developer (GUI) and software tester (ST). Each edge indicates friendship, *e.g.*, (Dave, Adam) indicates that Dave and Adam are friends.

To build up a team for software development, one issues a pattern graph Q depicted in Fig. 1(b) to find qualified candidates. The search intention asks team members to satisfy the following requirements: (1) with expertise: PM, PRG, DBA and ST; (2) meeting following friendship relations: (i) PM and PRG (resp. DBA) collaborated well before and are mutual friends; (ii) PRG and DBA have common friends, *i.e.*, they are connected within 2 hops; (iii) ST is a friend of PRG, but is possibly a direct (resp. indirect, with distance 2) friend of DBA.

It is often too restrictive to define matches as isomorphic subgraphs of Q. While if we extend Q with edge weights, more meaningful matches can be captured [5,7]. Indeed, this extension allows mapping from edges in Q to paths in G with distance constraint specified by edge weights, *e.g.*, edge (PRG, DBA) with weight 2 in Q can be mapped to a path between Adam and Mat in G. With the extension, match set $M(Q,G)$ includes following matches:

Match ID	Nodes in the match (PM, PRG, DBA, ST)	Match ID	Nodes in the match (PM, PRG, DBA, ST)
t_1	(Dave, Adam, Mat, Lucy)	t_2	(Phip, Dan, Mat, Rei)
t_3	(Phip, Dan, Carl, Rei)	t_4	(Bob, Bill, Carl, Emmy)
t_5	(Walt, Bill, Fred, Ruth)		

Observe that $M(Q,G)$ is (unnecessarily) too large to be inspected, while users may only be interested in top-k matches of Q that are as diverse as possible. It is hence unnecessary and too costly to compute the entire large set $M(Q,G)$. In light of this, an algorithm with the *early termination property* is desired, since it identifies top-k matches without inspecting the entire $M(Q,G)$.

To measure the quality of the top-k matches, one may consider the following criteria. (1) Tightness of connectivity [5]. Observe that matches t_2, t_4 and t_5 are connected more tightly than matches t_1 and t_3 due to shorter inner distances, and are considered more relevant to the query. (2) Social diversity [2,19]. Consider match set $\{t_2, t_3, t_4\}$, t_2 and t_3 share three common members, while t_2 and t_4 are quite "dissimilar" as they don't have members in common. Putting these together, when $k = 2$, $\{t_2, t_4\}$ makes a good candidate for top-k matches in terms of both relevance and diversity.   □

This example shows that *diversified Top-k graph pattern matching* may rectify the limitations of existing matching algorithms. To make practical use of it, however, several questions have to be answered. (1) What relevance and diversity functions should be used to rank the matches? (2) What is the complexity of computing top-k matches based on both of the functions? (3) How can we guarantee early termination by our algorithms for computing top-k matches?

Contributions. This paper answers these questions.

(1) We revise the traditional notion of subgraph isomorphism by supporting edge to path mapping, and define graph pattern matching with the revised notion, that's given Q and G, it is to compute the set $M(Q,G)$ of matches of Q in G, where each edge e_p of Q is mapped to a path in G with length bounded by the weight of e_p (Sect. 2).

(2) We introduce functions to rank matches of Q, namely, *relevance function* $w(\cdot)$ that measure the relevance of a match, and *distance function* $d(\cdot)$, which measure the "dissimilarity" of two matches. Based on both, we define a bi-criteria (balanced by a parameter λ) *diversification* function $F(\cdot)$, which aims to identify matches that are connected tightly and cover social elements as diverse as possible, simultaneously. We formalized *diversified top-k graph pattern matching problem* based on the diversification function $F(\cdot)$, and show that the decision version of the problem is NP-hard (Sect. 3).

(3) Despite hardness of the problem, we develop two approximation algorithms. One is in $O(|G|!|G| + \frac{k}{2}2^{|G|+1})$ time with *approximation ratio 2*, and the other one is in time $O((k \cdot |Q|)^{1/2}|G|!|G|)$, and in the meanwhile, not only preserves *the early termination property* but also has approximation ratio $\frac{|Q|k^2 - |Q|lk - |Q|k - lk^2}{k(2|Q|-l)(l+1)}$, where l indicates the level number when the algorithm terminates (Sect. 4).

(4) Using both real-life and synthetic data, we experimentally verify the performance of our algorithms (Sect. 5). We find that they effectively reduce excessive matches: when $k = 10$, our top-k matching methods only need to examine 12%–18% of matches in $M(Q, G)$ on average. Better still, our algorithms are efficient: with only 6.5 s to identify top-k matches on a graph with 4 million nodes and 53.5 million edges. In addition, they scale well with $|G|$ and $|f_e(e)|$, and are not sensitive to the change of k.

These results yield a promising approach to querying big social data.

Related Work. We categorize the related work as follows.

Top-k Graph Pattern Matching. Top-k graph pattern matching is to retrieve k best matches from the match set. There has been a host of work on this topic. For example, [21] propose to rank matches, *e.g.*, by the total node similarity scores [21], and identify k matches with highest ranking scores. [13] investigates top-k query evaluation for twig queries, which essentially computes isomorphism matching between rooted graphs. To provide more flexibility of top-k pattern matching, [5] extends matching semantics by allowing edge to path mapping, and proposes to rank matches based on their compactness. Instead of matching with subgraph isomorphism, graph simulation [15] is applied as matching semantic, and pattern graph is designated an output node in [8], then match result includes a set of nodes that are matches of the output node.

Diversified Graph Pattern Matching. Result diversification is a bi-criteria optimization problem for balancing result relevance and diversity [3,11], with applications in *e.g.,* social searching [2]. Following the idea, diversified graph pattern matching has been studied in, *e.g.,* [8,20]. [8] takes both diversity and relevance into consideration, and proposes functions to capture both relevance and diversity. In contrast, [20] considers diversity only, and measures diversity by the number of vertices covered by all the matches in the result. Our work differs from prior work in the following: (1) our matching semantic is quite different from that in [8,20], where [8] applies graph simulation as matching semantic, designates an output node in pattern graph, and treats matches of the output node as match result; and [20] adopts traditional subgraph isomorphism as matching semantic, a more strict semantic than ours. (2) [20] only considers match diversity, while ours considers both relevance and diversity.

2   Preliminary

In this section, we first define data graphs and pattern graphs. We then introduce graph pattern matching problem.

2.1   Data Graphs and Pattern Graphs

We start with notions of data graphs and pattern graphs.

Data Graphs. A *data graph* (or simply a graph) is an undirected graph $G = (V, E, L_v)$, where (1) V is a finite set of nodes; (2) $E \subseteq V \times V$, in which (v, v') denotes an edge between v and v'; and (3) L_v is a function such that for each v in V, $L_v(v)$ is a label from an alphabet Σ. Intuitively, the node labels denote *e.g.*, keywords, social roles [7].

We shall use the following notations. (1) A path ρ in a graph G is a sequence of nodes $(v_1, \cdots, v_n)$ such that (v_i, v_{i+1}) $(i \in [1, n-1])$ is an edge in G. (2) The length of the path ρ, denoted by $\mathsf{len}(\rho)$, is $n - 1$, *i.e.*, the total number of edges in ρ. (3) The distance $\mathsf{dist}(v, v')$ between v, v' is the length of the shortest path between them. (4) An edge weighted graph $G = (V, E, L_v, L_e)$ is a data graph, with V, E, L_v defined the same as its unweighted counterpart, but carrying weight $L_e(v, v')$ on each edge (v, v').

Pattern Graphs. A *pattern graph* is an undirected graph $Q = (V_p, E_p, f_v, f_e)$, where (1) V_p is the set of *pattern nodes*, (2) E_p is the set of *pattern edges*, (3) f_v is a function defined on V_p such that for each node $u \in V_p$, $f_v(u)$ is a label in Σ, and (4) f_e is a function defined on E_p such that for each edge (u, u') in E_p, $f_e(u, u')$ is either a positive integer k or a symbol $*$.

We denote $|V_p| + |E_p|$ as $|Q|$ (the size of Q), and $|V| + |E|$ as $|G|$ (the size of G).

2.2   Graph Pattern Matching Revised

We now introduce the graph pattern matching problem, denoted by GPM. Consider a data graph $G = (V, E, L_v)$ and a pattern graph $Q = (V_p, E_p, f_v, f_e)$.

Graph Pattern Matching. A *match* of Q in G is an n-ary node-tuple $t = \langle v_1, \cdots, v_n \rangle$, where $v_i \in V$ $(i \in [1, n])$, $n = |V_p|$, and there exists a *bijective function* h from V_p to the nodes in t such that (1) for each node $u \in V_p$, $f_v(u) = L_v(h(u))$; and (2) (u, u') is an edge in Q if and only if there exists a path between $h(u)$ and $h(u')$ with length no more than $f_e(u, u')$ in G.

The *answer* to Q in G, denoted by $M(Q, G)$, is the set of node-tuples t in G that matches Q. Abusing notations, we say a node v is a *match* of u, if v is in a match t and is mapped by $h(\cdot)$ from u. Intuitively, the node label $f_v(u)$ of u specifies search condition on nodes, and the edge label $f_e(u, u')$ imposes a bounded (resp. an unbounded) distance on the length of a path in G, that is mapped from edge (u, u') in Q, if $f_e(u, u')$ is not $*$ (resp. $f_e(u, u') = *$). Traditional pattern graphs are a special case of the patterns defined above with $f_e(u, u') = 1$ for all (u, u') in E_p.

Example 2. Recall G, Q and $M(Q, G)$ in Example. 1. One may verify that each pattern node carries search conditions, *i.e.*, a node v in G can match u only if $L_v(v) = f_v(u)$; and each pattern edge with weight specifies distance bound. Take t_1 as example, it is mapped via h from V_p, for each pattern edge (u, u'), there exists a path with length no more than $f_e(u, u')$ between $h(u)$ and $h(u')$, *e.g.*, pattern edge (PRG, DBA) is mapped to a path (Adam, Mary, Mat) with length 2, that is no more than f_e(PRG, DBA). $\qquad\qquad\qquad\qquad\qquad\qquad\qquad\qquad\square$

3   Diversified Top-k Graph Pattern Matching

In practice, the match set could be excessively large, while users are often interested in the best k matches that are not only relevant to the query, but also as diverse as possible. This suggests us to study the *diversified top-k graph pattern matching problem*.

In this section, we first propose two functions: *relevance function* for measuring the relevance of matches, and *distance function* to measure match diversity (Sect. 3.1). We next define a *diversification function*, a bi-criteria objective function combining both relevance and diversity, and introduce the *diversified top-k graph pattern matching problem* based on the function (Sect. 3.2).

3.1   Relevance and Distance Measurement

We start with a function to measure the *relevance* of matches of pattern graph.

Relevance Function. On a match t of Q in G, we define the relevance function $w(\cdot)$ as following:

$$w(t) = \frac{\sqrt{|E_p|}}{\sqrt{\Sigma_{(u,u') \in E_p, h(u), h(u') \in t}(\mathsf{dist}(h(u), h(u')))^2}}.$$

That is, the relevance function favors those matches that are connected tightly. The more tightly the nodes are connected in a match t, the more relevant t is to Q, as observed in study [5]. Thus, matches with high $w(\cdot)$ values are preferred for relevance.

We next introduce a metric for result diversity [17]. As observed in [2,19], it is important to diversify (social) search results so that groups identified can cover more elements (see Example 1).

Distance Function. To measure the "dissimilarity" of matches, we define a distance function as following. Given two matches t_1 with node set V_1 and t_2 with node set V_2, we define their *distance* $d(t_1, t_2)$ as following:

$$d(t_1, t_2) = 1 - \frac{|V_1 \cap V_2|}{|V_1 \cup V_2|}.$$

Intuitively, the distance between two matches indicates their social diversity, and the larger $d(t_1, t_2)$ is, the more dissimilar t_1 and t_2 are.

Observe that the function constitutes a *metric*. For any matches t_1, t_2 and t_3 of a pattern graph Q, (1) it is symmetric, *i.e.*, $d(t_1, t_2) = d(t_2, t_1)$, and (2) it satisfies the triangle inequality, *i.e.*, $d(t_1, t_2) \leq d(t_1, t_3) + d(t_2, t_3)$.

Example 3. Recall graph G, pattern Q in Fig. 1, the relevance values of the matches are $w(t_1) = \sqrt{\frac{5}{11}}$, $w(t_2) = \sqrt{\frac{5}{8}}$, $w(t_3) = \sqrt{\frac{5}{11}}$, $w(t_4) = \sqrt{\frac{5}{8}}$ and $w(t_5) = \sqrt{\frac{5}{8}}$, hence t_2, t_4 and t_5 are more relevant to Q than t_1 and t_3, since they are connected more tightly. On the other hand, the distance values between each pair of matches are $d(t_1, t_2) = \frac{6}{7}$, $d(t_1, t_3) = 1$, $d(t_1, t_4) = 1$, $d(t_1, t_5) = 1$, $d(t_2, t_3) = \frac{2}{5}$, $d(t_2, t_4) = 1$, $d(t_2, t_5) = 1$, $d(t_3, t_4) = \frac{6}{7}$, $d(t_3, t_5) = 1$ and $d(t_4, t_5) = \frac{6}{7}$, thus compared with t_2, t_3 is more dissimilar to t_1, as t_1 and t_3 correspond to two completely different groups of people.   □

3.2 Match Diversification

It is recognized that search results should be relevant, and at the same time, be as diverse as possible [11,19]. Based on the relevance and distance functions $w(\cdot)$ and $d(\cdot)$, we next introduce a diversification function.

Diversification Function. On a match set $\mathcal{S} = \{t_0, \ldots, t_k\}$ of a pattern graph Q, the diversification function $F(\cdot)$ is defined as

$$F(\mathcal{S}) = (1 - \lambda) \sum_{t_i \in \mathcal{S}} w(t_i) + \frac{2 \cdot \lambda}{k - 1} \sum_{t_i \in \mathcal{S}, t_j \in \mathcal{S}, i < j} d(t_i, t_j),$$

where $\lambda \in [0, 1]$ is a parameter set by users.

The diversity metric is scaled down with $\frac{2 \cdot \lambda}{k-1}$, since there are $\frac{k \cdot (k-1)}{2}$ numbers for the difference sum, while only k numbers for the relevance sum. The function $F(\cdot)$ is a minor revision of max-sum diversification introduced by [11]. It is a bi-criteria objective function to capture both relevance and diversity, and strikes a balance between the two with a parameter λ that is controlled by users, as a trade-off between the two [19].

Diversified Top-k Graph Pattern Matching Problem. With the diversification function $F(\cdot)$, we next state the problem of diversified top-k graph pattern matching, denoted by DivTopK. Given G, Q, a positive integer k, and a parameter $\lambda \in [0, 1]$, it is to find a set of k matches $\mathcal{S} \subseteq M(Q, G)$ such that

$$F(\mathcal{S}) = \arg\max_{\mathcal{S}' \subseteq M(Q,G)} F(\mathcal{S}'),$$

i.e., for all k-element sets $\mathcal{S}' \subseteq M(Q, G)$, $F(\mathcal{S}) \geq F(\mathcal{S}')$.

Example 4. Recall data graph G, pattern graph Q in Fig. 1. One can verify that when $0 < \lambda < 1$, $\{t_2, t_4\}$ or $\{t_2, t_5\}$ makes a top-2 diversified match set, since their diversification value is maximum among all 2-element subsets of $M(Q, G)$.   □

Theorem 1. *The* DivTopK *problem is* NP-*hard (decision problem).*

Proof Sketch: The decision problem of DivTopK is to decide, given Q, G, k, λ and bound B, whether a k-element subset $\mathcal{S} \subseteq M(Q, G)$ with $F(\mathcal{S}) \geq B$ exists. We show DivTopK problem is NP-hard, by reduction from the NP-complete *3-dimensional matching problem* (3DMP) [9]. An instance ϕ of 3DMP comprises three finite and disjoint sets W, X, Y, a subset $\tau \subseteq W \times X \times Y$, and an integer D. It is to determine if a 3-dimensional matching $M \subseteq \tau$ with $|M| \geq D$ exists.

Given any instance of 3DMP, we construct an instance of DivTopK as follows. (1) We construct a pattern graph Q, with three nodes u_w, u_x and u_y labeled by w, x and y, respectively, and three edges $\{(u_w, u_x), (u_w, u_y), (u_x, u_y)\}$. (2) We construct a data graph G as follows. (a) For each element o_w (resp. o_x and o_y) in set W (resp. X and Y), we construct a node v_w (resp. v_x and v_y), and insert into G. (b) For each matching $(o_{w_i}, o_{x_i}, o_{y_i})$ in τ, we insert edges (v_{w_i}, v_{x_i}), (v_{w_i}, v_{y_i}) and (v_{x_i}, v_{y_i}) in G. (3) We set $\lambda = 1$ in $F(\cdot)$, and $k = D$, $B = D$.

One may verify the following: the transformation above is in PTIME, and every matching in τ corresponds to a match of Q in G. One may further verify that the transformation is indeed a reduction, since there exists a matching of size at least D if and only if there exists a k-element matches with $F(\mathcal{S}) \geq B$. As 3DMP is NP-complete, the decision problem of DivTopK is hence NP-hard.   □

Despite hardness, we provide approximation algorithms for the problem (Sect. 4). Better still, one of the algorithms *preserves early termination property*.

4   Finding Top-k Diversified Matches

In spite of hardness, we develop approximation algorithms for the DivTopK problem, where one algorithm approximates diversification value with entire match set, while the other one applies *level-wise* strategy [20], which searches matches level by level, and terminates at certain level, *e.g.*, l, once k matches are identified. The main results of this section are as follows.

Theorem 2. *Given any instance of* DivTopK *problem, and assume that* $\mathcal{S}_{OPT}$ *is the optimal solution of the instance, the* DivTopK *problem can be solved by (1) an algorithm that finds a set* $\mathcal{S}$ *of* k *matches with* $F(\mathcal{S})$ *at least* $\frac{1}{2} \cdot F(\mathcal{S}_{OPT})$ *in* $O(|G||!|G| + \frac{k}{2} 2^{|G|+1})$ *time; and (2) another algorithm that preserves early termination property, and identifies match set* $\mathcal{S}$ *with* $F(\mathcal{S})$ *no less than* $\frac{|Q|k^2 - |Q|lk - |Q|k - lk^2}{k(2|Q| - l)(l+1)} \cdot F(\mathcal{S}_{OPT})$, *where* l *indicates the level number at which the algorithm terminates.*

We next provide two such algorithms and their detailed analysis as proofs of Theorem 2(1) and (2), respectively.

4.1   Approximating Diversification

Intuitively, the DivTopK problem can be divided into two subproblems: (1) *graph pattern matching*, which computes the match set $M(Q, G)$, and (2) *max-sum diversification* [3], that identifies a k-element subset $\mathcal{S}$ from $M(Q, G)$. Following the strategy, we provide an algorithm with approximation ratio 2 as a constructive proof of Theorem 2(1).

Algorithm. The algorithm, denoted as TopkApx (not shown), takes Q, G, k and λ as input, and works in three stages. (1) It computes an edge weighted graph $G' = (V', E', L'_v, L_e)$ from G, where V' is a subset of V, L'_v is defined the same as L_v in G, each edge $e_v = (v, v')$ in E' corresponds to an edge $e_u = (u, u')$ in Q, i.e., $L'_v(v) = f_v(u)$, $L'_v(v') = f_v(u')$, and L_e is a function defined on edges such that the edge weight $L_e(e_v)$ of e_v is no larger than $f_e(e_u)$. (2) It computes $M(Q, G)$ with the algorithm which revises VF2 [6] by combining constraints imposed via edge weights. (3) TopkApx applies the strategy given in [14] to select top-k matches. Specifically, TopkApx iteratively selects a pair of matches with maximum diversification value, and puts them in $\mathcal{S}$. After $\frac{k}{2}$ times selection, if k is odd, i.e., $|\mathcal{S}| = k-1$, TopkApx selects one more match t to maximize $F(\mathcal{S} \cup \{t\})$, and enlarges $\mathcal{S}$ with t. Finally, it returns $\mathcal{S}$ as diversified top-k matches.

Example 5. Given G and Q in Fig. 1, TopkApx first generates an edge weighted graph G', and identifies t_1-t_5 as matches of Q (see Example 1). Assuming $\lambda = 0.5$ and $k = 3$, TopkApx first selects $\{t_2, t_4\}$ to extend $\mathcal{S}$ since $F(t_2, t_4) = 1.79$ is maximum. It then enlarges $\mathcal{S}$ with t_5 as the diversification value can be maximized with t_5. TopkApx finally returns $\{t_2, t_4, t_5\}$ as the top-3 diversified matches.   □

Correctness and Complexity. TopkApx always terminates, and returns at most k matches when it terminates. To see the approximation ratio that TopkApx preserves, observe that an instance of DivTopK can be transformed to an instance of the *Maximum Dispersion problem* (MAXDisp) [14]. The MAXDisp problem is to find a subgraph G_k induced by a k-node subset V_k from a weighted complete graph G_c with the maximum sum of edge weights. Given a set of matches $M(Q, G)$, we construct a complete graph G_c as following: each node v_i in G_c represents a match t_i in $M(Q, G)$, and each edge (v_{i_1}, v_{i_2}) carries a weight $f_e(v_{i_1}, v_{i_2}) = \frac{1-\lambda}{k-1}(w(t_{i_1}) + w(t_{i_2})) + \frac{2\lambda}{k-1}d(t_{i_1}, t_{i_2})$. Then, the diversification value $F(\mathcal{S})$ of a k-element subset $\mathcal{S}$ equals to $F'(G_k) = \Sigma_{v_{i_1}, v_{i_2} \in G_k, i_1 < i_2} f_e(v_{i_1}, v_{i_2})$, i.e., total edge weights of G_k. One may easily verify that $\mathcal{S}$ is the top-k match set if and only if G_k maximizes $F'(\cdot)$. As TopkApx simulates the 2-approximation algorithm for MAXDisp to select matches, hence TopkApx approximates DivTopK with ratio 2.

To see the computational complexity, observe that, it takes TopkApx $O(|V|(|V| + |E|))$ time to construct graph G'. Since $|G'|$ is bounded by $|V| + |E|$, the match set computation is hence in $O(|V|!|V|)$ time. The cost for top-k match selection takes $O(\frac{k}{2}2^{|V|+1})$ time, as there may exist $2^{|V|}$ matches of Q in G. Thus, TopkApx is in $O(|G|!|G| + \frac{k}{2}2^{|G|+1})$ time in the worst case.

The analysis above completes the proof of Theorem 2(1).   □

4.2   Early Termination Algorithm

Though TopkApx preserves approximation ratio 2, it requires all the matches of Q to be computed, which may be inefficient on large graphs as there may exist exponentially many matches. To rectify this, we develop an algorithm that not only preserves performance guarantee, but also has *early termination property*.

In a nutshell, the algorithm applies the *level-wise* strategy [20] to discover best k matches. The discovery process starts from level 0, and may last for $|V_p|$ levels. At level i ($i \in [0, |V_p|]$), the algorithm identifies a set of matches, which are as relevant to Q as possible, and have i nodes in common with node set of existing matches. Once k matches are found, the algorithm terminates immediately without enumerating all the matches. We next show Theorem 2(2) by providing an algorithm and its detailed analysis as a constructive proof.

Input: Pattern Q, graph G, k, λ.
Output: A k-element set of matches of Q.

1. set $\mathcal{S}:=\emptyset$, $\mathcal{T}:=\emptyset$; integer $i := 0$;
2. construct an edge weighted graph G';
3. **while** $i \leq |V_p|$ **do**
4.   restore $\text{can}(u_s)$; pick u_s from V_p;
5.   **while** $\text{can}(u_s) \neq \emptyset$ **do**
6.   pick a node v_s from $\text{can}(u_s)$; $\text{can}(u_s) := \text{can}(u_s) \setminus \{v_s\}$;
7. $G_s = \text{Expand}(v_s, u_s, G', Q, \mathcal{T}, i)$; $\mathcal{S} := \mathcal{S} \cup \{G_s\}$; $\mathcal{T} := \mathcal{T} \cup V_s$;
8.   **if** $|\mathcal{S}| = k$ **then return** $\mathcal{S}$;
9. $i := i + 1$;
10. **return** $\mathcal{S}$;

Procedure Expand
Input: v_s, u_s, G', Q, $\mathcal{T}$, i.
Output: A match G_s of Q.

1. initialize an index $\mathcal{I} := \emptyset$, a stack $\mathsf{q} := \{u_s\}$, an empty graph G_s;
2. initialize pattern Q_o with node u_s, $\mathcal{I}(Q_o)$ with a match taking node v_s;
3. **while** $\mathsf{q} \neq \emptyset$ **do**
4. $u := \mathsf{q}.\text{pop}()$;
5.   **for each** unvisited edge (u, u') in Q **do**
6.   generate Q'_o by expanding Q_o with (u, u');
7.   **for each** match G_o in $\mathcal{I}(Q_o)$ **do**
8.   generate G'_o by extending G_o with (v, v');
9.   **if** G'_o is a match of Q'_o **then**
10. $\mathcal{I}(Q'_o) := \mathcal{I}(Q'_o) \cup \{G'_o\}$;
11.   **if** Q'_o is equivalent to Q **then**
12.   select G_s with $|V_s \cap \mathcal{T}| = i$ and largest $w(\cdot)$ from $\mathcal{I}(Q'_o)$; **break** ;
13. $\mathsf{q}.\text{add}(u')$ if u' is not visited before;
14. **return** G_s;

Fig. 2. Algorithm TopkET

Algorithm. The algorithm, denoted as TopkET and shown in Fig. 2, takes Q, G, k and λ as input. It first initializes two empty sets S and T to keep track of match set of Q and node set of all the matches in S, and sets level number $i=0$ (line 1). TopkET then computes an edge weighted graph $G' = (V', E', L'_v, L_e)$ in the same way as TopkApx does; as a byproduct, for each pattern node u, a set $can(u)$ that includes nodes in G' with the same node label as u is initialized (line 2). After G' is constructed, TopkET iteratively identifies matches of Q by applying the level-wise strategy, starting from level 0 (lines 3–9). Specifically, TopkET first restores $can(u_s)$ to its original value, and selects a pattern node u_s from V_p (line 4). It next repeatedly conducts the following to find matches: (a) pick a match candidate v_s of u_s for expansion. Once v_s is selected, it is removed from $can(u_s)$ (line 6); (b) employ procedure Expand to generate a new match G_s of Q by growing from v_s, and extends sets S and T with G_s and V_s, i.e., the node set of G_s, respectively (line 7). After extension, if $|S|$ already reaches k, TopkET terminates immediately, and returns S as final result (line 8); otherwise, it starts next round expansion from another match candidate. When all the candidates of u_s are used, i.e., $can(u_s) = \emptyset$, TopkET increases level number i by 1, and starts a new round evaluation (line 9). Finally, TopkET returns the set S as top-k matches of Q (line 10).

Procedure Expand. The procedure takes v_s, u_s, G', Q, T and i as input, and works as following. It first initializes an empty index $\mathcal{I}$ to keep track of the mapping between a sub-pattern Q_o of Q and its matches, and a stack q with u_s (line 1). It next initializes a pattern graph Q_o with node u_s, as a sub-pattern of Q, and initializes $\mathcal{I}(Q_o)$ with the match, that includes v_s (line 2). Procedure Expand then expands Q_o as well as its matches following the topological structure of Q (lines 3–13). Specifically, it first pops up the uppermost node u from the stack q (line 4). For each unvisited pattern edge (u, u') of Q, (a) Expand uses it to expand existing sub-pattern Q_o and generates a new sub-pattern Q'_o (line 6), (b) for each match G_o of sub-pattern Q_o, Expand identifies edges (v, v') in G' and uses them to extend G_o, where v is in G_o as a match of u in Q_o, and v' is the neighbor of v in G' (line 8). If the newly generated subgraph G'_o is a valid match of Q'_o, Expand enlarges index $\mathcal{I}$ by including G'_o in $\mathcal{I}(Q'_o)$ (lines 9–10). After all the matches of Q_o are processed, if Q'_o is already equivalent to Q_o, Expand selects one match G_s from $\mathcal{I}(Q'_o)$ such that the node set V_s of G_s has i nodes in common with T and the relevance value of G_s is maximum among all the matches, and breaks the **while** loop (lines 11–12). If the neighbor node u' of u was not visited before, Expand pushes it in stack q for further extension (line 13). When **while** loop terminates, Expand returns G_s as a match of Q (line 14).

Example 6. Recall Example 1. Given $\lambda = 0.5$ and $k = 3$, TopkET identifies diversified top-k matches starting following the level-wise strategy. Starting from level 0, it first selects a pattern node u_s with minimum candidate set, e.g.,DBA, initializes a pattern graph Q_o with u_s, and applies procedure Expand to generate matches. If a candidate Mat is selected in the first round iteration, two matches t_1 and t_2 are identified after expansion. Since t_2 has higher relevance value, it is included in set S. In the second round iteration, if Carl is chosen, two matches

t_3 and t_4 are found, and t_4 is added in $\mathcal{S}$ as its relevance value is higher. In the third round iteration, only Fred can be used for expansion, and the match t_5 is generated. One may verify that t_5 can not be chosen at level 0, since it shares node Bill with node set T. After all the candidates are processed, TopkET proceeds match identification at level 1, following the same way as before, and may select match t_1, which has one common node with $\mathcal{T}$, to enlarge $\mathcal{S}$. Thus, TopkET returns $\{t_1, t_2, t_4\}$ as diversified top-3 matches.   □

Analyses. To complete the proof of Theorem 2(2), it remains to verify the correctness and complexity of TopkET.

Correctness. To see the correctness of the algorithm, it suffices to show that when TopkET terminates, a set $\mathcal{S}$ including at most k matches of Q are returned, and $F(\mathcal{S})$ is no less than $\frac{|Q|k^2 - |Q|lk - |Q|k - lk^2}{k(2|Q|-l)(l+1)} F(\mathcal{S}_{OPT})$, where l is the level number.

(1) It can be easily verified that when TopkET terminates, either (a) a set $\mathcal{S}$ of k matches have been identified (termination condition in line 8), or (b) all the matches of Q are returned, since otherwise TopkET will not terminate.

(2) To see the approximation ratio TopkET preserves, observe the following. (a) $\frac{F(\mathcal{S})}{F(\mathcal{S}_{OPT})} \geq \frac{\text{LB}}{\text{UB}}$, where LB and UB refer to lower bound of $F(\mathcal{S})$ and upper bound of $F(\mathcal{S}_{OPT})$, respectively. (b) When $w(t_i) = 1$ ($i \in [1, k]$), and $d(t_i, t_j) = 1$ for $1 \leq i < j \leq k$, the diversification value of the optimal solution reaches its upper bound k, i.e., UB $= k$. (c) We show lower bound of $F(\mathcal{S})$, i.e., LB below.

We denote $\mathcal{S}_l$ as the increment of match set $\mathcal{S}$ at the l-th level, and $F_d(\mathcal{S}_l)$ as the increment of distance part of diversification value introduced by including $\mathcal{S}_l$ in $\mathcal{S}$. Assume that the algorithm terminates at level l, then by induction, when the algorithm terminates, $F_d(\mathcal{S}_l)$ is lower bounded by $(\Sigma_{j \in [0, l-1]} |\mathcal{S}_j| - \frac{l}{2|Q|-l}) |\mathcal{S}_l| + \frac{|\mathcal{S}_l|(|\mathcal{S}_l|-1)}{2}(1 - \frac{l}{2|Q|-l})$. Hence, $F(\mathcal{S}) = (1 - \lambda)\Sigma_{t_i \in \mathcal{S}} w(t_i) + \frac{2\lambda}{k-1}(F_d(\mathcal{S}_0) + F_d(\mathcal{S}_1) + F_d(\mathcal{S}_2) + \cdots + F_d(\mathcal{S}_l))$, which is lower bounded by $(1 - \lambda) * \Sigma_{t_i \in \mathcal{S}} w(t_i) + \frac{2\lambda}{k-1}(\Sigma_{i \in [0, l]}(|\mathcal{S}_i| * \Sigma_{j \in [i+1, l]} |\mathcal{S}_j|) - \Sigma_{i \in [1, l]} |\mathcal{S}_i| * \frac{i}{2|Q|-i} + \Sigma_{i \in [0, l]} \frac{|\mathcal{S}_i|(|\mathcal{S}_i|-1)}{2}(1 - \frac{i}{2|Q|-i}))$. Furthermore, it can be verified that $(1 - \lambda) * \Sigma_{t_i \in \mathcal{S}} w(t_i) \geq 0$, $\frac{2\lambda}{k-1}(\Sigma_{i \in [0, l]}(|\mathcal{S}_i| * \Sigma_{j \in [i+1, l]} |\mathcal{S}_j|) \geq \frac{k^2}{4} - (|\mathcal{S}_l| - \frac{k}{2})^2$, $\Sigma_{i \in [1, l]} |\mathcal{S}_i| * \frac{i}{2|Q|-i} \geq -\frac{l(k-|\mathcal{S}_0|)}{2|Q|-l}$, and $\Sigma_{i \in [0, l]} \frac{|\mathcal{S}_i|(|\mathcal{S}_i|-1)}{2}(1 - \frac{i}{2|Q|-i})) \geq \frac{1}{2}(1 - \frac{l}{2|Q|-l})(\frac{k^2}{l+1} - k)$. Putting these together, $F(\mathcal{S})$ is lower bounded by $\frac{|Q|k^2 - |Q|lk - |Q|k - lk^2}{(2|Q|-l)(l+1)}$, hence the approximation ratio $\frac{F(\mathcal{S})}{F(\mathcal{S}_{OPT})}$ is lower bounded by $\frac{|Q|k^2 - |Q|lk - |Q|k - lk^2}{k(2|Q|-l)(l+1)}$, where l is the level number.

Complexity. Observe that it takes TopkET $O(|V|(|V| + |E|))$ time to construct an edge weighted graph G' (line 2). The outmost **while** loop runs at most $\min(|V_p|, k) \leq (k \cdot |V_p|)^{1/2}$ times, for a single iteration, it is in $O(|G'|!|G'|)$ time to identify one match of Q in G', which is further bounded by $O(|G|!|G|)$ time. Thus, the worst time complexity of TopkET is $O((k \cdot |V_p|)^{1/2}|G|!|G|)$ time.

Early termination. Algorithm TopkET preserves the *early termination property*. In contrast to algorithms, *e.g.*, TopkApx, that identifies top-k matches from match set $M(Q, G)$, TopkET applies level-wise strategy to discover diversified top-k matches, and terminates as soon as top-k matches are identified. As will be verified in Sect. 5, TopkET is more efficient than TopkApx.

These completes the proof of Theorem 2(2).   □

5   Experimental Evaluation

We next experimentally verify the effectiveness and efficiency of our diversified top-k graph pattern matching algorithms, using real-life and synthetic data.

Experimental Setting. We used the following datasets.

(1) *Real − life graphs.* We used two real-life graphs: (a) *Pokec* [1], a social network with 1.63 million nodes of 269 types and 30.6 million edges. (b) *Google+* [12], a social network with 4 million entities of 5 types and 53.5 million edges.

(2) *Synthetic data.* We designed a generator to produce synthetic graphs $G = (V, E, L)$, controlled by the number of nodes $|V|$ and edges $|E|$, where L are assigned from a set of 15 labels. We generated synthetic graphs following the linkage generation models [10]: an edge was attached to the high degree nodes with higher probability. We use $(|V|, |E|)$ to denote the size of G.

(3) *Pattern generator.* We implemented a generator for pattern graphs $Q = (V_p, E_p, f_v, f_e)$, controlled by four parameters: $|V_p|$, $|E_p|$, label f_v from the same Σ, and an upper bound k for $f_e(e)$.

(4) *Implementation.* We implemented our algorithms TopkET, TopkET$_{\text{OPT}}$, which optimizes TopkET via selecting a pattern node u_s with minimum $|\text{can}(u_s)|$ for propagation, and TopkApx vs. DSQL [20], all in Java.

All the experiments were repeated 5 times on a Intel Core(TM)2 Duo 3.00 GHz CPU with 4 GB of memory and the average is reported here.

Experimental Results. We next present our findings.

Exp-1: Effectiveness. We first evaluated the effectiveness of our diversified top-k matching algorithms, *i.e.*, TopkET, TopkET$_{\text{OPT}}$ vs. TopkApx. We measured effectiveness by computing the ratio $\text{IR} = \frac{|M^i(Q,G)|}{|M(Q,G)|}$, where $|M^i(Q, G)|$ indicates the amount of matches identified by the algorithms when they terminate.

Varying $|Q|$. Fixing k=10, $\lambda = 0.5$, $f_e(e) = 1$ for all $e \in E_p$, we varied $|Q|$ from $(4, 6)$ to $(8, 16)$, and evaluated IR over two real-life graphs. The results shown in Figs. 3(a) and (b) tell us the following. (1) TopkET, TopkET$_{\text{OPT}}$ effectively reduce excessive matches. Taking patterns with size $|Q| = (4, 6)$ as example,

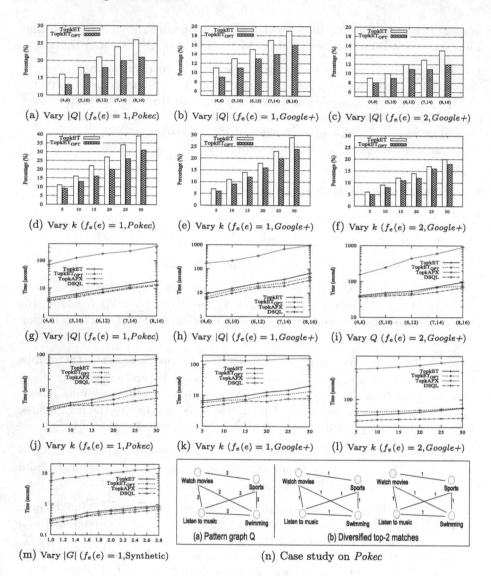

Fig. 3. Performance evaluation

TopkET (resp. TopkET$_{OPT}$) only inspected 21% and 17.6% (resp. 15% and 12.6%) matches, on average, on *Pokec* and *Google+*, respectively, when it terminates. (2) TopkApx identifies all the matches of Q, hence has IR = 100% (results not shown in figures). (3) TopkET$_{OPT}$ has lower IR than TopkET since its optimization strategy effectively reduces its search space.

Varying k. Fixing $|Q| = (4, 6)$, $f_e(e) = 1$ and $\lambda = 0.5$, we varied k from 5 to 30 in 5 increments, and reported IR on *Pokec* and *Google+*. As shown in

Figs. 3(d) and (e), the ratio IR of TopkET (resp. TopkET$_{OPT}$) increased from 11% (resp. 9%) to 39% (resp. 31%), and 7% (resp 6%) to 29% (resp. 24%) on *Pokec* and *Google+*, respectively, when k was increased from 5 to 30. The reason is that, for larger k, more matches have to be identified and examined by the algorithms.

Varying $f_e(e)$. We evaluated IR in the same setting as in Figs. 3(a) and (b), but with $f_e(e) = 2$ for pattern edges. Figure 3(c) shows the results on *Google+*, and tells us that (1) TopkET, TopkET$_{OPT}$ only inspect a small portion of matches, *e.g.*, only 9% and 8% of matches for patterns with $f_e(e) = 2$, and identify top-k matches. The results are consistent with Fig. 3(b); (2) our early termination algorithms are more effective for larger $f_e(e)$, *e.g.*, the IR of TopkET (resp. TopkET$_{OPT}$) on average is 11.8% (resp. 10.2%), which is smaller than IR when $f_e(e) = 1$. The results on *Pokec* are consistent with Fig. 3(c), hence are not reported. We still evaluated IR in the same setting as in Figs. 3(d) and (e), but using patterns with $f_e(e) = 2$. The results on *Google+*, shown in Fig. 3(f), tell us the following: our algorithms are more effective for larger $f_e(e)$, *e.g.*, the IR of TopkET$_{OPT}$ is on average 11.7%, less than 14.5% when $f_e(e) = 1$. As the results on *Pokec* are consistent with Fig. 3(f), hence are not reported.

As can be observed, our early termination algorithms work even better for larger $f_e(e)$. This is because larger $f_e(e)$ will result in larger match set, while our algorithms can identify top-k matches with less matches. Due to space constraint, for even larger edge weights, *e.g.*, $f_e(e) > 2$, we do not report results here, while we confirmed that the larger $f_e(e)$ is, the smaller IR is, which is consistent with the analytics given above.

Exp-2: Efficiency. We next evaluated the efficiency of the algorithms, in the same setting as in Exp-1. For comparison purpose, we set $\lambda = 1$, and only search diversified top-k matches without considering match relevance to favor DSQL.

Varying $|Q|$. Fixing $k = 10$ and $f_e(e) = 1$ for all pattern edges, we varied $|Q|$ from $(4, 6)$ to $(8, 16)$, and evaluated efficiency of the algorithms over real-life graphs. As shown in Figs. 3(g) and (h), (1) TopkET (resp. TopkET$_{OPT}$) takes only 5.5% (resp. 4.6%) and 5.4% (resp. 4.3%) time of TopkApx, on average, on *Pokec* and *Google+*, respectively. This verifies the effectiveness of the early termination property that TopkET and TopkET$_{OPT}$ preserves; (2) TopkET (resp. TopkET$_{OPT}$) spends extra 32.5% (resp. 9.6%) and 72.2% (resp. 26.9%) time than DSQL on *Pokec* and *Google+*, respectively, since they used a more costly strategy to identify matches to favor result relevance, even though relevance is ignored in the test; (3) TopkET$_{OPT}$ outperforms TopkET by 17.1% and 26% on *Pokec* and *Google+*, respectively, owing to its optimization strategy.

Varying k. Fixing $|Q| = (4, 6)$, $f_e(e) = 1$, we varied k from 5 to 30 in 5 increments, and tested efficiency of the algorithms on real-life graphs. The results are shown in Figs. 3(j) and (k). We find that (1) all the algorithms spend more time for larger k, which is consistent with the observations in Figs. 3(g) and (h); (2) TopkET, TopkET$_{OPT}$ and DSQL are more efficient than TopkApx, taking only 10.9%,8.2%,6.2% and 8.2%, 6.2%, 4.4% time of TopkApx on *Pokec* and *Google+*, respectively; (3) TopkET, TopkET$_{OPT}$ and DSQL are more sensitive to the increase of k than TopkApx, as a large part of time used by TopkApx is match set computation, which is not sensitive to the increase of k.

Varying $f_e(e)$. We evaluated efficiency of the algorithms, in the same setting as in Figs. 3(g) and (h) but with $f_e(e) = 2$. As DSQL only works for patterns with $f_e(e) = 1$, to rectify this, we slightly revise DSQL by integrating a preprocessing task, *i.e.*, extracting edge weighted graph G' from G, and take G' as input of DSQL. Results on *Google+* are shown in Fig. 3(i), and tell us following: all the algorithms spend more time on larger $f_e(e)$, *e.g.*, it takes TopkET$_{OPT}$ 27.5s (resp. 6.5s) to identify diversified top-k matches of Q with size $(4, 6, 2)$ (resp. $(4, 6, 1)$). The results on *Pokec* are consistent with Fig. 3(i) and are hence not reported. We also tested efficiency in the same setting as in Figs. 3(j) and (k), but with $f_e(e) = 2$. Figure 3(l) shows results on *Google+*, and tells us that it takes all the algorithms more time to identify diversified top-k matches, compared with patterns with $f_e(e) = 1$. We do not report results on *Pokec* as they are consistent with Fig. 3(l). The main reason that our algorithms spend more time for larger $f_e(e)$ is that all the algorithms need extra time to generate edge weighted graph G'. For larger $f_e(e)$, we do not report efficiency results due to space constraint, while we confirmed that the larger $f_e(e)$ is, the more costly our algorithms are, which is consistent with the analytics given above.

Exp-3 Scalability. Fixing $|Q| = (4, 6)$, $f_e(e) = 1$, we varied $|G|$ from $(1M, 2M)$ to $(2.8M, 5.6M)$ and evaluated the scalability of the algorithms with synthetic data. The results, shown in Fig. 3(m), tell us: (1) all the algorithms scale well with $|G|$; (2) the running time of TopkET, TopkET$_{OPT}$ and DSQL are less sensitive to $|G|$ than TopkApx.

Exp-4 Case Study. On *Pokec*, we manually inspected the diversified top-2 matches found by TopkET$_{OPT}$ for Q of Fig. 3(n) and (a). As shown in Fig. 3(n) and (b), the two matches found by TopkET$_{OPT}$ are not only relevant to Q, but also very dissimilar, since they are closely connected, *i.e.*, with $L_e(e) = 1$, and share no common element.

Summary. (1) Our algorithms are effective: top-10 matches can be found by TopkET when only 21% (resp. 17.6%) matches on *Pokec* (resp. *Google+*) are identified. (2) Our algorithms are efficient: it only takes TopkET 8.8s to find top-10 matches of Q with $|Q| = (4, 6)$, $f_e(e) = 1$. (3) Our optimization strategy is effective: TopkET$_{OPT}$ improves both IR and efficiency of TopkET. (4) Our algorithms scale well with $|G|$ and $f_e(e)$.

6   Conclusion

We have introduced and studied approximating diversified top-k graph pattern matching problem. We have revised pattern graphs by allowing edge to path mapping, defined functions to measure match relevance and diversity, and proposed diversification function as a bi-criteria objective function to capture both relevance and diversity. We have established the complexity of the problem, and provided approximation algorithms for computing diversified top-k matches with *early termination* property. As verified analytically and experimentally, our methods remedy the limitations of prior algorithms, by eliminating excessive matches and improving efficiency on big social graphs.

Acknowledgments. This work is supported by NSFC 71490722, Fundamental Research Funds for the Central Universities, and National Key R&D Program No. 2017YFA0700800, China.

References

1. Pokec social network. http://snap.stanford.edu/data/soc-pokec.html
2. Alonso, O., Gamon, M., Haas, K., Pantel, P.: Diversity and relevance in social search. In: DDR (2012)
3. Borodin, A., Lee, H.C., Ye, Y.: Max-sum diversification, monotone submodular functions and dynamic updates. In: PODS, pp. 155–166. ACM (2012)
4. Brynielsson, J., Högberg, J., Kaati, L., Martenson, C., Svenson, P.: Detecting social positions using simulation. In: ASONAM (2010)
5. Cheng, J., Zeng, X., Yu, J.X.: Top-k graph pattern matching over large graphs. In: ICDE, pp. 1033–1044 (2013)
6. Cordella, L.P., Foggia, P., Sansone, C., Vento, M.: A (sub)graph isomorphism algorithm for matching large graphs. TPAMI **26**(10), 1367–1372 (2004)
7. Fan, W., Li, J., Ma, S., Tang, N., Wu, Y., Wu, Y.: Graph pattern matching: from intractable to polynomial time. PVLDB **3**(1), 264–275 (2010)
8. Fan, W., Wang, X., Wu, Y.: Diversified top-k graph pattern matching. PVLDB **6**(13), 1510–1521 (2013)
9. Garey, M., Johnson, D.: Computers and Intractability: A Guide to the Theory of NP-Completeness. W. H. Freeman and Company, New York (1979)
10. Garg, S., Gupta, T. Carlsson, N., Mahanti, A.: Evolution of an online social aggregation network: an empirical study. In: IMC 2009
11. Gollapudi, S. Sharma, A.: An axiomatic approach for result diversification. In: WWW (2009)
12. Gong, N.Z., et al.: Evolution of social-attribute networks: measurements, modeling, and implications using Google+. In: IMC (2012)
13. Gou, G., Chirkova, R.: Efficient algorithms for exact ranked twig-pattern matching over graphs. In: SIGMOD (2008)
14. Hassin, R., Rubinstein, S., Tamir, A.: Approximation algorithms for maximum dispersion. Oper. Res. Lett. **21**(3), 133–137 (1997)
15. Henzinger, M.R., Henzinger, T.A., Kopke, P.W.: Computing simulations on finite and infinite graphs. In: FOCS (1995)
16. Papadimitriou, C.H.: Computational Complexity. Addison-Wesley, Boston (1994)
17. Qin, L., Yu, J.X., Chang, L.: Diversifying top-k results. PVLDB **5**(11), 1124–1135 (2012)
18. Terveen, L.G., McDonald, D.W.: Social matching: a framework and research agenda. ACM Trans. Comput. Hum. Interact. **12**(3), 401–434 (2005)
19. Vieira, M.R., et al.: On query result diversification. In: ICDE (2011)
20. Yang, Z., Fu, A.W., Liu, R.: Diversified top-k subgraph querying in a large graph. In: SIGMOD, pp. 1167–1182 (2016)
21. Zou, L., Chen, L., Lu, Y.: Top-k subgraph matching query in a large graph. In: Ph.D. workshop in CIKM (2007)

Boosting PageRank Scores by Optimizing Internal Link Structure

Naoto Ohsaka[1]([✉]), Tomohiro Sonobe[2,6], Naonori Kakimura[3],
Takuro Fukunaga[4], Sumio Fujita[5], and Ken-ichi Kawarabayashi[2,6]

[1] NEC Corporation, Tokyo, Japan
n-ohsaka@ak.jp.nec.com
[2] National Institute of Informatics, Tokyo, Japan
{tomohiro_sonobe,k_keniti}@nii.ac.jp
[3] Keio University, Tokyo, Japan
kakimura@math.keio.ac.jp
[4] RIKEN Center for Advanced Intelligence Project, Tokyo, Japan
takuro.fukunaga@riken.jp
[5] Yahoo Japan Corporation, Tokyo, Japan
sufujita@yahoo-corp.jp
[6] JST, ERATO, Kawarabayashi Large Graph Project, Tokyo, Japan

Abstract. We consider and formulate problems of PageRank score boosting motivated by applications such as effective web advertising. More precisely, given a graph and target vertices, one is required to find a fixed-size set of missing edges that maximizes the minimum PageRank score among the targets. We provide theoretical analyses to show that all of them are NP-hard. To overcome the hardness, we develop heuristic-based algorithms for them. We finally perform experiments on several real-world networks to verify the effectiveness of the proposed algorithms compared to baselines. Specifically, our algorithm achieves 100 times improvements of the minimum PageRank score among selected 100 vertices by adding only dozens of edges.

1 Introduction

Google's "PageRank" [6,28] determines the importance and measures the popularity of webpages based on the linkage structure of the web. The intuition behind PageRank is that *a webpage is important if important webpages point to it,* and the *PageRank score* of a vertex is defined as the solution of a system of linear equations. PageRank can be also interpreted as follows. Consider a random surfer who usually randomly follows its out-edge, but with a certain probability (e.g., 0.15), jumps to a uniformly chosen vertex. Then, the expected frequency of visiting a vertex is equal to its PageRank score. Thanks to its simplicity and generality, PageRank has been applied in a wide range of areas including chemistry, biology, recommender systems, and social network analysis [14].

© Springer Nature Switzerland AG 2018
S. Hartmann et al. (Eds.): DEXA 2018, LNCS 11029, pp. 424–439, 2018.
https://doi.org/10.1007/978-3-319-98809-2_26

In this paper, we study the following problem of "boosting" PageRank scores.

Given a graph and target vertices, extract a small number of missing edges that maximizes the minimum PageRank score among the targets.

Concretely, we consider certain scenarios, where we are required to

1. find a set of k missing edges such that in the resulting graph, the minimum PageRank score among the target vertices is maximized,
2. find the smallest number of missing edges such that in the resulting graph, the PageRank score of any target vertex is at least a threshold l, or
3. find a set of k missing edges such that in the resulting graph, the number of target vertices with a PageRank score of at least a threshold l is maximized.

We face such problems in various applications including the example below.

Optimizing Linkage Structure for Effective Web Advertising. Computational advertising has become increasingly important, e.g., approximately $137.53 billion dollars was spent on online advertising in 2014[1]. In computational advertising, there are three key parties [7]: *advertiser*, who "creates" advertising; *customer*, who "views" advertising; and *publisher*, who "sells" web banner spaces to advertisers. One of the main challenges is to design markets that simultaneously maximize values for advertisers, customers, and publishers. From the viewpoint of a publisher, it is desirable to sell the publisher's web banner spaces for the highest possible price to advertisers. To obtain such a high price, publishers need to provide some guarantee to advertisers; that is, users *frequently* visit (all of) the *target* webpages wherein web banner spaces are offered. However, the distribution of page visits [1] (and so PageRank scores [5, 13, 29]) is generally heavily skewed. Hence, it is desirable to have more webpages of high PageRank scores. In this manner, advertisers will have more chances to turn visitors of web banner spaces into their customers.

For this purpose, we can imagine the situation where a publisher owns the host network (i.e., webpages with the same hostname) and is given a set of target webpages with web banner spaces. Then, the goal is to guide visitors to any target webpages effectively by making "marginal" changes to its internal linkage structure. Here, it is plausible to allow the publisher to choose only a few hyperlinks because an excessively large number of hyperlink insertions may affect the browsing behavior of users and each insertion may incur some cost. Also, a set of candidate hyperlinks to be added to the host network is given since it might be impossible to create hyperlinks from certain webpages (e.g., the homepage). Hence, the goal is achieved by solving an instance of PageRank boosting problem, and the solution would help improve the effectiveness of advertising.

---

[1] https://www.emarketer.com/m/Article/Digital-Ad-Spending-Worldwide-Hit-36137 53-Billion-2014/1010736.

Contributions. In this paper, we present the study on PageRank boosting problem. Our main contributions are summarized as follows.

- **Problem formulations (Sect. 3):** We formulate discrete optimization problems of *PageRank score boosting*. Specifically, we introduce a problem of seeking for a set of k missing edges that maximizes the minimum PageRank score among a specified set of target vertices and further its two variants.
- **Complexity analysis (Sect. 4):** We prove that all of the three problems are NP-hard to solve exactly and indeed some problem is hard even to approximate under some complexity assumption.
- **Algorithms (Sect. 5):** To overcome the hardness result, we develop efficient heuristic algorithms for the problems. We propose to select edges missing in the current graph having the maximum contribution on PageRank boosting and add them to a solution. To this end, we propose a contribution-based approach for missing edge selection.
- **Experiments (Sect. 6):** We perform experiments on real-world webgraphs with up to millions of edges to verify the effectiveness of the proposed algorithms compared to baseline algorithms. Specifically, the proposed algorithm demonstrates 100 times improvements of the minimum PageRank score among selected 100 webpages by adding only dozens of hyperlinks.

2   Related Work

PageRank score boosting has the potential to various applications including link spam, search engine optimization, and measuring the PageRank error in incomplete networks.

Building Outlinks. Assume that a webmaster owns some webpages and wants to increase their PageRank scores by modifying his/her own webpages, i.e., by building their *out-edges*. Sydow [30] showed through computational experiments that carefully chosen out-edges to a single vertex increases its PageRank score. Avrachenkov et al. [3] theoretically analyzed a change of the PageRank score of a single vertex caused by adding its new out-edges and reported an optimal linking strategy. Subsequently, de Kerchove, Ninove, and Van Dooren [19] considered a generalization of [3]'s problem whose aim is of maximizing the *sum* of the PageRank scores of *multiple* vertices by changing their out-edges and provided an optimal linking strategy for this problem.

Link Building. In the literature of search engine optimization, *link building* aims to construct edges entering to target webpages (a.k.a. *backlinks*) with the purpose of increasing website ranking in search engine results. From theoretical aspects, a few hardness and approximation results were established such as [26, 27].

Inserting Edges Under Control. In this situation, we can insert any of hyperlinks under control to maximize the PageRank scores. Besides our motivation, this situation has been applied in the measurement of an error in PageRank scores calculated from incomplete graphs [17]. Because of technical reasons, e.g., server down or a crawling strategy, we often overlook some edges, which results in incorrect edges. The possible effect of such "fragile" edges on PageRank scores can be estimated as a solution for this problem.

From an algorithmic point of view, there have been several results on approximation and hardness. Csáji *et al.* [9,10] gave a polynomial-time algorithm for maximizing the PageRank score of a single vertex. Olsen [25] considered a problem of finding k missing (allowing multiple) edges that maximize the minimum PageRank score among a given set of vertices and proved its NP-hardness. Though our problem is closely related to Olsen's problem formulation, there exist some differences between them. Firstly, multiple edges are not allowed in our problem, and thus we cannot employ a trivial strategy which inserts multiple edges connecting from a webpage with very high PageRank score to a target webpage. Secondly, our problem is more general in a sense that we are given a candidate set for missing edges, while Olsen's problem assumes that every possible missing edge can be a candidate, i.e., a candidate set is *fixed* to be the set consisting of all possible missing edges. Note also that we prove that our problem is NP-hard even if the maximum degree of an input graph is bounded by a small constant.

Link Spam. *Link spams* aim to unethically increase the rankings of target webpages in search engine results by exploiting link-based ranking algorithms such as PageRank. Gyöngyi and Garcia-Molina [15] introduced link farms, where webpages exchange edges for mutual benefit. The best strategy to boost the PageRank score of a webpage is to have all webpages in the link farm link to that target. Baeza-Yates, Castillo, and López [4] empirically studied different link farm structures (star and ring topologies). Remark that the link farm involves webpage additions.

3   Problem Formulations

Notations. Let $G = (V, E)$ be a directed graph where V is a set of n vertices and E is a set of m edges. We assume that G is simple, i.e., it has no self-loops or multiple edges. For a graph G and a set F of edges not in E (i.e., $F \subseteq (V \times V) \setminus E$), the symbol $G \cup F$ stands for the graph obtained from G by inserting edges of F. If an edge set consists of a single edge, say, $F = \{(s, t)\}$, we simply write $G + (s, t)$ instead of $G \cup \{(s, t)\}$. We denote the in- and out-degrees of v by $d_G^-(v)$ and $d_G^+(v)$, respectively.

The *transition matrix* $\mathbf{P} = (P_{ij})$ of a graph G is defined as

$$P_{ij} = \begin{cases} 1/d^+(j) & (j, i) \in E, \\ 0 & \text{otherwise.} \end{cases} \tag{1}$$

The *PageRank* score $\mathbf{x}_G$ of G is a solution of the following linear equation [6, 28]:

$$\mathbf{x}_G = \alpha \mathbf{P} \mathbf{x}_G + (1 - \alpha)|V|^{-1}\mathbf{e}, \tag{2}$$

where $\mathbf{e}$ is the all-one vector and $\alpha \in (0, 1)$ is a decay factor, which is typically set to $\alpha = 0.85$.

Problem Descriptions. We here define PageRank score boosting problems. Given a directed graph $G = (V, E)$ (e.g., the host network of publishers), a set of target vertices T (e.g., each of which is a webpage with web banner spaces), and a set of missing edges $S \subseteq (V \times V) \setminus E$ under control, problems of PageRank score boosting are formalized as follows.

*Problem 1 (**M**inimum **P**ageRank **M**aximization; MPM).* Given a positive integer k, find k edges $F \subseteq S$ so that the minimum of $x_{G \cup F}(v)$ for v in T is maximized.

*Problem 2 (**Min**imum **P**ageRank **T**hreshold **C**overage; MinPTC).* Given a threshold number $l \in (0, 1)$, find the minimum number of edges $F \subseteq S$ so that $x_{G \cup F}(v) \geq l$ for every vertex v in T.

*Problem 3 (**Max**imum **P**ageRank **T**hreshold **C**overage; MaxPTC).* Given a threshold number $l \in (0, 1)$, and a positive integer k, find k edges $F \subseteq S$ so that the number of vertices in T with $x_{G \cup F}(v) \geq l$ for v in T is maximized.

4   Hardness Results

In this section, we show that the problems formulated in the previous section are NP-hard.

Theorem 1. *The MPM problem is NP-hard even if G is of maximum out-degree 3 and maximum in-degree 2.*

Proof. We reduce the vertex cover problem to MPM. In the *vertex cover problem*, we are given an undirected graph $H = (W, F)$ and a positive integer k, and the aim is to find a vertex subset X of size at most k such that every edge has an end vertex in X. The problem is known to be NP-hard even if H is a cubic graph (i.e., the degree of each vertex is three) [12].

Suppose that we are given an instance of the vertex cover problem, consisting of a cubic graph $H = (W, F)$ and a positive integer k. We construct a directed graph $G = (V, E)$ as follows. Set $V = W \cup F \cup F' \cup \{s, s'\}$, where F' is a copy of F. Two vertices $w \in W$ and $f \in F$ are adjacent if and only if w is an end vertex of f. Moreover, we add edges (f, f') and (f', f) for all $f \in F$, and (s, s') and (s', s). See Fig. 1. We define $T = F$ and $S = \{(s, w) \mid w \in W\}$.

Let d be the minimum of PageRank $x_G(v)$ among $v \in T$. We claim that H has a vertex cover of size at most k if and only if the instance G for MPM has a solution I such that $\min_{v \in T} x_{G \cup I}(v) > d$. We first observe that $x_G(v) = d$

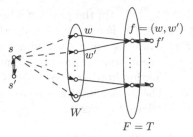

Fig. 1. Reduction from the vertex cover problem.

for each vertex $v \in T$. This is because the out-degree of w is the same for any $w \in W$, as H is cubic, and the in-degree of f is the same for any $f \in F$. If we add an edge (s, w) in T, then the PageRank score of a vertex in $\Gamma(w)$, where $\Gamma(w)$ is the set of vertices adjacent to w in T, increases. Therefore, the minimum of $x_G(v)$ among $v \in T$ becomes more than d if and only if there exists a vertex subset S in V such that $\bigcup_{v \in S} \Gamma(v) = T$. This means that these two problems are equivalent.

Corollary 1. *The MaxPTC problem is NP-hard.*

Proof. We show that the MPM problem can be reduced to the MaxPTC problem. Suppose we are given an instance (G, T, S, k) of MPM. Let d be the minimum of $x_G(v)$ among all target vertices $v \in T$. Notice that the MPM problem is equivalent to finding the maximum number $l \in [d, 1]$ such that a solution I for the MaxPTC instance (G, T, S, l, k) satisfies $x_{G \cup I}(t) \geq l$ for all $t \in T$. Thus, the MPM problem can be solved by repeatedly solving the MaxPTC problem with changing l by the bisection search on the interval $[d, 1]$. Cramer's rule implies that $x_G(v) = x_{G'}(v)$ or $|x_G(v) - x_{G'}(v)| \geq 1/|V|^{|V|+1}$ holds for any two graphs G and G' on the vertex set V. This fact means that the number of iterations in the bisection search is $O(|V| \log |V|)$.

The reduction from the MPM problem in the proof of Corollary 1 can be also adapted for MinPTC with a slight modification. Although this already implies the NP-hardness of MinPTC, we can prove a stronger approximation hardness of MinPTC by reducing the vertex cover problem to MinPTC directly. The following theorem describes this fact in detail.

Theorem 2. *It is NP-hard to approximate MinPTC within a factor of 1.3606. Moreover, if the unique games conjecture is correct, MinPTC admits no $(2 - \epsilon)$-approximation algorithm for any $\epsilon > 0$.*

Proof. We show that the vertex cover problem in a regular graph can be reduced to MinPTC. Indeed, the reduction is almost same as the one in the proof of Theorem 1. The only difference is that we have a threshold number l in the reduced instance of MaxPTC; we set l to $\min_{v \in T} x_G(v) + \epsilon$, where ϵ is a small positive number such that $0 < \epsilon \le 1/|V|^{|V|+1}$ (in the proof of Theorem 1, $x_G(v) = x_G(u)$ has been proven for any $u, v \in T$). Feige [12] showed that if the vertex cover problem with regular graphs admits an α-approximation algorithm for some $\alpha \ge 1$, then the problem with arbitrary graphs also admits an α-approximation algorithm. Thus, any approximation hardness result on the vertex cover problem can be adapted to MinPTC. The former hardness result in the statement follows from [11], and the later one follows from [20].

It should be noted that the reductions in Theorems 1 and 2 can be extended to the reductions from the vertex cover problem (or hitting set problem) on a k-uniform l-regular hypergraph, and the k-densest subgraph problem on a regular graph. Moreover, the reductions from these problems to MinPTC are approximation-preserving. A *hypergraph* G is denoted by $G = (V, \mathcal{E})$, where $\mathcal{E} \subseteq 2^V$. A hypergraph is k-*uniform* if each hyperedge in $\mathcal{E}$ has a size exactly k. Note that a 2-uniform hypergraph is just a graph. A hypergraph is l-*regular* if each node in V is contained in exactly l hyperedges, i.e., $|\{e \in \mathcal{E} \mid v \in e\}| = l$ for every $v \in V$. The *vertex cover problem on a hypergraph* is the problem that, given a hypergraph G, finds a minimum set S of vertices that covers every hyperedge, i.e., $S \cap e \ne \emptyset$ for every $e \in \mathcal{E}$. In the k-densest subgraph problem, we are given an undirected graph, and the problem seeks to find a set of k vertices that induces the maximum number of edges. Although we are not aware of any hardness results on these problems, no known algorithm attains a constant-approximation guarantee for these problems. Hence, the relationship between these problems and MaxPTC is a side evidence for the fact that it is hard to obtain a good approximation guarantee for MinPTC.

5   Proposed Algorithms

This section presents heuristic-based algorithms for our optimization problems.

Framework. As mentioned in the previous section, our problems include the vertex cover problem on a hypergraph. Because the problem is NP-hard, attempting to develop a polynomial-time algorithm is futile. However, this problem can be approximately solved via a greedy algorithm in an efficient manner. More specifically, the greedy algorithm begins with an empty solution set, and repeats adding a vertex that covers the maximum number of hyperedges not yet covered by the solution. This algorithm admits $O(\log n)$-approximation for hypergraphs with n vertices [8,18,23].

Inspired by the above-mentioned greedy algorithm, we here design a greedy heuristic for our problems. Our algorithm repeats adding an edge in a solution similarly to the greedy algorithm. For each iteration, we evaluate the "contribution" of a candidate edge and select one that maximizes the contribution.

In the following subsections, we present two strategies of contribution evaluation and devise the proposed algorithms.

Algorithm 1. Naive strategy for SELECTEDGE.

1: **procedure** SELECTEDGE($G = (V, E), T, S, F$)
2:     compute PageRank scores $\mathbf{x}_{G \cup F}$ for current graph $G \cup F$.
3:     $t \leftarrow \operatorname{argmin}_{v \in T} x_{G \cup F}(v)$.
4:     **for all** $s \in V$ in decreasing order of $\frac{x_{G \cup F}(s)}{d^+_{G \cup F}(s)+1}$ **do**
5:         **if** $(s, t) \in S \setminus (E \cup F)$ **then**
6:             **return** (s, t).

Algorithm 2. Contribution-based algorithm for MinPTC.

Require: a graph $G = (V, E)$, target vertices $T \subseteq V$, missing edges $S \subseteq (V \times V) \setminus E$, a threshold
    $l \in (0, 1)$.
1: $F \leftarrow \emptyset$.
2: **loop**
3:     compute PageRank scores $\mathbf{x}_{G \cup F}$ for current graph $G \cup F$.
4:     **if** $\min_{v \in T} x_{G \cup F}(v) \geq l$ **then**
5:         **break.**
6:     $S \leftarrow$ vertices with top-$|T|$ PageRank $\mathbf{x}_{G \cup F}$ scores.
7:     **for all** $s \in S$ and $t \in T$ **do**
8:         compute PageRank scores $\mathbf{x}_{G \cup F + (s,t)}$ for $G \cup F + (s, t)$.
9:         contrib$(s, t) \leftarrow \sum_{v \in T} \min(0, x_{G \cup F + (s,t)}(v) - l)$.
10:    $F \leftarrow F \cup \{\operatorname{argmax}_{(s,t) \in S \times T : (s,t) \notin E \cup F} \text{contrib}(s, t)\}$.
11: **return** F.

Naive Algorithms. The first contribution evaluation strategy is based on a
well-known fact that if an edge from u to v is inserted into a graph, then the
PageRank score of v will increase [2,16]. Because our objective is to increase
the "minimum" PageRank score, it is natural to select a vertex in T with the
"minimum" PageRank score as the *head* v of an edge to be added. We now
discuss which vertex is appropriate for the *tail* u of an edge to be added. From
the random-walk interpretation, inserting an edge leaving from a vertex with a
higher PageRank score is apparently more desirable. More precisely, we roughly
measure the effectiveness of a vertex u by its PageRank score divided by its
out-degree because the probability of a random walker moving to v through an
edge (u, v) is approximately equal to $x_G(u)/d^+(u)$ rather than $x_G(u)$.

To sum up, for a set F of already selected edges, we pick up a missing
edge connecting from a vertex s in V with the maximum value of $x_{G \cup F}(s)/$
$(d^+_{G \cup F}(s) + 1)$ to a vertex t in T with the minimum PageRank score
$\min_{t \in T} x_{G \cup F}(t)$. The procedure SELECTEDGE is shown in Algorithm 1.

Our naive algorithms for MPM, MinPTC, and MaxPTC start with an empty
set $F = \emptyset$ and continue the greedy selection according to SELECTEDGE until
k edges have been added into F (in the case of MPM and MaxPTC) or the
minimum PageRank score among T for a graph $G \cup F$ attains a given threshold
l (in the case of MinPTC).

Contribution-Based Algorithms. The disadvantage of the naive algorithms
is that it only evaluates a "local" influence of edge insertion on PageRank scores.
In reality, however, inserting a single edge not only increases the PageRank score

---

Algorithm 3. Contribution-based algorithm for MPM.

Require: a graph $G = (V, E)$, target vertices $T \subseteq V$, missing edges $S \subseteq (V \times V) \setminus E$, a solution size k.

1: low $\leftarrow \min_{v \in T} x_G(v)$ and high $\leftarrow 1$.
2: $F \leftarrow \emptyset$
3: **repeat**
4:     mid $\leftarrow$ (low + high)$/2$
5:     $F' \leftarrow$ solve MinPTC with parameters $G, T, S,$ mid.
6:     **if** $|F'| \leq k$ **then**
7:         low $\leftarrow$ mid and $F \leftarrow F'$.
8:     **else**
9:         high $\leftarrow$ mid.
10: **until** convergence
11: **return** F.

Algorithm 4. Contribution-based algorithm for MaxPTC.

Require: a graph $G = (V, E)$, target vertices $T \subseteq V$, missing edges $S \subseteq (V \times V) \setminus E$, a solution size k, a threshold $l \in (0, 1)$.

1: low $\leftarrow 0$ and high $\leftarrow |T| + 1$.
2: $F \leftarrow \emptyset$.
3: **repeat**
4:     mid $\leftarrow \lfloor$(low + high)$/2\rfloor$
5:     $F' \leftarrow$ select k edges according to the top-mid contribution.
6:     **if** $|\{v \in T \mid x_{G \cup F}(v) \geq l\}| \geq$ mid **then**
7:         low $\leftarrow$ mid and $F \leftarrow F'$.
8:     **else**
9:         high $\leftarrow$ mid.
10: **until** convergence
11: **return** F.

---

of its head but may also increase the PageRank scores of other vertices. The naive strategy, which only considers a local influence on the PageRank score, cannot capture such a "wide" influence and it may select ineffective edges.

To capture such a wide influence, we here introduce another kind of a measure to evaluate the contribution of an edge. For the MinPTC problem, we define the contribution of inserting an edge (s, t) to G as $\sum_{v \in T} \min(0, x_{G+(s,t)}(v) - l)$. This represents how close the graph $G + (s, t)$ is to the goal. Note that if all target vertices have PageRank score at least l, then this value takes zero.

Algorithm for MinPTC. Utilizing the contribution introduced above, we describe our contribution-based algorithm for the MinPTC problem. Starting with an empty set $F = \emptyset$, we evaluate the contribution of each edge in a candidate set and add one with the maximum contribution into F. We continue this until the minimum PageRank score among T attains a given threshold l. To reduce a candidate set, we only evaluate edges whose tail has a high PageRank score and whose head is in T. Pseudocode is shown in Algorithm 2.

Algorithm for MPM. We cannot directly use the above-mentioned contribution to MPM because a threshold l is not given. One might suggest a variant of Algorithm 2 which selects an edge with maximum $\min_{v \in T} x_{G+(s,t)}(v)$ (we denote this strategy by *Contrib-simple*). This algorithm, however, does not take into account the increase in the PageRank scores of all vertices excluding $\text{argmin}_{v \in T} x_{G+(s,t)}(v)$. In fact, *Contrib-simple* will be shown to be less effective than the naive strategy in our experiments. We then use the reduction from MPM to MinPTC that is presented in the proof of Corollary 1. That is, we repeatedly solve the MinPTC problem with the bisection search on the interval $[d, 1]$, where d is the minimum of $x_G(v)$ among $v \in T$, to obtain the maximum number l such that the answer of MinPTC is at most k. Pseudocode is shown in Algorithm 3.

Algorithm for MaxPTC. For the MaxPTC problem, the goal is somewhat different from MPM, i.e., maximizing the *number* of the target vertices with PageRank score at least l, though both the problems ask for selecting k edges. We again claim that a variant of Algorithm 2, which selects an edge with maximum $\sum_{v \in T} \min(0, x_{G+(s,t)}(v) - l)$ is not so effective similarly to the case of MPM (we denote this strategy by *Contrib-simple*). Hence, we again rely on the bisection search, that is, we repeatedly decide whether or not inserting k missing edges can increase the PageRank scores of r target vertices to l. To this end, instead of maximizing the sum of $\min(0, x_G(v) - l)$ for *all* target vertices v, we aim at greedily maximizing the sum of $\min(0, x_G(v) - l)$ for target vertices v having the *top-r* PageRank scores in T. It is clear to see that if at least r target vertices have PageRank score at least l, then the sum of contributions is zero. Pseudocode is shown in Algorithm 4.

Efficient Update of PageRank Scores. As mentioned so far, contribution-based algorithms choose the most effective edge from a set of candidate edge in each iteration. However, it requires computing the PageRank scores of a graph obtained by inserting each candidate edge to evaluate its contribution. This would be computationally expensive if we simply apply static algorithms.

In each iteration of the contribution-based algorithms, we have already the PageRank scores of a graph $G \cup F$, and we are asked to compute the PageRank of $G \cup F + (s, t)$ for all candidate edges (s, t). This situation is similar to the dynamic network setting when a graph is evolved, which allows us to use an incremental algorithm for tracking PageRank scores on dynamic graphs proposed by [24]. Their algorithm can manage both the addition and deletion of several edges. Thus we can compute the PageRank scores $\mathbf{x}_{G \cup F + (s,t)}$ for each candidate edge (s, t), given $\mathbf{x}_{G \cup F}$.

The algorithm is proven to perform efficiently if edges are *randomly* inserted [24]. Unfortunately, this is not the case in our scenario because we insert a single edge leaving from a vertex with a high PageRank score. Nevertheless, our experimental results indicate the scalability of our proposed algorithms.

6   Experimental Evaluation

We conducted experiments on several webgraphs to demonstrate the effectiveness of the proposed algorithms. All experiments were conducted on a Linux server with an Intel Xeon E5540 2.53 GHz CPU and 48 GB memory. All algorithms were implemented in C++ and compiled using g++ 4.8.2 with the −O2 option.

Datasets. We use three webgraph datasets: Google network with 12,354 vertices and 164,046 edges, which is a webgraph from google.com domain, from Koblenz Network Collection [21], and Stanford network with 150,532 vertices and 1,576,314 edges, which is a webgraph from stanford.edu domain, and Berk-Stan network with 334,857 vertices and 4,523,232 edges, which is a webgraph from berkely.edu and stanford.edu domains, from Stanford Network Analysis Project [22]. For each graph, we extracted the subgraph induced by the largest strongly connected component.

Selection of Target Vertices. Because the structural properties of T affect the performance of each algorithm, we use the following different methods for selecting T.

- RANDOM: randomly selects 100 vertices from V.
- 2-HOP: (1) initialize $r = 10,000$ and $T = \emptyset$, (2) pick up a vertex v having the r-th highest PageRank score, (3) add the 2-hop neighbors of v into T, (4) if $|T|$ is less than 100, then increase r by one and return to (2); otherwise, (5) randomly select 100 vertices from T.

Note that vertices selected by 2-HOP are expected to be close to each other.

Algorithms. We compare the following four algorithms:

- *Degree*: a baseline algorithm that repeatedly selects an edge from a vertex s with the maximum value of $d^-(s)/(d^+(s) + 1)$ to a target vertex t with the minimum in-degree. Note that this method does not take into account PageRank scores.
- *Naive*: the naive algorithm with Algorithm 1.
- *Contrib-simple*: the contribution-based algorithm without the bisection search for MPM and MaxPTC.
- *Contrib*: the contribution-based algorithm for MPM, MinPTC, and MaxPTC. For MPM, the bisection search is repeated 30 times.

A parameter ϵ, which specifies the accuracy of PageRank estimation, is set as $\epsilon = 10^{-7}$.

Solution Quality. We first examine the effectiveness of each algorithm. We compute the PageRank scores of the graph obtained from an original graph by inserting edges in the solution produced by each algorithm using power iteration.

Results for MPM. Table 1 shows the minimum PageRank score among target vertices in the graph obtained by inserting missing edges selected by each algorithm. We set the solution size k as 20, 40, and 80. In the 2-HOP setting, *Contrib* gives more effective missing edges compared to *Contrib-simple* and *Naive*. Inserting

Table 1. Experimental results for MPM.

Settings		Minimum PageRank score [$\times 10^6$]				Run time [s]			
Dataset	k	Contrib (Algorithm 3)	Contrib-simple	Naive	Degree	Contrib (Algorithm 3)	Contrib-simple	Naive	Degree
Google	0	12.97	12.97	12.97	12.97	–	–	–	–
(Random)	20	**13.94**	13.57	13.71	12.97	3,803.6	120.2	0.1	0.1
	40	**18.09**	17.64	18.03	14.12	7,820.4	236.7	0.1	0.1
	80	**36.43**	27.74	31.54	16.00	13,901.9	501.2	0.2	0.2
Google	0	13.12	13.12	13.12	13.12	–	–	–	–
(2-hop)	20	**267.99**	78.70	152.70	18.86	3,219.6	93.3	0.1	0.1
	40	**439.04**	205.75	370.33	77.66	5,861.4	184.5	0.1	0.1
	80	**1,379.17**	235.57	1,279.54	129.61	11,840.2	354.7	0.2	0.2
Stanford	0	1.03	1.03	1.03	1.03	–	–	–	–
(Random)	20	**1.09**	1.05	1.06	1.03	7,006.0	228.4	1.1	0.9
	40	**1.50**	1.43	1.47	1.06	12,600.5	410.9	1.4	1.3
	80	**4.01**	3.81	**4.01**	1.06	22,291.1	729.0	2.3	2.3
Stanford	0	1.47	1.47	1.47	1.47	–	–	–	–
(2-hop)	20	**187.93**	114.28	27.58	6.09	5,833.2	213.3	1.0	0.9
	40	**347.97**	163.44	237.69	6.22	10,388.6	379.4	1.4	1.3
	80	**556.98**	200.51	386.48	6.38	18,756.4	729.2	2.2	2.2
BerkStan	0	0.46	0.46	0.46	0.46	–	–	–	–
(Random)	20	0.46	**0.47**	**0.47**	0.46	4,723.8	155.8	2.0	1.9
	40	**0.68**	**0.68**	0.63	0.47	8,684.8	284.2	2.8	2.7
	80	1.56	**1.57**	1.49	0.47	16,243.7	512.2	4.6	4.4
BerkStan	0	0.56	0.56	0.56	0.56	–	–	–	–
(2-hop)	20	**41.88**	25.62	24.70	1.24	4,418.9	125.9	1.9	2.0
	40	**69.54**	39.16	42.97	1.24	7,865.5	232.6	2.8	2.9
	80	**178.94**	58.51	170.14	6.61	14,298.6	431.4	4.6	4.4

only dozens of edges significantly improves the PageRank scores of target vertices, e.g., the minimum PageRank score for Stanford improves from 0.0000015 to 0.00018. In the RANDOM setting, however, edge insertions hardly increase the PageRank scores of target vertices, and there are no significant differences between Contrib and Naive. Note that Degree does not improve the minimum PageRank scores at all. In fact, it mostly selected missing edges with the same head.

Let us investigate the change of PageRank scores of target vertices caused by edge insertion. Figure 2 illustrates a subgraph of Google induced by T (2-HOP) in which each target vertex is colored according to its PageRank score (red for higher values, blue for lower values). We can observe that vertices in T are connected to each other, as expected from the construction of T. In the original graph, target vertices take PageRank scores from 0.000013 to 0.072. In the resulting graph obtained by adding 80 missing edges chosen by Contrib, most of the target vertices are colored in green (PageRank score at least 0.0014). In other words, this edge addition improves the minimum PageRank score by 100 times. We also note that red vertices in the middle of the figures actually have the highest and second-highest PageRank scores among all the vertices.

Table 2. Experimental results for MinPTC.

| Settings | | # inserted edges | | Run time [s] | |
Dataset	l	Contrib (Algorithm 2)	Naive	Contrib (Algorithm 2)	Naive
Google	0.0001	**90**	93	494.4	0.2
(Random)	0.0002	**95**	97	510.9	0.3
	0.0004	120	**98**	652.9	0.3
	0.0008	181	**98**	988.8	0.3
Google	0.0001	**7**	9	31.7	0.1
(2-hop)	0.0002	**10**	30	46.2	0.1
	0.0004	**37**	43	165.9	0.1
	0.0008	**54**	56	238.1	0.2
Stanford	0.0001	101	**100**	840.4	2.6
(Random)	0.0002	103	**100**	861.5	2.5
	0.0004	**233**	379	1,530.1	7.9
	0.0008	**780**	1,027	4,576.9	20.0
Stanford	0.0001	**11**	32	100.3	1.1
(2-hop)	0.0002	**21**	38	168.6	1.3
	0.0004	**52**	78	406.5	2.5
	0.0008	**148**	233	1,019.1	5.0
BerkStan	0.0001	118	**98**	699.0	4.9
(Random)	0.0002	**254**	287	1,299.3	12.4
	0.0004	**574**	675	2,698.3	26.7
	0.0008	**1,890**	2,136	7,175.1	81.3
BerkStan	0.0001	**60**	69	324.7	3.7
(2-hop)	0.0002	**93**	136	467.6	6.3
	0.0004	**244**	313	1,059.2	13.3
	0.0008	**795**	1,083	2,900.1	42.6

Results for MinPTC. Table 2 shows the number of inserted edges with each threshold value l for each algorithm. The threshold value l is set as $l = 0.0001$, 0.0002, 0.0004, and 0.0008. Note that we did not run *Degree* because it may not produce a reasonable number of missing edges, as expected from the results for MPM. When target vertices are selected by 2-hop, *Contrib* outperforms *Naive* under almost every setting. Particularly, *Contrib* requires only 10 edges to ensure that every PageRank score in T reaches 0.0004 for Google, whereas *Naive* requires 30 edges for the same. On the other hand, both *Naive* and *Contrib* demonstrate poor performance when target vertices are selected by Random. Note that both *Naive* and *Contrib* select at least $100 = |T|$ missing edges for this setting. This is because *randomly* selected target vertices are very far from each other, and thus,

Table 3. Experimental results for MaxPTC.

Settings		# vertices with PageRank score $\geq l$				Run time [s]			
Dataset	k	Contrib (Algorithm 4)	Contrib-simple	Naive	Degree	Contrib (Algorithm 4)	Contrib-simple	Naive	Degree
Google	0	4	4	4	4	–	–	–	–
(Random)	20	**30**	29	24	24	695.7	109.3	0.1	0.1
$l = 0.0001$	40	**50**	49	44	40	1,394.8	229.3	0.1	0.1
	80	**90**	90	87	51	2,441.0	454.2	0.2	0.2
Google	0	79	79	79	79	–	–	–	–
(2-hop)	20	100	100	100	97	125.2	94.0	0.1	0.1
$l = 0.0001$	40	100	100	100	99	125.7	186.6	0.1	0.1
	80	100	100	100	100	109.1	364.2	0.2	0.2
Stanford	0	0	0	0	0	–	–	–	–
(Random)	20	**21**	21	20	20	1,311.0	194.7	1.0	1.0
$l = 0.0001$	40	**41**	41	40	39	2,569.3	397.2	1.4	1.3
	80	**81**	78	80	55	3,791.2	748.9	2.3	2.2
Stanford	0	11	11	11	11	–	–	–	–
(2-hop)	20	100	100	56	33	516.0	183.5	1.0	0.9
$l = 0.0001$	40	100	100	100	41	511.0	344.0	1.4	1.3
	80	100	100	100	51	458.5	638.4	2.2	2.2
BerkStan	0	2	2	2	2	–	–	–	–
(Random)	20	**22**	22	22	22	953.3	141.9	1.9	1.8
$l = 0.0001$	40	41	**42**	**42**	21	1,796.3	277.8	2.9	3.1
	80	80	**82**	**82**	39	3,386.0	520.3	4.5	4.3
BerkStan	0	15	15	15	15	–	–	–	–
(2-hop)	20	**52**	51	41	33	990.5	165.8	2.0	1.9
$l = 0.0001$	40	**73**	72	62	42	1,554.8	241.3	2.9	2.8
	80	100	100	100	56	1,996.8	428.8	4.5	4.4

both *Naive* and *Contrib* have no other choice but to insert an edge connecting a vertex with a high PageRank score to every target vertex.

Results for MaxPTC. Table 3 shows the number of vertices with PageRank score at least $l = 0.0001$ for each algorithm. *Contrib* produces quite effective missing edges in some settings, e.g., adding 40 missing edges to BerkStan increases the number of target vertices (selected by 2-hop) with a specified condition from 15 to 73, whereas *Naive* increases it to 62. *Contrib-simple* performs slightly worse than *Contrib*. For Random, the number of target vertices with PageRank score at least $l = 0.0001$ is roughly equal to the number of inserted edges. This also comes from the fact that vertices chosen by Random are far from each other.

Scalability. We then study the scalability of each algorithm. Tables 1, 2 and 3 report the run time of each algorithm for MPM, MinPTC, and MaxPTC. *Degree* is naturally the fastest in most cases. *Naive* is highly scalable for all settings; it only requires at most 100 s. *Contrib-simple* requires at most 1,000 s. Although *Contrib* is the slowest among all algorithms, it even scales to BerkStan with millions of edges. Note that *Contrib* requires a longer time for MPM compared to MinPTC because it solves MinPTC 30 times for the bisection search.

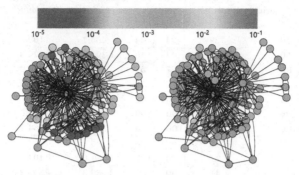

(a) Before edge addition. (b) After 80 edge addition.

Fig. 2. Subgraph of Google induced by T (2-HOP). Each vertex is colored according to its PageRank score. (Color figure online)

7 Conclusion

In this paper, we have considered the three graph optimization problems of boosting PageRank scores. We have proven the NP-hardness and then proposed heuristic-based algorithms inspired by the greedy strategy. Through experiments on real-world graph data, we have verified the effectiveness of the proposed algorithms compared to baseline algorithms.

References

1. Adar, E., Teevan, J., Dumais, S.T.: Resonance on the web: web dynamics and revisitation patterns. In: CHI, pp. 1381–1390 (2009)
2. Avrachenkov, K., Litvak, N.: Decomposition of the Google PageRank and optimal linking strategy. Technical report (2004). http://doc.utwente.nl/80247/
3. Avrachenkov, K., Litvak, N.: The effect of new links on Google PageRank. Stoch. Model. **22**(2), 319–332 (2006)
4. Baeza-Yates, R.A., Castillo, C., López, V.: PageRank increase under different collusion topologies. In: AIRWeb, pp. 25–32 (2005)
5. Becchetti, L., Castillo, C.: The distribution of PageRank follows a power-law only for particular values of the damping factor. In: WWW, pp. 941–942 (2006)
6. Brin, S., Page, L.: The anatomy of a large-scale hypertextual Web search engine. Comput. Netw. ISDN Syst. **30**(1), 107–117 (1998)
7. Broder, A., Josifovski, V.: MS&E 239: Introduction to Computational Advertising (2011). http://web.stanford.edu/class/msande239/
8. Chvatal, V.: A greedy heuristic for the set-covering problem. Math. Oper. Res. **4**, 233–235 (1979)

9. Csáji, B.C., Jungers, R.M., Blondel, V.D.: PageRank optimization in polynomial time by stochastic shortest path reformulation. In: ALT, pp. 89–103 (2010)
10. Csáji, B.C., Jungers, R.M., Blondel, V.D.: PageRank optimization by edge selection. Discrete Appl. Math. **169**, 73–87 (2014)
11. Dinur, I., Safra, S.: On the hardness of approximating minimum vertex cover. Ann. Math. **1**, 439–485 (2005)
12. Feige, U.: Vertex cover is hardest to approximate on regular graphs. Technical report (2004)
13. Fortunato, S., Boguñá, M., Flammini, A., Menczer, F.: Approximating PageRank from in-degree. In: WAW, pp. 59–71 (2006)
14. Gleich, D.F.: PageRank beyond the web. SIAM Rev. **57**(3), 321–363 (2015)
15. Gyöngyi, Z., Garcia-Molina, H.: Link spam alliances. In: VLDB, pp. 517–528 (2005)
16. Ipsen, I.C.F., Wills, R.S.: Mathematical properties and analysis of Google's PageRank. Bol. Soc. Esp. Mat. Apl. **34**, 191–196 (2006)
17. Ishii, H., Tempo, R.: Computing the PageRank variation for fragile web data. SICE J. Control Measur. Syst. Integr. **2**(1), 1–9 (2009)
18. Johnson, D.S.: Approximation algorithms for combinatorial problems. J. Comput. Syst. Sci. **9**(3), 256–278 (1974)
19. de Kerchove, C., Ninove, L., Van Dooren, P.: Maximizing PageRank via outlinks. Linear Algebra Appl. **429**, 1254–1276 (2008)
20. Khot, S., Regev, O.: Vertex cover might be hard to approximate to within 2-epsilon. J. Comput. Syst. Sci. **74**(3), 335–349 (2008)
21. Kunegis, J.: KONECT - the Koblenz network collection. In: WWW Companion, pp. 1343–1350 (2013)
22. Leskovec, J., Krevl, A.: SNAP Datasets: Stanford Large Network Dataset Collection, June 2014. http://snap.stanford.edu/data
23. Lovász, L.: On the ratio of optimal integral and fractional covers. Discrete Math. **13**(4), 383–390 (1975)
24. Ohsaka, N., Maehara, T., Kawarabayashi, K.: Efficient PageRank tracking in evolving networks. In: KDD, pp. 875–884 (2015)
25. Olsen, M.: The computational complexity of link building. In: COCOON, pp. 119–129 (2008)
26. Olsen, M.: Maximizing PageRank with new backlinks. In: CIAC, pp. 37–48 (2010)
27. Olsen, M., Viglas, A., Zvedeniouk, I.: A constant-factor approximation algorithm for the link building problem. In: COCOA, pp. 87–96 (2010)
28. Page, L., Brin, S., Motwani, R., Winograd, T.: The PageRank citation ranking: bringing order to the web. Technical report, Stanford InfoLab (1999)
29. Pandurangan, G., Raghavan, P., Upfal, E.: Using PageRank to characterize web structure. In: COCOON, pp. 330–339 (2002)
30. Sydow, M.: Can one out-link change your PageRank? In: AWIC, pp. 408–414 (2005)

Finding the Most Navigable Path in Road Networks: A Summary of Results

Ramneek Kaur[1]([✉]), Vikram Goyal[1], and Venkata M. V. Gunturi[2]

[1] IIIT-Delhi, New Delhi, India
{ramneekk,vikram}@iiitd.ac.in
[2] IIT Ropar, Rupnagar, India
gunturi@iitrpr.ac.in

Abstract. Input to the Most Navigable Path (MNP) problem consists of the following: (a) a road network represented as a directed graph, where each edge is associated with numeric attributes of cost and "navigability score" values; (b) a source and a destination and; (c) a budget value which denotes the maximum permissible cost of the solution. Given the input, MNP aims to determine a path between the source and the destination which maximizes the navigability score while constraining its cost to be within the given budget value. This problem finds its applications in navigation systems for developing nations where streets, quite often, do not display their names. MNP problem would help in such cases by providing routes which are more convenient for a driver to identify and follow. Our problem is modeled as the arc orienteering problem which is known to be NP-hard. The current state-of-the-art for this problem may generate paths having loops, and its adaptation for MNP, that yields simple paths, was found to be inefficient. In this paper, we propose two novel algorithms for the MNP problem. Our experimental results indicate that the proposed solutions yield comparable or better solutions while being orders of magnitude faster than the current state-of-the-art for large real road networks. We also propose an indexing structure for the MNP problem which significantly reduces the running time of our algorithms.

1 Introduction

The problem of finding the most navigable path takes the following as input: (a) a directed graph representation of a road network where each edge is associated with a cost (distance or travel-time) value and a navigability score value; (b) a source and a destination and; (c) a budget value. Given the input, the objective is to determine a path between the source and the destination which has the following two characteristics: (i) the sum of navigability score values of its constituent edges is maximized and, (ii) the total length (in terms of distance or travel-time) of the path is within the given budget. In other words, MNP is a constrained maximization problem.

This problem finds its applications in navigation systems for developing nations. Quite often, streets names in developing countries are either not

© Springer Nature Switzerland AG 2018
S. Hartmann et al. (Eds.): DEXA 2018, LNCS 11029, pp. 440–456, 2018.
https://doi.org/10.1007/978-3-319-98809-2_27

displayed prominently or not displayed at all. In such cases, it becomes difficult to follow the conventional navigation instructions such as *"Continue on Lal Sai Mandir Marg towards Major P Srikumar Marg"*. In such nations, it is desirable to travel along a path which is "easily identifiable" by a driver. For instance, consider the example shown in Fig. 1. Here, Path 1 is the shortest path between KFC in Cannaught Place (point A) and the Rajiv Chowk metro station (point B). The path has a travel time of 10 min. However, this route is potentially confusing due to lack of prominently visible sites and easily identifiable turns. To navigate on this route, one would have to heavily rely on a very accurate GPS system (potentially expensive) operating over a good quality map with no missing roads, both of which may be non-trivial for a common traveler in developing countries.

In contrast, consider Path 2 from A to B in Fig. 1 with a travel time of 12 min. This route involves going past some popular sites like the hotel Sarvana Bhavan and the hotel Radisson Blu Marina, and then taking the first right turn that follows. Given the challenges associated with transportation and navigational infrastructure in a developing nation setting, one may choose Path 2 even if it is 20% longer than the shortest path. This is because it is easier to describe, memorize, recall and follow. This option would be even more amenable if the driver is not well versed with the area. Furthermore, travelers who are not comfortable using navigation systems while driving, senior citizens for instance, generally look up the route suggestions before starting their journey. Such travelers tend to memorize the route based on the sites en-route to their destination. These travelers would thus be benefited by our concept of most navigable paths which are potentially easier to memorize.

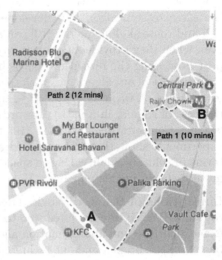

Fig. 1. Problem illustration

The concept of most navigable paths would also help drivers of two-wheelers (predominant in developing countries) by suggesting routes which are easier to follow, as it can be difficult to follow step-by-step instructions on a screen while driving two-wheelers. To the best of our knowledge, both Google maps (www.maps.google.com) and Bing maps (www.bing.com/maps) do not have the option of most navigable paths.

Computational Challenges. Finding the "Most Navigable Path (MNP)" is computationally challenging. The MNP problem is formalized as the Arc Orienteering Problem (AOP) (a maximization problem under constraints) which is known to be an NP-hard combinatorial optimization problem [2,7]. The AOP problem can easily be reduced to the Orienteering Problem (OP) [15] which is also NP-hard [7]. Another factor which adds to the complexity of the MNP problem is the scale of real road networks, which typically have hundreds of thousands of road segments and road intersections.

Challenges in Adapting a Minimization Problem. It is important to note that a maximization problem such as the MNP problem *cannot be trivially reduced into a minimization problem* by considering the inverse of the navigability scores. Even without the budget constraint, the MNP problem involves maximizing the sum of navigability scores of the output path. Mathematically, it is equivalent to maximizing the sum of n parameters, $s_1, s_2, \ldots, s_n$ (s_i denotes the score of edge e_i in a path); *which is not equivalent* to minimizing the sum of their inverses, $\frac{1}{s_1}, \frac{1}{s_2}, \ldots, \frac{1}{s_n}$. Thus, it is non-trivial to generalize algorithms developed for minimization problems (e.g., [6,8,9]) for the MNP problem. For the same reason, algorithms developed for k shortest loop-less paths problem [12] and route skyline queries [10] can also not be adapted for the MNP problem even when they provide an opportunity to incorporate the budget constraint by post-processing the returned k paths based on their total cost. We compare the performance of our algorithms with that of an algorithm that optimally solves the formulation: $Minimize \sum_{e_i \in path} \frac{1}{s_i}$, subject to $\sum_{e_i \in path} c_i \leq budget$ (c_i denotes the cost of edge e_i in a path). Towards this end, we adapt the Advanced Route Skyline Computation algorithm [10] to optimally solve this formulation. Our experiments show that the proposed algorithms outperform this adaptation in terms of solution quality. Due to space constraint, the results of this experiment are not shown in this paper.

Limitations of Related Work. The existing algorithms for the AOP problem (or the OP as AOP can be reduced to OP) can be divided into three categories: exact, heuristic and approximation algorithms. An exact algorithm for the AOP based on the branch-and-cut technique has been proposed in [2]. However, the algorithm would not be able to scale up to real-world road networks as it takes up to 1 h to find a solution for a graph with just 2000 vertices.

There have been several works which proposed heuristic algorithms for the AOP [14,16] and OP [4,13] problems. A core requirement of these algorithms is pre-computation of *all-pairs shortest paths* of the input graph. This pre-computation step is necessary for ensuring their scalability (as also pointed out by Lu and Shahabi [11]). However, it is important to note that in any realistic scenario, urban networks keep updating frequently, e.g., roads may be added, closed or heavily congested (due to repair or accidents), etc. Thus, any real-life system for the MNP problem working on large-scale urban road maps cannot use these techniques which require frequent computation of all-pairs shortest paths.

To the best of our knowledge, the only heuristic algorithm that does not pre-compute shortest paths is [11]. However, it generates paths with loops. We adapted their solution for the MNP problem. Our experimental results indicated a superior performance of the algorithms proposed by us. Our understanding is that their algorithm is most suitable when there are very few edges with a non-zero navigability score value.

Gavalas et al. [7] proposes an approximation algorithm for the OP problem where edges are allowed to be traversed multiple times, a relaxation not suitable for our problem as it would be pointless to drive unnecessarily in a city to reach a destination. Similarly, approximation results were proposed in [3,5]. But they would also need to pre-compute all-pairs shortest paths for efficiency, which as discussed previously is not suitable for the MNP problem in typical real-world scenarios.

Our Contributions. This paper makes the following contributions:

1. Proposes the novel problem of finding the most navigable path. This problem has a potential to add value to the current navigation systems so that they can be easily used in developing nations.
2. Proposes a novel indexing structure (*navigability index*) for road networks which can estimate (in constant time) the potential of any given segment (a sequence of edges in a path) for navigability score improvement.
3. Presents two algorithms for the MNP problem which use a novel *Weighted Bi-directional Search (WBS)* procedure and the previously described *navigability index*.
4. Extensively evaluates the proposed algorithms on three real-road network datasets, and compares their performance against that of the current state-of-the-art algorithm, ILS(CEI) [11].
   Our experiments demonstrate that the proposed algorithms yield comparable or better solutions while being orders of magnitude faster than the state-of-the-art.

Outline. The rest of the paper is organized as follows. In Sect. 2, we cover the basic concepts and formally present the problem definition. In Sect. 3.1, we describe the WBS algorithm. The proposed algorithms for MNP are presented in Sects. 3.2 and 3.3. In Sect. 4, we discuss the experimental evaluation. Finally, we discuss our conclusions and future work in Sect. 5.

2 Basic Concepts and Problem Definition

Definition 1. *Road network:* *A road network is represented as a directed graph $G = (V, E)$, where the vertex set V represents the road intersections and the edge set E represents the road segments. Each edge in E is associated with a cost value which represents the distance or travel-time of the corresponding road segment. Each edge is also associated with a score value (≥ 0) which represents the navigability score of the corresponding road segment. We refer to an edge with a score value >0 as a navigable edge.*

Definition 2. *Path:* *A path is a sequence of connected edges $<e_1\, e_2\, ...\, e_n>$. For this work, we consider only simple paths (paths without cycles).*

Definition 3. *Segment of a path:* *A sequence of connected edges, $<e_i\, e_{i+1}\, ...\, e_j>$, denoted as S_{ij}, is a segment of the path $P = <e_1\, e_2\, ...\, e_n>$ if $1 \leq i \leq j \leq n$. $S_{ij}.score$ denotes the sum of scores of all the edges in S_{ij}. Likewise, $S_{ij}.cost$ denotes the sum of costs of all the edges in S_{ij}. $S_{ij}.start$ and $S_{ij}.end$ denote the first and last vertices of S_{ij}.*

2.1 Problem Definition

Input consists of:

(1) A road network, $G = (V, E)$, where each edge $e \in E$ is associated with a non-negative cost value and a navigability score value.
(2) A source $s \in V$ and a destination $d \in V$.
(3) A positive value *overhead* which corresponds to the maximum permissible cost allowed over the cost of the minimum cost path from s to d. In this paper, we refer to the term (*overhead* + *cost of the minimum cost path from s to d*) as the *budget*.

Output: A path from s to d
Objective function: *Maximize path.score*
Constraint: *path.cost* $\leq$ *budget*.

2.2 Practical Considerations While Using MNP

While using MNP in real life one would have to assign navigability scores to the road segments. Though the score values may be subjective, a rule of thumb could be followed. The values could be assigned over a range (e.g., 1–15) where higher values (e.g., 10–15) could be given to edges with unique/popular sites like a well-known temple or a prominent building, and lower values (e.g., 1–5) to edges with sites like petrol pumps and ATMs.

3 Proposed Approach

Our proposed algorithms for the MNP problem primarily consists of the following two steps. Firstly, we compute a shortest path from s to d which optimizes only on the cost attribute of the path. This path is referred to as the *initial seed path*. In the second step, we iteratively modify portions of this seed path with the goal of improving the navigability score of the solution. While this is being done, we ensure that the total cost of the resulting path is within the budget value. A key procedure used for improving the navigability of the seed path is our proposed *weighted bi-directional search*. We first describe the *weighted bi-directional search* procedure in Sect. 3.1. Following this, in Sects. 3.2 and 3.3, we propose two novel algorithms for the MNP problem which use this search procedure on "segments" of the initial seed path to improve its navigability score.

3.1 Weighted Bi-directional Search (WBS)

Input to the WBS algorithm consists of the input road network, a specific segment (S_{ij}) of the initial seed path, and a budget value (B'). The goal of the algorithm is to determine a replacement (S'_{ij}) for the specified segment such that the following criteria are satisfied:

1. The new segment (S'_{ij}) has a higher score value than the input segment.

2. The resultant path from the source to the destination, obtained after replacing S_{ij} with S'_{ij}, is simple (i.e., no loops).
3. The total cost of the new segment S'_{ij} is within B'.

The WBS algorithm employs a bi-directional search to determine the replacement S'_{ij}. The forward search starts from the first vertex of the input segment S_{ij}, whereas the backward search starts from the last vertex of S_{ij}. In each iteration of the WBS algorithm, the forward search determines the *best-successor* of the current tail node (denoted as F_{tail}) of the partial segment it is developing.

In contrast to the forward search, the backward search determines the *best-predecessor* of the current tail node (denoted as B_{tail}) of the partial segment it is developing. This is done by processing the incoming edges at B_{tail}. At the beginning of the algorithm, F_{tail} is initialized to the $S_{ij}.start$, whereas B_{tail} is initialized to $S_{ij}.end$. We now provide details on computation of the best-successor and best-predecessor.

Determining the *Best-Successor* of F_{tail}. Given the current tail node of the forward search frontier F_{tail}, and the target node (current B_{tail}), the algorithm computes the *Forward Navigability Potential* (Γ^f in Eq. 1) of all the *outgoing neighbors u of F_{tail}*. Following this, the neighbor with the highest Γ^f is designated as the best-successor of F_{tail}. Algorithm 1 details this process.

$$\Gamma^f(u, F_{tail}, B_{tail}) = \frac{1 + score(F_{tail}, u)}{cost(F_{tail}, u) + D_E(u, B_{tail})} \qquad (1)$$

In Eq. 1, D_E denotes the Euclidean distance[1] between the outgoing neighbor u (of F_{tail}) and the current tail node of backward search B_{tail}. As per Eq. 1, neighbors of F_{tail} which are closer to B_{tail} (i.e., lower Euclidean distance), and involve edges with high navigability score values and low cost values, would get a higher value of Γ^f. Algorithm 1 chooses the neighbor which has the highest value of Γ^f.

Algorithm 1. Best-Successor of F_{tail} (G, F_{tail}, B_{tail})

Input: A road network G, F_{tail} and B_{tail}
Output: Best-Successor of F_{tail} and its gamma value (Γ^f_{best})
  1: **for all** OutNeighbors u of F_{tail} **do**   ▷ only the unvisited Outneighbors
  2:     Compute $\Gamma^f(u, F_{tail}, B_{tail})$
  3: **end for**
  4: Best-Successor of F_{tail} ← OutNeighbor u of F_{tail} with highest Γ^f

---

Determining the *Best-Predecessor* of B_{tail}. Analogous to the computation of Γ^f, the *Backward Navigability Potential* (Γ^b) can be computed using Eq. 2.

---

[1] If the edge costs represent travel-times, then a lower bound on the travel time may be used. This can be computed using the upper speed limit of a road segment.

Here, we consider the incoming edges of B_{tail}. The neighbor with the highest Γ^b is designated as the best-predecessor of B_{tail}.

$$\Gamma^b(v, B_{tail}, F_{tail}) = \frac{1 + score(v, B_{tail})}{cost(v, B_{tail}) + D_E(v, F_{tail})} \tag{2}$$

It is important to note that both the forward and the backward searches have "moving targets." In other words, the value of B_{tail} in Eq. 1 (and F_{tail} in Eq. 2) would change as the algorithm proceeds. To this end, the WBS algorithm employs a design decision to help in terminating quickly (while not sacrificing on the navigability score).

Algorithm 2. Weighted Bi-directional Search (G, P, S_{ij}, B')

Input: A road network G, a path P, a segment S_{ij} of P, a budget value $B' = B - P.cost + S_{ij}.cost$, **Output:** A new segment S'_{ij} to replace S_{ij}

1:   $F_{tail} \leftarrow S_{ij}.start$, $B_{tail} \leftarrow S_{ij}.end$
2:   Mark F_{tail} as colored by forward search and B_{tail} as colored by backward search
3:   **while** Forward and backward searches do not color a common node **do**
4:      **if** F_{tail} and B_{tail} are connected via a 1-hop or 2-hop path **then**
5:         $P_{cand} \leftarrow S_{ij}.start \rightsquigarrow F_{tail} \rightsquigarrow B_{tail} \rightsquigarrow S_{ij}.end$
6:         **if** $P_{cand}.cost \leq B'$ **then**
7:            Save P_{cand} in the set (Ω) of candidate segments
8:         **end if**
9:      **end if**
10:     Compute Best-Successor of F_{tail} & its Gamma, Γ^f_{best} (using Algorithm 1)
11:     Compute Best-Predecessor of B_{tail} & its Gamma, Γ^b_{best}
12:     **if** $\Gamma^f_{best} > \Gamma^b_{best}$ **then**
13:        $F_{tail} \leftarrow$ Best-Successor of F_{tail} ▷ Move the forward search
14:        Compute Best-Predecessor of B_{tail}
15:        $B_{tail} \leftarrow$ Best-Predecessor of B_{tail} ▷ Move the backward search
16:        Mark F_{tail} and B_{tail} as colored by their respective searches
17:     **else**
18:        $B_{tail} \leftarrow$ Best-Predecessor of B_{tail} ▷ Move the backward search
19:        Compute Best-Successor of F_{tail} (using Algorithm 1)
20:        $F_{tail} \leftarrow$ Best-Successor of F_{tail} ▷ Move the forward search
21:        Mark F_{tail} and B_{tail} as colored by their respective searches
22:     **end if**
23:     **if** $(S_{ij}.start \rightsquigarrow F_{tail}).cost + (B_{tail} \rightsquigarrow S_{ij}.end).cost > B'$ **then**
24:        Break
25:     **end if**
26: **end while**
27: **if** Forward and backward searches have colored a common node **then**
28:     $P_{cand} \leftarrow$ Reconstructed path between $S_{ij}.start$ and $S_{ij}.end$ by following the Best-Successors/Best-Predecessors
29:     Save P_{cand} in Ω
30: **end if**
31: Return the segment in Ω with highest navigability score

The algorithm first moves the frontier (forward or backward) whose next node to be added[2] comes in with a higher value of Γ. The rationale behind this design decision is the following: a node with higher Γ can imply one or more of the following things: (1) closer to the current target; (2) higher navigability; (3) lower edge cost. Needless to say that all these circumstances are suitable for the needs of the WBS algorithm. After advancing the selected search, WBS re-computes the next node to be added to the other search frontier before it is advanced. For instance, if in the first step the node added by the forward search had higher Γ, then WBS would first advance the forward search frontier by updating its F_{tail} to its best-successor. Following this, best-predecessor of the backward search is re-computed based on the new value of F_{tail}. After that, the backward search is also advanced by updating its B_{tail} to its best-predecessor. Note that in any particular iteration of WBS, both the forward and the backward searches are advanced.

Putting Together Forward and Backward Searches. Algorithm 2 puts together our proposed forward and backward searches along with a *termination condition* and a mechanism to *collect candidate solutions* during the execution. We now describe both these aspects of the WBS algorithm.

As one can imagine, a natural termination condition for the WBS algorithm would be meeting of the forward and the backward searches, i.e., both the searches color the same vertex. Algorithm 2 uses this as the primary termination condition as indicated in the while loop on line 3 of the pseudo-code. In addition to this, the algorithm also terminates, if at any stage, the total cost of the partial paths constructed so far by the forward and the backward searches happens to be greater than the available budget B'. This termination clause is indicated in lines 23–25 of Algorithm 2.

During the course of the algorithm, it collects several candidate solutions in a set called Ω (lines 4–9 and lines 27–29 in Algorithm 2). In the end, WBS returns the solution having the highest navigability score. The primary reason to collect these candidate solutions being that the forward and the backward searches may not always meet during the course of the algorithm. WBS collects the candidate solutions in the following two ways:

1. At any time during the exploration, if the current F_{tail} and B_{tail} are connected through either a direct edge or two edges then, the segment formed by concatenating this direct edge (or two edges) with the current partial segments formed by forward and backward searches is saved as a candidate solution in the set Ω. This is done only if the total cost of this candidate solution is less than the budget B'. This case is illustrated in lines 4–9 of the algorithm.
2. Trivially, if the two search frontiers meet, the partial segments formed by the forward and the backward searches are concatenated to create a candidate solution between the first and last nodes of the original segment S_{ij}.

---

[2] Best-successor in case of forward search and best-predecessor in case of backward search.

Time Complexity Analysis. The number of vertices visited by WBS is $O(|V|)$. No vertex is visited twice, and each vertex v that gets colored leads to computation of $degree(v)$ number of navigability potential values. Thus, the time complexity of WBS is $O(|E|)$.

3.2   MS(WBS): Multiple Segment Replacement Algorithm with WBS

As described earlier, the core idea of both our algorithms is to compute an initial *seed path* (P) first, and then replace segments of this seed path to improve its navigability score. In this section, we describe an algorithm which efficiently chooses a set of segments of P, which when replaced by the output of WBS, lead to a high increase in the navigability score. Note that we may not be able to replace all the segments due to the budget constraint.

For the sake of brevity, we describe only the most crucial part of the algorithm, which is to determine the set of segments of the initial seed path for replacement. One can trivially put this along with the previously described WBS algorithm and the construction of the initial seed path to create a full pseudocode.

In the MS(WBS) algorithm, we first execute an instance of WBS for each of the possible segments of the initial seed path. While calling an instance of WBS for a segment S_{xy} in the initial seed path, we pass a budget value $B' = B$ - cost of seed path $(P.cost) + S_{xy}.cost$. Here, B is the budget value given in the problem instance ($B = overhead + P.cost$).

Following these calls to the WBS algorithm, we would have a pair $<score\,gain, cost\,gain>$ for each of the possible segments S_{xy} in the initial seed path. Here, *score gain* of a segment S_{xy} is defined as the difference in the navigability score values of S_{xy} and its replacement S'_{xy}. $S_{xy}.sgain = S'_{xy}.score - S_{xy}.score$. *Cost gain* is also defined in an analogous way: $S'_{xy}.cgain = S'_{xy}.cost - S_{xy}.cost$

Selecting a set of segments to replace from a given seed path is non-trivial. This is because of the following three reasons: (a) segments chosen for replacement may have common edges, (b) the budget constraint and, (c) replacements of the chosen segments may overlap.

As an instance of challenges (a) and (b), refer to Table 1. The table illustrates a sample scenario on replacing segments of an initial seed path $<e_1\,e_2\,e_3\,e_4>$. The ten possible segments of this path are shown in Table 1 along with their sample *sgain* and *cgain* values. In the table, an entry $(0, 0)$ implies that no solution was found by WBS for that segment. In this example, we can either replace the segment $<e_2\,e_3>$ or the segment $<e_1\,e_2\,e_3>$. Replacing both would not be possible as e_2 is common to both segments.

In addition, if the allowed overhead was 15 then, the combination $<e_2\,e_3>$ and $<e_1>$ can be collectively replaced. Whereas, segments $<e_1\,e_2>$ and $<e_3>$ cannot be collectively replaced, as their total cost gain is 20. We now formalize this idea using the concept of a feasible and disjoint set of segments corresponding to a path.

Table 1. Set of all segments of path $<e_1\, e_2\, e_3\, e_4>$

Segment	$(sgain, cgain)$	Segment	$(sgain, cgain)$
$<e_1>$	(7, 5)	$<e_2\, e_3>$	(16, 10)
$<e_2>$	(7, 5)	$<e_3\, e_4>$	(15, 10)
$<e_3>$	(5, 8)	$<e_1\, e_2\, e_3>$	(15, 18)
$<e_4>$	(0, 0)	$<e_2\, e_3\, e_4>$	(0, 0)
$<e_1\, e_2>$	(15, 12)	$<e_1\, e_2\, e_3\, e_4>$	(10, 15)

Feasible and Disjoint Set of Segments (FDSS). FDSS of a given path (initial seed path P) is a set of segments such that no two segments in the set share an edge, and $P.cost + \sum_{\forall S \in FDSS} S.cgain \leq B$. In our previous example, $<e_2\, e_3>$ and $<e_1>$ forms an FDSS. As expected, the central goal would be to determine an FDSS which results in the highest increase in navigability score.

Computational Structure of the FDSS Problem. The FDSS problem can be seen as an advanced version of the 0/1 knapsack problem where certain items are not allowed together (i.e., segments having common edges). For solving FDSS, one can first enumerate all sets of disjoint segments and then, run an instance of 0/1 knapsack on each set of disjoint segments. Basically, each set is generated by cutting a path at k unique locations ($0 \leq k \leq$ #edges in the path - 1). However, this technique would have two computational challenges: (i) knapsack problem is known to be NP-hard, (ii) a path with n edges would have 2^{n-1} unique sets of disjoint segments. Given these reasons, this paper proposes to drop the feasibility constraint of the FDSS definition. This gives us the concept of a Disjoint Set of Segments (DSS).

Disjoint Set of Segments (DSS). DSS of a given path is a set of segments such that no two segments in the set share an edge. Our algorithm attempts to compute a DSS which results in the highest increase in navigability score (optimal DSS) without considering the budget constraint.

Dynamic Programming (DP) Based Solution for Computing Optimal DSS. We observe that the DSS problem exhibits the optimal substructure property. We exploit this property to design a DP based solution to compute the optimal DSS which takes $\theta(l^3)$ time and consumes $\theta(l^2)$ space (for a seed path with l edges). The central idea in this DP formulation is to consider the DSS problem analogous to that of the rod cutting problem. Our initial seed path P becomes the "rod" being cut. We aim to "break up this rod" (i.e., initial seed path) into pieces (i.e., disjoint segments) such that the total score gain obtained by replacing these disjoint segments is maximum. Understandably, the rod is assumed to be made up of edges, and we are allowed to place cuts at the head or tail node of edges.

$$f_{ij} = \max_{i \leq k \leq j} (S_{ik}.sgain + f_{(k+1)j}) \tag{3}$$

Equation 3 represents the underlying recurrence equation for our DP based solution. Here, f_{ij} denotes the optimal solution for the sub-problem having edges numbered i through j. The optimal solution for the initial seed path P containing n edges (numbered 1 through n) is given by f_{1n}. Variable k denotes the edge after which the first cut in the optimal solution is assumed to be placed. For each possible first cut (i.e. for each value of k), we check the sum of *sgain* values of S_{ik} (solution found by WBS for segment towards the left of the cut) and $f_{(k+1)j}$. $f_{(k+1)j}$ is the optimal break-up computed by this algorithm for the sub-problem having edges numbered $(k + 1)$ through j. The option for having no cut at all is considered when $k = j$. The base conditions for this recurrence equation are: $f_{ii} = S_{ii}.sgain$ (subproblem of size 1) and $f_{(i+1)i} = 0$. Algorithm 3 presents a bottom up procedure for computing this recurrence equation.

Algorithmic Details of MS(WBS). This algorithm has three primary steps. In the first step, *sgain* values are computed for all possible segments of the initial seed path using the WBS algorithm. After this, in the second step, we compute the optimal DSS of the initial seed path using Algorithm 3. Next, WBS is invoked on all pieces in the optimal DSS in decreasing order of their *sgain* values, and the path is updated after each invocation of WBS. The reason for invoking WBS again is to ensure that the path remains free from loops as it is updated. The segments are extracted in decreasing order of their *sgain* values to maximize the improvement in the score as replacing all segments within the budget may not be feasible.

Algorithm 3. DP algorithm for computing optimal DSS

Input: *sgain* values for all segments of seed path, **Output:** Optimal DSS

1: **for** $spsize = 2$ to n **do** ▷ Subproblem size
2:    **for** $i = 2$ to $n - spsize + 1$ **do** ▷ First edge of segment
3:       $j \leftarrow i + spsize - 1$ ▷ Last edge of segment
4:       $f_{ij} \leftarrow 0$ ▷ Initializing optimal solution for S_{ij}
5:       **for** $k = i$ to n **do** ▷ Edge after which first cut is placed
6:          **if** $S_{ik}.sgain + f_{(k+1)j} > f_{ij}$ **then**
7:             $f_{ij} \leftarrow S_{ik}.sgain + f_{(k+1)j}$
8:          **end if**
9:       **end for**
10:    **end for**
11: **end for**

Time Complexity Analysis. For the time complexity analysis of MS(WBS), we do not consider the complexity of initial seed path computation. Given that we are using a shortest path from s to d as our seed path, this cost is upper bounded by the cost of Dijkstra's algorithm $O(|E| + |V| \log |V|)$.

The first step of MS(WBS) estimates the *sgain* values for all segments of the seed path using WBS. For a path with l edges, this takes $O(l^2|E|)$ time since the total number of segments in the path is $\theta(l^2)$. A DSS is then selected in the second step using our DP solution, in $\theta(l^3)$ time. Step three involves updating the path by calling WBS on all segments in the selected DSS. Since the maximum number of segments in a DSS can be l, this step requires $O(l|E|)$ time. This gives a time complexity of $O(l^2|E|)$ for MS(WBS).

3.3   VAMS(WBS): Vicinity Aware Multiple Segment Replacement Algorithm with WBS

Recall that MS(WBS) computes the estimates of score gain values for all the possible segments of the initial seed path. In other words, given an initial seed path with l edges, MS(WBS) calls WBS algorithm $\theta(l^2)$ times to get the score gain of each of the possible segments. Following which, it determines the optimal set of segments (DSS) for replacement. Invoking $\theta(l^2)$ instances of WBS may not be computationally scalable. To this end, this section proposes a novel metric called *vicinity potential* (VP) of a segment. The VP values serve as a proxy to the score gain values of the $\theta(l^2)$ segments in the initial seed path. The algorithm (VAMS(WBS)) proposed in this section uses VP values instead of the score gain values. We now provide details on computing the VP value of a segment.

Given a segment S_{ij} and a budget value B', the vicinity of S_{ij} is defined as the area bounded by the ellipse with focal points as $S_{ij}.start$ and $S_{ij}.end$. The length of its major axis is B'. The properties of an ellipse allow us to claim the following: any path between $S_{ij}.start$ and $S_{ij}.end$ of length $\leq B'$ would not include any edge lying completely outside or intersecting this ellipse.

Vicinity Potential (VP) of a Segment. The VP value of a segment S_{ij} is defined as the average of navigability scores of all the navigable edges lying in the *vicinity* of S_{ij}. Equation 4 presents this formally[3]. Recall that only the edges having a navigability score value > 0 are referred to as navigable edges (denoted in Eq. 4 as $'ne'$).

$$VP(S_{ij}, B') = \frac{\sum_{ne \in ellipse(S_{ij}.start, S_{ij}.end, B')} ne.score}{Count(ne \in ellipse(S_{ij}.start, S_{ij}.end, B'))} \qquad (4)$$

This definition is based on the intuition that more the number of highly navigable edges in a segment's vicinity, the higher would be the probability of finding a segment to replace it.

Navigability Index for Computing the VP Values. To compute the VP value of a segment, one needs to obtain the set of edges that are contained in its vicinity. This computation can be made efficient using our *navigability index*.

Navigability Index is similar to a regular spatial grid. Each cell in this index stores two numeric values: *sum* and *count*. The sum value of cell (x, y) is set to the sum of scores of all navigable edges contained in the rectangle bounded between cells $(0, 0)$ and (x, y). The count value of cell (x, y) is set to the number

---

[3] If edge costs represent travel-times, then the travel-time based budget can be converted to a distance based budget using the upper speed limit of a road segment.

of navigable edges contained in this rectangle. Given these, the sum and count values of any rectangle in the grid, bounded between cells (i, j) and (m, n), can be computed using Eqs. 5 and 6 (refer Fig. 2). Here, $sum(x, y)$ and $count(x, y)$ respectively denote the sum and count values of cell (x, y).

$$Sum = sum(m, n) - sum(i - 1, n) - sum(m, j - 1) + sum(i - 1, j - 1) \quad (5)$$

$$Count = count(m, n) - count(i - 1, n) - count(m, j - 1) + count(i - 1, j - 1) \quad (6)$$

The proposed index structure computes the VP value for any segment in $O(1)$ index lookups, irrespective of the order of the grid index used. This is done as follows: to compute the VP value of S_{ij}, we take the spatial coordinates of $S_{ij}.start$ and $S_{ij}.end$ as the foci of the ellipse with the length of major axis= B'. Next, we compute the grid aligned minimum bounding rectangle (MBR) of this ellipse. The VP value of S_{ij} can then be computed by plugging the bottom-left (i, j) and upper-right (m, n) coordinates of this MBR in Eqs. 5 and 6. This makes the computation much faster. The idea of this index structure was inspired by the work of Aly et al. [1].

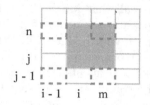

Fig. 2. Navigability index illustration

Algorithmic Details of VAMS(WBS). VAMS(WBS) is similar to MS(WBS) with exceptions in the first and the second steps. Given the initial seed path, the first step of VAMS(WBS) involves computing the VP values of all segments of the seed path. In the second step, the optimal DSS is computed based on the VP values of segments. Next, the VAMS(WBS) invokes the WBS algorithm to determine the actual replacements for the segments in the optimal DSS.

Time Complexity Analysis. The time complexity of the first step of VAMS(WBS) is $\theta(l^2)$, since VP values are computed for $\theta(l^2)$ segments, and each such computation takes $O(1)$ time. The complexity of the remaining steps is the same as the steps of MS(WBS). Thus, the time complexity of VAMS(WBS) is $O(l|E| + l^3)$.

4   Experimental Evaluation

Performance of the proposed algorithms was evaluated through experiments on three real-road networks of different sizes (refer Table 2). Datasets 1 and 3 were obtained from OpenStreetMap (http://www.openstreetmap.org). Dataset 2 was obtained from Digital Chart of the World Server (https://www.cs.utah.edu/~lifeifei/SpatialDataset.htm). All datasets constituted directed spatial graphs with vertices as road intersections and edges as road segments. Cost of each edge corresponded to the metric of distance. Navigability score values of edges were generated synthetically. We assumed that 40% of the total edges in each dataset have no prominently visible site, and assigned them a score value of zero. The remaining 60% of the edges were assumed to be navigable. These edges were assigned a non-zero score value. The navigable edges were distributed uniformly across space, and were assigned a random integral score value in the range 1–15.

Table 2. Description of datasets

Dataset	Place	Vertices	Edges
1	Delhi	11,399	24,943
2	California	21,048	39,976
3	Beijing	55,545	95,285

Baseline Algorithm. Given the score gain values of all segments of the initial seed path, a naive algorithm would be to replace the segment with the highest *sgain* value. We call this algorithm the Single Segment replacement algorithm with WBS (SS(WBS)), and use it as the baseline algorithm.

Experimental Setup. All algorithms were implemented in Java language on a machine with a 2.6 GHz processor and a 32 GB RAM. The shortest paths were computed using the A* algorithm. For the VAMS(WBS) algorithm, the order of the grid index built for datasets 1, 2 and 3 was 300×300, 500×500 and 1000×1000 respectively. The idea was to create $1\,km \times 1\,km$ grid cells in each navigability index. To study the effects of change in the available overhead, we set the overhead as a fraction of the length of the shortest path. An overhead of x% implies that the total length allowed for the resultant path is: shortest path length + x% of the shortest path length. The statistics reported for an overhead of 0% represent the values for the shortest path. We report the average statistics for 100 random query instances for all three datasets.

4.1 Comparative Analysis of Proposed Algorithms

Effect of Increase in Overhead on the Score of a Path. Figure 3 illustrates the results of this experiment for queries where shortest path length was 40 kms. MS(WBS) gives paths with higher navigability score values than SS(WBS) and VAMS(WBS). This is because MS(WBS) actually computes the score gain values of all possible segments (unlike VAMS(WBS)). In addition, it computes the optimal DSS (unlike SS(WBS)) using these score gain values.

Effect of Increase in Overhead on the Running Time. Figure 4 shows the results of this experiment for queries where shortest path length was 30 kms. The running time for overhead of 0% marks the time taken to compute the shortest path. In general, VAMS(WBS) takes less time because it avoids the expensive repetitive invocation of WBS algorithm to compute the score gain value of each segment in the path. As the overhead is increased from 10% to 40% the increase in running time of all three algorithms is steady. The comparative performance of the algorithms is in accordance with their time complexities.

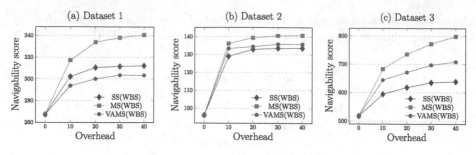

Fig. 3. Effect of increase in overhead on score of a path

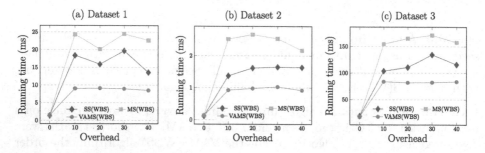

Fig. 4. Effect of increase in overhead on running time

Effect of Increase in Path Length on the Running Time. Figure 5 shows the results of this experiment. The results are shown for an overhead of 30%. We observe that VAMS(WBS) performs the best, followed by SS(WBS) and MS(WBS), in that order. Also, the rate of increase in running time for VAMS(WBS) is steady as compared to the other two algorithms.

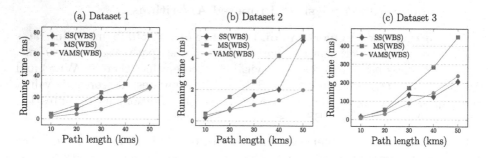

Fig. 5. Effect of increase in path length on running time

4.2   Comparative Analysis of Proposed Algorithms and ILS(CEI)

Score Value and Running Time as a Function of Overhead. Figure 6(a) shows the results of this experiment on Dataset 2 for queries with a shortest path length of 40 kms. For this experiment, we implemented an adaptation of ILS(CEI) [11] that yields a simple path. Further, the original ILS(CEI) repeats its iterations till the running time is less than some threshold value. We implemented a version which executes a single iteration of ILS(CEI) and terminates, similar to the working of our algorithms.

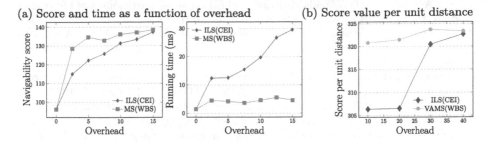

(a) Score and time as a function of overhead   (b) Score value per unit distance

Fig. 6. Comparative analysis of proposed algorithms and ILS(CEI)

Our results show a superior performance of MS(WBS) over ILS(CEI) in terms of both solution quality and running time. However, the difference between the solution quality keeps decreasing with increase in overhead. For overhead values higher than 15%, the solution quality of ILS(CEI) exceeds that of MS(WBS). In contrast to this, the difference in running times of both algorithms becomes increasingly significant as the overhead increases, making ILS(CEI) impractical for higher values of overhead. The change in running time of MS(WBS) is steady. Our results for other datasets followed a similar trend. Note that, in real-life, a traveler would generally prefer lower values of overhead, implying that MS(WBS) is a superior algorithm for the MNP problem.

Score Value per Unit Distance as a Function of Overhead. In this experiment, we compared the score value per unit distance for the solutions given by VAMS(WBS) and ILS(CEI). Figure 6(b) shows the results of this experiment on Dataset 1 for queries with a shortest path length of 30 kms. Overhead values were varied in this experiment. VAMS(WBS) has higher score value per unit distance. The gap between the algorithms reduces as the overhead is increased.

5   Conclusions and Future Work

In this paper, we introduced the Most Navigable Path (MNP) problem and proposed two novel algorithms for MNP: MS(WBS) and VAMS(WBS). We also proposed an indexing structure for MNP which can be used to estimate (in constant time) the gain in navigability score of a path on replacement of some segment of the path. We demonstrated, through experimental analysis, the superior performance of our algorithms over the state-of-the-art heuristic algorithm for AOP. In future, we plan to design the iterative versions of our algorithms which shall improve the solution quality until some threshold on running time.

Acknowledgements. This work was in part supported by the Infosys Centre for Artificial Intelligence at IIIT-Delhi, Visvesvaraya Ph.D. Scheme for Electronics and IT, and DST SERB (ECR/2016/001053).

References

1. Aly, A.M., et al.: AQWA: adaptive query workload aware partitioning of big spatial data. Proc. VLDB Endow. **8**(13), 2062–2073 (2015)
2. Archetti, C., Corberán, A., Plana, I., Sanchis, J.M., Speranza, M.G.: A branch-and-cut algorithm for the orienteering arc routing problem. Comput. Oper. Res. **66**(C), 95–104 (2016)
3. Bolzoni, P., Helmer, S.: Hybrid best-first greedy search for orienteering with category constraints. In: Gertz, M., et al. (eds.) SSTD 2017. LNCS, vol. 10411, pp. 24–42. Springer, Cham (2017). https://doi.org/10.1007/978-3-319-64367-0_2
4. Bolzoni, P., Persia, F., Helmer, S.: Itinerary planning with category constraints using a probabilistic approach. In: Benslimane, D., Damiani, E., Grosky, W.I., Hameurlain, A., Sheth, A., Wagner, R.R. (eds.) DEXA 2017. LNCS, vol. 10439, pp. 363–377. Springer, Cham (2017). https://doi.org/10.1007/978-3-319-64471-4_29
5. Chekuri, C., Pal, M.: A recursive greedy algorithm for walks in directed graphs. In: Proceedings of the 46th Annual IEEE Symposium on Foundations of Computer Science, FOCS 2005, pp. 245–253 (2005)
6. Delling, D., Goldberg, A.V., Nowatzyk, A., Werneck, R.F.: Phast: hardware-accelerated shortest path trees. J. Parallel Distrib. Comput. **73**(7), 940–952 (2013)
7. Gavalas, D., Konstantopoulos, C., Mastakas, K., Pantziou, G., Vathis, N.: Approximation algorithms for the arc orienteering problem. Inf. Process. Lett. **115**(2), 313–315 (2015)
8. Jing, N., Huang, Y.W., Rundensteiner, E.A.: Hierarchical encoded path views for path query processing: an optimal model and its performance evaluation. IEEE Trans. Knowl. Data Eng. **10**, 409–432 (1998)
9. Kanoulas, E., Du, Y., Xia, T., Zhang, D.: Finding fastest paths on a road network with speed patterns. In: 22nd International Conference on Data Engineering (ICDE 2006), p. 10, April 2006
10. Kriegel, H.P., Renz, M., Schubert, M.: Route skyline queries: a multi-preference path planning approach. In: 2010 IEEE 26th International Conference on Data Engineering (ICDE 2010), pp. 261–272 (2010)
11. Lu, Y., Shahabi, C.: An arc orienteering algorithm to find the most scenic path on a large-scale road network. In: Proceedings of the 23rd SIGSPATIAL International Conference on Advances in Geographic Information Systems, pp. 46:1–46:10 (2015)
12. Martins, E., Pascoal, M.: A new implementation of Yen's ranking loopless paths algorithm. Q. J. Belg. French Ital. Oper. Res. Soc. **1**(2), 121–133 (2003)
13. Singh, A., Krause, A., Guestrin, C., Kaiser, W., Batalin, M.: Efficient planning of informative paths for multiple robots. In: Proceedings of the 20th International Joint Conference on Artificial Intelligence, IJCAI 2007, pp. 2204–2211 (2007)
14. Souffriau, W., Vansteenwegen, P., Berghe, G.V., Oudheusden, D.V.: The planning of cycle trips in the province of east flanders. Omega **39**(2), 209–213 (2011)
15. Vansteenwegen, P., Souffriau, W., Oudheusden, D.V.: The orienteering problem: a survey. Eur. J. Oper. Res. **209**(1), 1–10 (2011)
16. Verbeeck, C., Vansteenwegen, P., Aghezzaf, E.H.: An extension of the arc orienteering problem and its application to cycle trip planning. Transp. Res. Part E: Logist. Transp. Rev. **68**, 64–78 (2014)

Load Balancing in Network Voronoi Diagrams Under Overload Penalties

Ankita Mehta[1], Kapish Malik[1], Venkata M. V. Gunturi[2(✉)], Anurag Goel[1], Pooja Sethia[1], and Aditi Aggarwal[1]

[1] IIIT-Delhi, New Delhi, India
[2] IIT Ropar, Rupnagar, India
gunturi@iitrpr.ac.in

Abstract. Input to the problem of Load Balanced Network Voronoi Diagram (LBNVD) consists of the following: (a) a road network represented as a directed graph; (b) locations of service centers (e.g., schools in a city) as vertices in the graph and; (c) locations of demand (e.g., school children) also as vertices in the graph. In addition, each service center is also associated with a notion of *capacity* and an *overload penalty* which is "charged" if the service center gets overloaded. Given the input, the goal of the LBNVD problem is to determine an *assignment* where each of the demand vertices is allotted to a service center. The objective here is to generate an assignment which minimizes the sum of the following two terms: (i) total distance between demand vertices and their allotted service centers and, (ii) total penalties incurred while overloading the service centers. The problem of LBNVD finds its application in the domain of urban planning. Research literature relevant to this problem either assume infinite capacity or do not consider the concept of "overload penalty." These assumptions are relaxed in our LBNVD problem. We develop a novel algorithm for the LBNVD problem and provide a theoretical upper bound on its worst-case performance (in terms of solution quality). We also present the time complexity of our algorithm and compare against the related work experimentally using real datasets.

1 Introduction

The problem of Load Balanced Network Voronoi Diagram (LBNVD) takes the following three inputs: (1) a road network represented as a directed graph $G = (V, E)$; (2) a set of nodes V_d ($V_d \subset V$) designated as the *demand nodes* and; (3) a set of nodes V_s ($V_s \subset V$) designated as the *service centers*. With each service center $s_i \in V_s$, we have information on its allowed capacity and overload penalty.

Given the input, LBNVD outputs an assignment $\mathcal{R}$ from the set of demand vertices V_d to the set of the service centers V_s. In other words, for each demand vertex $v_{d_i} \in V_d$, LBNVD determines a "suitable" service center $s_j \in V_s$.

The objective of the LBNVD problem is to determine an assignment $\mathcal{R}$ which minimizes the sum of the following two quantities: (i) sum of distances from the

© Springer Nature Switzerland AG 2018
S. Hartmann et al. (Eds.): DEXA 2018, LNCS 11029, pp. 457–475, 2018.
https://doi.org/10.1007/978-3-319-98809-2_28

demand vertices to their allotted service centers and, (ii) total penalty incurred across all the overloaded service centers.

LBNVD finds its applications in the area of urban planning. Examples include the problem of defining the zones of operation for schools in a city (school catchment area [1]). Here, the school going children (the demand) are to be divided across a set of schools (service centers). For such a situation, it is important to consider both the capacity of a school (e.g., in terms of #teachers) and a notion of penalty as overloading the school too much (e.g., by not maintaining appropriate student:faculty ratio) may decrease the quality of education.

Limitations of Related work: The current state of the art most relevant to our work includes the work done in the area of network voronoi diagrams without capacities [2,3], network voronoi diagrams under capacity constraints [4–6], weighted voronoi diagrams [7] and optimal location queries (e.g., [8–10]).

Work done in the area of network voronoi diagrams without capacities [2,3] assume that the service centers have infinite capacity, an assumption not suitable in many real-world scenarios. On the other hand, work done in the area of network voronoi diagrams with capacities [4–6] did not consider the notion of "overload penalty." They perform allotments (of demand nodes) in an iterative fashion as long as there exists a service center with available capacity. In other words, the allotments stop when all the service centers are full (in terms of their capacity). The problem of LBNVD is different in the sense that it allows the allotments to go beyond the capacities of the service centers. After a service center is full, LBNVD uses the concept of the *overload penalties* for guiding the further allotments and load sharing.

Weighted voronoi diagrams [7] are specialized voronoi diagrams. In these diagrams, the cost of allotting a demand vertex x to a service center p is a linear function of the following two terms: (i) distance between x and p and, (ii) a *real number* denoting the weight of p as $w(p)$. LBNVD problem is different from weighted voronoi diagrams. Unlike the weighted voronoi diagrams, our "$w(p)$" is a *function* of the number of allotments already made to the service center p. And it would return a non-zero value only when the allotments cross beyond the capacity. Whereas in [7], $w(p)$ is assumed to play its role throughout.

Optimal location queries (e.g., [8, 10]) focus on determining a suitable location to start a new facility while optimizing a certain objective function (e.g., total distance between clients and facilities). Whereas, in LBNVD, we already have a set of facilities which are up and running, and we want to load balance the demand around them.

Our Contributions: This paper makes the following 5 contributions:

(1) Define the problem of a Load Balanced Network Voronoi Diagram (LBNVD).
(2) Propose a novel *Local Re-Adjustment based approach* (LoRAL algorithm) for the LBNVD problem. The proposed approach adopts a *very prudent* approach towards optimization. Before allotting any demand vertex to a service center, it checks if there exists a bounded number of local re-adjustments which can be made to the current partially built network voronoi diagram

to reduce its objective function value. These bounded number of local re-adjustments are decided by exploring a bigger (but bounded) superset of potential local re-adjustments. Such an approach helps in getting an overall lower objective function values without increasing the execution time too much.

(3) Provide a theoretical bound on the worst-case performance (in terms of final objective function value) of the proposed LoRAL algorithm.

(4) Present an asymptotic time complexity analysis of the proposed approach.

(5) Experimental evaluation (on a real dataset of New Delhi road network) which includes a comparison against the related work.

Outline: The rest of this paper is organized as follows: In Sect. 2, we provide the basic concepts and the problem statement. Section 3 presents our proposed approach. Sections 4 and 5 provide a detailed theoretical analysis and the time complexity analysis of our proposed approach. Section 6 provides an experimental evaluation of our proposed approach and the related work on real datasets.

2   Basic Concepts and Problem Statement

Definition 1. *A **Road Network** is represented as a weighted directed graph $G(V,E)$, where V is the set of vertices and E is the set of edges. Each vertex represents a road intersection. A road segment between two intersections is represented as a directed edge. Each edge is associated with a cost $w(u,v)$ which represents the cost to reach vertex **v** from vertex **u**.*

Definition 2. *A **Service Center** is a vertex in the road network representing a public service unit of a particular kind (e.g., schools, police stations and hospitals). A set of service centers is represented as $V_s = \{s_1, ..., s_{|V_s|}\}$, where $V_s \subset V$ and $|V_s|$ is the number of service centers.*

Definition 3. *A **Demand Vertex** is a vertex in the road network representing the location of a unit population which is interested in accessing the previously defined service center. A set of demand vertices is represented as $V_d = \{v_{d_1}, ..., v_{d_{|V_d|}}\}$, where $V_d \subset V$ and $|V_d|$ is the number of demand vertices. Whenever the context is clear we drop the subscript d_i from $v_{d_i}s$' to maintain the clarity of text.*

Definition 4. *Capacity of a service center s_i (c_{s_i}) is the prescribed number of unit demand that a service center s_i can accommodate.*

Definition 5. *Penalty function of a service center s_i ($q_{s_i}()$) is a function which returns the extra cost that must be paid for a new allotment to the service center s_i which has already exhausted its capacity c_{s_i}. This extra-cost is added to the objective function as a "penalty" for undertaking this assignment. If the allotment is done within the capacity of a service center, then no penalty needs to be paid. Examples of penalty cost in real world include cost to add additional infrastructure and/or faculty in a school.*

The penalty function of s_i $(q_{s_i}())$ takes a parameter j $(1 \leq j \leq |V_d|)$ as an input. Here, the value j implies that the service center s_i has already been assigned $j - 1$ demand vertices, and now we are attempting the jth allotment. $q_{s_i}()$ returns 0 when $j \leq c_{s_i}$. Otherwise, it returns a positive value denoting the extra-cost which must be added to the objective function (as a "penalty") for undertaking this assignment. $q_{s_i}()$ returns only positive values and is monotonically increasing over j $(1 \leq j \leq (|V_d| - c_{s_i}))$. The intuition behind monotonically increasing penalty being: one may need to add increasingly more resources to a school (or a hospital) as the overload keeps increasing.

2.1   Problem Statement

Input:

- A road network G(V,E), where each edge $e \in E$ has a positive cost.
- A set of service centers $V_s = \{s_1, ..., s_{|V_s|}\}$ where $V_s \subset V$.
- A set of demand vertices $V_d = \{v_{d_1}, ..., v_{d_{|V_d|}}\}$ where $V_d \subset V$.
- A set of positive integer service center capacities $C = \{c_{s_1}, ..., c_{s_{|V_s|}}\}$.
- A set of penalty functions for the service centers $Q = \{q_{s_1}, ..., q_{s_{|V_s|}}\}$.

Output: An assignment $\mathcal{R} : V_d \rightarrow V_s$. Each demand vertex is allotted to only one service center.

Objective Function

$$Min \left\{ \sum_{\substack{\forall \ Service \\ Centers \ s_i}} \left\{ \sum_{\substack{\forall \ demand \ vertices \ v_{d_j} \\ allotted \ to \ s_i}} Dist(v_{d_j}, s_i) \right\} + Total \ Penalty \ across \ all \ s_i \right\}$$

(1)

Here, $Dist(v_{d_j}, s_i)$ denotes the shortest distance between the demand node v_{d_j} and the service center s_i. We assume that for every demand vertex in V_d, there exists a path to reach at least one service center in V_s. Also, we consider the case where the total capacity of all service centers is less than the total demand. Apart from this, there can be two other cases. First, the total capacity is equal to the total demand. And second, the total capacity is greater than the total demand. In both these cases, some service centers can still get overloaded as a demand vertex can choose to forcibly go (as it lead to lower objective function value) to a nearby service center (which is already full) and pay the penalty instead of going to a far-off "free" service center.

2.2   Using LBNVD in Real Life

Consider again the problem of defining the zone of operation for each school (school catchment area [1]). For this problem, the unit of measurement for the objective function (Eq. 1) could be the total cost of "operation" for one day in Rupees (or Dollars). For this objective function, the edge costs in our road network should be the fare amount spent while traveling that edge. And if the number of teachers in a school dictate its *capacity*, then penalty could be defined using metrics like faculty - student ratio. For ease of interpretation one can assume this penalty to be a constant number for each new allotment beyond the capacity. Note that this would still adhere (in a strictly mathematical sense) to the definition of penalty given in Definition 5. In this setting, if a faculty-student ratio of 1:F is maintained in a school, then its overload penalty would be Faculty-annual-salary/(F × 365) for each allotment over the capacity of the school. For Indian scenarios, this fraction comes out in the range of 40–60 for government schools. Note that one can easily create more complex notions of capacities and penalty functions by incorporating other parameters like library-books-student ratio, lab-equipments-student ratio, etc. It should be noted that proposed algorithm is oblivious of these implementation intricacies of the penalty functions as long as they adhere to Definition 5.

3 Proposed Approach - LoRAL Algorithm

3.1 Key Idea: Cascade of Local Re-adjustment

Cascade of local re-adjustments to the partially constructed voronoi diagram is the central idea of our proposed approach. We use this key technique in our algorithm to bring down the objective function value. We first describe this idea at a high level in this section and then detail its operationalization in Sect. 3.2. We now describe the idea of local re-adjustment through an illustrative example.

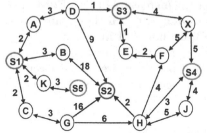

Service Center	Capacity
S1	4
S2	1
S3	3
S4	2
S5	1

Fig. 1. A sample transportation network.

Given a LBNVD problem instance, the first step is to determine the closest service center for each demand vertex $v_{d_i} \in V_d$. For this step, the algorithm internally computes the shortest distance between all pairs of demand vertices and service centers. Following this, we sort the $(v_{d_i}, closest\ s_i)$ pairs in the increasing order of the shortest distance to the closest s_i. After this, the algorithm would process the pairs $(v_{d_i}, closest\ s_i)$ in the increasing order of the shortest distance. Consider a stage when v_{d_i} is being processed. The algorithm would **try** to allot v_{d_i} to its corresponding closest s_i. At this stage, one of the following two cases can happen. **Case 1:** s_i is not yet full (in terms of its capacity), in which case, the allotment goes through. **Case 2:** s_i is already full and we need to "pay a penalty" for this assignment.

For addressing case 2, we propose the concept of *Cascade of Local Re-Adjustments* of the current partially built network voronoi diagram. It is important to note that once a demand vertex v_{d_i} is allotted to a service center s_i it need not be removed from s_i, unless it is required during the

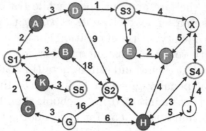

Service Center	Capacity
S1	4
S2	1
S3	3
S4	2
S5	1

Fig. 2. Partially constructed LBNVD for network shown in Fig. 1 (best it color).

local re-adjustment. And if removed from s_i, another demand node (since the algorithm deemed it as a better option) would be associated with s_i. This implies that as the algorithm progresses, the total penalty paid on any service center either increases or remains the same (i.e., monotonically increases).

Concept of Single Local Re-adjustment of Partial Voronoi Diagram: Figure 1 shows a sample road network with 5 service centers $(S1, S2, \ldots, S5)$ and 11 demand vertices. Figure 2 illustrates a partially constructed network voronoi diagram on the road network shown in Fig. 1. In the figure, the demand vertices which are allotted to a service center are filled using the same color as that of their allotted service center. Vertices which are not yet allotted are shown without any filling. Furthermore, nodes where a service center is located are assumed to have zero demand. In the problem instance shown, the first few <demand vertex- closest service center> pairs (in increasing order of distance to the closest service center) have already been processed. All these pairs were processed through case 1, and $S1$, $S2$ and $S3$ are now full in capacity. Now, consider the case when the pair $<X, S3>$ is being processed. Ideally the demand vertex X should be assigned to service center $S3$ as per the nearest service center criteria. However, since $S3$ is full we need to consider one of the following 9 options for *Local Re-Adjustment*.

1. Allot X to $S3$ and pay the penalty of overloading $S3$;
   Total increase in objective function: $4 + q_{S3}()$ (penalty paid as 1st insertion beyond the capacity of $S3$).
2. Allot X to $S3$, but push the vertex D to another service center.
   Total increase in objective function to associate X to $S3$ and then transfer D to:
   - $S1$: $4 + (5 - 1) + q_{S1}()$ (penalty paid as 1st insertion beyond the capacity of $S1$)
   - $S2$: $4 + (9 - 1) + q_{S2}()$ (penalty paid as 1st insertion beyond capacity in $S2$)
   - $S4$: $4 + (10 - 1)$ (No penalty since under-loaded)
   - $S5$: $4 + (10 - 1)$ (No penalty since under-loaded)

3. Allot X to $S3$ but push the vertex F to another service center.
   Total increase in objective function to associate X to $S3$ and then transfer F
   to:

   – $S1$: $4 + (42 - 3) + q_{S1}()$ (penalty paid as 1st insertion beyond capacity in $S1$)
   – $S2$: $4 + (21 - 3) + q_{S2}()$ (penalty paid as 1st insertion beyond capacity in $S2$)
   – $S4$: $4 + (10 - 3)$ (No penalty since under-loaded)
   – $S5$: $4 + (47 - 3)$ (No penalty since under-loaded)

The algorithm would evaluate the cost of each of these options (in terms of increase in objective function value) and then choose the minimum. Note that while considering options for allotted demand nodes to be pushed out at $S3$, we only consider what we refer to as the *boundary vertices* of $S3$. In Sect. 4, we prove this formally in Lemma 1. Following is a formal definition of the *boundary vertices* of a service center s_i.

Definition 6. *Set of Boundary Vertices of a service center* s_i (B_{s_i}): $B_{s_i} = \{b_{s_i}^1, b_{s_i}^2, \ldots, b_{s_i}^m\}$ *for a service center* s_i *is a set of demand vertices allotted to the service center* s_i *such that each vertex in* B_{s_i} *has at-least one of the following three properties: (a) an outgoing edge to a vertex allotted to a different service center* s_j, *(b) an outgoing edge to a different service center* s_j *or, (c) an outgoing edge to an unprocessed demand vertex. In Fig. 1, D and F were currently the only boundary vertices of S3. Whenever the context is clear we drop the superscript m from* $b_{s_i}^m s$ *in our text.*

Cascade of Local Re-adjustments: Continuing our previous example, in options (2) and (3), $S1$, $S2$, $S4$ and, $S5$ would in-turn also consider local re-adjustment after receiving a boundary vertex from $S3$. This would happen only if they themselves are overloaded. This process continues leading to what we refer to as *cascade of local re-adjustments*. One such case could arise if X was allocated to $S3$ and D was pushed to service center $S1$, thereby overloading it by 1. Now, $S1$ can in-turn choose between the following two decisions:

Option 1. Accept D and overload itself.
Option 2 Accept D and in-turn push one of its boundary vertices to another service center (for e.g., K to $S5$).

More details on this are covered in the upcoming sections. It is important to note that in our actual operationalization of the previously described idea, the proposed algorithm first *simulates a set of k cascades* and then *chooses* the cascade which results in lowest increase of objective function.

3.2   Best of K Cascades LoRAL Algorithm

This section details the operationalization of our previously discussed idea of local re-adjustments. In our proposed algorithm we use the following heap.

MinDistance Heap: a binary min-heap data structure which is ordered on the distance between a demand vertex and its closest service center.

Our algorithm starts by first computing the shortest distance between all pairs of demand vertices and service centers. This paper uses shortest path algorithm as a black box. As a proof of concept in our implementation, we used the Floyd-Warshall [11] algorithm for computing the shortest distances. One can easily replace Floyd-Warshall algorithm with other shortest path algorithms (e.g., [12–14]).

Following the shortest distance computation, we insert the pairs <Demand vertex, closest s_i> in the *MinDistance heap*. Demand vertex to be processed next is picked from the MinDistance heap using the extract-min operation. Initially all vertices are unprocessed and a *while* loop is executed till the *MinDistance heap* is not empty. This ensures that all demand vertices are processed.

Let the result of extract-min operation be a demand vertex v_{d_i} and its closest service center s_i. Currently v_{d_i} is unprocessed, and as per the closest distance metric it should go to the service center s_i. However, before allotting a demand vertex to a service center, the algorithm checks if s_i can accommodate the demand or not. In case the service center s_i has the required capacity, then v_{d_i} is allotted to s_i. Following this, the objective function is incremented by the shortest distance between s_i and v_{d_i}.

However, had s_i been full or overloaded, the algorithm would have reached the scenario of *local re-adjustment*. The concept of local re-adjustment is operationalized in the following way. We first simulate the total increase in objective function in k *cascades of local re-adjustments*. Then, the algorithm chooses the cascade which has the lowest increase in the objective function. The algorithm then implements the local re-adjustments of this chosen cascade in the partially constructed voronoi diagram **provided it is cheaper** (i.e., lower increase in objective function value) than just accepting additional overload at s_i. Following are details on simulating the k cascades.

Structure of One Simulated Cascade: Consider again the previous scenario of v_{d_i} and its closest service center s_i. Each of the simulated cascades start at s_i and progresses in the following way: the service center s_i accepts its current assignment (i.e., the demand node v_{d_i}) and pushes out one of its boundary vertices to another service center, say s_j. Now, s_j evaluates the following two options: (a) just accept the s_i's boundary vertex or, (b) accept and then push out one of its own boundary vertices to another service center, say s_k. s_j would proceed with the option which leads to a lower increase in objective function value. The same process repeats at s_k. This cascade of local re-adjustments keeps rolling until one of the following three things happen:

(1) At any stage, s_j to which a boundary vertex is transferred is under-full.
(2) Total #re-adjustments in this simulated cascade becomes greater than a certain threshold.
(3) At any stage, s_j to which a boundary vertex is transferred assess that it is cheaper (i.e., lower increase in objective function value) to simply just accept the vertex being forced upon rather than accepting and in-turn rejecting one

of its own boundary vertices. In other words, s_j chooses option (a) described in the previous paragraph. We now provide details on the simulating k cascades.

Simulating k Cascades: Consider again the previous scenario of v_{d_i} and its closest service center s_i. B_{s_i} is the set of boundary vertices of s_i.

Firstly, the algorithm calculates the sum of the following four terms for every pair of boundary vertex $b_{s_i} \in B_{s_i}$ and service center $s_j \in \{V_s - s_i\}$: (1) shortest distance between v_d and s_i, (2) shortest distance between b_{s_i} and service center s_j, (3) penalty value of the service center s_j (if any) and, (4) negative of distance between boundary vertex b_{s_i} and service center s_i. Note that at this stage we don't need to worry about the penalty that was paid by b_{s_i} to s_i as instead of b_{s_i} we would be adding another demand vertex (v_{d_i}) to s_i. In other words, the total number of demand vertices allotted to any service center will only increase (or remain the same) as the algorithm progresses. Following this, we choose the *top* k pairs of b_{s_i} and their respective s_js' which had the lowest sum. These k pairs

---

Algorithm 1. Best of **k** Cascades LoRAL Algorithm

Input: (a) Number of Cascades k, (b) Allowed length of cascade l (must be $\leq |V_s| - 1$)
Output: LBNVD with Objective Function value Δ

1: Compute the shortest path distance between every demand vertex in V_d and every service center in V_s.
2: For each $v_{d_i} \in V_d$ determine the closest service center and create a MinDistance heap with all <demand vertex-closest service center> pairs.
3: **while** MinDistance heap is not empty **do**
4:     $< v_{d_i}, s_i > \leftarrow$ extract-min on MinDistance heap.
5:     **if** s_i has vacancy **then**
6:         Allocate v_{d_i} to s_i, decrement capacity of s_i, increment Δ by $dist(v_{d_i}, s_i)$
7:     **else**
8:         Set $\Omega \leftarrow$ top k boundary vertices (b_{s_i}s') and their corresponding service centers (s_js') for re-adjustment.
9:         **for** Each boundary vertex b_{s_i} in Ω **do**
10:             Simulate a cascade starting with s_i accepting v_{d_i} and pushing out b_{s_i} to s_j
11:             $Cascade^i \leftarrow$ resulting cascade
12:             $\beta^i \leftarrow$ Increase in objective function value after $Cascade^i$
13:         **end for**
14:         Set cost of overload: $\delta_1 \leftarrow dist(v_{d_i}, s_i) + q_{s_i}()$
15:         Set cost of transfer: $\delta_2 \leftarrow Min_{\forall i \in [1,k]}\{\beta^i\}$
16:         **if** $\delta_1 < \delta_2$ **then**
17:             Allot v_{d_i} to s_i with penalty, and $\Delta = \Delta + \delta_1$.
18:         **else**
19:             Allot v_{d_i} to s_i. Implement the cascade which had the lowest β value
20:             $\Delta = \Delta + \delta_2$.
21:         **end if**
22:     **end if**
23: **end while**

seed the k cascades of the local re-adjustments whose cost would be simulated by the algorithm.

After this, the algorithm takes forward each of the k cascades individually and unravels them in a manner similar to that explained previously. Note that in each of these cascades, at any stage, a service center s_j ($\neq s_i$) rolls forward the cascade only along the *best boundary vertex* (along its respective s_k ($\neq s_j$)). Basically, a service center s_j ($\neq s_i$) compares the increase in objective function seen after the re-adjustment of each of its boundary vertices $b_{s_j} \in B_{s_j}$ (to each of the service centers $s_k \neq s_j$) and then chooses the option with minimum increase to roll forward. This is in contrast to the first service center s_i which chooses *top* k boundary vertices and rolls forward each of them. For each of the k simulated cascades, we note the final increase in objective function value at the end of cascade. The cascade which generates the minimum increase in objective function overall is chosen as the *best cascade*. Note that in each of these k cascades, no service center is allowed to repeat. Finally, the algorithm chooses between performing overload at s_i or implementing the set of local re-adjustments as per the *best cascade*. Algorithm 1 shows the pseudo-code for best of k cascades LoRAL algorithm.

3.3   Execution Trace

Figure 3 illustrates a partially constructed LBNVD on a road network containing 5 service centers $(S1, S2, \ldots, S5)$ and 11 demand vertices. Figure 3 also illustrates the capacities and the penalty functions of the service centers. For ease of understanding, in this example, penalty functions are assumed to be just constants. Note these also adhere to the formal definition of penalty functions stated in Definition 5. As mentioned earlier, one can use any monotonically increasing function as a penalty function in the LoRAL algorithm. In Fig. 3, the demand vertices which are allotted to a service center are filled using the same color as that of their allotted service center. Vertices which are not yet allotted are shown without any filling. Nodes where a service center is located are assumed to have zero demand.

In the problem instance shown, the first few <Demand vertex, closest s_i> pairs (in increasing order of distance to the closest s_i) have already been processed. All these pairs were processed as direct assignment as their respective service centers were under-full at that stage. $S1$, $S3$ and $S4$ are now full in capacity. Now, consider the case when the pair $<X, S3>$ is being processed. $S3$ is full, and thus we reach the scenario of re-adjustment. Assume that the max number of cascades in the algorithm (refer Algorithm 1) is set to 2. $S3$ has only two boundary vertices D and F, so both would be chosen for simulating cascades. Now for both D and F, we compute the "best service center" (a service center which leads to lowest increase in objective function). For node D, this would be $S1$ with a total increase of 18 $(4 + 5 - 1 + 10)$ in the objective function value. Whereas for node F, the best service center is $S4$ with a total increase of 19 $(4 + 10 - 3 + 8)$. Now the pairs $<F, S4>$ (Cascade 1) and $<D, S1>$ (Cascade 2) become the seeds for the two cascades simulated by the algorithm.

Cascade 1 (Seed <F, S4>): Given that S4 is also full, it would consider between the following two options: (a) accept F and pay the penalty. The overall increase in objective function value in this case would be 19. (b)

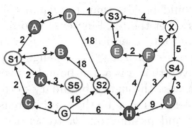

Service Center	Capacity	Penalty
S1	4	10
S2	2	10
S3	3	10
S4	1	8
S5	1	6

Fig. 3. Illustrating best of K cascades LoRAL algorithm (best it color).

Accept F and push out its only boundary vertex J^1 to its corresponding "best service center", which happens to be S2. The overall increase in this case be 18 $(4 + 10 - 3 + 10 - 3)$. One can observe that option (b) leads to lower increase in objective function and thus would be chosen to roll forward the cascade beyond S2. However, this cascade would not roll beyond S2 as S2 was under-full.

Cascade 2 (Seed <D, S1>): Given that S1 is also full, it would consider between the following two options: (a) accept D and pay the penalty. The overall increase in objective function value in this case would be 18. (b) Accept D and push out one of its boundary vertices to another service center. For this option it would evaluate all pairs of its boundary vertices (A, B, K, C) and service centers $(S2, S3, S4$ and $S5)$ to determine the best boundary vertex and its corresponding best service center to roll forward. For our example, this happens to be the pair <K, S5>. The total increase in this case would be 9 $(4 + 5 - 1 + 3 - 2)$. Option (b) leads to lower increase in objective function and thus would be chosen to roll forward the cascade beyond S5. However, this cascade would not roll beyond S5 as S5 was under-full.

Now, the LoRAL algorithm would compare the increase in objective across the three options: (a) force allotment of X to S3, (b) Cascade 1 and, (c) Cascade 2. In our example, Cascade 2 leads to lowest increase and thus would be implemented by the algorithm. This process continues until all demand nodes are allotted.

4   Theoretical Analysis

Lemma 1. *Consider a full service center s_i where a new demand vertex v is being inserted. For a local re-adjustment operation on s_i, it is sufficient to consider only the boundary vertices of s_i (refer Definition 6).*

Proof. Let s_x be any service center and u be any non-boundary demand vertex of s_i. We would prove that there would exist at-least one boundary vertex b_{s_i} of s_i such that total increase in objective function obtained by pushing out b_{s_i}

---

[1] In case it had more boundary vertices, it would have considered the boundary vertex which lead to lowest increase in objective function.

to a service center s_x would be less than (or equal to) the increase observed while pushing out u to s_x. Let the total increase in objective function value by accepting v and pushing out u to a service center s_x is Θ. Thus, $\Theta = Dist(v, s_i) + Dist(u, s_x) - Dist(u, s_i) + Penalty\,at\,s_x$ (no change in penalty paid to s_i as v got in and u got out). Given that u is a non-boundary vertex of s_i, any shortest path from u to s_x must pass through a boundary vertex of s_i. W.l.g assume that b_i is this boundary vertex. Let Φ be the total increase in objective function obtained by accepting v and pushing out b_i to service center s_x. Thus, $\Phi = Dist(v, s_i) + Dist(b_i, s_x) - Dist(b_i, s_i) + Penalty\,at\,s_x$. We can ignore the terms corresponding to $Dist(v, s_i)$ and the penalty paid at s_x as they are common on both sides. Thus, we focus only on the terms $\Theta_{new} = Dist(u, s_x) - Dist(u, s_i)$ and, $\Phi_{new} = Dist(b_i, s_x) - Dist(b_i, s_i)$. Now, we need to establish that $\Theta_{new} \geq \Phi_{new}$. $\Theta_{new} - \Phi_{new} = Dist(u, s_x) - Dist(b_i, s_x) + Dist(b_i, s_i) - Dist(u, s_i)$. By triangle inequality of shortest paths: $Dist(u, s_i) \leq Dist(u, b_i) + Dist(b_i, s_i)$. Therefore, $\Theta_{new} - \Phi_{new} \geq Dist(u, s_x) - Dist(b_i, s_x) + Dist(b_i, s_i) - Dist(u, b_i) - Dist(b_i, s_i)$. Since, $Dist(u, s_x) = Dist(u, b_i) + Dist(b_i, s_x)$. We have, $\Theta_{new} - \Phi_{new} \geq Dist(u, b_i) + Dist(b_i, s_x) - Dist(b_i, s_x) + Dist(b_i, s_i) - Dist(u, b_i) - Dist(b_i, s_i)$. Therefore $\Theta_{new} - \Phi_{new} \geq 0$. Thus, b_i becomes the chosen boundary vertex to push out instead of u.

4.1 LBNVD Problem and Min-Cost Bipartite Matching

The LBNVD problem can be reduced to min-cost bipartite matching [15]. The first step of this reduction (postulated in [15] from a theoretical perspective) is creation of a bipartite graph from the demand vertices and service centers. In the first set of vertices of the bipartite graph, we put $|V_d| \times |V_s|$ number of vertices. In the second set (of vertices), we put one vertex for each of the demand vertex given in the input. Now, we make the number of vertices same in both the sets by creating dummy vertices and adding them to the set corresponding to the demand vertices. Following this, we add edges between the two sets. We would be adding an edge between all pairs of nodes in the bipartite graph. Cost of an edge between the sets is defined as follows:

(a) **If Vertex in the second set is not a dummy vertex:** Edge cost (vertex, j_{th} copy of s_i) = ShortestDistance(vertex, s_i) + penalty paid for being the "j-capacity of s_i" (if j > capacity of s_i) insertion beyond the capacity of s_i. Note that the penalty term is not added if $j \leq$ capacity of s_i.
(b) **If Vertex in the second set is a dummy vertex:** Edge cost (vertex, j_{th} copy of s_i) = Infinite.

A perfect matching of minimum weight on this bipartite graph would be a solution to our LBNVD problem instance. Note that we would have to remove the cost contributed by the dummy vertices in the final objective function value. Also the edges need to be directed in a certain way as explained next.

4.2   Bounding the Performance of LoRAL Algorithm

Consider again the bipartite graph representation of the LBNVD problem. In the optimal algorithm [16], we first add a super source s and super destination node t. Following this, we add directed edges from super source to all the unmatched (initially all) demand vertices and edges from copies of the unmatched (initially all) service centers to the super destination. During the course of the algorithm, as the nodes get matched, we remove the edges incident on s and t. Also, as the matching progresses, edges which are in the matching are oriented from the copies of service centers to the demand nodes and vice-versa if otherwise. In addition, costs of the edges oriented from set of demand vertices to service centers are deemed to be positive. Whereas, the costs of edges directed from service centers to demand nodes are deemed to be negative. Note that it is possible to work with only positive edge costs but in such a case, the algorithm also assigns a potential (≥ 0) to each node (refer [16]) and works with both edge costs and node potentials.

In each iteration of the optimal algorithm, we compute the shortest path between s and t and augment the matching. The edges which are directed from the set of demand vertices to the copies of the service centers get added to the matching and the ones directed otherwise are removed. It has been theoretically proven [16] that, at given stage of the matching, augmenting along the shortest path between s and t increases the size of the matching by one, i.e., one more demand node gets added to the matching. And the increase in the objective function value is the least possible. Furthermore, continuing this way leads to optimal answer. Note that LoRAL algorithm also increases the size of matching by one in each iteration of the main loop (line number 3).

It should be noted that the "cascade of local re-adjustments" created while processing a demand node (at line 4) is same as the described s-t path on this bipartite graph. Also note that, we are referring to "cascade of local re-adjustments" in a general sense. It includes both direct allotment (line 6) and "cascade of re-adjustments" (lines 8–21 in Algorithm 1). Therefore, the primary difference between the optimal algorithm and our proposed LoRAL algorithm turns out to be the difference in total cost of the $s - t$ path chosen in each iteration. We use this difference to bound the worst case approximation of our LoRAL algorithm.

Lemma 2. *Given an instance of the LBNVD problem. Let c_{min} be the smallest capacity, i.e., $c_{min} = \min\{c_{s_1}, ..., c_{s_{|V_s|}}\}$ and p_{max} be the maximum possible penalty that could be paid by any particular allotment to any service center $s_j \in V_s$. Let $cost_{min}$ be the smallest distance between any demand vertex and any service center, i.e., $cost_{min} = \min_{\forall v_{d_i} \in V_d \wedge s_j \in V_s} dist(v_{d_i}, s_j)$. Let $cost_{max}$ be the largest increase in objective function possible while inserting any demand node, i.e., $cost_{max} = \max_{\forall v_{d_i} \in V_d \wedge s_j \in V_s} dist(v_{d_i}, s_j) + p_{max}$. The worst case deviation in final objective function value between the LoRAL and the optimal algorithm is bounded by: $(|V_d| - c_{min}) \times (cost_{max} - cost_{min})$.*

Proof. We prove this using the value of the worst case increase in objective function value after processing each demand vertex. Note that both LoRAL and the optimal algorithm would in general process the demand vertices in different order, but one could derive a bound using the net increase in objective function value after a unit increase in the size of matching (and number of demand vertices is same across both algorithms). It is important to note that, (at-least) the first c_{min} demand vertices processed by both the algorithms would be same. This is because of the greedy nature of algorithms where they choose only the lowest cost (as determined by them) augmenting paths. And these first c_{min} demand vertices would be processed as direct allotment (by both the algorithms) to the closest service center as all the service centers are guaranteed to have free space then. For each of the first c_{min} demand vertices, both algorithms are guaranteed to have the same amount of increase in objective function value. After this, for each new addition (though different demand vertices), the LoRAL algorithm could in worst case have $cost_{max}$ increase in objective function value, i.e., none of the paths of local re-adjustments helped. Whereas the optimal algorithm could (theoretically) lead to only $cost_{min}$ increase in objective function value. Therefore, the worst case deviation in final objective function value between the LoRAL and the optimal algorithm is bounded by: $(|V_d| - c_{min}) \times (cost_{max} - cost_{min})$. Note that this bound is just a mathematically possible upper bound.

5   Asymptotic Complexity

The input graph G(V,E) has n vertices and m edges, out of which we have $|V_d|$ demand vertices and $|V_s|$ service centers.

Time Complexity: The shortest path distance between all pairs of demand vertices and service centers is computed using the floyd-warshall algorithm [11] which takes $O(n^3)$. Using the result of the floyd-warshall, we determine the closest service center for each demand vertex which takes $O(|V_d| \times |V_s|)$. In step 2, we construct MinDistance heap which takes $O(V_d)$ time. In step 3, the algorithm enters into a *while* loop which runs for each of the $|V_d|$ demand vertices. In step 4, each extract-min operation takes $O(log(|V_d|))$ time. Steps 5–7 take $O(1)$ each. After this, steps 8–21 simulate increase in objective function value in k cacades of local re-adjustments starting at the service center s_i. For one local-adjustment, the algorithm has to see #boundary vertices $\times$ $|V_s|$ pairs of boundary vertices and service centers. This can take at most $O(|V_d| \times |V_s|)$ time (in worst case). Therefore, simulating increase in objective function in k sequences of local re-adjustments where length of each sequence in upper bounded by $|V_s|$ (as no service center is allowed to repeat) would take $O(k|V_s||V_d||V_s|)$. Given steps 8–13 are inside the while loop on step 3 (which would run for $|V_d|$ times), the total running cost of the LoRAL algorithm becomes: $O(n^3) + O(|V_d||V_s|) + O(|V_d|) + O(k|V_d||V_s||V_d||V_s| + |V_d|log(|V_d|))$. If $|V_s| \ll |V_d|$, the total running cost of the LoRAL algorithm becomes: $O(n^3) + O(k|V_d|^2 + |V_d|log(|V_d|))$. It is important to note the term $k \times |V_d|^2$ is an absolute mathematical upper bound and can happen only in highly contrived datasets where each demand node has a direct edge to all the $|V_s|$ service centers. Only in such cases, the cost of one re-adjustment becomes $O(|V_d| \times |V_s|)$. For any realistic road networks, the #boundary-vertices $\ll |V_d|$.

Also, the $O(n^3)$ time for computing the shortest distance between $|V_d| \times |V_s|$ pairs can be significantly reduced in practice by employing techniques like A*, contraction hierarchies, etc.

6   Experiment Results

Algorithms were tested on a real road network dataset of New Delhi containing **65902 vertices and 144410 edges** obtained from OpenStreetMaps [www. openstreetmap.org]. Some vertices (chosen randomly) of this graph were designated as the service centers and other vertices as the demand vertices (each with unit demand). Algorithms were implemented in Java 1.7 and the experiments were carried out on a Intel(R) Xeon(R) CPU machine (2.50 GHz) with 96 GB of RAM and Cent OS 6.5. We pre-computed the closest service center to each demand vertex using floyd-warshall as a black box. In our experiments, we measured run-time and final value of objective function as different parameters were varied. We compared the performance of the following 5 algorithms.

(1) **Best of k Cascades LoRAL:** Algorithm 1 while upper bounding the values of k to: (a) number of demand vertices and, (b) number of demand vertices/2 (denoted as "best of k/2 cascades" in the plots). By upper bounding we mean that each service center would explore Min{its #boundary vertices, k} number of cascades. l was set to #service centers - 1.
(2) **Bounded LoRAL:** Algorithm 1 with $k = 1$ and $l = $ #service centers/2.
(3) **Unbounded LoRAL:** Algorithm 1 with $k = 1$ and $l = $ #service centers - 1.
(4) **Min-Cost Bipartite Matching algorithm** given in [15].
(5) **CCNVD** algorithm proposed in [4].

Experiment 1: Comparing Our Algorithms with the Optimal Algorithm: Figures 4 and 5 shows the results of this experiment. In this experiment, the number of service centers in each network was set according to the following scheme: 1 service center for every 400 demand vertices (denoted as service center ratio 400:1). Capacities of all service centers were chosen randomly (integers) from the range 250–350. For sake of ease of interpretation, we took each penalty function as constant a random integer from the range 50–80. Note that this still follows our formal definition of penalty functions given in Definition 5.

Our experiments showed that LoRAL algorithms were much faster than the optimal algorithm while maintaining comparable values of the final objective function. Best of k cascades algorithm was closest to the optimal algorithm in terms of the final value of the objective function. The optimal algorithm did not scale up with the increase in the size of the input network (Fig. 4) and thus, was excluded from further experiments with networks of size 65000 vertices.

Experiment 2: Effect of Number of Service Centers: Figure 6 shows the result of this experiment. In this experiment, the number of service centers was varied by changing the ratio of the # demand vertices to the # service centers. A smaller value of this ratio indicates a larger number of service centers. Capacities

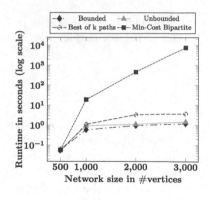

Fig. 4. Exp1 - run-time analysis.

Fig. 5. Exp1 objective func. analysis.

of all service centers were chosen randomly (integers) from the range 250–350. For sake of ease of interpretation, we took each penalty function as constant a random integer from the range 50–80.

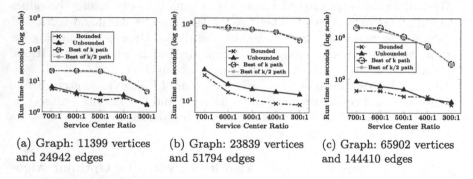

(a) Graph: 11399 vertices and 24942 edges

(b) Graph: 23839 vertices and 51794 edges

(c) Graph: 65902 vertices and 144410 edges

Fig. 6. Effect of number of service center.

Experiment 3: Effect of Total Capacity of Service Centers: In this experiment, penalty costs were in the range of 50–80 units and the number of service centers were according to the ratio of 300:1 (Figs. 7a and b) and 500:1 (Fig. 7c). Our experiments showed that as the parameter total capacity/# demand vertices decreased, runtime increased as more local re-adjustments were made and value of the objective function increased as more penalty was being paid.

Experiment 4: Effect of Penalty Costs: In this experiment we took each penalty function as constant a random integer from the set range (x-axis in Fig. 8a and b). Ratio of total-capacity/#demand-vertices was set to 0.6 and the number of service centers were set according to the ratios 300:1 (Fig. 8a and b) and 500:1 (Fig. 8c).

Experiment 5: Comparing LoRAL Algorithms with CCNVD: Table 1 shows the results of this experiment. Note that since CCNVD [4] does not have a notion of penalty, we focus on the "obj func value - total penalty paid" values

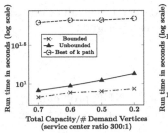

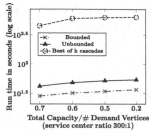

Capacity Ratio	Bounded	Unbounded	Best of k paths
0.7	8688072	8679187	8664776
0.6	8887163	8881991	8871237
0.5	9113295	9109435	9098743
0.2	10127435	10126311	10123650

(a) Graph: 23.8K vertices   (b) Graph: 65.9K vertices   (c) Graph: 65.9K vertices (Objective Function Value)

Fig. 7. Effect of total capacity of service centers.

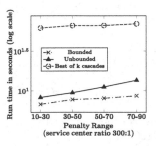

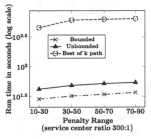

Penalty Range	Bounded	Unbounded	Best of k paths
10-30	7798550	7798225	7798106
30-50	8646181	8644853	8643786
50-70	9512618	9511183	9505962
70-90	10371867	10368714	10361819

(a) Graph: 23.8K vertices   (b) Graph: 65.9K vertices   (c) Graph: 65.9K vertices (Objective Function Value)

Fig. 8. Effect of penalty costs.

Table 1. Experiment 5 - comparing LoRAL algorithm with CCNVD

Network size	Cost type	Unbounded	Bounded	Best of k cascades	CCNVD
1000 vertices	Final obj func value	91910	91910	91910	168688
	Total penalty	15400	15400	15400	NA
	Obj func value - Total penalty paid	76510	76510	76510	168688
2000 Vertices	Final obj func value	184197	186197	184197	263585
	Total penalty	37924	37924	37924	NA
	Obj func value - Total penalty paid	146273	146273	146273	263585

for comparison. One can observe that CCNVD [4] had very large values on this metric. This is because CCNVD [4] aims for service area contiguity.

7   Conclusions

This paper proposed a novel LoRAL algorithm for the LBNVD problem. Approach was evaluated both theoretically and experimentally, using real datasets of the New Delhi road network. Our experiments indicated that the LoRAL algorithm was significantly more scalable than the optimal algorithm. In future, we would like to work on other generalizations of LBNVD.

Acknowledgement. We would like to thank Prof Sarnath Ramnath, St. Cloud State University and the reviewers of DEXA 2018 for giving their valuable feedback towards improving this paper. This paper was in part supported by the IIT Ropar, Infosys Center for AI at IIIT Delhi and DST SERB (ECR/2016/001053).

References

1. Wikipedia: Catchment area. http://en.wikipedia.org/w/index.php?title=Catchment%20area
2. Okabe, A., et al.: Generalized network voronoi diagrams: concepts, computational methods, and applications. Int. J. GIS **22**(9), 965–994 (2008)
3. Demiryurek, U., Shahabi, C.: Indexing network voronoi diagrams. In: Lee, S., Peng, Z., Zhou, X., Moon, Y.-S., Unland, R., Yoo, J. (eds.) DASFAA 2012. LNCS, vol. 7238, pp. 526–543. Springer, Heidelberg (2012). https://doi.org/10.1007/978-3-642-29038-1_38
4. Yang, K., et al.: Capacity-constrained network-voronoi diagram. IEEE Trans. Knowl. Data Eng. **27**(11), 2919–2932 (2015)
5. U, L.H., et al.: Optimal matching between spatial datasets under capacity constraints. ACM Trans. Database Syst. **35**(2), 9:1–9: 44 (2010)
6. U, L.H., et al.: Capacity constrained assignment in spatial databases. In: Proceeding of the International Conference on Management of Data (SIGMOD), pp. 15–28 (2008)
7. Aurenhammer, F.: Power diagrams: properties, algorithms and applications. SIAM J. Comput. **16**(1), 78–96 (1987)
8. Yao, B., et al.: Dynamic monitoring of optimal locations in road network databases. VLDB J. **23**(5), 697–720 (2014)
9. Xiao, X., Yao, B., Li, F.: Optimal location queries in road network databases. In: Proceedings of the 27th International Conference on Data Engineering (ICDE), pp. 804–815 (2011)
10. Diabat, A.: A capacitated facility location and inventory management problem with single sourcing. Optim. Lett. **10**(7), 1577–1592 (2016)
11. Cormen, T.H., Stein, C., Rivest, R.L., Leiserson, C.E.: Introduction to Algorithms, 2nd edn. McGraw-Hill Higher Education, New York (2001)
12. Delling, D., et al.: PHAST: hardware-accelerated shortest path trees. J. Parallel Distrib. Comput. **73**(7), 940–952 (2013)
13. Bast, H., et al.: Fast routing in road networks with transit nodes. Science **316**(5824), 566 (2007)

14. Jing, N., et al.: Hierarchical encoded path views for path query processing: an optimal model and its performance evaluation. IEEE Trans. KDE **10**(3), 409–432 (1998)
15. Bortnikov, E., Khuller, S., Li, J., Mansour, Y., Naor, J.S.: The load-distance balancing problem. Networks **59**(1), 22–29 (2012)
16. Kleinberg, J., Tardos, E.: Algorithm Design. Pearson Education, London (2009)

Author Index